WOMEN SCIENTISTS
IN AMERICA

WOMEN SCIENTISTS IN AMERICA

Before Affirmative Action
1940–1972

MARGARET W. ROSSITER

The Johns Hopkins University Press

Baltimore and London

This book has been brought to publication with the generous assistance of the
National Endowment for the Humanities.

The Johns Hopkins University Press
2715 North Charles Street
Baltimore, Maryland 21218-4319
The Johns Hopkins Press Ltd., London

ISBN 0-8018-4893-8

Library of Congress Cataloging-in-Publication Data
will be found at the end of this book.
A catalog record for this book is available from the British Library.

CONTENTS

ILLUSTRATIONS

Plates

Following page 148:
Doris Mable Cochran
Cathleen Morawetz and her family
Three women graduate students of Lipman Bers
Ann Chamberlain Birge and Robert Birge
Marguerite Thomas Williams
Gerty and Carl Cori
Nun chemists
Barbara McClintock
Katharine Burr Blodgett

Following page 332:
Grace Murray Hopper
Ruth Benerito
Mary Bunting
Special award designed for Mary Shane
Local newspaper article about mother-scientist Maria Goeppart Mayer
Maria Goeppart Mayer
Essie White Cohn and Sarah Ratner
Lise Meitner with President Harry S. Truman
Symposium on Women in Science and Engineering at MIT

Figures

TABLES

ACKNOWLEDGMENTS

*I*t is a very great pleasure at the end of a long project to be able to thank publicly the many persons and institutions that helped so much along the way. Grants from the then History and Philosophy of Science Program and the Visiting Professorships for Women Program, both at the National Science Foundation, and fellowships from several other institutions, including the John Simon Guggenheim Memorial Foundation, the then Gender Roles Program of the Rockefeller Foundation, and the MacArthur Fellows Program of the John D. and Catherine T. MacArthur Foundation, all kept me at my task when jobs were scarce. Yet none of these institutions can be held responsible for this book's contents, for which I alone am accountable. I am also glad to acknowledge the many archivists and librarians who helped to locate material and suggested still more. Although the libraries—at first the University of California at Berkeley, then Harvard, and finally Cornell—have all undergone great technological changes in the past decade, a scholar still needs (perhaps now even more than ever) knowledgeable reference and other librarians to find his or her way in their great holdings. There is a wealth of information available to those who can get at it. In addition, anyone working on obscure persons and with rare materials cannot get along without interlibrary loan, at which both the Berkeley and Cornell staffs excel. If something existed, they would find it.

One institution and its staff deserve especial thanks—the American Academy of Arts and Sciences, where John Voss and Alexandra Oleson took on the task of administering a series of NSF grants over nine years and even provided office space in their new building outside Harvard Square for three. This was timely, invaluable, and far above and beyond any expectation. It enabled me to press on, wherever I was located; I still marvel at their thoughtfulness and generosity.

Among fellow scholars special gratitude goes to Geraldine Joncich Clifford for a timely referee report and to Sally Gregory Kohlstedt, who in a conversation in a basement lunchroom at Simmons College, where she was teaching in the 1970s, suggested that I ought to expand an article on women scientists in the United States before 1920 into a larger study, *even if it took two volumes*. At first this seemed far-fetched, but it grew more plausible as the material for the study ballooned out of control and threatened to burst out of the pages of an earlier volume (*Women Scientists in America: Struggles and Strategies to 1940*, published in 1982). One just could not start World War II and however much might come after it on page 500. This second, rather different volume is the sequel. Over the years Sally has kept in touch with the project, and despite her own researches and increasingly heavy tasks and duties, she has read at least two drafts. Her pertinent, pointed queries have marked each page and have made this a much better book.

At the Johns Hopkins University Press, Henry Tom and Barbara Lamb made the big decisions, such as to include a bibliographical essay, a rarity nowadays, and to assign Joanne Allen to the project. She then spent the better part of a memorable year painstakingly transforming a lengthy manuscript into the handsome book you have before you. At Cornell University, Patricia S. Dean facilitated my move into the world of word processors in 1991, and Lillian Isacks showed extraordinary patience with the endless details and even learned new software to get the figures and tables just right. Then, too, one's relatives deserve thanks, particularly my mother and younger twin Charles, for gently inquiring how the book was coming and not groaning too loudly when I said I had found another archive.

The following persons and institutions have given permission to cite material printed here: the American Society of Microbiology, the Bancroft Library of the University of California at Berkeley, the California Institute of Technology, the International Council of Psychologists, the Historical Collections and Labor Archives at Pennsylvania State University, the Schlesinger Library of Radcliffe College, the Society of Women Engineers, and Enid and Doris Wilson.

INTRODUCTION

Although by all accounts the period 1940–72 was a golden age for science in America, it has generally been considered a very dark age for women in the professions.[1] How could this have been? Were women not an integral part of American science by 1940? Why, then, in a period of record growth in almost every aspect of American science that one could count—money spent, persons trained, jobs created, articles published, even Nobel prizes won—were women so invisible? What had happened? Was this not at first the period of World War II, when women were told that they could do anything and were even recruited for certain scientific and technical projects? Was this not followed by the Cold War, when, because of the heightened manpower needs of a highly technological military-industrial complex, officials launched a campaign for "woman-power," urging bright women to seek training in nontraditional areas such as science and engineering? Was this not a period when record numbers of women were earning doctorates in scientific and technical fields? What had happened to them? Were they congregated in areas or fields that did not grow as much as the men's? Did the scientific job market work differently for them? Did marriage as then constituted (or, more broadly, marital status in general) have a differential impact on their scientific careers? Why were there these limits to the women's opportunities?

Unfortunately for the trained women of the 1940s through the 1960s, the evidence indicates that the growth and affluence of the period that could have made room for more and better-trained scientists of both sexes did not benefit the two equally; in fact, it generally unleashed certain forces that hastened the women's exit and subsequent marginalization and underutilization, which could then be cited to justify denying further training for their successors. Evidence presented here indicates that most of the women's traditional employers, such as women's colleges, teachers colleges, and colleges of home economics, were closing their doors to them, and they were not included on the faculties of the growing and new coeducational institutions. Much of this new exclusion was tied to marriage, as in the reinstatement of the antinepotism rules on most campuses after World War II. But even single women were ousted or just not hired in these pronatalist years, often for fear that they might later get married. Prestigious universities had never had many women in visible places, but in the affluent 1950s and 1960s less well regarded institutions began to aspire to become prestigious themselves. Thus the pattern was deliberate and grew widespread. Formerly ridiculed and poorly supported, these institutions began to seem ripe for upgrading and colonization; for the first time they could afford to pension off the older women and employ more men.

Large numbers of often older women also seemed to spur, though this was never stated publicly, a lot of ageism, sexism, and possibly homophobia. It was portrayed as a sign of progress to get rid of the old girls, raise salaries, reduce teaching loads, hire more Ph.D.'s, rename the school a state university, and urge the new faculty to get on with their research, all of which would forcibly upgrade the school's level of prestige. Indeed, much of the discussion in these decades of the need for higher salaries at the liberal arts colleges and elsewhere, usually by foundation officials and academic administrators, was a kind of code language for the masculinization of formerly female-dominated areas. After the mid-1950s this "chill" spread beyond employment to levels of recognition that earlier women scientists had achieved, such as election to certain professional offices and selection for coveted prizes. Thus, at a time when young male scientists faced enhanced opportunities at every turn, the young women were supposed to be home with the children, whether they had them or not or whether they wanted to be there or not.

For many years there was little consciousness that these attitudes and practices might constitute something as ugly as discrimination. It was just the way it was. Even respectable people behaved this way, and they were often proud of it. The prevailing assumption, considered so obvious that hardly anyone expressed it in the 1950s, was that though some women were present in science, they were at best invisible and at worst an embarrassment. They could be, and were, blithely omitted as either unimportant or anomalous from accounts and thinking about the profession. Thus, at the same time that some governmental officials were urging young women to greater efforts (see ch. 3), previously trained women scientists were finding it hard to get hired or promoted or even to be taken seriously. Whether single or married, they had been defined as obstacles to overall progress, and their removal was a desired goal. There did not seem to be any way to stop the juggernaut. Any protests or complaints were unlikely to change anything, and they were likely to be dismissed as special pleading not worthy of consideration.

The leaders of a few women's organization were aware of the closing doors and spoke out, though with limited success. Unfortunately, those women's organizations whose leaders saw what was happening and might have protested more successfully than they did on their behalf were themselves compromised in various ways, starting in the late 1940s. Liberals complained that the American Association of University Women (AAUW) had let some of its chapters refuse to accept Negro members, while the right wing retaliated with accusations that a few AAUW leaders and staff held Communist sympathies. Intimidated by such public censure, leaders of both the AAUW and, to a lesser extent, the National Federation of Business and Professional Women's Clubs took a less outspoken stand on the rising tide of discrimination than they might have otherwise. As a whole, organizations turned cautious and quiet, and the women scientists, one by one, either persevered with difficulty or turned to other things. (In fact, as will be discussed in chapter 5, statistical surveys of the 1950s and 1960s found "other" a rather frequent designation of women's activities.)

A few women scientists noticed what was happening. Some saw instances such as quotas but were reluctant to see a pattern. Most women and scientists of the time lacked the vocabulary (e.g., "sexist" or "male chauvinist") and the civil-rights concepts to recognize systematic patterns, identify the responsible parties, and plan how to correct the situation. A few individuals and even small groups did see a pattern and reported it, but because they were either grateful for their current status or desperate not to lose even that, they were reluctant to criticize the powerful and successful, especially when so little seemed likely to be gained from it. The few who did something to bring attention to the situation or even to try to correct it were ineffectual for various reasons. Mildred Mitchell published an article in 1951 that provoked such a strong counterblast from a Harvard psychologist that the topic dropped out of sight for at least a decade. Then, in the mid-1950s, at the very time when so much rhetoric claimed that trained women were needed more than ever, a committee of Radcliffe trustees and Harvard faculty glimpsed and discussed the limits to women's opportunities in academia, its own final report, reflecting many compromises, was hopelessly muddled and even counterproductive. When one committee of the AAUW did discuss discriminatory practices in the 1950s and even took a few retaliatory steps, AAUW members voted it out of existence in 1963. Occasional individuals, such as Myra Sampson at Smith College, Frances Clayton at Brown University, and Jessie Bernard at Pennsylvania State University, collected data that showed that the women's colleges were no longer hiring many women but evidently preferred men. Yet for various reasons none was able to use this material effectively. In short, even those women who were most inclined to take an interest in the professional status of scientific and academic women were themselves too much a product of the thinking and politics of the time to raise much "consciousness" or effect change.

Yet one suspects from the few actions that were taken that if the data had been interpreted otherwise, they would not have been published, or if they had been published, they would not have been taken seriously. It is more likely that they would have been ignored, as Sampson's data were, or dismissed as the ideas of radical or crazy women. Certainly federal legislation regarding employment or the internal workings of universities would have seemed intrusive and detrimental at a time of McCarthyism. From this it took even the most politically astute women a minimum of about fifteen years to recover.

Although the prevailing mindset made it nearly impossible for the women of the time, including several social scientists, to glimpse the larger reality, suddenly in the early 1960s Betty Friedan and others began to put enough pieces together into a radically new but recognizable pattern. Alice Rossi, a well-trained sociologist of the 1950s, devoted several years in the early 1960s to rereading, rethinking, and reformulating the prevailing wisdom. Perhaps her new view was possible then because in the context of the civil-rights movement she could begin to see a pattern of oppressive attitudes and practices, that is, that the "victims" were a class that did not "deserve" their fate and should not be blamed for it; that society could and should be changed; that laws would have to be passed by a reluctant Congress and then, what would be even harder, the executive branch would have to be pressured into enforcing them. As the women's numbers increased in the late 1960s (especially in the biological sciences) and their "consciousness" began to rise, the continuing discriminatory practices grew increasingly outrageous and intolerable. Hundreds of women and other feminists, becoming aware for the first time of the enormity of their marginalization and exploitation and the scale of the resources denied to them, were energized to seek change immediately. "Sisterhood" was exhilarating, and within a few months many of them began to fight back. Finally, about 1969 this anger coalesced into a movement, which called for the innumerable "status of women" reports of 1969–72. These in turn, as they documented and brought into full view the totality of the women's exclusion, angered even more persons. By 1970 these reports had become prime evidence in federal hearings on sex discrimination on campus and in the work force, which led to landmark legislation—equal pay and affirmative action in academia—by 1972.

Thus, women scientists were there in record numbers in the 1950s and 1960s, though one might have to look rather hard to find them. Trained to advanced levels, they were, to use some military terms of the period, "camouflaged" as housewives, mothers, and "other" and "stockpiled" in cities and college towns across America (where many still remain), ready but uncalled for the big emergency that never came.

WOMEN SCIENTISTS
IN AMERICA

1

World War II

Opportunity Lost?

World War II is generally termed a "total war" because it affected the daily lives of almost all Americans. Not only did it cause many thousands of casualties on battlefields around the world but it wrought tremendous economic and social changes at home; it restored high employment and long-awaited prosperity, disrupted family life, and, since waging war is a governmental function, greatly increased the responsibilities and size of the federal government. In addition, government's support and interest in science and engineering grew dramatically during wartime, and its manpower policies affected many women scientists, who suddenly came to be considered a rare and precious national resource and were flooded with recruitment literature and special training programs. But even at the height of their demand, women scientists found that, despite the rhetoric, they were not welcome in all types of work or at high levels. Instead, they were considered temporary employees, "keeping the seat warm" for men assigned to other, higher-priority wartime duties, or else they were channeled into lower-level openings as aides and assistants to men who had been promoted. Thus, although many women scientists made important contributions to winning the war, they remained temporary and subordinate supplements to an essentially all-male labor force. They showed how valuable they could be in a crisis, but they also lost the best opportunity yet to create the kind of leverage that might have carried their wartime gains into the postwar economy.

Most textbooks report that the U.S. involvement in World War II began in December 1941, with the unexpected attack by the Japanese on the American fleet at Pearl Harbor, and ended with the dropping of the atomic bomb on Hiroshima and Nagasaki in August 1945, but this wartime chronology is deceptive. Since the war had started in Europe as early as September 1939, with the German march on Poland, which was followed by the Nazi invasions of Denmark, Norway, and Belgium, the fall of France, and the air attacks on Great Britain, the American government had been aware for more than two years before Pearl Harbor that it might soon be entering the war. In fact, it had used that prewar period to take several quiet but important steps to increase its scientific readiness by creating a great many of the agencies and programs that would later play a major role in its victory.[1]

The timing of America's delayed entrance into the war had some important consequences for women's involvement in it. Since Depression conditions continued within the American economy into the early years of the war, there was initially no lack of trained men to fill most of the senior positions. Women scientists, as psychologist Gladys Schwesinger later aptly described it, were "told to be good girls and advised to wait until plans could be shaped up to include [them]."[2] By 1943, when serious shortages had begun to develop (or were said to be developing) and special efforts were being made to recruit women scientists, the positions offered to most of them were chiefly at the low levels, for the wartime job market accorded quite different opportunities to the two sexes. After several thousand men (and perhaps a hundred women) got the top-priority governmental assignments and industrial positions, those women known to the male leaders or identified by the National Roster of Scientific and Specialized Personnel, set up in June 1940, were invited to fill the newly vacated positions. University teaching, for example, which was considered a lower-priority task, was left to those men who were unfit for the military or unable to get war jobs and to the thousands of women scientists who were available after all. A third category was made up of young women hastily trained to become temporary "aides" or assistants to men in industrial jobs. Thus, the women's contributions to World War II were different and less visible than the men's.

The most important war work done by women scientists was in the area of nutrition, which is often said to be as essential to a war effort as armaments and medicine. Accordingly, when urged by the President's National Defense Advisory Council in 1940 to consider possible upcoming war problems, the National Research Council (NRC) set up its Food and Nutrition Committee (later Board). As early as November 1940 this committee, comprising twenty-one of the nation's top biochemists, nutritionists, and physicians, realized that if the nation were to feed its armies properly and perhaps put its civilians on food rationing (a remote possibility in 1940), some important decisions would have to be made. But even this committee of experts was ill prepared for the task, for, as they knew all too well, hardly any work had been done in this area. Although the field of nutrition had developed greatly since World War I and a whole alphabet of vitamins were now known, there was still appalling ignorance about how to apply this knowledge to daily life. In particular, very little was known about the amount of each nutrient the average person needed daily. Even less was known about the requirements of such subgroups as the "average fighting man," pregnant women, or even children. Lydia Roberts of the University of Chicago, one of four women on this committee, has recounted, perhaps in jest, how the committee worked out its suggested levels for human-consumption requirements. Some of the women (Roberts, Helen Mitchell, and either Martha M. Eliot or Icie Macy Hoobler) stayed in the hotel at night and worked out tentative drafts for the next day's discussion, while the men presumably spent the evening "on the town." The next day all regrouped to discuss the proposals together. After several such meetings and extensive correspondence with many other researchers, the committee came up with its first "recommended daily (or dietary) allowances" (RDA), which were submitted to the National Nutrition Conference held at the White House in May 1941 and voluntarily accepted by manufacturers and governmental agencies. As more knowledge became available, the allowances were modified, first in 1944 and several times thereafter.[3]

Once the RDA had been approved, the Bureau of Human Nutrition and Home Economics, which was reorganized in 1943, began to recast its recipes to suggest menus that utilized these new recommendations. Hazel Stiebeling, who became chief of the bureau in 1944, was apparently the first to adapt her menus for four income levels, thus giving each segment of the population a nutritious diet that it could afford. The bureau's major wartime task was, of course, in extension work, or "nutrition education," advising housewives, businesses, and other institutions how to reduce purchases and make substitutions. To accomplish this, many home economists held a variety of positions in other governmental agencies. For example, Mary Barber of the W. K. Kellogg Company of Battle Creek, Michigan, was assigned to the Army's Quartermaster Corps, where she designed menus for its 1.4 million servicemen. Her counterpart in the Navy was Ina Lindman, who took a leave from the United Fruit Company, where she was head of its test kitchens (and its only woman executive), in order to revise the official Navy cookbook for wartime use. So great was the demand for nutritional advice for civilians (and so weak the demand for marine biologists) that even Rachel Carson, an associate aquatic biologist for the U.S. Fish and Wildlife Service, a low-priority agency during the war, was pressed into writing food-conservation bulletins on how to substitute local fish and other domestic seafoods in family menus.[4] Meanwhile, one of the country's best-known women anthropologists, Margaret Mead, whose husband, Englishman Gregory Bateson, had enlisted in the Royal Air Force in 1939, was in New York chafing for action. In December 1941 she left her two-year-old daughter with friends and moved to Washington, D.C., to serve as executive secretary of the new NRC Committee on Food Habits, which was trying to devise ways to help populations change their food habits and perhaps accept rationing. If this seems a rather odd assignment for one of the world's experts on the Pacific Islands, it at least allowed Mead time to lecture widely in the United States and England and to maintain contact with the numerous other social scientists in Washington during the war.[5]

Other women anthropologists were also able, after some initial difficulty, to convince governmental officials of their usefulness in dealing with "strange people and languages," as Gladys Reichard put it in 1942. A few even had unusual opportunities for important research. Best known of the war work by women anthropologists was that of Ruth Benedict for the Office of War Information (OWI), which commissioned her in 1943 to come to Washington and write an interpretive report on Japan and the Japanese mind for wartime and postwar use. Although she had never even been to Japan, Benedict examined numerous historical and literary sources and wrote an insightful report about Japan's firm traditions of aestheticism and militarism. This later became the best-selling book *The Chrysanthemum and the Sword* (1946).[6] Others, like Lucy Talcott, an archaeologist of Greece, and Cora DuBois, an expert on the California Indians, started new careers in the Office of Strategic Services (OSS, forerunner of the Central Intelligence Agency), where many professors wrote position papers on social and economic conditions abroad. DuBois, who had struggled to find an academic post in the difficult 1930s, seemed to thrive at the OSS, first as chief of its Indonesia section in Washington in 1942–44 and then as chief of its research branch in Ceylon in 1944–45. Several other young women anthropologists, such as Rosalie Wax and Katharine Luomala, were part of a research team under sociologist Dorothy Swaine Thomas of the University of California that studied the effects of internment on the 130,000 Japanese-Americans held in concentration camps on the West Coast during the war. (Elizabeth Colson, a Harvard graduate student at the time, did a similar study of internees in Arizona for the War Relocation Authority in 1942–43.)[7]

Although the Office of Scientific Research and Development (OSRD) was one of the largest and best-known employers of male scientists during the war (because of the influential men it attracted and its own eight-volume history), it called relatively few women to desirable assignments on its prestigious projects, which included radar, explosives, the proximity fuse, antimalarial drugs, blood transfusions on the battlefield, penicillin, and even the atomic bombs. The OSRD seems to have followed the classic "old-boy" recruitment patterns: the top men appointed men they knew already, and they in turn recruited their teams from among their own university and personal acquaintances. Since many of these projects were located at the very universities where women held only subordinate positions in the 1930s, it is not surprising that only a few women scientists worked on OSRD projects. In fact the OSRD's leaders fought even the slightest hint of open recruitment or other regulations that the White House was beginning to apply to "federal contractors," on the grounds that OSRD contracts were temporary and had highly technical specifications.[8]

So invisible were the women scientists at the OSRD that it may come as a surprise that there were any there at all. They were rarely mentioned in the standard histories of the projects on which they worked; the few women who were cited were administrative workers praised for their efficiency in cutting through red tape and obtaining special clearances for their male bosses. Even the postwar article "Science in Petticoats," by "technical aide" Virginia Shapley, which saluted women's contributions to the OSRD, unfortunately omitted their names. Thus, to comprehend the position of women scientists at this one agency, one has to piece together items from a wide variety of other sources.[9] Although about thirty women scientists have been located in this way (and no claim is made here for completeness), they constitute a relatively small group compared with the tens of thousands of men working on OSRD projects.

At least eleven women worked on what turned out to be the most glamorous, prestigious, and subsequently influential of the OSRD projects, the "Manhattan District," or the atomic bomb project. Physics graduate student Leona Marshall (later Libby) was the only woman to work on the first part of the bomb project, Enrico Fermi's first controlled nuclear chain reaction at the University of Chicago in December 1942, where she served as a "neutron counter." Later, when many of the luminaries of this phase of the project had moved to Los Alamos, New Mexico, physicist Rose Mooney (later Slater) of Sophie Newcomb College became the associate chief of the x-ray structure section of the Manhattan District's Metallurgical Laboratory in Chicago and thus one of the highest-ranking women on the project. At Columbia University in 1942–43, chemist Harold Urey had at least three women scientists on his project to separate uranium 235 from U-238 by gaseous diffusion: chemistry teacher Lotti Grieff and the brilliant physicists Maria Goeppart Mayer and C. S. Wu.[10]

At least three other women scientists worked on the calculations of the bomb's radiation damage to humans. Pioneering in this field of "health physics" were Elizabeth Graves, a 1940 Ph.D. at the University of Chicago, Elda Anderson at Los Alamos, and Edith Quimby in New York City.[11] Among the other contributors to the Manhattan District were two women geologists, mineralogist Helen Blair Barlett, who took a leave from her job at the A. C. Spark Plug Division of General Motors to join the ceramics branch of the project at MIT, where she developed a new nonporous porcelain, necessary for the interior of the bomb, and Margaret Foster, a geochemist at the U.S. Geological Survey (USGS) in Washington, D.C., who developed two new ways to separate thorium from uranium.[12] Meanwhile Karla Dan von Newmann (the wife of mathematician John von Neumann), who was apparently self-taught in mathematics, was one of the first persons to tackle the programming of the project's first computer.[13] All these "Manhattan" women seem to have been recruited informally: some (like Graves and von Neumann) were talented wives on location whose unexpected skills were pressed into service; others (like Libby, Mayer, and Wu) were students or former associates of the men running the projects; still others, like the geologists, though probably unknown to any of the principal figures on the bomb project, were experts in the precise specialties needed at the time.

There were a few women scientists among the thousands of professionals busy on other OSRD projects during the war. At least five women worked on the radar project at various locations: Jenny Rosenthal (later Bramley) worked with the Signal Corps; Pauline Morrow Austin, a recent Ph.D., and Monica Healea, of the Vassar physics department, both worked on the radar project at MIT; Margaret Nast Lewis (daughter of cytologist Margaret Reed Lewis) worked with a branch group at the University of Pennsylvania; and Elizabeth Laird, recently retired after thirty-five years of teaching at Mount Holyoke, worked on radar for the Canadian government for "the duration," 1941–45.[14] At least four women scientists contributed to the OSRD's large penicillin project during the war. Microbiologist Dorothy Fennell was at the Northern Regional Research Center of the U.S. Department of Agriculture (USDA) in Peoria, Illinois, where most of the work was done, and Elizabeth McCoy served on a related contract at the University of Wisconsin. Engineer Margaret Hutchinson, the first woman to become a member of the American Institute of Chemical Engineers, helped to design one of the nation's first commercial penicillin plants. (Earlier she had been one of the few women to work on the government's aviation fuel program.) Once there was enough penicillin for testing, immunologist Gladys Hobby of Columbia's College of Physicians and Surgeons, was, with Dr. Karl Meyer, reportedly the first to use it on a patient. (She was later a co-discoverer of Terramycin, an early antibiotic.)[15] Three women chemists, Professors Mary Sherrill and Emma Perry Carr of Mount Holyoke College and Dorothy V. Nightingale of the University of Missouri, were among the few women who worked on the OSRD's antimalarial drug project, which had contractors at more than thirty campuses.[16]

Despite the presence of these few women on important projects, the impression remains that the OSRD grossly underutilized its women scientists. For example, several women Ph.D.'s found themselves working as "technical aides," librarians, or other assistants in charge of paperwork. Thus, psychologist Helen Peak found herself, despite her Yale doctorate and professorship at Randolph-Macon Woman's College, the harassed administrative assistant under several ineffective male professor-bosses on an OSRD psychophysics project on Long Island. She left after just a few months for a better job at the OWI. Another kind of "women's work" also persisted at the OSRD: chemists Louise Kelley of Goucher College and Hoylande D. Young of Chicago were employed as "chemical librarians," Kelley at headquarters and Young on the toxic gas project.[17]

In particular, hardly any women were given any authority on OSRD projects. The highest-ranking women at the OSRD seem to have been physicists Gladys Anslow of Smith College, who had worked on E. O. Lawrence's cyclotron before the war and now served as chief of the communications and information section at OSRD headquarters, in charge of liaison with civilian scientists attached to the armed forces, and Dorothy Weeks of Wilson College, a Ph.D. from MIT, who was in charge of the British reports section from 1943 to 1945. Even mathematician Mina Rees, on leave from Hunter College, and the only woman scientist at the OSRD to move into a major science policy position after the war (see ch. 13), was listed as a "technical aide" and "executive assistant to the chief of the applied mathematics panel" in 1943–46.[18] Only three women seem to have supervised OSRD projects: Maria Telkes, a research associate in metallurgy at MIT, who directed a project to use solar energy to purify water on life rafts; Agnes Fay Morgan, a nutritionist at the University of California at Berkeley, whose OSRD project tested dehydrated foods for their vitamin content; and pathologist Virginia Frantz, who ran a project to develop her new cellulose compound that would quickly stop bleeding.[19] Yet all three of these women worked alone on their projects, a most unusual pattern given the large teams on other OSRD projects. All that these three women were "managing" was themselves!

Besides the nutritional, anthropological, and OSRD agencies, several other old and new governmental units hired women scientists to meet their expanding responsibilities. In particular, many geologists and geographers worked for the Military Geology Unit of the USGS, established jointly with the Army Corps of Engineers in June 1942 and greatly expanded (to comprise sixty geologists) after that. There women geologists such as Julia Gardner of the USGS and Dorothy Wyckoff of Bryn Mawr College did important secret work, such as preparing maps of battle zones in North Africa, Southeast Asia, and the Pacific Islands. Zoologist Mary Sears, on loan from Radcliffe College to the Navy's Hydrographic Office, turned her hand to "military oceanography" and headed a unit that prepared extensive reports on enemy waters, especially any conditions that might affect Allied submarine operations or amphibious landings. Geologist Grace Stewart did geographical work during the war for the highly secret OSS, and Louise Boyd, the San Francisco socialite, geographer, and Arctic explorer, investigated magnetic and radio phenomena in Greenland and the North Atlantic for the National Bureau of Standards (NBS) in 1941. Later she became a "technical aide" in the War Department, preparing so many maps and other reports of conditions in the polar regions that she was given a medal. Other women geologists were busy in the petroleum industry, and a few even entered the small and previously all-male field of meteorology, with jobs at the U.S. Weather Bureau or in private firms, especially commercial airline companies. By the end of the war, therefore, many new doors seemed to have opened at least temporarily for women in the earth sciences.[20]

Another new kind of job for women scientists was participation in the newly created military services for women, the Navy's WAVES, the Army's WAACS (later WACS), and the Coast Guard's SPARS. When these women's auxiliary forces were formed in early 1942, they needed immediately not only thousands of enlisted women but also hundreds of women officers, a totally new profession for college women. These jobs turned out to be some of the best positions open to younger women scientists during the war, since here they could receive the advancement denied to them elsewhere and become "veterans," with all the postwar benefits that implied. A young woman scientist would have been well advised in 1942 or 1943 to sign up for officers' training school (held at Smith and Mount Holyoke colleges). The WAVES developed strong ties to the women's colleges, several of whose junior faculty members became some of its first officers, for example, Lt. Cmdr. Frederica de Laguna, an anthropologist at Bryn Mawr, and Capt. Harriet Creighton, a botanist at Wellesley. (Lt. Grace Hopper, a mathematician at Vassar, was, by contrast, one of the few women in the Navy itself. While stationed at Harvard University in 1944, she programmed Mark I, the first large-scale computer, as part of her duties for the Bureau of Ordnance.)[21]

Among other women scientist-officers were thirty-eight psychologists, twenty-six of whom were in the WAVES. One of these, Dorothy C. Stratton of Purdue, was appointed the first head of the Coast Guard's SPARS in November 1942. Ensign Kathleen Lux, a graduate of Purdue University's School of Engineering (the largest producer of women engineers), became the first woman officer in the Navy's Civil Engineer Corps in 1943, and physical anthropologist Mildred Trotter of Washington University in St. Louis joined the WAVES and was assigned to help the Army identify bodies from partial skeletons, a very practical wartime application of the relatively "pure" science of anthropology. A few officer-scientists, such as Capt. Mildred Mitchell, a Radcliffe Ph.D. and clinical psychologist at several state hospitals before joining the WAVES in 1942, enjoyed the Navy so much that they decided to make their careers in it after the war, when the women's services were merged with the regular armed forces. Meanwhile, physical chemist Cmdr. Florence van Straten, a Navy meteorologist during the war, stayed on afterwards as a civilian in the Naval Weather Service. As one of the very few who could predict Pacific storms and, in the 1950s, the pathway of atomic fallout, she merited a "supergrade" position.[22]

Civilian women psychologists also applied their professional skills to a variety of wartime assignments. For example, seven, including the full professors Florence Goodenough of the University of Minnesota and Jean Macfarlane of the University of California at Berkeley, were designated "Expert Consultants to the Secretary of War" in the spring of 1942 and dispatched to WAAC recruiting areas to test candidates for officers' training school. Mary Vanuxem, recently retired from directing a home for mentally defective women, assisted in the supervision of the Medical Survey Program for the Selective

Service System. Mary Hayes, director of vocational guidance for the National Youth Administration in Washington, moved easily to a similar job at the new War Relocation Authority in 1942. And Grace Fernald switched from directing a clinic at UCLA for children with learning disabilities to running a program for illiterate soldiers at several Army bases in California.[23]

Still other women psychologists held a series of staff positions in Washington, D.C. Ruth Tolman, a clinician from Los Angeles (who was in the capital with her husband, a prominent Cal Tech physicist), served first as a "social science analyst" at the USDA, then as a "public opinion analyst" at the OWI, and finally as a clinical psychologist at the OSS, where, along with Eugenia Hanfmann, she served on a "psychological evaluation team" that selected and trained spies and other personnel for work overseas. Similarly, Helen Peak, after brief stints at the OSRD and the OWI, became a "special adviser" at the War Production Board and finally, in 1945–46, a "bombing analyst" for the U.S. Strategic Bombing Survey. Many other women psychologists too numerous to mention used their professional skills on the home front, in local draft board counseling, child care projects, psychiatric social work, or the mental health activities designed to help servicemen and their families overcome or adjust to the traumas and tragedies around them.[24] Some were recruited to these positions by a subcommittee on women's services formed in late 1941 by the combined Emergency Committee of several psychological associations, and others may have gotten them through the newly established and independent National Council of Women Psychologists (NCWP), set up the day after Pearl Harbor to ensure that women got some share of the war work.[25]

Meanwhile, even more women scientists, including many mathematicians and chemists who were not working on other war projects, were called to teach at the nation's colleges and universities. Some universities rescinded their prewar antinepotism rules for the duration in order to employ those local spouses they had formerly spurned. Science faculties were badly depleted when many men (who had held most of the academic jobs in the 1930s) left suddenly for governmental science projects or, if young enough, the military. Their substitutes and those who stayed behind worked harder than ever; enrollments decreased sharply, but the remaining students were taking more science courses than in peacetime, there was pressure to accelerate courses and graduate students in three or three and one-half years, and many professors also taught extra courses on war-related topics for military training or civilian programs, such as the Engineering, Science and Management War Training program (see below). Some women scientists, like astronomer Frances Wright on the staff of the Harvard College Observatory, taught navigation to Harvard students, including Army and Navy V-12 candidates, from 1942 to 1945, and under the Armed Services Training Program anthropologist Hortense Powdermaker of Queens College, in New York City, taught exotic languages and wrote handbooks for servicemen headed for the Pacific Islands and Southeast Asia. Some switched fields to help out; others moved from smaller schools to larger universities where they could teach their specialty (such as physicist Melba Phillips, who left Brooklyn College for the University of Minnesota, thereby getting her first taste of teaching in a major, though depleted, department); and many other women apparently entered university teaching for the first time or reentered it after an absence. Some of the older women at coeducational institutions, such as chemists Clara deMilt at Tulane and Leonora Bilger at the University of Hawaii, even became temporary department chairpersons, which probably would not have happened in peacetime.[26]

In fact, according to figures collected by the Women's Bureau (WB), the movement of women scientists onto college and university faculties during the war was so widespread that it constituted their principal contribution to the war effort, dwarfing their occasional appointments to specific war projects like those mentioned above. There special contacts or particular specialties were required, but in college teaching, practically any person with a doctorate (as well as many without) was in demand. The shortage of university teachers was apparently so great in 1943–45 that even all-male colleges and engineering schools were forced to accept women instructors. The WB found that the number of women teaching at American colleges and universities more than tripled between 1942 and 1946: in 1942, 2,412 women scientists were teaching at 1,573 schools; in 1946, 7,712 were teaching at 1,749 institutions (see table 1.1). Although the WB did not present any data on the number of men on these faculties in 1946, it did say that women constituted 12.1 percent of the total of 20,000 faculty members in 1942. From this, one can calculate that if the number of men dropped by one-third in the next four years and all were replaced by women, the faculty would be about 40 percent female, a probably intolerable situation once conditions returned to "normal."

This feminization of the faculty received far less publicity than did the entrance of women into low-level chemical and engineering jobs in industry. No one seemed eager to herald the women's arrival on the faculty as the opening of a new era. It was a temporary expedient best kept under wraps, and it was not clear whether these jobs were temporary or possibly permanent. The WB expected that the universities would retain all these women faculty members and even add more, since they would be getting record enrollments after the war.[27] Yet many large universities simply declared a moratorium on tenure decisions and postponed all such long-range planning until after the war, when the men would have returned. Thus, for most of the women stand-ins this meant that no matter what heroic feats they accomplished in their temporary jobs, they would be expected to leave without a fuss once victory was assured. Apparently they accepted this situation and served patriotically and without protest in these temporary jobs. At least there was no sign in the *AAUW*

Table 1.1
Women on Science Faculties, 1942 and 1946

Field	1942	1946	% Change
Biological sciences	782	2,587	+230.8
Mathematics	686	1,459[a]	+112.7
Chemistry	485	1,585	+226.8
Physics	178	411[b]	+130.9
Geography	157	115	−26.8
Engineering	50	53	+6.0
Geology	46	93	+102.2
Meteorology	28	16	−42.9
Miscellaneous		590	
More than one science		635	
Science with nonscience		202	
Total	2,412	7,746	+221.1

Sources: The Outlook for Women in Science, WB Bulletin 223-1 (Washington, D.C., 1949), 20 (tables 3, 4).

[a]Includes 44 in statistics.

[b]Includes 64 in astronomy.

Journal (AAUWJ) which would have been interested, of any attempt at an organized challenge.

On occasion, certain civilian agencies of the federal government had to find a woman scientist to fill the place of a man away on other duties. Although the USDA had had no position for her when she applied in 1933, plant pathologist Cynthia Westcott received a telegram in 1943 (perhaps as the result of registering with the National Roster) telling her to report to a USDA installation in Alabama. Although her assignment, which was to develop a cure for a disease in the azalea industry, hardly counts as a pressing war problem, the Southern florist industry had strong enough representation in Congress to get what it wanted, even if only from a woman, in wartime. Because Westcott was familiar with other sclerotinic plant diseases, she quickly discovered the cause of this epidemic, which had baffled her predecessor for years. She thus had the satisfaction of a job well done, but that was about all, since when the man she replaced returned from military service, she had to leave—with no chance at another governmental job, no promotion (which her achievement must have merited), and no veteran's pension such as the returning servicemen would receive from a grateful citizenry.[28]

But even this massive reallocation of the nation's scientific talent was apparently not enough, and by late 1942, if we are to believe official statements of the time, the shortages were worse than ever. (Authorities as eminent as James Bryant Conant and Secretary of Labor Frances Perkins doubted, as did many Southern male scientists not called to war projects, that there really was a shortage of scientists.)[29] Manpower agencies began to authorize new training programs for scientists and engineers, and by early 1943 women and to a lesser extent blacks, the two reserve labor forces of the country, were being specially recruited and trained for jobs in industry. Formerly spurned by chemical companies and engineering firms, they were now eagerly sought out and told they constituted a precious, national resource. The result was a sudden change of rhetoric, a flood of recruitment literature, and a spate of new programs to identify scientific talent and train it in time to contribute to the war effort. In a barrage of propaganda the War Manpower Commission (WMC) and the OWI glorified these essentially industrial positions into "scientific" ones that paid higher salaries than did clerical work. Thus in one of the OWI's first films, *Women in Defense* (1942), actress Katharine Hepburn narrated a script written by Eleanor Roosevelt that urged women to go to work in governmental or industrial science projects. In 1942 the award-winning Greer Garson showed how glamorous and heroic a woman scientist could be when she starred in the movie *Madame Curie*. Here the subject and her role united the patriotic themes of support for France and sympathy for Poland with the romantic theme of love for Pierre and the wartime need for women scientists.[30] As the war dragged on and the recruitment became more intense, bright high school students were sought out, publicized, and urged to major in science in college with such lures as scholarships in the Westinghouse National Science Talent Search (established in 1943) and the Bausch and Lomb Science Talent Search (1944). Both of these had several women among their early winners, and the Westinghouse program even required that there be the same percentage of women among the winners as among the entrants.[31]

Probably more effective in immediate recruitment was the torrent of books and articles that appeared between 1942 and 1945 urging women to pursue careers in science and engineering. Their usual approach was to laud women's past contributions to science, extol their current ones, and then paint a bright future for women in these fields. One of the most responsible of these books was Edna Yost's *American Women of Science* (1943), which contains short biographies and interviews of twelve such women and thus now has a certain historic value. It is probably also indicative of the low visibility of women scientists even during World War II that Yost, a member of the American Association of University Women (AAUW) and a former editorial worker at both the American Public Health Association and the American Society of Mechanical Engineers, expressed surprise in her preface that women had already done so much in science: "But I had not the slightest idea that American women of science had achieved even a fractional part of what they have actually accomplished. I was completely unprepared for what a little specialized research began to uncover. So, it seems, was everybody else. When we signed the contract, both my publisher and I knew there was material available for a pretty good book but we had no idea the achievements recorded would be of the caliber they actually are."[32]

It was not long before the women discovered that most of these much-publicized new opportunities were merely for young women in entry-level jobs. Employers were far more willing to promote the men they had on hand and hire young women as their "aides" and assistants than they were to give even mid-level positions to women already trained (often with doctorates). The government's all-male WMC readily accepted this employer bias rather than try to implement any kind of fair employment regulations for women. (James O'Brien, executive officer of the National Roster, admitted to a conference on recruitment, quoted in the *New York Times* in September 1942, that such resistance was "a real fact" in hiring women scientists.)[33] Faced with employers' intransigence, the government was forced to provide extensive and perhaps unnecessary training (at universities at federal expense) to assure industry a steady supply of the freshly trained beginners whom it would hire. Thus, despite all the presumably pressing needs for trained persons in industry, certain imbalances began to appear; some women "scientists" were sought after, and others, far more qualified, were allowed to remain unutilized. Evidently no matter how badly off the government and industry really were in 1942 and 1943, they were not so desperate that they had to hire women, even highly qualified ones, for mid-level positions.

Actual or expected labor shortages were thus only one part of the scientific manpower problem during World War II. Prevailing attitudes and practices played a much larger part than the availability of trained personnel in determining where women scientists were allowed to work during and after the war. A rich, or at least freespending, country could buy its way around its prejudices; rather than undertake the very difficult task of changing during wartime its attitudes toward occupational roles, it could find and train new personnel that fit its stereotypes. In fact American gender roles changed surprisingly little during World War II, even in the face of the unprecedented demands for technical manpower.

This channeling of women scientists into acceptable (i.e., lower-level) positions was evident in the government's own civil-service recruitment literature. A 1943 booklet, *Civilian War Service Opportunities for College and University Students,* provided descriptions of approximately one hundred jobs, with comments about the employability of women in each. These ranged from the encouragement offered prospective statisticians ("*Women* with statistical training and appropriate supporting experience are in great demand") to the grudging acceptance offered women engineers ("While women with engineering training are usually qualified for junior engineer positions rather than for the higher grades listed here, women who are qualified under this announcement are urged to apply"). Women physicists were channeled into lower grades:

"Women should be encouraged to enter the field of physics. There are a large number of openings for women, particularly in the lower grades." But the greeting to women in the agricultural fields was not even this lukewarm. For the position of "junior soil conservationist" they were warned, "*Women* have not held many positions of this type," and for the higher-level "agricultural specialist" there was the brief statement, "These positions will be filled primarily by men."[34]

One male-dominated field where the shortages were so great that women were vigorously recruited during World War II was that of engineering. There had been very few women engineers before World War II, but the field now seemed to be offering women unprecedented opportunities. The *American Men of Science (AMS)* had listed eight such women in 1938 (0.4 percent of its total women and 0.2 percent of its approximately 3,500 engineers), and the National Roster found 144 in 1941, but this was still only 0.3 percent of its almost 50,000 engineers. But these relatively few trained women were quickly absorbed by war projects (for example, Doris Wills, who had been at home with her twin daughters, became a control system engineer with the Minneapolis-Honeywell Company in Philadelphia during the war). The shortage was perceived as so great that twenty-nine engineering schools that had hitherto excluded women, including the Carnegie Institute of Technology, Columbia University's School of Engineering, and Rensselaer Polytechnic Institute, began to admit them between 1940 and 1945. By then the number of women engineers on the Roster had nearly tripled, to 395, but even this increase was far smaller than it was for men counted there, and most of it took place at or below the bachelor's degree level.[35]

But these numbers were minuscule when compared with the numbers of women who became "engineering aides" during World War II. When the call went out for still more women "engineers" in 1942 and 1943, civil engineer Elsie Eaves of the *Engineering News-Record* warned women to be aware that recruiters used the term "engineer" to describe two very different jobs: those few women with degrees in the field could expect beginning professional positions, but most others, including college graduates with a few additional courses in drafting and machine testing as well as women without degrees, were being used in subprofessional jobs as "engineering aides" or temporary assistants to men who had recently been promoted from lower positions within the company. There were plenty of openings for women in both kinds of jobs—even the federal government changed its longtime policy and began hiring women engineers in late 1942—but a young woman reading articles glorifying the contributions of women "engineers" to the war effort, such as "Co-ed Engineers Take Men's Places" (at Curtiss-Wright Aircraft Company in 1943), should realize that they were just talking about these temporary aides, or "cadettes," whose postwar prospects were uncertain.[36]

So great was the anticipated demand for this kind of subordinate personnel that the federal government set up a special program as early as October 1940 to subsidize their training. Administered by the U.S. Office of Education (USOE) and funded by a special appropriation to its vocational education budget for "defense training," this program was one of the first federal efforts to increase and train scientific manpower. At first this Engineering, Science, and Management War Training program (ESMWT) provided special training only at four-year technical schools. But as the shortages increased, it was extended to a total of 227 colleges and universities across the nation, including several women's colleges (Hunter, Simmons, Smith, Wellesley, Vassar, Bryn Mawr, and Wilson colleges), where courses were taught in elementary engineering, mathematics, chemistry, physics, and safety engineering.

This program turned out to be an open door of opportunity for many women and some minority group members. Because of the provision in its appropriation that "no trainee . . . shall be discriminated against because of sex, race, or color" and its special recruitment of women, their percentage in the program rose from less than 1 percent the first year to a surprising 21.8 percent in 1943–44. The program's final report revealed that of its almost 1.8 million enrollees between 1940 and 1945, there had been more than 280,000 women (15.7 percent) and 25,000 blacks (1.4 percent). This report was also very enthusiastic about the women's performance not only in the courses but also on the job, where their work was often considered better than that of men who had been in the program. Practically all of the graduates went to work immediately in industrial jobs, which usually had been arranged ahead of time, and they thus joined "Rosie the Riveter," the famed metalworker and poster woman, in increasing American production of essential war equipment. Some industries were so desperate for these additional trained workers that they cooperated with the ESMWT in establishing joint programs, such as that between Cal Tech and the southern California aircraft companies. Others, such as the joint Goodyear–University of Cincinnati program, even paid the women "cadettes" eleven dollars per week during their training. Based on its own final report, this program would seem to have been a model for showing what young women could do in a highly "masculine" field when public opinion favored it and government and industry gave them training and jobs.[37]

Yet these "aides" were both praised and blamed. In a poll of one group of the women, the WB found that 93 percent had liked their jobs most of the time, felt well paid, and found the experience worthwhile. Most also indicated, however, that the jobs they were given were so far beneath their level of training that they became bored with the work after two months and frequently left for better jobs or returned to college. As usual, the employers did not accept any responsibility for this situation but blamed the women for the high turnover, that is, for requesting transfers, expressing dissatisfaction with their assignments, and "expecting special treatment." They also blamed the women for their youthfulness, their lack of mechanical know-how, and their lack of work experience, which meant that they had required extra supervision at a time when supervisors were very scarce. In short, employers complained that the women substitutes were not their usual industrial employees. From the employers' postwar perspective, the program was thus only a minor success and not worth continuing, despite the women's low wages.[38]

Another field for which large numbers of women "scientists" were vigorously recruited and specially trained during World War II was industrial chemistry. So great was the demand for women in this field that recruiters toured the women's colleges in January and February 1942 to sign up the June graduates even before they had begun to seek work, an abrupt change since the Depression. Several articles appeared in the *Chemical and Engineering News (C&EN)* and the *Journal of Chemical Education (JCE)* in 1942–45 describing enthusiastically how well women "chemists," often without a college degree, were fitting into low-level jobs replacing men in the very industries—rubber, distilling, pharmaceuticals, petroleum, and metallurgical testing—that had systematically spurned them in the 1930s. Of the 2,900 women "chemists" added to the Roster by 1945, over 90 percent held a bachelor's degree or less.[39]

There seems to have been almost no industrial demand for fully trained women chemists. Unlike in engineering, there was a considerable supply of experienced women chemists in the United States at the start of the war. Almost 1,700 women chemists, of whom 395 held doctorates, were listed on the Roster in 1941. In wartime, and with such widespread shortages, these accomplished women should have been in great demand for responsible positions. Yet the experience of women chemists repeatedly demonstrated the ceiling or limits placed on women scientists in industrial employment. Most of the experienced and doctoral women chemists seem to have remained at their teaching posts, probably expecting that, having completed a questionnaire for the National Roster or taken other steps to seek war work, they would soon be called to an important project. Relatively few women chemists ever got such an assignment, however. Aside from those at the OSRD mentioned above, only three others seem to have held positions worth noticing, and one of them may not even have been a scientist after all. Gladys Emerson, one of Herbert Evans's research associates at the University of California, joined the Merck Company in Rahway, New Jersey, to work on vitamin preparation in 1942; Mary Rolland, who had held several possibly secretarial positions in the chemical industry in the 1930s, was put in charge of production priorities for petrochemicals at the War Production Board in Washington, D.C.; and Elizabeth Aldrich Bridgeman, who had joined the NBS in 1928, worked on fuels and lubricants, wrote numerous reports, and "directed the research of 10 to 30 assistants" there.[40]

This lack of demand for advanced women chemists must have been a factor in both fueling and then undercutting their rather limited protests during World War II. As they had before the war, women chemists continued to meet annually and give each other rather frank practical advice on how to behave and hold onto their jobs in industry. Thus, in a paper presented to the Women's Service Committee of the American Chemical Society (ACS) in September 1942 (and later published in the *JCE*), Lois Woodford of the American Cyanamid Company in Connecticut used her position as a chemical secretary to collect management's negative views of women chemists and to give traditional advice on how to overcome their many perceived weaknesses. In the expansive mood of a national buildup in 1942, the belief that individual merit would lead to higher pay and advancement was still credible, and even Icie Macy Hoobler, the outstanding liberal nutritionist, wrote Woodford that such self-discipline was the only answer.[41]

But a year later, at the September 1943 ACS meeting in Detroit, the women's mood had changed drastically. The Women's Service Committee sponsored an evening symposium entitled "Opportunities for Women in Chemistry Now and After the War," which was followed, as a report put it euphemistically, "by an active and rather heated discussion on woman's position today in the field of chemistry and her prospects for future success in the field."[42] The perennial problems of hostile employers, low pay (half that of men), lack of advancement, and the high turnover of married women all seem to have been discussed in animated terms. Yet no action was suggested. About all that these women chemists could think to do, even in 1943, at the height of the wartime demand for trained scientists (which was apparently close to a fiction in their eyes), was to reiterate what must have seemed barely tenable, namely, the importance once again of good work and a positive attitude: "In all cases, the success of a woman in chemistry depends upon her qualifications and her attitude. If a woman is absolutely sincere, there is no reason why she cannot succeed. If her employer shows symptoms of prejudice, she may help to overcome this attitude by doing just a little more than is expected of her. She can go just as far as she wants to, as long as she puts everything she has into her job."[43]

They had no real evidence that this strategy would work any better then than it had in the past. About all that they could do was hope that future generations would be able to solve the problem. The women's complaints still did not make any noticeable impact on management. Even those men who were trying to recruit college women into industrial employment, who might therefore have been expected to read the article from which the above quotation is taken, still spent most of their time telling the women what unstable and unreliable employees most of them would make! Thus, all Walter J. Murphy, editor of *Industrial and Engineering Chemistry,* could say at a college-industry symposium in May 1944 to those women who thought they should be getting equal pay was that they should "adopt a definitely realistic attitude." They should realize that women, in general, had so many weaknesses that an employer would find it hard to take them seriously. Almost any male chemist was preferable to any woman, since the women were known to be unable to do teamwork as well as (all) men; unwilling to work under the supervision of other women; lacking in aggressiveness (which seemed to mean originality, perseverance, dedication, and eagerness to get ahead) and in the focused type of imagination that (all) men have. In addition, of course, women would leave to get married and had probably attended a woman's college, where, Murphy said to their faces, they would have received an inferior kind of training! (No one, apparently, spoke up to challenge him on this blatant distortion of reality.) Accordingly, women who expected to be "equal" should "make adjustments" and lower their expectations. Nevertheless, Murphy concluded by expressing the vain hope that if women could "prove themselves" by their wartime experience, their postwar world looked "bright." In short, the women chemists had to expect to have better qualifications than male chemists in order to have any chance at "equality." Yet even this was unlikely, since a "woman chemist" was by definition too unstable to be a "good chemist" or even a reliable employee. Certainly nothing had changed in management's mind since the 1920s, and all that stoicism and overqualification that women chemists had been advocating for more than twenty years had not improved women's situation in the chemical industry.[44]

By September 1944 the women chemists seem to have become quite resigned to their plight. Lois Woodford was back with more advice on how to overachieve and adjust and even blamed the women for the high turnover in the entry-level jobs to which most of them were restricted. Other speakers reiterated the theme that a positive attitude was a woman chemist's only chance for any kind of advancement. In fact the mood of the 1944 meeting seems to have been quite somber, with one speaker advising the women to retain their positive attitude even to the point that "if demotion follows, it should be accepted cheerfully."[45]

It probably did young women no harm to be told that they would not be in as great a demand after the war ended as they were now and that they should start as soon as possible to upgrade themselves by taking night courses or enrolling for an advanced degree if they wished to hold on to their jobs. But this very cautious advice to the young, which was the product of the World War I and Depression experience, was no help to the hundreds of more advanced women who had already proven themselves and who might have wished to use their sudden leverage to some better effect. This kind of advice not only ignored their existence but also perpetrated industry's tired bromides about why it could not promote women. Even if it was unrealistic to expect industrial leaders to change their views, one might have expected some more venturesome strategy from the women's leaders themselves. Having worked in their chemical societies and "fraternities" for several decades, they might have been expected to have a new strategy for changing public opinion. At the very least the leaders of these women's organizations might have challenged management's views that no women were worth promoting, as engineer Lillian Gilbreth did in 1943 (see below), and used such occasions as Murphy's talk in 1944 to criticize this industrial mentality and push for some top-level appointments for women. Although biochemist Florence Seibert had written to the president of Iota Sigma Pi in 1942 urging some action—"In spite of all the miseries and setbacks due to a war I am sure there will be a forward step for women chemists, and our ranks will be greater and more important than ever. Iota Sigma Pi can become a great influence in this progress"[46]—there is no sign that any women's chemical organizations took any effective action to address the issue. Nor were any chemists visible on the committees of the National Federation of Business and Professional Women's Clubs (NFBPWC) and the AAUW that were protesting in Washington (see below). In fact, not only were the women leaders in chemistry unwilling to press for new opportunities during World War II but they were very quick to beat down any signs that their conservative strategy of extra qualifications and stoicism was inadequate and that a more vigorous, aggressive alternative might be needed.

Such a new view was presented in 1946 by two discouraged young women chemists at the Hercules Powder Company of Delaware. Eleanor F. Horsey and former Sarah Berliner Fellow Donna Price pointed out in an article entitled "Science Out of Petticoats," in the *AAUWJ*, that no women were promoted at chemical companies, whatever their accomplishments, and that the only possible solution was for management to change its attitudes. "The remedy to these problems is one that can be applied only from above by men at the head of management: regard the woman chemist as a chemist rather than as a woman."[47]

Although it was not clear in 1946 how this change might be brought about (governmental legislation on the issue was almost unthinkable), other women chemists do not seem to have applauded this fresh viewpoint. Rather than being praised and supported for shifting the blame away from the women's presumed lack of qualifications and onto industry's job structures and management's promotion policies, the two young women were sharply rebuked a few months later by none other than Emma Perry Carr, longtime professor of chemistry at Mount Holyoke College and author of a book of career advice for women chemists after World War I. Carr's letter to the editor in the next issue reiterated her belief that women chemists' worst enemies were other less qualified and uncommitted women, whose low standards of work and rapid turnover in industry hurt them all. Truly professional women chemists would stick to "a rather higher quality of work than might be expected of a man whose choice of profession is assured" and, through such self-discipline, "*within our*

group . . . keep open the door of opportunity for women in the postwar era"
(Carr's italics).[48]

Carr and others at Mount Holyoke College were still convinced of the value
of this strategy in 1946, despite all the evidence that it had failed to improve
women's lot in industry and had certainly not brought them any advancement
even in wartime. She continued to fear that any protest would hurt them far
more than it could help, that it was much more likely to lead to reprisals and less
employment for women at all levels than it was to bring about any improvements. She may have understood that the younger generation now wanted
more than just marginal entry-level positions in the chemical industry; they
wanted advancement, and were upset and insulted when a less qualified man
was advanced ahead of them and then, as Horsey and Price had described it,
had to ask their advice about how to do his job. Carr had no remedy for this
problem, except to say that a highly professional attitude by the women chemists would impress managers and lead to improvement eventually. From her
attitude and the reluctance of most women chemists to speak up and fight for
greater opportunities, one can see that neither the leading women in chemistry
nor the men running the companies had changed their thinking about wom-
en's advancement in industrial chemistry as a result of their wartime work.
Thus these conservative academics lacked the whole feminist perspective of the
unjustness of women's second-class status. They seemed to believe that women
scientists would not deserve equal employment until sometime in the distant
future when, despite their restricted opportunities, their collective performance had equaled (or bettered) the men's statistically.

The women chemists failed to use the wartime situation, in which for the
first time in decades they had some leverage, to improve their status and
opportunities. The leaders lacked confidence, commitment, and an effective
strategy to deal with the new demand and settled instead for a continuance of
the marginal status they had held for several decades. Not even the leaders of
Iota Sigma Pi or Sigma Delta Epsilon, who were used to helping the women
scientists adjust to discriminatory employment, seem to have offered any new
thoughts on the issue of women's place in postwar science. They were, in fact,
even more pessimistic than the recruitment literature of the time. Rather than
expecting that women's wartime achievements would impress employers and
lead to expanded wartime employment as the vocational guidance literature led
young women to expect, these older, experienced leaders apparently feared a
backlash like the one many of them had experienced after World War I.

More visible and more vigorous but equally ineffective were the efforts
throughout the war of several leaders of the NFBPWC and the AAUW, both
headquartered in Washington, D.C., to assure women's participation in the
expanding wartime employment. If a qualified woman could not get a mid- or
upper-level position in wartime, when there were presumably pressing short-

ages in several fields, what hope was there for any woman later on? When an unnamed former AAUW fellow wrote to the editors of the *AAUWJ* in 1942 that despite her doctorate in chemistry and weeks of looking in Washington, she could not find a war job, she got a sympathetic hearing. She wrote that "there are numerous positions open for clerks, secretaries, and laboratory technicians with various degrees of moderately specialized training. That is all a woman is expected to offer. Actually it seems that a woman is not wanted for a position entailing any real administrative responsibility. It is just the old cry of 'wolf.'" This confirmed the editors' suspicions. They agreed that "the men in authority are more likely to accept women welders than women scientists, more inclined to hire women bus drivers than women administrators."[49]

Similarly, a series of articles entitled "Womanpower 4-F," in the NFBPWC's *Independent Woman (IW)* in 1943, published the angry letters of many rejected by the very industrial companies that were presumably in dire need of workers.[50] Although none of these women were scientists, their criticisms did reveal a sharp split between the types of workers that the government was wooing (female, black, and older ones) and the ones that employers really wanted (experienced men in excellent health and under 40—or even 30). In fact this split between the government's rhetoric and occupational realities, so evident and seemingly outrageous in wartime, when ability and skills were touted as of greater importance than age and experience, was the start of a longtime credibility gap between the government's rhetorical conception of reality and the women's experience of it.

Yet what could the women do to protest and try to change such employment policies and attitudes? Fortunately, at least two leaders, neither of them scientists—Minnie Maffett, a Dallas physician, who was president of the seventy-six-thousand-member NFBPWC in 1941–44, and Sarah T. Hughes, a Dallas judge, who chaired the AAUW's Committee on the Economic and Legal Status of Women in 1941–47—were, along with Eleanor Roosevelt in the White House, among the strongest and most outspoken voices of protest during the war. Although they would have been the first to admit that they were largely unsuccessful in their efforts, they did win a few important victories, and even their presence and attempts were a big advance over women's position in World War I.[51]

Since the basic goal of these women was the appointment of women to high-level positions in the government, they utilized a variety of political tactics. Not only did Maffett urge those members of the NFBPWC with scientific training to register with the government's National Roster but, foreseeing that governmental officials might not realize that women had as many skills as they did, in 1941 she had the federation start its own listing of members who might be of use on wartime projects. Then whenever a new commission or other war agency was started, Maffett and the NFBPWC could recommend a suitable list of appointees. Maffett's major success in this direction was a partial one. After she spent months urging chairman Paul McNutt to appoint a woman member of the WMC, he responded by making the NFBPWC's own vice president, Margaret Hickey of St. Louis, chairman of a subordinate Women's Advisory Committee to advise the commission on women's affairs but not vote on them. It could no longer be said that women lacked the necessary qualifications for high public office.[52]

When, however, as usually happened, the women's nominations were ignored or otherwise passed over, Maffett and others resorted to a second, more public tactic: they took the offensive and proceeded to create their own publicity. For example, in 1942 Maffett issued an "open letter" to Paul McNutt, chairman of the WMC, protesting the small number of women being appointed to major positions in Washington and printed it on the inside cover of the *IW*. (She also published his evasive reply a month later.) Since Maffett also kept track of what major appointments were being made in Washington, in her frequent addresses to the membership she mentioned how few women were being given these positions. In September 1943 she issued a full report on the subject showing that only 8 of the 660 (1.2 percent) high-level jobs in new wartime federal agencies had gone to women. This report received such good press coverage (including in the *New York Times,* where, however, it was buried in the women's ["Society"] section) that it led to an even larger publicity-generating event, the White House Conference on How Women May Share in Policy-Making in June 1944.[53] The NFBPWC and the AAUW sponsored this meeting, but it must also have had the strong support of Eleanor Roosevelt herself. It in turn may have helped get a few women appointed to some postwar commissions, such as the commission to create the United Nations, which first met in San Francisco in April 1945.[54]

But suggesting names and then complaining later when they were not appointed was not the women's only tactic. A second approach, perhaps more effective in terms of consciousness-raising, to use a modern phrase, than in terms of actual response was confrontation or direct challenges, whenever the chance arose, to the vague and evasive statements of governmental and industrial spokesmen. Thus, engineer Lillian Gilbreth, an educational adviser to the OWI and a member of the education subcommittee of the WMC, and one of the most outspoken women critics of the government's manpower policy, spoke directly to the issue of women's advancement at a February 1943 college-industry conference at the AAUW headquarters in Washington. In her talk, entitled "At What Levels in Industry Are Women's Colleges Reservoirs for Employment?" Gilbreth deliberately raised such questions as, Were women in industry to be given promotions, and had any gotten them already? Were they to be given the assistants and aides that men were regularly assigned? Or were they all to be secretaries, and the women's colleges then berated for turning out poorly trained ones at that? Unfortunately, the responses to Gilbreth's address were not reported in any of the several published accounts of the conference.[55] One can assume, however, that if there were any, they were evasive. The "spokesmen" that were usually sent to such meetings were probably instructed to pacify any women critics but make as few commitments as possible. If the women were learning the tactic of confrontation, the male administrators were old hands at defusing it.

Probably the women's most effective tactic, and one that allowed them to advocate positive new programs rather than just protest past omissions, was that of lobbying and testifying before Congress, often against the military, in favor of legislation that would open new opportunities for women. Thus the NFBPWC, the AAUW, and other women's groups repeatedly spoke up for the Women's Bureau's budget and sponsored much new legislation, such as that creating the women's branches of the armed forces, commissioning women physicians for the military, and admitting women to the specialized officers' training schools (e.g., those for transportation and military government). They also advocated more promotions for women in the foreign service and postwar jobs for women veterans. In addition, they supported numerous bills relating to women's economic and legal status, including social security benefits, income tax status (opposing mandatory joint returns for married women), and the whole struggle for equal pay. Although the more conservative AAUW continued to oppose the Equal Rights Amendment, the NFBPWC succeeded in getting both the Democrats and the Republicans to endorse it in their party platforms in 1944. The women did not win immediately on most of these economic issues, but they were laying the framework for much of the progressive women's legislation of the next thirty years.[56]

Yet Maffett herself was disappointed in how little progress her efforts won. In mid-1944, upon stepping down from the presidency of the NFBPWC, she stated that most of these efforts had come to far less than she and others had hoped: there were still hardly any women in top positions in any part of the war effort.[57] Why this was so is not really clear. There were hundreds, perhaps thousands, of new executive jobs and, one suspects, not enough good men to fill them, yet the men in charge were unwilling to let the women have even one vote on the WMC, which controlled *millions* of women's jobs. Nor was public support for women strong enough to force them to do so. Even the women on the WMC's Women's Advisory Committee did not, according to Eleanor Straub, who has examined their history, push very hard for a stronger role; they tended to be apologetic when they asked for a stronger voice concerning even those issues that affected women most strongly. Nor, as Lillian Gilbreth observed, were most women really convinced that they deserved an equal role in the war effort. All too many, she found, had to battle the widespread belief that men should automatically get the best jobs.[58] Thus, despite the greatly increased employment of women at certain levels during World War II, resistance to them at higher levels and especially in the professions remained strong even, apparently, among the women themselves.

Perhaps in wartime the citizens felt safer in men's hands. Certainly protest was more difficult then, since critics of governmental policy could all too easily be accused of unpatriotic behavior. In any case, even at the height of the war, when gender roles were being stretched and changed, traditional American antifeminism remained very strong and perhaps even intensified. Professional women were learning how few friends they had politically, despite their important contributions to the war effort. During the Depression it had seemed that the biggest obstacle to women's advancement was the economy, but now, when the economy was booming, basic social attitudes remained quite conservative, despite all the oft-proclaimed "shortages." The demand for qualified persons in high places could never have been stronger, and yet the antifeminism was so strong and so persistent at high levels that it held back women in good times as well as bad. A tremendous change in public opinion, as well as full employment, would be necessary to uproot this basic belief.[59]

In assessing the war's overall impact on women's role in science, one is tempted to conclude that despite all the activity and publicity about their advances and wartime "firsts," the women came out of the war only slightly stronger than they had gone into it. Some women scientists, perhaps about 100, had held temporary jobs on governmental and industrial war projects. Many others, who had been unemployed in the 1930s, had apparently found temporary positions, especially in college teaching, during the war. Potentially more important, the numbers of women scientists had more than doubled in just four years, increasing by the hundreds if one counts only women with higher degrees, and by the thousands if one includes those with a bachelor's degree or less (see table 1.2). But owing to the overall wartime recruitment of both sexes, as well as improved recordkeeping and a broadened definition of what constituted a scientist, the total number of scientists on the Roster also more than doubled in these years, from 131,440 in 1941 to 324,145 in 1945. The result was that, despite all the talk about women's advances in science during the war, their percentage of the total on the Roster rose almost imperceptibly from 4.0 percent in 1941 to just 4.1 percent in 1945. Even then most of the women were added mostly at the lower levels, and layers of bureaucracy were added above them.

Yet the main reason why the women scientists failed to capitalize on their wartime demand as effectively as they might have has to be the prevailing public opinion about women's role in American society. To a large extent, the women scientists shared this ambivalence themselves. Although some women's leaders were out front challenging industry spokesmen, testifying in front of

Table 1.2.
Scientists on the National Roster, by Sex and Level of Education, 1941 and 1945

Level of Education	Total		Women		Women as % of Total
	N	%	N	%	
December 1941					
Doctorate	23,782	18.1	1,817	34.1	7.6
Master's	25,403	19.3	1,798	33.8	7.1
Bachelor's	64,118	49.5	1,529	28.7	2.3
Other	17,137	13.0	179	3.4	1.0
Total/average	131,440	99.9	5,323	100.0	4.0
December 1945					
Doctorate	27,210	8.4	1,958	14.6	7.2
Master's	41,074	12.7	2,930	21.9	7.1
Bachelor's	195,022	60.2	7,403	55.2	3.8
Other	60,849	18.8	1,117	8.3	1.8
Total/average	324,145	100.1	13,408	100.0	4.1

Sources: National Roster of Scientific and Specialized Personnel, *Report to the National Resources Planning Board, June 1942* (Washington, D.C.: GPO, 1943), app. D; idem, "Distribution of Roster Registrants by: 1. Professional Field, 2. Sex, 3. Age, 4. Extent of Education, December 31, 1945" (Washington, D.C., mimeographed), tables 2, 4.

governmental committees, and fighting those in power for larger opportunities for women, others, especially among scientists and academics, shrank from this role. They were not convinced that they were equal and deserved better jobs, and they were unsure whether, or how hard, to push for them. Their leaders, senior academic women chosen for the most part because of their success in a conservative profession, tended to be cautious and more inclined to dampen the fires of protest than to stoke them. They saw little value or possible gain in political activism and were more comfortable telling the other women to work harder to measure up to higher levels of discipline and achievement. Yet it is debatable what they could have accomplished if they had organized a unified onslaught against the pervasive discrimination; the barriers and villains were everywhere, while the allies were few and halfhearted. Nor did this very unwillingness of so many women to protest and push for advancement in wartime, when they were at least thought to be needed, bode well for their advancement in the postwar world.

By 1945, then, the women who had seemed in such demand and performed so well in wartime knew how weak and short-lived their welcome really was. Regardless of how well they performed, they were not promoted, and most expected to be ousted from their temporary positions. They were, as a whole, still the victims of social and economic forces they could not control, the reserve labor force that could be called up and channeled into certain levels of work as occasion required. They had no leverage or way of consolidating even these minimal wartime gains, and if all the expected layoffs took place, they might be as badly off in the late 1940s as they had been in the 1930s. Their only hope was that in science the highly publicized shortages would continue into the postwar world. Under such conditions of "full employment," the women might continue to be in as much demand as in wartime. How real these shortages were and what would really happen after the war, however, remained to be seen.

2

Postwar "Adjustment"

Displacement and Demotion

Long before the actual end of the war in September 1945, planning the postwar world had become a favorite American pastime. Starting as early as 1942 and reaching a crescendo in 1944–46, articles on the subject abounded in popular and professional journals. A central concern in these prescriptions for utopia was economic prosperity: there should be no more depressions like the last one, and to that end the federal government should guarantee full employment, by deficit spending if necessary. The place of women in this postwar world was as problematical as it always is in utopias. Some persons thought that women should be allowed to work if they wanted to or had to. Others thought that they should all be at home raising children. What female scientists should do was even less clear. Although it was a common (and accurate) assumption that the postwar economy would be a highly technological one, requiring even more scientists and engineers than had been needed during the war, women and scientists continued to be two separate, almost mutually exclusive populations in most postwar thinking. By 1946–47, however, the women scientists' contributions to the war effort, highly publicized just a few years before, had been completely forgotten. They were eclipsed by the returning veterans, who, aided by the famed GI Bill, itself a measure to guard against another depression, overwhelmed the nation's better colleges and universities. Aided by quotas and other enrollment restrictions, male veterans soon displaced female applicants and, before long, female staff and faculty. But the women did not protest very much or very effectively. A few spoke out, but in the late 1940s public opinion, ever ambivalent about the role of women, was far less sympathetic to the plight of the achieving woman than it had been earlier. Even groups that traditionally had been supportive of academic and professional women had been so undermined from within and intimidated from without that they were unable or unwilling to speak out forcefully. Thus, women's wartime accomplishments, rather than justifying an increased role for women in the postwar world, were quickly forgotten. The postwar pronatalism displaced and demoted them and put them on the defensive, just as the nation was gearing up for the largest expansion in its history, which, unlike the war, was not expected to be temporary. The postwar period would have lasting and detrimental impact on women's role in science for many years to come. In women's history there can be major steps backward.

The best source on women's situation and prospects in this transitional period, for both its data and its other information, is the series of eight employment bulletins on women scientists prepared by Marguerite Zapoleon, chief of the Employment Opportunities Section of the WB, and her staff in 1946–49. Inspired by contemporary concern about potential shortages of scientists, especially by Vannevar Bush's *Science, the Endless Frontier* (1945), which she cited on occasion, Zapoleon was well qualified to mastermind this first governmental study of women in science. She held a bachelor's degree in engineering, had done graduate work in economics and social work, and had been a vocational guidance counselor in the Cincinnati public schools, where she worked for a time with psychologist Helen Thompson Woolley, before moving to Washington, D.C., in 1935. In the vocational guidance tradition, her bulletins provided thorough treatments of the past experience, current situation, and future prospects of women in chemistry, biological sciences (botany, bacteriology, and zoology), mathematics and statistics, architecture and engineering, physics and astronomy, geology, geography, and meteorology, and "occupations related to science."[1] In order to estimate the current number of women in science, she used several diverse but readily available sources of data, such as the 1940 census, the *AMS,* 7th edition (1944), the wartime Roster, and membership counts of various professional associations. Thus, her total of 11,510 to 12,460 women scientists and engineers, distributed as shown in table 2.1, was lower than the 20,212 of the final (1945) wartime Roster, which had included many women with bachelor's degrees, but much higher than the nearly 3,000 given by the *AMS,* 8th edition (1949), presented in table 2.2, which included few engineers and a preponderance of Ph.D.'s. Her overall percentage of women in "science," 7.4 percent, dropped to a mere 2.7 percent when she included the heavily male-dominated field of engineering.[2]

What made Zapoleon's bulletins unique and historically valuable was that she and her staff not only collected numerical data on the current supply of women scientists but also visited hundreds of employers and telephoned others to ascertain as accurately as possible the probable future demand for women trained in the sciences. While in the field, they made note of what types of jobs women had held earlier, what types they held then, and, based on employers' comments, what types they probably would hold in the future. Thus, Zapoleon's estimate that in 1946–47 the employment of women scientists was down about 10 percent in most fields from the wartime peak in late 1944 but still far above the prewar level of 1940 seems as accurate an assessment of the

Table 2.1.
Scientists and Engineers, by Field and Sex, 1946–47

Field	Total	Women	Women as Percentage of Field
Chemistry	77,000	5,400	7.0
Mathematics[a]	10,200	2,050	20.1
Bacteriology	4,000	1,000	25.0
Physics	18,450	900	4.9
Zoological sciences	7,840	610	7.8
Biology, general, other	3,200	600	18.8
Botanical sciences[b]	10,000	350	3.5
Geology	11,000	330	3.0
Geography	800	140	17.5
Astronomy	600	100	16.6
Meteorology	2,800	30	1.1
Total sciences	145,890	11,510	7.4
Engineering	317,000	950	0.3
Total	462,890	12,460	2.7

Source: The Outlook for Women in Science, WB Bulletin 223-1 (1949), 5.

[a] Does not include statistics.

[b] Includes forestry.

situation as we are likely to get. Likewise, her expectation that the demand for them would continue strong for a few years at least, since the economy was expanding, but that most of the women added would be in traditionally feminine areas seems cautiously realistic. She also noted strong geographic variations in the employment of women in science, the best prospects being in college towns and northeastern urban areas. In addition, four groups of women scientists—those who were over forty, married, black, or handicapped—faced special obstacles. They could expect to be employed at levels far beneath their actual training or skills and should aim for library or editing jobs.[3] This attitude typified the one-sided "realism" of the time: women had to be exceptional and push themselves to the utmost to merit equal treatment, but employers had to be accepted as they were. Thus, Zapoleon advised women to plan their future in order to avoid or minimize this very real discrimination. Zapoleon's message was thus a mixed one: be something of a pioneer and go into science, but conform and adjust to the prevailing prejudices there. She did not tell young women that they were equal. Nor did she suggest any strategy for their getting beyond these traditional areas of "women's work." More of the same was her safest prediction, and yet even as she, the government's expert on the subject, wrote, historically unique events were already unfolding on the nation's college and university campuses.

One of the main reasons why the postwar employment demand for women

Table 2.2.

Scientists and Engineers, by Field and Sex, 1948

Field	Total	Women	Women as % of Total
Biology	6,915	807	11.7
Botany	(1,908)	(245)	12.8
Bacteriology	(1,690)	(238)	14.1
Zoology	(1,142)	(176)	15.4
General biology	(1,196)	(121)	10.1
Entomology	(979)	(27)	2.8
Chemistry	12,649	579	4.6
Psychology	1,892	385	20.3
Fields related to medicine[a]	3,171	269	8.5
Mathematics	1,993	196	9.8
Nutrition and foods	624	188	30.1
Medicine[b]	3,402	149	4.4
Other[c]	2,483	131	5.3
Earth sciences	2,487	108	4.3
Physics and electronics	4,094	108	2.6
Astronomy	252	25	9.9
Statistics	305	25	8.2
Agriculture	2,889	15	0.5
Engineering	5,840	8	0.1
Metallurgy	677	5	0.7
Total/average	49,673	2,998	6.0

Source: AMS, 8th ed., 1949, as calculated from data on sample of "about 84%" presented in *Employment, Education, and Earnings of American Men of Science,* BLS Bulletin 1027 (Washington, D.C., 1951), 7, 36.

[a] Includes clinical medicine, neuropsychiatry, obstetrics, ophthalmology, pediatrics, public health, and radiology.

[b] Includes anatomy, dental medicine, pathology, physiology, pharmacy, and veterinary medicine.

[c] Includes architecture, military applications of science, manpower resources, and all other.

scientists looked even as promising as it did in the late 1940s, despite all the initial fears of a postwar depression, was that rather than seeking immediate employment, more than one million veterans went to college or graduate school in 1946 and 1947 on the GI Bill. The educational provisions of the Serviceman's Readjustment Act of 1944, passed by Congress two weeks after the Normandy invasion, offered those honorably discharged veterans whose schooling, college, or professional education had been interrupted by military service up to five years of full tuition plus a subsistence allowance at the college of their choice, an important provision. This was part of a lavish package of veterans' benefits—pension rights, medical care, unemployment benefits, mustering-out pay, low-cost insurance, mortgages and loans, job training for the disabled, and social security credit for time served—that a grateful citizenry showered upon its heroes and heroines. (The Veterans' Preference Act, passed a week later, gave them an additional lifetime advantage in most governmental jobs.) The nation's four hundred thousand women veterans were also eligible for these benefits, but uneasily so, since they were frequently overlooked by groups that assumed that all veterans were male, and they were even omitted from some publications by the Veterans Administration (VA) itself.[4]

Early surveys of the veterans' postwar plans greatly underestimated their eagerness to return to college and invest in further training. No one predicted that an overwhelming 7.8 million World War II veterans, according to the calculations of a later governmental commission, would choose to take advantage of the educational provisions of the GI Bill to attend the nation's top colleges. These institutions, which had been underutilized during the war, now found that they did not have adequate staff or space, especially housing, for all their thousands of male applicants. Accordingly, male students took over the women's dormitories, and when this was not enough, the universities sought federal aid for additional temporary housing in trailer parks, Quonset huts, even nearby former military or naval bases or hospitals. Those told to wait or to attend other institutions (i.e., the teachers colleges and the smaller liberal arts schools, which the veterans were bypassing for more prestigious institutions) were the nonveterans, especially women. Cornell University, for example, where women had constituted more than 50 percent of wartime enrollments, quickly pared women's enrollments back to 20 percent of the 1946 total. State universities, such as Michigan and Wisconsin, where women had accounted for 64 percent and 70 percent, respectively, of wartime enrollments, now refused to accept applications from out-of-state women. (Later, when the deluge continued and officials realized that some veterans were women, they revised and extended the ban to all out-of-state nonveterans, still mostly women.)[5] Before long the percentage of women students on the bulging campuses had dropped to just 25%–35 percent, at which level it remained into the 1950s.

Even some women's colleges eagerly accepted male veterans in 1946. Pressed by Governor Thomas Dewey to take in more veterans, even if it meant lowering standards, several private and city-run women's colleges in New York State, such as Sarah Lawrence, Elmira, Vassar, Russell Sage, Adelphi, and Hunter (at its uptown campus in the Bronx) accepted modest numbers of male veterans in 1946.[6] Most dramatic of all, the heretofore all-male campus of the University of Florida at Gainesville was so crowded that the state not only sent male veterans temporarily to the Florida State College for Women at Tallahassee but even remade that into the permanently coeducational Florida State University. A decade later an official there reported on the great developments that had taken place since the changeover to coeducation—a big increase in the state appropriation, more students and more air-conditioned buildings, a larger and better-trained faculty, higher salaries, more publications, and even graduate work. But what he failed to say was that most of the newly hired faculty, who were given far better resources and lighter teaching loads than their predecessors, were men.[7] Both the plentiful new resources and the self-congratulatory tone were harbingers of what was to come.

Not only were the men taking over certain parts of the women's colleges but, aided by a rising pressure for "adjustment," they were retaking the "masculine" domain of engineering. Because some women veterans, wartime "engineering aides," and others trained in elementary engineering skills under the ESMWT had attended engineering school after the war to get the degrees that they lacked, there was a brief bulge in the number of females graduating in engineering in the late 1940s: 191 in 1947–48 and 175 in 1949–50. By 1951–52, however, the number was down to 60, and in 1952–53 it dropped sharply to just 37.[8] Although part of this decline was due to the waning of GI enrollments at engineering schools in general, the dwindling numbers of women were a sign of diminished tolerance for such pioneering (or in the new language of social adjustment, "deviant") women.

As the first enrollment restrictions and quotas became known in early 1946, the leaders of the AAUW, which had long represented college women's interests, publicized and protested the limitations in ominous tones, noting in their journal, "Doors Closing for Women Students." The most outspoken critic of the limitations was Alice Lloyd, dean of women at the University of Michigan, whose "Women in the Postwar College" appeared in the *AAUWJ* in the spring of 1946. Lloyd deplored the restrictions, whose full impact was not yet clear, as a threat to all minorities and urged the AAUW to take strong action. Colleges, she insisted, did not have to accept every veteran who applied; the large number of applicants could be cited as a reason to raise admission standards on most campuses.[9] College administrators denied that they had a "men first" policy or meant to squeeze women out but insisted that they had a duty to accept as many veteran applicants (97 percent male) as possible, even, apparently, if they were less qualified academically than many nonveterans (usually women). As Herman Wells, the president of Indiana University, put it, since these men would soon be the leaders of the state, it was in the best interests of the university to accept as many as possible.[10]

Meanwhile, the NFBPWC, at its biennial convention in the summer of 1946, passed a resolution opposing percentage restrictions on women's enrollments in colleges and universities "not only as discriminatory, but also as curtailing leadership where it is most needed in the years ahead."[11] Yet neither of these two large groups—the AAUW with more than 90,000 members and the NFBPWC with nearly 120,000—was able to rouse public opinion on this issue in the key year 1946, when eagerness to help veterans was paramount in

American society. Hardly anyone else seemed to care that thousands of young-er women might be losing their opportunity for higher education and future professional positions, least of all the young women themselves, who, accord-ing to those interviewed by the *New York Times* in a May 1946 radio forum, were eager to make sacrifices for the veterans. Lloyd and the AAUW leaders could only hope that in time the quotas would be dropped and that someday women's enrollments would again reach the prewar level of 41 percent.[12]

The quotas on women's enrollment at the graduate level in these years were even more restrictive than at the undergraduate level. In some cases the wom-en's numbers, as well as their percentages, actually dropped, as departments and whole graduate schools reserved their places for male candidates. At Rad-cliffe Graduate School, the female part of the Harvard Graduate School of Arts and Sciences, the total enrollment was 400 in 1945–46, but a strict postwar quota pushed that back to fewer than 300 in 1946, while the male graduate enrollment at Harvard soared from 1,088 in 1946 to 1,960 in 1947. Moreover, Radcliffe's female enrollment level was not allowed to reach 400 again until 1957, despite a great increase in the number of applicants, especially married ones. Radcliffe authorities apparently did not protest this administrative policy publicly, though some students were upset by the continuing lack of fellowship support and rules discouraging married or part-time students. Similarly, at Johns Hopkins, where graduate admissions were controlled entirely by the departments, the total female enrollment fluctuated between 120 and 155 each year in 1946–51 (always 21–25 percent of the totals, as if a percentage quota were being applied).[13]

Despite these caps on female admissions, the number of doctorates earned by women continued to rise. The number of science doctorates granted to women by American universities rose slowly but steadily from about 120 per year in 1940 to about 290 per year in 1954 (see fig. 1). Because the men's performance was far more volatile, however, reflecting the wartime plummet to about 700 in 1945 and its dramatic postwar resurgence to more than 4,500 per year by 1954 (with about 500 per year in engineering alone), the women were now earning only about 6 percent of the total. The remasculinization of science in the 1950s and 1960s was under way with a vengeance.

Not only women students were being pushed aside in the spring of 1946; pushed aside as well were many women faculty and deans of women, often-times the highest-ranking women administrators. One of the most prominent members of the National Association of Deans of Women (NADW), as it was then named, psychologist Kate Heuvner Mueller, dean of women at Indiana University, described in her autobiography how she was demoted in the spring of 1946. Suddenly, without any warning or advance consultation, she was ousted from her office and given the lesser title "assistant dean of students" by the new dean, a former military officer who had run training programs on

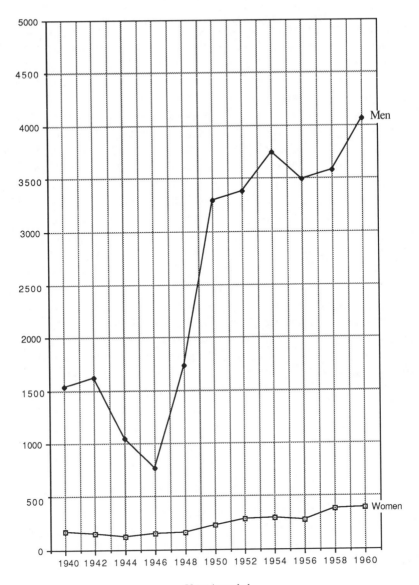

Fig. 1. Number of science doctorates awarded per year, by sex, 1940–1960. Data presented here include engineering, psychology, and anthropology. Based on Lindsay R. Harmon and Herbert Soldz, comps., *Doctorate Production in United States Universities, 1920–1962,* NAS Publication No. 1142 (Washington, D.C., 1963), 50–53.

campus during the war. (The next year she was given a still more anomalous title: "educational adviser for women students.") Mueller, who had spent eight seemingly successful years in her position, had attained a national reputation, and possessed far better professional credentials (a doctorate in psychology and a diploma in counseling from the American Board of Examiners in Psychology) than did the ex-colonel who superseded her, was never given an explanation for the change.[14] The move was a continuation of a prewar trend to supplant unspecialized deans of men and women with a group of specialists headed by a (male) "dean of students" or "director of the office of student affairs." Nor does it seem that Mueller would have received much support from either her president or the students had she chosen to protest. Feminist historian Alma Lutz complained to the AAUW in the late 1940s that the women students of the time did not understand why women were discontent being assistants rather than deans (or deans rather than presidents). A properly subordinate position seemed fine to them.[15] Again only Alice Lloyd, herself a beloved past president of the NADW, called attention to this "disquieting" organizational trend that was already afflicting women in academic administration. The result, she predicted, would be that before long hardly any women would be included in the top levels of university administration.[16]

Given the record student enrollments on most campuses in the late 1940s, the women faculty who had filled in during the war years might have expected to be kept on a while longer, but they too were pushed out in various ways. On some campuses the trustees reinstated the antinepotism rules that had been temporarily relaxed during the war. Thus, spouses (usually wives) who had been sought out in the emergency were now dismissed. At the University of Kansas, for example, the antinepotism rules were reinstated in 1949 for wives in the same department as their husbands and then a year later for all wives.[17] Similarly, women faculty who married men who were also on the faculty immediately lost their jobs, as did assistant professor Margaret Kuenne (Harlow) in psychology at the University of Wisconsin in 1948.[18] Other women faculty married graduate students—not all of the veterans were young—and then quit their tenure-track and even tenured positions to follow their husbands when training or a job took them (and so whole couples) elsewhere. Thus Gretchen Ann Magaret, at age thirty-two a full professor of psychology at the University of Wisconsin, left after her marriage in 1948 for a series of jobs in Illinois, Nebraska, and finally a professorship in medical psychology at the University of Oregon Medical School.[19] Similarly, Mary D. Salter, an assistant professor of psychology at the University of Toronto, left in 1949 when her new spouse sought further training in England.[20]

Yet when such vacancies occurred, they were not filled by other women, not even by women with doctorates who were wives of newly hired or other faculty at the university or by other, even eminent women already in the area. Applied mathematician Hilda Geiringer von Mises, for example, discovered this barrier both at the new Brandeis University in 1946 and at the restructuring Tufts University in 1947. Ever eager to have a position near her husband in Cambridge, Massachusetts, von Mises, upon hearing in late 1946 about the proposed new Jewish university to be built outside Boston, wrote Albert Einstein, who was involved in its planning and whom she had known since the 1920s in Berlin, to inquire about a position in the university's new mathematics department. Unfortunately, however, before he could do much on her behalf, Einstein had some differences of opinion with the fledgling board of trustees and resigned his advisory post. (They then named the new school for Louis Brandeis rather than Einstein.) Thereupon the dean of science decided that the new mathematics department should not hire established persons, as most departments at Brandeis were doing. He wanted to fill the department with bright young men who had pioneered the new specialties developed during World War II. Accordingly, he then hired ten such men. This move smacks a bit of what would later be termed "ageism" as well as "sexism," for von Mises, despite her age (she was born in 1893), was still doing pioneering work in applied mathematics, and many of the university's other early appointees of whom President Abram Sachar was very proud were persons over sixty who, like her, had fled the Nazis in the 1930s.[21]

But von Mises did not give up. A year later, when she heard that mathematician John Barnes was leaving nearby Tufts University for a position at UCLA (his wife, also a mathematician, later got a job at Occidental College), von Mises had a professor at Swarthmore make inquiries for her. He did, and he passed on to her the rather direct reply from a professor at Tufts, "I am quite sure that President Carmichael will not approve of a woman to replace Barnes. We have Wm. Graustein's widow on our staff, and Ralph Boas' wife, so it is not merely prejudice against women, yet it is partly that, for we do not want to bring in more if we can get men."[22] Evidently the same Leonard Carmichael who had headed the National Roster during World War II and was to serve on virtually every major scientific manpower commission in the 1950s and 1960s, the very ones that would urge that women be "encouraged," felt that a third woman in the Tufts mathematics department, whatever her ability and eminence, would be detrimental to its reputation. Yet von Mises, who was one of the more accomplished applied mathematicians in the country at the time, could only thank her friend at Swarthmore politely for his efforts and add stoically that she felt "pretty discouraged about this rather open discrimination. I hope there will be better conditions—for the next generation of women. In the meantime one has to go on as well as possible."[23]

Meanwhile, a slightly positive sign that women were being considered as science faculty in the postwar world appeared in the 1947 Steelman Report to President Truman on immediate and future scientific needs of the nation, but it apparently led nowhere. In April 1947 staff members preparing the report sent a three-page questionnaire to 250 science department chairmen, 5 at each of the top 50 doctorate-producing institutions (192 responded), asking, among other things, "Do you think it would be desirable to have more women teachers of science at the college and university level? If so, what steps could be taken? What about other segments of the population?" Of 22 mathematics department chairmen responding, 9 (40.9 percent) professed indifference to the sex of the teacher, claiming that competence alone mattered, but 5 (22.73 percent) were "definitely opposed." Of the eleven geologists who responded, most were "generally indifferent," but 2 (18.2 percent) claimed that they would be "severely handicapped because of the necessity of geological field work." Appended to a summary of these responses in the final report was another, more positive one: "In many families there are wives with excellent college training and teaching experience who are barred from active teaching. Doubt the desirability of this." Yet very few later reports would make even this modest recommendation that trained and even married women could be used in the upcoming expansion.[24]

Some women faculty protested as they felt the walls closing in around them in the postwar world. Alice Lloyd was not alone when she deplored the continuing low status of women on college faculties in her 1946 article; several other concerned women published even stronger protests in professional journals in the next few years. Geraldine Hammond of the University of Wichita, in "And What of the Young Women?" in the *AAUP Bulletin* in 1947, questioned the discriminatory hiring practices that resumed at universities after the war: Women were hired as instructors rather than as assistant professors, were paid less, and were hardly ever promoted. They were also subjected to a constant barrage of negative criticism and psychological harassment; at the very least they were informed that the department had not wanted to hire a woman anyway, denounced as emotionally unstable, blamed for past difficulties with other faculty women, criticized either as poor risks who would leave soon to get married or as "old maids" if they did not, and, after a few years, belittled for complaining or becoming bitter when they were not promoted. It was not enough, Hammond said, for a chairman to exonerate himself by saying, "It isn't fair, but what am I to do?" All academia must change.[25]

Psychologist Jane Loevinger of the Washington University Medical School in St. Louis complained to the editor of the *American Psychologist (AP)* in 1948 that department chairmen were using married women scientists as scab labor, offering them salaries that they would be embarrassed to offer to similarly qualified men. What was even more galling, they managed to make such a faculty wife "feel indebted to the university for the opportunity to serve." Loevinger recommended that the profession try to enforce some semblance of equal pay and create some "respectable parttime jobs."[26] A year later Marguerite Fisher, longtime assistant professor of economics at Syracuse University and an active member of the NFBPWC, published a study entitled "Economic Dependents of Women Faculty" in the *AAUP Bulletin*. In it she presented data on 158 faculty women at 20 institutions in the Northeast. More than half (53.8 percent) were supporting dependent relatives, especially mothers. Nearly a third had two or more full or partial dependents. Since the women's academic salaries were low and only 6 percent reported any additional income, they were carrying a serious financial burden. Fisher hoped that her report would add fuel to the NFBPWC's ongoing campaign for federal legislation for equal pay.[27] Thus, as it became clear that women's postwar problems in academia would be no different than they had been before the war, anger was rising in some quarters at least. But these first calls for attention were lone voices in the wilderness.

Any efforts on behalf of women scholars at a time of such overwhelming antifeminism merit interest. In the late 1940s two such projects were able to open doors for women at Harvard and the University of Michigan, respectively, by using the old strategy of "creative philanthropy." In 1947 Samuel Zemurray, founder of the United Fruit Company, whose son had been killed during World War II and whose daughter was a Radcliffe alumna, trustee, and Harvard-affiliated anthropologist in Central America, donated $250,000 to Radcliffe to endow the Zemurray Stone–Radcliffe Professorship for "a distinguished woman scholar in any field of knowledge" on the Harvard faculty. This was quite a novelty at Harvard, which after more than three hundred years in existence had only just tenured a woman on its faculty of arts and sciences, Sirarpie Der Nersessian, in Byzantine art at Harvard's Dumbarton Oaks branch in Washington, D.C. Once the Radcliffe trustees had accepted the money, with its controversial restriction, discussions at both Harvard and Radcliffe revolved around which departments might be interested in such an addition to their ranks and which women, worldwide, might measure up to Harvard's exacting standards. Several seemed possible, including in the sciences psychoanalyst Anna Freud of London, economist Joan Robinson of Oxford, physicist Lise Meitner of Stockholm (dismissed early as too old), anthropologists Dorothy Lee of Vassar and Ruth Benedict of Columbia, and Harvard's own astrophysicist Cecilia Payne-Gaposchkin, a Radcliffe Ph.D., long eminent on the staff of the Harvard College Observatory.[28] But because the tiny astronomy department was then fully staffed, whereas the history department had had a recent retirement, the choice in 1947 was the Englishwoman Helen Maud Cam of Girton College, Cambridge University, a specialist on medieval English legal history. Upon her retirement in 1954, she was succeeded by anthropologist Cora DuBois, who had earlier lost a position at the University of California at Berkeley by refusing to sign its regents' controversial loyalty oath, and she was succeeded in 1969 by Wellesley archaeologist Emily Vermeule.[29]

Besides bringing a senior woman to the faculty of arts and sciences, the Zemurray contribution slightly modified the generally hostile climate surrounding the selection and retention of other women faculty members at Harvard. For example, psychologist Eugenia Hanfmann later recounted in a short autobiography that the 1947 discussions at Harvard about the Zemurray Stone–Radcliffe Professorship created enough interest in women on the faculty that after she had held a one-year lectureship, her chairman dared to recommend her for a three-year position. But because this arrangement would permit her the hitherto banned privilege of attending faculty meetings, the dean disapproved, recommending instead three successive one-year appointments. A year later, however, as the Zemurray Stone discussions continued, he approved a three-year appointment for her after all. (She was feisty enough to go to one faculty meeting just to confirm her suspicion that they were not worth attending.)[30] Yet the ferment did not help another psychologist also there at the time, Thelma Alper, one of the best graduate students Gordon Allport of the Harvard faculty had ever had. A lecturer there in social relations, she should have been pleased with her offer of an associate professorship at Clark University, where she was to be its first woman faculty member. Yet as she reported the incident years later in an autobiography, she was so angry at having no future at Harvard that she found it very fitting to leave the Harvard Faculty Club that day not by the women's side door but by the front door, knowing full well that this was, as she put it, "GANZ VERBOTEN" (totally forbidden).[31]

Meanwhile, at the University of Michigan a ten-thousand-dollar bequest made in 1899 by one Catherine Naife Kellogg for a professorship for "a woman of acknowledged ability" had slowly been accruing interest and adding alumnae contributions. In the late 1940s, when the total reached about one hundred thousand dollars, university officials began to search for a suitable woman.[32] In 1950 they finally appointed psychologist Helen Peak, then of Connecticut College for Women, as the first Catherine Naife Kellogg Professor. Peak had received her doctorate from Yale University in 1930 and had been professor and chairman at Randolph-Macon Woman's College from 1933 to 1943. A wartime analyst at the U.S. Strategic Bombing Survey, she had been one of the very few women in the new field of survey research and had met many of the men who would be important in this growing area after the war, especially at Michigan. In 1946 she had moved to New London, where she was a full professor as well as the recording secretary of the APA, the highest office held by a woman in that organization in decades. Thus, Catherine Kellogg's gift to the university back in 1899 enabled one accomplished woman scientist five decades later the opportunity to leave the faculty of a woman's college for that of one of the largest doctoral programs in psychology in the nation, one that also trained many women in that field, and become one of the most visible and most honored women in psychology since Florence Goodenough's retirement from the University of Minnesota in 1947. Unfortunately, the one chosen was one of the least active in championing the cause of women in the field.[33]

Meanwhile, at its annual meetings between 1946 and 1948 the AAUW passed a series of resolutions condemning quotas in undergraduate, graduate, and professional admissions and discriminatory employment practices. In 1949 its Committee on the Status of Women urged the WB to respond to recent critics of women's abilities to be upper-level administrators by preparing a bulletin on the topic.[34] In addition, the committee prepared jointly with the Committee on Standards and Recognition a public "statement of principles" in early 1949 urging that "action was needed in particular to counteract a movement threatening the holding by women of policy-making and administrative positions in colleges and universities. An equally important aspect of this problem is that eliminating women from higher positions deprives girls of the advice and counsel of women, and conditions them to feel that there is a limitation upon their own position as responsible members of society."[35] But such resolutions and statements had no noticeable impact.

Unfortunately, the AAUW's only means of pressuring universities was still what it had been decades earlier, the relatively weak one of public embarrassment: it could urge its Committee on Standards and Recognition to review the eligibility of certain universities for AAUW membership for their graduates. Many, it was suspected, no longer had enough women on the faculty or a dean of women of sufficient rank to meet AAUW membership requirements. Accordingly, at the AAUW's biennial convention in June 1949 the delegates voted to have the committee begin just such a review. But the AAUW's leaders were of mixed minds about the wisdom of such a militant stand. The committee was already overburdened with its review tasks; and AAUW leaders did not want to upset national accreditation groups by undoing previous decisions. What was perhaps most important, the AAUW itself was not exactly guilt-free in the area of discrimination in 1949. It was still in the midst of a difficult period of dealing with those southern and border state chapters that had refused to accept Negro members. (In 1950 AAUW vice-president Dorothy Kenyon, who held decidedly liberal views on a variety of subjects, would be the first person to testify before Senator Joseph McCarthy's Foreign Relations Subcommittee.)[36] Thus, the AAUW leaders were nervous about taking on the universities that were so patriotically helping the veterans; they feared that theirs would not be a popular cause in most quarters and that any adverse publicity might hurt the organization further. Even the bland 1955 article "AAUW Standards Re-applied," which summarized nineteen of the committee's first reviews of twenty-six institutions, was deliberately as vague as possible. Before the manuscript was finally cleared for publication several AAUW officials had carefully removed from it any mention of specific universities, especially that of one of the better-known universities in the country.[37] Thus, the little that the AAUW could or would do in the early 1950s to fight sex discrimination in universities would be limited and late.

Meanwhile, the real target was not the single career woman (that pathetic "old maid" who all agreed had to work) but the married woman who chose to stay on the job. She was far more threatening to the happy vision of rigid sex roles. She might be "feminine," but if she did not have any children, she could still be faulted for being derelict in her duty to perpetuate the race (the fear of overpopulation had yet to emerge). If she did have children but continued to work, she was obviously a negligent mother. Every reader of Dr. Benjamin Spock knew that a normal child needed the full-time attention of his (or in later editions, her) loving mother, and any woman who did not provide that full-time attention was not only not doing her duty but also jeopardizing her child's future well-being. This proper maternal role might involve considerable self-sacrifice, but it was her duty, and others were justified in pointing out her "negligence" to her. (Some mothers, it was admitted, might overdo it and smother their children with too much attention ["Momism"], but they were assumed to represent abnormal cases.)[38]

Since apparently not enough women scientists were heeding the growing chorus of male and female psychoanalysts as to "what women want," California housewife and educator Olive Lewis took aim at them in 1948 in one of the most vicious articles ever to appear about women scientists. Lewis was convinced that (all) scientist-mothers were "more necessary in the home than in the laboratory." Her chief "evidence" was her impression that they were always so preoccupied with their own domestic problems that they did not do much good work of their own and even often upset other scientists' experiments. It was typical of the time that Lewis did not suggest any ways to help these hard-pressed women scientists, such as perhaps a part-time schedule or reliable daycare to relieve their minds of so much worry about their children. Instead, in an article in *Hygeia,* a popular health magazine published by the American Medical Association and thus prominently placed in doctors' waiting rooms all across the nation, where young mothers were likely to see it, she self-righteously blamed them for trying to make the best out of what was for them a difficult situation.[39] Yet at the very same time hardly any women's or popular magazines featured or even noted what surely was one of the big stories of 1947 about women in science, namely, that biochemist Gerty Cori of the Washington University Medical School in St. Louis, a wife and even mother, was the first American woman to win the Nobel prize in any science. She and her husband Carl, both born and trained in what later became Czechoslovakia, had shared (with an Argentine) the Nobel prize in medicine. Surely this was a story for *Life,* the women's magazines, the Sunday supplements, and all the rest, make of it what they would. Yet outside the science press it did not get the attention it deserved. Possibly Gerty Cori, afraid of the intrusions and the opprobrium of an Olive Lewis, shied away from the media and turned down all interviews.[40]

In 1948 and 1950 two children of Lillian Gilbreth, that beloved psychologist-turned-engineer, were able to capitalize on the popularity of motherhood and domesticity to recount their childhood adventures in two sentimental bestsellers of the time, *Cheaper By the Dozen* and *Belles on Their Toes,* both of which were made into movies. They managed to obscure the fact that their mother, who had run the Engineers for Hoover campaign in 1932 and confronted Army officials during World War II, was an able professional who had managed to raise eleven children while practicing her career only because of her considerable domestic assistance (up to six persons). They portrayed her instead as a young widow to whom all things came easily. The moral of her tale might have been a more realistic one, namely, that with the proper help working women scientists (even those who were single parents) could do anything, even raise eleven healthy children![41]

But it was Pauline Beery Mack, a chemist-nutritionist at Texas Woman's University, who in 1952 confronted and ridiculed one of the media's standard stereotypes about achieving women. Usually interviews with or feature stories about a nontraditional woman ended with the reassurance that despite it all she still loved to cook. Thus, when a reporter from *Nation's Business* concluded an interview with Professor Mack by asking the usual question, she gave a most unexpected reply: "Frankly . . . I never liked cooking, and after investing all my time and thousands of dollars in a scientific education, it never made any sense to me to do it personally. Any businessman would understand this."[42]

In the federal government, where advancement had always been rare for women, conditions worsened after the war, when many career employees who had held temporary supervisory positions were returned to their prewar ranks as budgets were cut drastically and wartime promotions terminated. The case of Doris Mable Cochran, long of the Division of Reptiles and Batrachians at the Smithsonian Institution, is illustrative. The first woman hired as a "herpetological aid" as soon as the Civil Service Commission (CSC) had opened the examination to women in 1919, Cochran then graduated from George Washington University and assisted the already sixty-nine-year-old herpetologist Leonhard Stejneger. Promoted to assistant curator in 1927, she earned a doctorate from the University of Maryland in 1933. Upon the death of Stejneger in 1943 (a special act of Congress had exempted him from retirement rules), Cochran was promoted to associate curator and appointed verbally only "acting curator in charge" of the division, of which she was now the only professional staff member and which she had in effect already been running for many years. When, however, she learned in 1945 that newly hired scientists were being paid the same salary she was being paid after twenty-six years, she started a long protest to have her grade raised to GS-13 and her title officially changed to full curator. Although she had the full support of her nominal supervisor, Waldo Schmitt, he could not defend her successfully to his superiors, Remington Kellogg, the head of the Museum of Natural History, and Alexander Wetmore, secretary of the Smithsonian, who complained about her clerical work and cited some possible misidentifications in some of her recent publications. Schmitt noted on a page headed "Kellogg's complaints": "Though girl has developed + its responsibility of supervisor to see that she does; they continue to want to rate her down. Wetmore is prejudiced against career women."[43]

Though she was denied the promotion, Cochran continued to run the division (with more supervision from Schmitt, who was chastised for his laxity) until finally in 1956 she was promoted to full curator. Years later her obituary in *Copeia,* the official journal of the American Society of Ichthyologists and Herpetologists, did not mention her personnel difficulties but praised her generosity and kindness to generations of local schoolchildren, who had brought in their collected snakes and lizards for professional identification. Unlike Stejneger, who had been known as the "child eater," she took the time to talk to and encourage these budding collectors, several of whom did become herpetologists. Thus Cochran, a victim of prejudice as well as changing professional standards at the Smithsonian, was able to rise above her own mistreatment to encourage others to take up her science.[44]

Yet had an official or staff member from an outside group such as the NFBPWC, the main group concerned with civil-service women, which frequently passed resolutions calling for the appointment of more women to top-level posts and opposing discrimination in federal employment, taken some interest or intervened in some practical way, perhaps she would have had her promotion years earlier. Nor was the AAUW, which had passed a similar resolution back in 1941, much help for women scientists facing demotion or even dismissal.[45] The NFBPWC was reluctant to charge discrimination even if individual cases of it were occurring because, according to a subcommittee reporting on the issue in 1951, there was "insufficient" evidence on how widespread the pattern was. To obtain such evidence, a sample survey, preferably by the WB, would be needed, but the subcommittee recommended against any action, since "the times are not normal" and thus "any survey made under present conditions would fail to present a true picture of women's opportunities or lack of opportunities in the fields of labor, business, or the professions."[46] Thus, although NFBPWC officials knew that almost all the top jobs in government were going to men, including those ranked lower on their civil-service rosters than women on their separate list, they felt unable to act.

In this repressive climate the strongest protest of the period, a report on the status of women by Navy psychologist and diehard feminist Mildred Mitchell, delivered to the International (formerly National) Council of Women Psychologists in 1950 and published in the *AP* in 1951, was remarkable. This final blast grew out of a series of postwar events within the NCWP, which had been formed in December 1941 and whose members were in the postwar world, as Alice Bryan put it, "groping toward a function."[47] When some of its original leaders became interested in international problems, especially the care of children abroad, the group changed its name in 1947 to the International Council of Women Psychologists (ICWP) to reflect this new emphasis. In 1948 the group applied for divisional status within the newly reorganized APA but was rejected on the grounds that any all-female group was inherently discriminatory.[48] (Although the ICWP reportedly did have a few male associate members, it did not make the modern side step of calling itself the International Council *for* Women Psychologists.)[49]

Part of of this postwar "groping" was a study of the fourteen hundred women in the *APA Yearbook,* undertaken in 1947, submitted to the *Journal of Social Psychology* in 1948 but not published until 1950. In it ICWP members Harriett Fjeld of Douglass College and Louise Bates Ames of the child guidance clinic at Yale University Medical School broke down their 393 usable responses both by occupational groupings—those employed in universities, clinics, public schools, and miscellaneous (government, business, and other)—and by type of response to the question, "In what ways has the fact that you are a woman affected your career?" (see table 2.3). This revealed that there were two kinds of women psychologists: those working in "men's work" (universities and miscellaneous), almost half of whom were dissatisfied because men were hired more often than women, promoted more often, and, of course, paid much more; and those working in "women's work" (clinics and public schools), who least often felt hindered by their gender and most often felt helped by it. They reported that their employers actually preferred to hire women, feeling that they could get along better with children or even male prisoners, who resented male psychologists. Thus, as one might have expected, women in "women's work" felt more confident and secure, since they could expect to advance and grow in stature over the years. Women in "men's work" could not, however, and felt increasingly isolated, unhappy, and anomalous as they grew older without advancement. Yet after reporting these and other differences in the women psychologists' attitudes, Fjeld and Ames concluded rather blandly that "the respondents, taking full account of the difficulties, indicated, with some dissenting votes, that they had found psychological work a satisfying profession."[50]

Shortly after this article was published in early 1950, NCWP founder Gladys Schwesinger cited it in the society's newsletter as evidence that now at last the ICWP had completed its original purpose of improving the status of women in psychology.[51] At this the dissenters erupted. They were not willing to let their complaints be so easily minimized and even dismissed and the society dis-

Table 2.3.
Perceived Impact of Gender on Women's Careers in Psychology, by Type of Employer, 1947

Perceived Effect	All Groups (N = 393)	University (N = 158)	Clinical (N = 131)	Public Schools (N = 53)	Miscellaneous (N = 51)
Hindered	40%	47%	36%	21%	49%
No effect	29	25	28	40	33
Helped	18	15	22	26	6
Marriage hindered	8	9	9	6	10
No answer	5	4	5	7	2

Source: Harriett A. Fjeld and Louise Bates Ames, "Women Psychologists: Their Work, Training, and Professional Opportunities," *Journal of Social Psychology* 31 (1950): 88 (table 12).

banded, for they were convinced that full equality had certainly not yet been achieved. In particular Schwesinger's statement provoked a strong rebuttal from the still discontented feminist Mildred Mitchell, who presented her own report on the status of women psychologists at the ICWP's next annual meeting. She claimed that since the status of women within the field, and especially in the APA, remained low, the group's original purpose of improving women's status within the field had not ended with V-E and V-J days. Members listened "with great interest" to her report documenting women's continuing underrepresentation in APA offices: They made up one-third of the membership but only 21 percent of the fellows and 10 percent of the national officers. The APA had not had a woman president since 1921, and she had been only the second to serve since the association started in 1893. On the other hand, five of the six members of national committees were women, making this new level of participation one area of improvement since 1940. A second was in the new position of divisional secretaryships, 24 percent of which had gone to women since the 1946 reorganization had created such offices. Thus, Mitchell felt that the women's protests since 1940 had had some impact at the committee and divisional level but that they had hardly produced equality. Mitchell's audience of ICWP members was delighted with her report and recommended it for publication. It appeared in the *AP* in June 1951.[52]

When Edwin Boring, Mitchell's former Harvard and Radcliffe professor, saw the article, he seized the opportunity to submit a stinging reply. Boring was eager to write an article on "The Woman Problem," as he called it, because a few years earlier, in 1943–46, he and psychologist Alice Bryan of the Columbia School of Library Science had collaborated uneasily and inconclusively on a series of three articles on the status of women in psychology. Bryan, a feminist and founder of the NCWP, had been convinced that women were not being given the positions and recognition in the field that they deserved, but Boring was more skeptical. He evidently believed that they were already being treated equally and that their minimal contributions to the field to date were adequate proof that they were not entitled to any more recognition than they were already receiving. Despite their almost irreconcilable political differences, Boring was intrigued enough by the problem to suggest to Bryan that together they write a series of joint articles based on the thousands (4,580) of wartime questionnaires completed by psychologists of both sexes for the NRC's Office of Psychological Personnel. (They later added a followup study of 880 male and female doctorates.) Though some of Bryan's more militant friends warned her against associating with such a "dangerous person" as Boring, she had accepted in late 1943.[53]

Bryan and Boring differed over how to interpret the differences the data revealed between men and women psychologists, who less often held doctorates, were paid strikingly less, did less research, held fewer offices in the APA, and more often worked part-time. Part of the explanation was in the separate labor markets open to men and to women, the traditional division into "men's work" and "women's work" in psychology. But Bryan and Boring went beyond this to probe the psychologists' attitude toward their plight. Boring's interpretation of the wartime women's professed widespread, almost cheerful acceptance of the phenomenon and contentment with it was to minimize the unfairness of the discrimination and even to see positive elements of "cultural adaptation" and "realistic adjustment" in it. The most Bryan could do was to insert occasional demurrals or cautionary notes. Finally in November 1946 Bryan could no longer accept Boring's attitude and broke off the collaboration, apparently accusing him of being biased. (Some key letters in the Boring-Bryan correspondence in the Harvard University Archives [HUA] are missing at this point.) Puzzled and hurt, Boring turned to his less militant former student Helen Peak, a staunch antifeminist, who assured him that he was not biased (or "unobjective," as he put it) and encouraged him to write out his own ideas by himself. He did so and then put them away until 1951, when he was provoked by Mildred Mitchell's outburst of feminism.[54]

In Boring's "Woman Problem," published in the *AP* in 1951, even more than in the earlier set of joint articles, one catches rare and revealing glimpses into what later came to be called the "male chauvinist" mentality. With all the tact and understanding of the haves telling the have-nots why they are poor, Boring blithely ignored all the occupational data he and Bryan had collected over several years, which formed the basis of the feminist attack. He considered it naive and irrelevant for "the ICWP or anyone else to think that the problem can be advanced toward solution by proving that professional women undergo more frustration and disappointment than professional men, and by calling then on the conscience of the profession to right a wrong." This approach was "to fail to see the problem clearly in all its psychosocial complexities." Instead, Boring had a novel explanation. The problem turned, he said, "on the mechanisms for prestige, which leads to honor and greatness and often to the large salaries, [and this] is not with any regularity proportional to professional merit or the social value of professional achievements."[55]

But rather than condemning a system that did not reward merit and was blind to certain kinds of professional contribution, Boring went on to describe it in admiration. Success in the field, he explained, was not based on naive notions of merit; it was based on quantities of prestige, which certain personality types accumulated over long years in the profession. These successful people, of whom he was proud to be one, were the compulsive personalities who were utterly devoted to the field and contributed 168, or at least 80, hours per week to it. If no women had been elected to the presidency of the APA in the last thirty years, it was because women did not have this personality or level of "job concentration." Such people, he gladly admitted, were not well adjusted in the usual sense—they made poor parents, bad spouses, and unstable human beings—but they contributed greatly to the field and were highly rewarded for it. Although some women did start out well and were "charmingly enthusiastic" for a few years, they failed as they grew older to maintain the intensive commitment to the field that final success required. Apparently even unmarried women were too distracted by family concerns or "conflicts" of other sorts to keep up this rarefied level of "professional fanaticism." His advice to aspiring women was that fewer of them should become deans and that they should write more books on general subjects and worry less about the helpless and downtrodden of society; he overlooked a recent study by his own research assistant showing that women already wrote proportionally more books and tests than men, though fewer articles.[56] Boring also overlooked all the occupational advantages that his Harvard professorship had contributed to his high productivity and even omitted the contributions of his psychologist wife, Lucy Day Boring, Ph.D. (whose professional judgment he had admitted elsewhere to Mary Calkins was often better than his own).[57] Nor was he willing to imagine possible counterexamples to his prejudices, such as that (1) some of the highly honored men might not be living up to these high standards or (2) some women might be working as hard as he was and yet be unrecognized by the field. (Experimentalist Alberta Gilinsky, a nonfaculty researcher at Columbia University, must have had this article in mind when she closed a letter to him chidingly in 1954, "I have resumed the 80-hour week. . . .")[58]

Boring, however, sincerely believed that the present system, especially the APA elections, was eminently just and a legitimate validator of one's worth as a psychologist. The rules were fair, but men and women were not equal in the field of psychology. Therefore, though the outcome might seem lopsided, he saw no reason to think that anything needed changing: "I do not believe that sex prejudice operates against women in APA elections to top level offices. I can not prove this faith, but I think that on the average and given everything else equal, a male psychologist will vote for a woman in preference to a man—or for a member of any minority group that he thinks is underprivileged or discriminated against. Everything else is, however, not often equal and women are usually not preferred for the top level jobs because some of their male competitors have more prestige." Society was to blame, or as he insisted, "It is not the APA which keeps women down, but the universities, industry, the government, the armed services," and these were such immutable pillars of American society in the 1950s that to think of changing them bordered on the subversive. Women should be "realistic" and accept their situation. With this, the debate subsided (without even a letter of protest to the editor) and did not revive until the mid-1960s.[59]

Boring's thundering response to Mitchell's report served to hasten the declining feminism within the ICWP. In 1951 it voted to admit men "quietly," though the emotion over the decision was so strong that some of the members' letters about the matter were still restricted decades later. When in 1953 the ICWP formed a new committee to examine the status of women within the field, the group decided, after considerable difficulty, that it was not prudent or politic to investigate such potentially explosive issues any further and described in a final report the resistance it had met:

> During the spring quarter [of 1954], Dr. [Virginia] Sanderson [professor of speech education at Ohio State University] contacted six people in Ohio, hoping to get the study of that state started; all six stated that they were opposed to such a study, believing that no matter how carefully it was undertaken it would be misinterpreted, create antagonism, seem to be an attempt to complain about sex differences. (One head of a department whom I know well but who does not want to be identified said that if women were really capable nothing could hold them back, but that he felt the effort to support women psychologists, just because they were women, was a mistake.)[60]

By then, "feminism" had become such an embarrassment that even sympathetic sentiments were being prefaced with disclaimers, as in this moderate statement by Harriet O'Shea, associate professor of child psychology at Purdue University, in 1954:

> Although I do not like a feminist approach to things, nevertheless, I do think I see a kind of overweighting of professional advantages on the side of men in our own field as well as in other fields. Women lawyers and women doctors have told us at different times in the past that they felt that their women's organizations, without doing anything belligerent, had a little value in just evening up the thinking of the profession so that women are not quite as much forgotten as they might otherwise be. Therefore, it is possible that even in the USA we would feel that an organization of women psychologists is worthwhile.[61]

In retrospect, one can see that there were issues aplenty for the group to consider if it had dared or wanted to. It could, for example, have pursued Margaret Wooster Curti's warning in 1951 that the treatment of women "is changing for the worse in graduate schools and elsewhere" and would bear investigation. But the slide continued, and in 1959 the group voted overwhelmingly to omit the word "women" from its name. In 1961 it quietly dropped any mention of the furtherance of women's concerns from the official statement of purpose in its newsletter.[62]

In the postwar era any hint that American society might have flaws, that it might be functioning unfairly or hurting some groups within it unduly, was unacceptable not only in the political arena, where McCarthyism held sway, but even, as Ravenna Helson has shown, in psychological journals.[63] The absence of complaints made it all the harder for the few remaining reformers to fight discrimination, as Alice Bryan had found among the women psychologists in 1946. No one was willing to come forward, force the issue, or be a test case. Discrimination was just as severe as ever, perhaps worse, but everyone ignored the problem and tried to forget it.[64] Although some later investigators would fallaciously interpret this lack of protest as evidence that the problem had evaporated or been "solved," it persisted and worsened until it was confronted by a revived civil-rights movement in the late 1950s and 1960s and the "women's liberation" movement after that. Postwar "adjustment" had smothered protest and postponed change, but it had not really solved anything.

Thus by the early 1950s, women scientists and their traditional supporters in the AAUW, the NFBPWC, the NCWP, and elsewhere were losing their will to fight their continuing and even worsening second-class treatment. They had lost ground in employment and advancement in academia, government, and industry, and yet, except for a few individuals, they were too afraid even to protest. The result was that women scientists, who had held some leverage in wartime, just a few years earlier, were left without a voice in manpower circles just as the age of federal spending and tremendous expansion were coming to American science and higher education in the 1950s.

3

"Scientific Womanpower"

Ambivalent Encouragement

Besides being told to be "well adjusted," American women were also being warned in the years after 1950 of the importance, even the necessity, of their contributing to the nation's defense, if the United States was to survive the Cold War. Yet just what this role might be shifted over the course of the so-called defense decade. At first women were to defend or protect the home, the community, and spiritual values, but as major governmental officials and a growing number of agencies and commissions charged with overseeing the nation's scientific manpower publicized the serious impending shortages of trained technical personnel, their vision of women's role broadened to include other tasks outside the home. Even before the Soviet launching of the first Sputnik in October 1957 these planners had unleashed a torrent of vocational articles urging bright young women to study for careers in science and engineering. By the end of the decade the first programs in "continuing education" were training mothers to reenter the work force, and several governmental fellowship programs were helping others to earn doctorates in defense-related fields. Nevertheless, it was often unclear just what jobs this new kind of "womanpower," trained in physics rather than the basket weaving recommended by advocates of "adjustment," was supposed to fill. Although it often sounded as if women were to staff the defense industries while men served in the military forces, by the mid-1950s their envisioned role was often restricted to such traditionally womanly jobs as technical librarians, laboratory technicians, and science teachers in the expanding public school systems. In time it became clear that bright women were to be trained only for "readiness" and "preparedness" purposes and then, like precious minerals and other natural resources, "stockpiled" for a future emergency. Despite the frustrations of such overtraining and later underuse, this inclusion of women in plans for "partial mobilization" was to have greater impact on both women and science than even World War II, for such steps were no longer thought to be mere temporary expedients but would be needed into the "foreseeable future."

At first it was thought that the onset of the Cold War in the late 1940s and the Korean War in June 1950 in particular would have important long-lasting consequences for women in science. As the draft quota reached fifty thousand men per month in September and October 1950, the conscription of women became briefly a distinct possibility, to which Dorothy Stratton, the wartime head of the U.S. Coast Guard's SPARS, now about to become executive secretary of the Girl Scouts of America, recommended serious consideration in an article entitled "Our Great Unused Resource—Womanpower" in the *New York Times Magazine*. Meanwhile, the chief reaction of many male scientists was to secure deferments and possibly even total exemptions from military duty for those graduate and undergraduate students who they claimed were needed more either in school or in the so-called critical occupations. After much debate all though the spring of 1951, a draft policy was worked out that exempted all women and deferred certain male college students in the top quarter of their class.[1]

By then, however, the subject of manpower had grown far beyond military needs and immediate deferments to long-term civilian recruitment, for in the fall of 1950, after the Red Chinese had joined the war, President Harry S. Truman declared a national emergency. By January 1951 he was recommending that the United States prepare for a standing army of 3.5 million men to deter aggression not only in Korea but everywhere for the foreseeable future.[2] By early 1951, therefore, a certain panic was felt by those in the government and out who were concerned with future supplies of what were now termed "human resources," especially highly trained ones such as scientists and engineers. Although the nation would now need both soldiers and engineers to win the Cold War, there might not be enough able-bodied men available to fill both jobs. The shortage of trained engineers was particularly acute because of the low birth rate of the 1930s and the recent drop in engineering enrollments (attributed to highly publicized unemployment in the field in 1949 and 1950). So few engineers would be graduating from college before 1953 that the nation's very survival could be imperiled.

In this atmosphere the founding in May 1950 and the early meetings of the Society of Women Engineers (SWE) were reported in the *New York Times* and elsewhere. The group had grown out of a series of local meetings in the Northeast in 1949–50 of two types of women engineers: the pioneers, who had been trained before or during World War II, and those who after working as "aides" during the war had since returned to college to earn an engineering degree.[3] SWE membership grew rapidly in the 1950s, reaching 350 by 1953 and surpassing 500 by 1958. SWE's first president and longtime leader was thirty-one-year-old Beatrice A. Hicks of the Newark (N.J.) Controls Company, run by her father.[4]

At this point Eugene Rabinowitch, a biophysicist at the University of Illinois who had worked on the wartime Manhattan District and was now editor of the liberal *Bulletin of Atomic Scientists,* publicized the idea in a February 1951 editorial, "Scientific Womanpower," that women, who could not as yet be drafted, should be encouraged to become scientists and engineers instead. The federal government might even provide fellowships to help them get the necessary training. A few might even become latter-day Madame Curies. This idea was considered so novel, just six years after the end of World War II, that even *Newsweek* reported it in a column entitled "Help Wanted: Woman."[5]

Yet the idea was not widespread, as was made evident at an American Council on Education (ACE) conference, "Women in the Defense Decade," in New York City in September 1951. Although the nine hundred persons who attended the conference, mostly women educators and representatives of women's organizations (including Lillian Gilbreth of the newly formed SWE), by all accounts had left euphoric that there was, after all, something that women could do in the crisis, the meeting broke no new ground. It might have gone differently had Anna Rosenberg, Assistant Secretary of Defense for Manpower, spoken. Having earlier favored the drafting of women into the armed services, she might have brought the war a little closer to the women's colleges, but she withdrew at the last minute. Instead the speakers present, unable to envision or articulate innovative responses to impending wartime conditions, especially the dawning manpower shortages in technical fields, voiced traditional opinions, simply urging women to defend the home ("the source of the nation's strength"), to protect their community, to preserve spiritual values (including free speech in a time of hysteria), and to support those women who chose to join the armed services. They also recognized the need for some educated women to work in the growing defense industries. If this need became overwhelming, some easing of the double burden of housework plus a job would have to be arranged, but the need was not yet urgent, and suburban life was not expected to be jeopardized.[6]

Meanwhile, the only up-to-date summary of current manpower thinking, the September 1951 issue of the *Scientific American,* devoted to "The Human Resources of the United States," sold out immediately and went quickly through three reprintings. One of its articles, "Mobilization," by Arthur S. Flemming of the new Office of Defense Mobilization (ODM) in the Executive Office of the President, stressed the increasingly official view that women were needed as scientists and engineers. At age forty-six, Flemming, the assistant to the director for manpower of the ODM, had already served on several governmental boards (including the CSC and the WMC) and was currently on leave from his post as president of Ohio Wesleyan University, his alma mater. Almost immediately after his arrival at the ODM in early 1951, he had formed the (all-male) Committee on Specialized Personnel (CSP) to make recommendations about the then pressing issue of deferments in critical fields and to make long-range plans for partial and then full mobilization.[7] This was very difficult in 1951, since the existing data were scattered, out of date, and rarely broken down by gender. The recently revived wartime Roster was frantically putting together its first fresh data, and the newly established National Science Foundation (NSF) had not yet moved into the area of manpower statistics.

Despite this muddle, the CSP considered ways to recruit more young persons into science and to ensure that those once trained were, as the jargon put it, fully "utilized." In this context Flemming emerged as one of the very few feminists of the time. He sincerely believed, as he said several times in his *Scientific American* article, that women could and should be scientists and engineers. His continued forcefulness on this issue reportedly startled both the representatives of seven women's colleges who called upon him in late 1951 and an audience of engineering deans whom he addressed in late 1953. Moreover, when in June 1952 the CSP considered briefly the recommendation of Fred C. Morris, professor of civil engineering at Virginia Polytechnic Institute, that women be trained as engineering aides as in World War II, Flemming rejected it on the grounds that the "plan implies that women are suitable only for subprofessional work in the engineering field and does not encourage them to achieve full professional status."[8] Thereafter other members of the committee and its consultants either strongly deplored management's unrealistically low expectations for women engineers or were apologetic about a suggestion in 1954 that a corps of women (only) be trained to be science teachers.[9] Flemming's speeches and attitude also inspired the WB to prepare another vocational bulletin, *Employment Opportunities for Women in Professional Engineering* (1954), in rebuttal of Morris's plan, which he had published in the *Journal of Engineering Education (JEE)*. For this report the WB surveyed six hundred persons suggested by the SWE to discover what women could do in the field if they were given the chance.[10]

The chief result of the ODM's efforts was Defense Manpower Policy No. 8, issued in September 1952, which established for the nation its future goals in the training and utilization of scientific and engineering manpower. Flemming's mark on it is evident, and the preface contains the following strong statement: "Throughout this document all references to scientists and engineers make no distinction between the sexes or between racial groups; it being understood that equality of opportunity to make maximum effective use of intellect and ability is a basic concept of democracy."[11] In addition, among the policy's twelve recommendations to employers of scientists and engineers, the eleventh urged them "to reexamine their personnel policies and effect any changes necessary to assure full utilization of women and members of minority groups having scientific and engineering training."

This strong but vague recommendation was as far as anyone was willing to go in the 1950s, and further than most. The memory of World War II controls was still so vivid that none of the labor or management representatives who had to approve ODM policies would agree to recommend any such governmental involvement, not even if "full mobilization" were necessary. Nor did anyone seriously consider the gentler concept of governmental incentives such as tax credits to bring about the desired full "utilization" of womanpower. Thus, throughout the 1950s the government's official policy was to encourage women to enter scientific and technical fields and to urge employers, including the federal government itself, to hire and utilize them fully. But it would not back these recommendations up with any kind of federal incentives or enforcement. As William D. Carey, then at the Bureau of the Budget, put it in 1956, in reviewing a spate of still more such recommendations, "Who is going to guide the 'better utilization' of scientists and engineers? This is formidable indeed, and I don't see how it can be done except under conditions of extreme crisis and manpower control."[12] Because few were willing to endorse such controls in the 1950s, all that aspiring women scientists were going to get from the federal government for the next decade or more was this official rhetoric of encouragement, along with, in time, better statistics and a chance for a federal fellowship.

Meanwhile, the officers of the Rockefeller Foundation were also worried about the implications of the nation's voracious appetite for trained experts. Even before the Korean War they had been concerned that there might not be enough bright, trained people in the nation to do all the things that were projected for the postwar world. In 1947 they determined to publicize the impending shortages by hiring someone to write a major report on current and future manpower needs in the humanities and social sciences as well as the physical and biological sciences. After a lengthy search, Dael Wolfle, executive secretary of the recently reorganized APA and already a member of several manpower advisory groups in Washington, agreed to direct the project. Starting in October 1950 and continuing into early 1954 Wolfle quietly directed his staff and wrote a major report that eventually cost the foundation $240,000.[13]

By the time his report appeared, in the spring of 1954, the Korean War had been over for almost a year, and manpower interests had reverted to the primarily civilian and educational topic that interested the Rockefeller Foundation officials most: the future needs of higher education in a technological society. Wolfle's report, *America's Resources of Specialized Talent: A Current Reappraisal and a Look Ahead,* was a complete inventory, reportedly the first ever, of the nation's future supply of and demand for educated specialized personnel. Its chief theme was an extension of Vannevar Bush's vision of 1945 that "educated manpower" was a critical "national resource." In a complex society such as the United States had become, the education of future specialists was too important to be dependent on individuals' choices. The nation's survival depended on having enough people well trained in certain fields. The public should be aware of its vulnerability in this area and begin to take steps both to broaden the "potential supply" of such specialized persons and to improve the "utilization" of those already trained. Wolfle's recommendations in this area included providing more financial aid for those men and women who were bright enough to benefit from college but could not afford to attend; increasing their desire for more college education (primarily through publicity and vocational guidance); improving education in various ways; and increasing the likelihood that a student, once enrolled, would graduate. Data available on the ability of those students who did not go on to college (two-thirds of whom were women) indicated that a great increase in the numbers of persons completing college would not lead to any lowering of the quality of graduates. Far too many bright people were being "wasted."[14]

Wolfle also recognized that a major root of the serious undertraining and later underutilization of both women and minority students was the systematic discrimination they faced at all levels. Since they were not expected to hold major jobs, they had little incentive for prolonged or advanced training. Wolfle discussed this problem sympathetically (he was perhaps influenced by his wife, who was also a trained psychologist), but he did not think the situation was likely to change very soon and therefore did not recommend, for example, special fellowships for women or minority students. He did urge that ways needed to be found to alleviate this problem. In addition, he broached the idea that in order for there to be adequate manpower statistics in the future, the major professional societies should be asked to take on a larger role and, if necessary, be supported by a federal subsidy.

Thus, Wolfle's report, three and a half years in the making, became part of a growing chorus articulating a liberal vision of the importance of lessening discrimination in both education and employment and strengthening higher education across the nation primarily by nongovernmental means, such as raising awareness, reforming vocational guidance, and modernizing curriculums. Federal aid in the form of undergraduate scholarships and for maintaining better manpower statistics would also be helpful. Women's important place in this vision was spelled out to those who cared to notice, but in general reaction to the report was rather muted, and most reviews were short summaries of its contents.[15]

Meanwhile, there were faint but positive signs that more women were going into science, even into engineering. The numbers of women enrolled for a bachelor's degree in engineering more than tripled between the fall of 1951 and the fall of 1957 (from 561 to an all-time high of 1,661); similarly the number of degrees awarded to them increased fourfold, from 37 in 1952–53 to 145 in 1959–60. These increases may have been partly owing to the decision at some formerly all-male engineering schools, such as the Georgia Institute of Technology in 1953 and Clemson University, in South Carolina, in 1955, to admit women, although thirty-two others, mostly private ones, still refused to accept them in the early fifties.[16] But because the men's enrollments also nearly doubled between 1951 and 1957, the women's percentage of both figures still remained well under 1 percent.[17]

But if only a few persons were able in the early 1950s to picture women contributing to the military or becoming engineers, many more could support the idea of their becoming science teachers in the expanding public schools. Starting around 1954–1955 there was a sudden explosion in the number of articles urging women to study science, primarily in order to become laboratory workers or public school teachers. So popular was this channeling that by the end of the decade programs had been set up at several colleges and universities to train or retrain those women who wished to enter these occupations. In June 1954 Bryn Mawr College hosted the small Conference on the Role of Women's Colleges in the Physical Sciences, chaired by Walter C. Michels, professor of physics and a longtime advocate of training more women in this field, and President (and psychologist) Katherine B. McBride. Attended by representatives of seventeen women's colleges, twelve corporations, and eight other educational and governmental institutions, the conference participants promoted the idea that "with some moral support and financial aid," the women's colleges could "contribute much toward increasing the nation's supply of scientific and technical personnel." In particular they concluded that "women can increasingly fill the need for teachers of science in our secondary schools," for which the need was "tragically acute."[18] The conference also stimulated additional publicity about the alleged need for women in science, for in one month, March 1955, at least four articles referring to it appeared.[19]

Typical of a new genre of articles that started off asserting that the study of science was not at all unwomanly but ended up declaring that women were especially needed as laboratory technicians and science teachers was the editorial "Women in Physics and Chemistry," in the *Electrochemical Society Journal* for August 1954. It first applauded the new source of technical manpower that trained women represented but then immediately specified and minimized the roles that they might play. Thus, though the unnamed author asserted that "industrial laboratories are interested in training and ability regardless of sex," he also felt that "women are most successful and appear happiest in the position of technical aide to scientists and engineers engaged in research and development programs. They are more willing than men to do routine repetitive work. They are patient, faithful, and dexterous. They are subjective rather than objective in attitude. They get satisfaction in the approbation of colleagues."[20]

Repeating these sentiments was the 1956 article "Courting Distaff Employees," in *Chemical Week*, which praised employers who met the manpower shortage by hiring women despite their well-known drawbacks, "such as short job duration, low job motivation, and everything covered by the phrase 'female temperament.'" These employers knew, however, "that women have outstanding qualities of patience, neatness, manipulative skill, and color perception, which eminently qualify them for such exacting jobs as microchemistry, chromotography, laboratory analysis, and repetitive projects of one sort or another." Therefore, such enlightened employers were hiring women for jobs "where production continuity is not seriously jeopardized, where particularly feminine traits can be utilized," including technical writing and editing, laboratory benchwork, product development, technical literature search, and patent searching.[21]

Meanwhile, Howard Meyerhoff's "Science Careers for Women," in the new journal *Science Counselor*, for teachers in Catholic schools, urged women to become teachers. (It was also one of the first articles to refer to the large number of Russian women engineers.) By 1956 Meyerhoff, executive director of the Scientific Manpower Commission, a nonprofit organization established in 1953 and supported by seven major professional societies, was noting in his annual report that of late there had been a "keen interest" in both scientific careers for women and vocational guidance for science teachers, two topics that were no doubt linked in the popular mind.[22] Similarly, in 1957 Otto Kraushaar, president of Goucher College, in his article "Science and the Education of Women," deplored the great waste of talent among girls and suggested that "our colleges for women are in an excellent position to recruit and educate teachers of science for elementary and secondary schools, for whom there is such a desperate need."[23] Thus, by 1956 and 1957 some opinion-makers in colleges and industry were beginning to focus on some expanded but still highly segregated accommodation of women in science. Any increased numbers of "scientific womanpower" could be safely channeled into traditionally feminized roles, where they would reinforce rather than threaten existing sex segregation in the work force.

Elsewhere, however, impatience was increasing with all the vague exhortation to society in general and vocational guidance and employers in particular to "encourage" women to go into science, and pressure was mounting to move onto something more specific. Even Lee DuBridge, president of the men-only California Institute of Technology and a member of the scientific advisory committee of the ODM, stressed in a 1956 article entitled "Scientists and Engineers: Quantity Plus Quality," in *Science,* that past rationalizations of the failure to interest women in science on the grounds that they would grow up to be mothers was insufficient: "Millions of women do work in spite of home duties and motherhood; indeed they work so that they can have better homes and more children. Junior science fairs have uncovered some very able girl scientists. Why not a nation-wide effort to attract girls into technical interests?"[24]

In April 1956 President Eisenhower, urged by ODM director Flemming, took the small step of setting up a National (later President's) Committee on the Development of Scientists and Engineers (PCSE) to serve as a clearinghouse for all the nongovernmental efforts springing up everywhere to train more scientists and engineers. It is significant that Eisenhower appointed nineteen men, and no women, to this committee, overlooking such obvious ones as representatives of the SWE, the AAUW, the women's colleges, and all other women's groups. Two women presidents of the National Education Association (NEA) did serve as ex officio members of the committee before it disbanded in December 1958. Their appointments would have reinforced the prevailing impression that the only women prominent in science were schoolteachers. The omission of other women lends credence to the suspicion that despite all the talk and exhortation, some important people did not really want or expect women to become full-fledged scientists or engineers.[25]

Although some of the committee's staff members were interested in holding a meeting about women in scientific careers, their records show how painfully ignorant they were of both the personnel and the issues involved. At one point they hired Marguerite Zapoleon of the WB as a consultant, and at another point a staff member helped journalist Helen Hill Miller with an excellent article on women in science for the *Atlantic Monthly*. Among the fresh practical suggestions that the PCSE did not follow up on were two put forth by others in 1956–57. Allaire U. Karzon, a thirty-one-year-old Buffalo lawyer and mother of two, suggested that if more womanpower really was needed in the labor force, as one heard repeatedly, then the federal tax code ought to be reformed to remove some of the economic disincentives that working mothers currently faced. Specifically they ought to be allowed to deduct a realistic amount of the cost of childcare from their own or joint income tax returns. Nursing, teaching, and professional women's groups were already behind the idea; if the scientific manpower lobby, which was constantly bewailing the "loss" of trained women, supported it too, such a bill might pass.[26] The second idea came, possibly from a student, in a letter to the editor of *Science Digest* in 1957 complaining that despite all the efforts to attract young women into science and engineering, there were still no girls' science clubs in high schools, only those for homemakers (which, ironically, despite the stereotypes about home economics, might well have been urging the women students to take more science courses).[27]

The committee's lukewarmness on women's issues was evident in an exchange of correspondence between its vice-chairman, Eric A. Walker, the dean of engineering at Pennsylvania State University, and committee consultant Donald S. Bridgman. Apparently convinced that it was unrealistic and perhaps even undesirable to recruit many women into engineering, Walker criticized a recent talk by Bridgman before the American Society for Engineering Education. He wanted numbers, not of the total persons available, as given in Wolfle's book, but just of the available bright men. Educators, he apparently felt, would do better by focusing their recruiting efforts on these desirable young men.[28] Ironically, however, forcing Bridgman to break his data down by gender, even with the immediate intention of excluding women from calculations of future engineers, would have unforeseen beneficial consequences later on when they were presented to a governmental committee that did have women among its members.

Similarly, in March 1957, when the National Manpower Council (NMC) of Columbia University, of which Eisenhower had been president, released its report, *Womanpower*, it barely hinted that some women might be or might become scientists or engineers. This report had grown out of a long effort going back to 1953, led by the economist Eli Ginzberg and supported by the Ford Foundation (but only at the strong urging of Arthur S. Flemming), to study women's place in the American economy and society. Comprising the results of sixteen "brainstorming" conferences held around the country, *Womanpower* was a mélange of pedestrian recommendations and innovative ideas (a repetitious compendium, Eleanor Dolan of the AAUW called it in a review in *Science*). Generally about women and work, its ideas—that because women were in the work force in large numbers despite the back-to-the-home movement of a few years ago, they would need better planning and counseling and even institutional supports, such as childcare and innovative programs at colleges—had been under discussion elsewhere for years. Because the Ford Foundation wanted a maximum amount of favorable publicity for its venture, the report also shied away from (as its records reveal) controversial issues, such as equal pay for equal work, job discrimination, and women in nontraditional areas. For example, in the chapter "Shortages in Highly Trained Personnel," Ginzberg and the council talked mostly of schoolteachers and nurses, giving hardly any hint that some women might undertake scientific or technical careers, although he noted at one point that the number of women working toward advanced degrees was higher than ever before.[29]

In fact the strategists at the NMC turned *Womanpower*'s very lack of originality into a political advantage. By reflecting a kind of cautious blue-ribbon consensus on some women's issues, the report succeeded in separating them from those that were still too controversial for foundation action. Because of the NMC's close ties to Eisenhower, its own expertise, the endorsement, or at least participation, of hundreds of major figures across the nation (listed in eighteen pages of acknowledgments), and its massive favorable publicity, *Womanpower* gave these topics a certain importance and legitimacy that helped to strengthen an emerging conservative women's movement, which could work for women's "special needs" without embarrassment. It might also embolden some foundations to support proposals for programs to help womankind.

Also emerging on the public scene in 1957 was Mary Ingraham Bunting, microbiologist, widow and mother, and dean of Douglass College, who was soon to become a familiar figure in national policy circles and champion and personification of the cause of women in science. A graduate of Vassar College (in physics), she had earned a doctorate in agricultural bacteriology at the University of Wisconsin in 1934. After teaching at Bennington College in Vermont for a year, she had married Henry Bunting, a fellow student at Wisconsin. When he became a resident physician in Baltimore two years later, she taught physiology and hygiene at Goucher College. In 1938 he joined the faculty of the Yale Medical School, and she became a research assistant there until she retired to motherhood (they had three children) during World War II. This paraspouse pattern resumed in 1946, however, when she accompanied her husband to Harvard for a year and taught part-time at Wellesley College. Upon their return to New Haven, she was a full-time research assistant and lecturer and might have continued in such traditional accompanying roles much longer. But in 1954 her husband died suddenly, and in 1955, through a lucky "old girl" contact—a Vassar classmate was married to the president of Rutgers—she became the new dean at Douglass College. By the fall of 1957, she was known well enough to be invited to join Katherine McBride on the NSF's Divisional Committee for Scientific Personnel and Education, an eye-opening assignment in the post-Sputnik days.[30]

It was at one of her advisory committee meetings that Mary Bunting came across a later version of Donald Bridgman's data for the PCSE on who was dropping out of science broken down by gender. When Bunting raised questions about the wide discrepancies, she realized that certain NSF officials were deliberately hiding such data, lest the public realize that all the persons trained to fill these shortages would be women. As she later put it,

> It was a single statistic that first sparked my interest in women as such. It seemed important as we checked over the flow of talent into science to know, along with many other inquiries, how many able high school graduates did not continue to college. Foundation staff analyses of data from three sources indicated that of all high school graduates scoring in the top ten per cent by ability tests, at least 97 per cent and perhaps 99 per cent of those who did *not* go on to college were female. This statistic surprised me. I expected girls to be in the majority, but not by such a margin. How sad, I thought. What a waste. Who were they? What had stopped them?
>
> But what really alerted me to the possibility that women *hadn't* yet been liberated was the fact that the statistic was suppressed. (I was reminded of how difficult it had been for a Wisconsin colleague to publish research on the nutritional equivalency of oleomargarine and butter.) I never received an explanation of the suppression of the statistic on college entrance, but my best guess was, and is, that the Foundation staff believed that if America knew that all the bright boys were going to college, no one would think there was a problem in the schools. It was then that I first sensed that in this country we did not expect women to contribute anything really important to science or—I soon realized—to any intellectual frontier.[31]

Radicalized by this revealing NSF suppression, Bunting determined to publicize this systematic waste of women. Before long she would be asking why, if there really was a shortage of scientists, the government (or even the AAUW) did not do more about providing fellowships for women, including part-time ones for married students, and why, if John Gardner really wanted "excellence," as he proclaimed in a best-selling book of 1961, the Carnegie Corporation, which he headed, was not supporting more projects for women? She became a leader in the movement for "continuing education" for reentry women, starting immediately at Douglass College[32] and then, after 1960, at Radcliffe College.

Just as this portion of the "womanpower" problem was beginning to be solved with programs to retrain "mature" women (over thirty-five usually) for jobs as schoolteachers and computer programmers (see ch. 12), the Soviet launching of Sputnik I in October 1957 added a new tone of urgency and stridency to the talk of recruiting women scientists. "Mature" women were no longer enough; more younger women would have to go into science as well if the nation were to best the Russians either in outer space or here on earth. As Congress discussed increases in science budgets and possible new fellowship programs, the *Wall Street Journal (WSJ)* published on its front page on January 16, 1958, a skeptical article by staff reporter Arthur Lack, "Science Talent Hunt Faces Stiff Obstacle: 'Feminine Fallout'; Officials Fear Many Federal Scholarships Will Go to Girls—Who'll Shun Careers." This article revealed that unnamed federal officials expected at least one-third of their new scholarships to go to women, many of whom would then marry, have children, and "at least interrupt their careers, if not end them altogether." Lack added a little extra ridicule: "Hence it's inevitable that some Government money will go to train scientists who experiment only with different household detergents and mathematicians who confine their work to adding up grocery bills." He then explained that the government could not, as some (also unnamed) private fellowship groups did, place a quota on the number of grants going to women,[33] as this "probably would embroil the Government in a great controversy with the many 'equal rights' advocates among the ladies," a testament that the women's movement was not dead in 1958.

In order to estimate how many women graduate students would be dropping out, Lack cited a recent report by the AAUW on its first seventy-five years of fellowships for its statistic that 55 of its 323 responding fellows, or as he put it, "at least one sixth," were not working. (Luckily, no one referred him to a 1956 Radcliffe report on its Ph.D.'s and its statistic of 78 of 321 unemployed.) Yet no one that he interviewed was alarmed at this. Eleanor Dolan (a Radcliffe Ph.D.) of the AAUW staff, Esther Lloyd-Jones of Teachers College and formerly on the American Council on Education's Commission on the Education of Women (ACE-CEW), and even Ralph Flint, director of higher-education programs of the Department of Health, Education and Welfare (HEW) and thus in charge of the new National Defense Education Act (NDEA) fellowships, all minimized the problem, and Flint was quoted as saying that "most of the girls who enter professions 'don't let marriage interfere.'" But instead of dropping the whole issue (or non-issue), Lack went on to discuss other recent hearsay "evidence" of women's poor showing in science and mathematics: although women made up 55 percent of the high school students taking the then new National Merit Scholarship examinations, only 28 percent of the winners were women. No one at the National Merit Scholarship Program commented on this, but Lack quoted Nancy Duke Smith (actually Lewis), dean of Pembroke College at Brown University, as saying that "the girls were eliminated largely because of 'their almost complete failure to measure up on the mathematical questions.'" Lack also cited an unidentified NMC report *(Womanpower)* that said that boys did better on college entrance examinations than did girls. Nevertheless, he had to admit that girls tended to get better grades and were a majority of the honor students in high school. He also cited Eleanor Dolan's reference to a recent HEW study by Robert Iffert that had shown that women were more likely than men (40.5 percent compared with 38.5 percent) to graduate from the college at which they had first enrolled. But rather than thinking that it was unfair that these women would be only one-third of the federal scholarship holders, Lack persisted in his belief that even that proportion was dangerously high.[34]

Two weeks later the *Wall Street Journal* printed a letter to the editor from Susan Spaulding of New York City, an AAUW member and executive assistant to the president of New York University. She severely criticized Lack's "confusing barrage of extraneous statements," which "proves nothing, but unfortunately contrives to give the impression that scholarships to such girls would represent a loss." This she challenged on several grounds. She had recently visited the physics department of a woman's college, where the male chairman had told her that even if the women students married, they tended to go on in science in one way or another. They generally married other scientists, found it easy to keep up with the field, and "as soon as circumstances permit they return to the laboratory." Spaulding then cited some statistics about the Institute of Mathematical Science at her own university. On its staff of 190 were 23 women, of whom 11 were married, including seven with children. One of the mothers (applied mathematician Cathleen Morawetz, who eventually had four children) was even on the faculty. Neither marriage nor motherhood had deterred these women. Spaulding went on to question the whole idea of "fallout" and "loss." Was it a "loss" (and not a gain?) to the supply of future scientists that some mothers were trained scientists? One reason why women had to spend so much time in mothering was that fathers spent such long hours away from the suburbs. Certainly the government was already making great efforts to identify scientific ability at a young age, even in the elementary schools, and to give it the fullest opportunities to develop. If this was one of the nation's goals, then science fellowships for future mothers "could add up to the most effective recruitment of future talent that we could devise and support."[35]

Finally, despite the criticism and after many months of debate, prolonged partly because some critics wanted educational reforms *before* voting massive federal aid, Congress passed the monumental NDEA in August 1958. Linking higher education clearly to national defense, the act was the culmination of the manpower analysts' message. The act boldly asserted:

> The Congress hereby finds and declares that the security of the Nation requires the fullest development of the mental resources and technical skills of its young men and women. . . .
>
> We must increase our efforts to identify and educate more of the talent of our Nation. This requires programs that will give assurance that no student of ability will be denied an opportunity for higher education because of financial need; will correct as rapidly as possible the existing imbalances in our educational programs which have led to an insufficient proportion of our population educated in science, mathematics, and modern foreign languages and trained in technology.[36]

Thereupon the Congress established a total of ten new programs, including a federal student loan program[37] and, most important for our purposes here, a new graduate fellowship program (Title IV) that was to be even larger and broader than that at NSF. Awarding a thousand new fellowships in 1959 and fifteen hundred each year from 1960 through 1964 (before shooting up to seventy-five hundred in 1967 and 1968, its boom years), the program seemed a genuine attempt to reach its stated goals of alleviating "an existing and projected shortage of qualified college teachers" by awarding three-year fellowships for full-time graduate study.[38]

Although the act did not talk of "equal educational opportunity" or of rectifying the evils of discrimination, but only of shortages and national security, a few observers did make the link between its greatly expanded pool of future college teachers and increased employment prospects for women as well as men. For example, the NEA's report *Teacher Supply and Demand in Universities, Colleges, and Junior Colleges,* released in June 1959, applauded the new NDEA programs and raised the rhetorical question, "Will this expanding movement bring more women into the university and college classroom?"[39] Two years later Charles Warnath, assistant professor of psychology and director of counselor training at the University of Oregon, published in the *Vocational Guidance Quarterly* a short article on the subject, provocatively entitled "Is Discrimination against Talented Women Necessary?" He wondered whether women trained to high levels under the new NDEA would be willing upon completion of their studies to be restricted to "women's tasks," as was still the custom. He recounted local tales of well-educated women whose careers had been hampered by antinepotism rules and other sex-linked restrictions. If the woman was unmarried, she would have to be "much better than the male applicants in order even to be considered for a position," since "many college departments simply ignore female applicants for positions." He thought that "some serious thinking" ought to be done about the problem. His solution must have been at least partly tongue-in-cheek:

> Either some discrimination will have to be built into programs such as NDEA so that girls begin to feel discouragements early in their academic life (and only the toughest survive to fight for their place in the male world) [as if this were not already the case] or new ways of looking at the development and use of talented women will have to be considered. Early encouragement followed by later discouragement can only lead to tremendous frustration on the part of many in this group—a group which will be by its very nature a superior and articulate group.

He then recommended that the NDEA sponsor studies of ways to change attitudes toward talented women. Perhaps, he said, "a full-scale public relations job needs to be done soon."[40]

Although by 1958 one AAUW leader had put into words what others had been groping toward—that it was no longer enough to talk of encouragement and reforming vocational guidance: actual programs were needed to help women plan their lives to fit modern demands[41]—doubts persisted about whether women were really wanted or needed in science and engineering. Authors of articles that seemed to exhort them to become scientists and engineers oftentimes maladroitly blamed them instead for not already doing more. When recruitment efforts cost money, surprisingly, despite the billion dollars for the new NDEA programs, there was none available. When by 1958 individuals and groups identified specific institutional barriers or obstacles, such as antinepotism rules at universities, governmental officials acted helpless. Women were needed, or so some were saying, but hardly any actual steps were taken to mobilize them effectively. Instead the falsity of most of the rhetoric and the limits to reform became all too evident.

In recruitment literature, for instance, the high proportion of women in the Soviet scientific labor force had often been mentioned, almost as a curiosity, in American articles about women scientists in 1955–58, even before Sputnik. Starting in 1958, however, some journalists began to use this high percentage as a reason why more American women should become scientists and engineers as well. But the ploy was so often flawed in its execution that one wonders if it was almost deliberately counterproductive. For example, after presenting all the statistics on Soviet women engineers, a journalist would point out (usually emphasizing the point with a photo) how unattractive (plain and stocky) these Soviet women were by Western standards. They were never shown with children, although statistics also showed that most were married and mothers. Thus the journalists, female as well as male, usually ended up showing not that a normal American girl should want to become an engineer like Olga but that she should be relieved that her country was not yet so badly off that she had to![42]

The urgent question that remained was how the bright young women who had not planned to go on to college would respond to this now almost patriotic duty to become scientists and engineers? Newspapers and magazines such as the *Saturday Review*, the *New York Times Magazine*, the *AAUWJ*, and even *Mademoiselle* carried a flood of articles in 1958–61—under such titles as "Woman's Place Is in the Lab, Too," "Science for the Misses," "Bright Girls: What Place in Society?" and "Plight of the Intellectual Girl"—addressing and denying the purported conflict between science and femininity.[43] At a time when the country reportedly badly needed more trained scientists and engineers and when Soviet women constituted one-quarter of that nation's engineers (but fewer than 1 percent of those in the United States), it was almost unpatriotic for bright young American women to follow traditionally feminine wiles and hide their brains. They must come forward, show their intelligence, and prepare for graduate school. More thoughtful, because it did not deny the existence of the conflict but accepted its reality and tenacity, was Diana Trilling's "Female-ism: New and Insidious," which appeared in *Mademoiselle* in June 1960. Trilling realized that it might seem to the young women of the time who preferred to marry young and remain at home that "femininity" and competence outside the home were antithetical, but this was a rather recent notion and a political luxury. It was certainly not true in embattled countries such as Israel or the Soviet Union or in pioneer ones as America had been until recently. If America's liberty was really in danger, as the country's leaders kept saying, then it was definitely the job of women, especially those who were well educated, to defend it wherever they were needed.[44]

In this charged atmosphere it was perhaps inevitable that women's mixed response would be distorted and deplored whatever they were doing. Thus, in 1959, when the WB completed its third annual survey of the postcollegiate experiences of the class of 1957, it chose to criticize the seemingly low percentage going directly into scientific employment. Seventy-nine percent of the class had found immediate employment, mostly in traditionally "feminine" jobs, including 58 percent as schoolteachers and 7 percent as nurses. But rather than being pleased or relieved that so many were employed, as *Womanpower* said was the current pattern, the account in *Science* deplored the comparatively low proportion—a mere 5 percent—going directly into scientific work. Only 1 percent of the class of 1957 had become chemists, 1 percent mathematicians or statisticians, and 3 percent "technicians, biological," a catchall category that included 48 percent of the biological science majors and 51 percent of those in the "health fields other than nursing." Both the WB and *Science* concluded that the percentage employed in science was not high enough and that young women needed more vocational guidance. (The survey's positive statistics—that 10 percent of the women were going on to postgraduate study, including an encouraging high of 21 percent of the physical science majors and 17 percent of those in the social sciences, including psychology—were ignored, as was the fact that a record number of women had earned Ph.D.'s in science in 1957.)[45]

Immediately the WB, to which NSF staffers routinely sent all inquiries from young women interested in a future in science, released several new bulletins: *Careers for Women in the Physical Sciences* (1959), *Careers for Women in the Biological Sciences* (1961), and *Careers for Women as Technicians* (1961). Yet when director Alice Leopold asked the NSF to help the WB pay for this unexpectedly large drain on its budget, it refused, despite its very large budget for science education. When she then appealed to James Killian, the president's science adviser, for funds to reimburse the costs of her recent bulletins on women in science, he recommended that she apply to the private Alfred P. Sloan Foundation, but that too failed. This must have seemed like a strange economy at a time when science budgets were being tripled and there was presumably a pressing national shortage of womanpower.[46] Did they want change or not? And if they did, why were they not willing to pay for it?

Nevertheless, the publicity accorded these new bulletins provides some hint that some teenagers were already far ahead of those very governmental officials who were publicly but lukewarmly urging them to enter science. Although the *New York Times*'s Fred Hechinger in January 1960 reflected the WB's pessimism, saying that "science in America is still a man's world, trying to put out a welcome mat for women," Maxine Cheshire of the *Washington Post* took a more hopeful view. When she talked to some high school girls identified by their principal as some of the "most industrious" seniors, she found that although none had seen the WB bulletins, one had already been admitted to MIT, and others planned to major in physics, mathematics, and political science. They were already "sold" on a career in science, even without the latest official vocational guidance.[47]

Yet problems, even pathological hostility, awaited some of them, for traditional exclusions did not necessarily just fade away. In 1960 Bernard Berelson, a former program director in the "behavioral sciences" (a term he popularized) at the powerful Ford Foundation, published *Graduate Education in the United States,* a major review of the nation's graduate schools and their needs. Three years in the making, supported by a grant of $100,000 from the Carnegie Corporation of New York, and based on consultations with forty-one male academics (listed in an appendix) and questionnaires completed by many of the deans of ninety-two graduate schools (omitting Radcliffe and Bryn Mawr), this presumably was an authoritative report. Although Berelson claimed not to speak for any foundation or to represent anyone other than himself, his background, source of support, and methods would lead one to expect him to reflect pretty accurately the prevailing views of the male graduate deans and the foundation world. Therefore it is important to note the brutally negative expectation that Berelson held out for women's future in academia. At a time when governmental committees were talking about "encouraging" women and Congress was appropriating millions of dollars to alleviate shortages of trained personnel, Berelson strongly asserted the graduate deans' contrary belief in a brief paragraph:

> So much is being said these days about the possibility of solving the "college teacher shortage" with women Ph.D s. that it is perhaps worth noting that they constituted 10% of the doctorates in 1910 and not quite 11% last year. Except for the war years, they have never accounted for a much larger segment than that. Furthermore, as the N[ational] O[pinion] R[esearch] C[enter] study shows, only about 15% of women graduate students prefer a full-time professional career for the first five years after completing their work. When these facts are combined with the reluctance of academic employers, not to mention the rules against joint husband-wife employment, they do not leave much room, in my judgment, for reliance on this "solution." It may be, as a foundation officer put it, that "we are losing half our brains" in this way, but it is hard to see much that can be done about it.[48]

Thus, Berelson was not only using incorrect percentages about previous doctorates (and thus trivializing the important fact that women were earning record numbers of Ph.D.'s in the 1950s) but, even worse, twisting the intent of the women respondents to one of the very large (and also foundation-supported) "opinion" or "survey research" projects of the time. Because those women graduate students polled "preferred" to be both married and pursuing a career, something that was only rarely feasible at the time, Berelson refused to believe that they or any women in graduate school would ever make any significant contribution to academia! It is hard to calculate the damage that such a devastating assessment by such an authoritative source must have done to the burgeoning interest in adding women to college faculties.

Another flagrantly defiant flaunting of the "womanpower" message occurred in 1962 at a session ostensibly designed to recruit women into engineering. When in the early 1960s, despite a decade of outreach, the percentage of engineers who were women was still below 1 percent, the SWE cooperated with the Office of Civil and Defense Mobilization of the Executive Office of the President as well as several universities on a series of two-day "utilization conferences" that urged guidance counselors, teachers, and others to encourage more women to take up engineering. Usually these occasions were just more of the same pollyanna "womanpower" messages of exhortation and denial—there are no problems, you are needed, you can do anything. Yet the published proceedings of the first such conference (at the University of Pittsburgh in April 1962), which included questions from the audience, indicated that only the speakers, and not even all of them, were particularly optimistic. There was plenty of anecdotal and statistical evidence of cultural and institutional discouragement. One speaker, psychologist Donald Ford (later dean of human development) of Penn State University, said there was a need for "a female John Glenn," referring to the recent celebrity of the astronauts of the National Aeronautics and Space Administration (NASA). Another, Dean Erwin Steinberg of Margaret Morrison College of Pittsburgh's Carnegie Institute of Technology, doubted the sincerity of the manpower analysts and the "womanpower" message, for if there were a real shortage, there would have been more change and action by 1962. A third, Donald Feight, manager of industrial relations at U.S. Steel in Pittsburgh, was insulting and brutally negative. He started out by saying that he had played with a doll until he was nine years old, trying for six years to smash its unbreakable head, and he added, "Maybe that's the reason I don't like women too well." He then waved two papers, one by himself on women engineers and the other by Eric A. Walker of Penn State University entitled "Women Are NOT for Engineering," which he described as showing that though a few women (Edith Clarke and Lillian Gilbreth) had made careers in the field, this was not the path for most women. In working on his own paper he had asked the men in his company's research laboratories, "Does U.S. Steel Corporation hire and train women with the same degree of enthusiasm as it does men?" The answers, he said, were negative. During the question period, Feight's response to one query was that although U.S. Steel would be hiring about six hundred engineers that year, none of them would be females.[49] Assuming that he accurately reflected the view that U.S. Steel chose to communicate to the federally sponsored recruitment panel, one can glimpse the depth of the hostility facing women engineers in some major places.

Thus, despite a lot of rhetoric "encouraging" the eager young women of the 1950s and the 1960s to become scientists and engineers, there were signals that the attitudes of employers and the atmosphere at the nation's graduate schools might not be even rhetorically encouraging.

4

Graduate School

Record Numbers Despite It All

*I*f it had become part of federal manpower policy in the 1950s that bright women were now a "precious national resource" and that more of them should be encouraged to earn doctorates, that message was strongly resisted at the nation's graduate schools. In fact many of the universities that just a few years earlier had coped so flexibly, even heroically, with the influx of veterans continued to exclude women (e.g., Princeton University and Texas A&M) or admitted them with the expectation that they would soon drop out. The nation's graduate deans, including two women in the 1950s, were determined to uphold the value of the Ph.D., especially in a period of expansion, and did not wish to cheapen the degree by making it easy to get. They were convinced that a high dropout rate was a sign of high standards and were utterly opposed to such arrangements as part-time study or childcare as unbecoming of "serious" students at prestigious institutions. Only a few leaders of graduate schools took even such basic action as constructing new dormitories for women students. Most interpreted women's recently increased propensity to marry while in graduate school as still another sign that they would drop out before finishing their degrees. Accordingly, rather than devising ways to help them to stay on and complete their degrees, they expected them to quit and then blamed them when, laboring under poverty as well as the usual isolation and uncertainty of graduate studies, they did drop out. Then their case could be used to justify denying aid or discouraging future female applicants. Unfortunately, when increased amounts of federal fellowship became available in the late 1950s, they were generally in the fields where the women were not: a great deal in engineering and the physical sciences, much less (and later) in the social sciences.

It is thus quite remarkable that despite such a discouraging atmosphere, record numbers of women scientists did enroll, persevere, and complete their degrees in the 1950s and 1960s. Yet what data are available on these women indicate that they were not distributed uniformly across all fields and all graduate schools. Even at this preemployment stage, "science" was not, for them anyhow, one big homogeneous aggregate though which they might move at will; instead it was a kaleidoscope of very different subfields, in a great many of which they were, despite all the "womanpower" rhetoric, unwelcome and unwanted. Similarly, among institutions there were myriad patterns and sub-patterns, as some departments and particular professors proved more welcoming and more supportive than others. Mentors, more than money, may have been the missing ingredient.

Although many women scientists in the 1970s published reports of their unhappy graduate school experiences in the fifties and sixties, there were few contemporary accounts. Yet these few confirm their later view that graduate study was indeed a rather grim prospect, or at least it seemed so to the occasional reporter from a woman's magazine who tried to romanticize graduate school and make it attractive to readers. Nancy Lynch's "The Young American Graduate Student," in the September 1956 (i.e., pre-Sputnik) issue of *Mademoiselle*, was downright discouraging: all that the hundred graduate students she interviewed did was work and worry. Although there were plenty of single men around, the women's idea of an interesting date was "talk (*serious* talk)" or a concert or a lecture. The only women who seemed happy in graduate school were those who were in love with their subject, and they were, to put it euphemistically, rare beings who were "able to take and even savor the—yes—one-sided life at graduate school." Most "nice bright girls" would probably be better off spending the same money on a year of travel abroad, an impression quickly confirmed by a look at the seemingly low marriage statistics of graduate women. Few readers of *Mademoiselle* in 1956 would find graduate school at all attractive, Lynch decided.[1]

A later report by journalist David Boroff was more detailed, more insightful, and even a bit optimistic. Published in 1961, his chapter "The University of Michigan: Graduate Limbo for Women" in his *Campus USA: Portraits of American Colleges in Action* found the graduate women "academically motivated but professionally confused." They had done well in college and, despite the discriminatory admissions policy (admitted to by one unnamed psychology professor) and the very low income from their teaching fellowships, embarked on higher studies. This allowed for few luxuries, so that though a few were strikingly attractive, most were "that most sexless of creatures: an academic drone, indifferently dressed." They worked hard, dated the men in the department (though most complained that they were "either married—or 'impossible!'"), and found "considerable nervous excitement in the search for, and discovery of the academic father, a distinguished professor." Over time they grew tired of their straitened circumstances, but they were slow to finish their degree because they were not sure what they would do next. They were reluctant to take jobs in small college towns and feared most of all "entombment in a small *woman's* college—no fate more harrowing than that." Yet Boroff was confident that soon the problem would solve itself: "The demand for college teachers during the next decade will be so intense that the barriers will topple. As more women join the ranks of graduate students, it will no longer seem a heresy when they devote themselves to the intellectual life. The gates of the ghetto will swing open."[2]

One step toward improving or humanizing the living conditions for these graduate women was a kind of "female philanthropy" of the fifties, the construction of new dormitories. For years women graduate students had found what rooms they could in boardinghouses in Cambridge, New Haven, and elsewhere, but as enrollments increased and as expectations rose that the social experience of meeting other students could be a valuable part of one's graduate school years, the vision and goal of a dormitory for graduate women took shape. Thus Bernice Brown Cronkhite, the longtime dean of the Radcliffe Graduate School, who, egged on by an empty lot in a prime location, had envisioned a "graduate center" as early as the 1920s, intensified her fund drive in the fifties. Despite much skepticism from potential donors about the need for graduate education for women, by 1955 she had raised enough ($200,000) to lay the cornerstone and in 1957 to open the almost luxurious Radcliffe Graduate Center, later renamed Cronkhite Hall. It was built in installments, as money became available, and finally housed 139 students, mostly in single rooms, since Cronkhite felt, perhaps recalling Virginia Woolf, that "every graduate student must have a room of her own."[3]

Similarly at Yale University, when the female enrollments at the graduate and professional schools reached about two hundred in the mid-fifties, President A. Whitney Griswold decided that it was time for a new women's dormitory. When in 1956 Yale was deemed eligible for a million-dollar federal loan at 2.75 percent interest, he pressed on without special donors and opened a dormitory for more than two hundred women in 1959. Named Helen Hadley Hall in honor of the wife of James Hadley, the president who had admitted women to Yale Graduate School in 1892, the building proved both too expensive for many women students and too plain and ugly for undergraduates nearby in visibly more comfortable "colleges." This served to reemphasize the vastly different financial resources behind the graduate women at Yale and the undergraduate men. In particular, the women's dorm had no central dining room or formal meal plan, just kitchenettes on each floor.[4]

The first dormitory for graduate women in the South was Rayzor Hall at Texas Woman's University. Originally an apartment building near campus, it was purchased by the university in 1960 with the help of Mr. and Mrs. J. Newton Rayzor of Houston. Once renovated and equipped Texas-style, with air conditioning and a parking lot for student cars, the five-story building was able to house fifty-three graduate students.[5] Yet such dormitories, comfortable and convenient, were not necessarily the morale boosters their builders had envisioned. As Cronkhite's successor at Radcliffe reported in 1966, the rooms were all occupied, but the women in them were discouraged and depressed and needed much counseling.[6]

Many of them were poor. Despite the rapidly growing number of AAUW and federal fellowships, there were still too few to go around. Nevertheless, it would be hard to find a better example of the power of "female philanthropy" in the fifties and sixties than the AAUW fellowship programs; as Bernice Brown Cronkhite put it, "The AAUW was quite wonderful."[7] Since the 1880s, the AAUW had been raising funds to support women's graduate work. In the 1920s it had set the ambitious goal of increasing its endowment funds to a million dollars, which it attained in 1953. When the organization's master fund-raiser Dorothy B. Atkinson Rood of Minneapolis urged the members to double this amount by 1981, they impressed even her by accomplishing this by 1961. Perhaps they were spurred on by reports of skeptical public officials, such as Arthur Lack's notorious front-page article in the *Wall Street Journal* on January 16, 1958.[8] If the federal government, with all its money and highly publicized new programs for scientific manpower, were to withhold fellowships from many women, the need for the AAUW fellowships was greater than ever. Seasoned members must redouble their efforts to raise funds, which they did again and again.

For example, in 1956–57 the AAUW awarded eighty-four fellowships with an average stipend of $2,430, or a total of $204,000. In 1958, in order to meet new Internal Revenue Service rulings on nonprofit organizations, the AAUW set up a separate AAUW Educational Foundation to handle all the moneys, and it built a new headquarters building in Washington, D.C., to house it. Despite this expense, the new foundation was still able to increase the number and size of its awards to ninety-seven at $3,320 each (or $322,000) in 1965–66. This sharp increase over previous decades is shown in figure 2. Among the winners were crystallographer Isabella Karle, whose 1943 AAUW award was especially welcome because at that time the chemistry department at the University of Michigan did not give fellowships to women,[9] chemist Elga Wasserman at the Radcliffe Graduate School in 1947, geographer Ann Larimore at the University of Chicago in 1956, economist Marina von Neumann Whitman at Columbia University in 1961, and Vivienne Malone Mayes, the first black woman to be awarded a Ph.D. in mathematics from the University of Texas, in 1964.[10]

Although some members of the AAUW fellowships committee feared that married women applicants would represent a greater gamble than single women, since more of them might drop out, as a whole they felt that the risk was worth taking. Nevertheless, the committee did hedge its bets by asking all the married applicants to detail their plans for childcare during the fellowship year.[11] Yet even the AAUW continued to require that all of its fellows be full-

Number of Awards

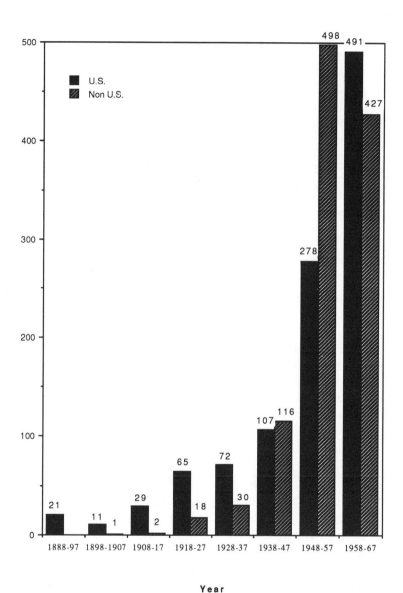

Fig. 2. AAUW awards to women of the United States and other countries, 1888–1967. These 2,166 awards do not include the 25 recipients of the AAUW Achievement Award. Based on data in *Idealism at Work, Eighty Years of AAUW Fellowships: A Report by the American Association of University Women* (Washington, D.C.: AAUW, 1967), 6.

time students. Mary Bunting of Radcliffe fought this for many years, arguing that if the AAUW, which specialized in the needs of women scholars, did not pioneer in supporting part-time study, no federal agency would. Nevertheless, the restriction long remained in effect.[12]

A related form of "female philanthropy" in the 1960s was a special fund to provide graduate fellowships for Catholic nuns. This grew out of the Sister Formation Movement, which had started after an address by Pope Pius XII to the First International Congress of Teaching Sisters in Rome in 1951. One goal was to improve and upgrade the education of teaching nuns at all levels, including those at the many Catholic colleges that were seeking accreditation and thus needed to have more Ph.D.'s on their faculty. Although many orders of nuns and their local bishops managed to provide the funds for some nuns' advanced degrees, for others a fellowship was needed. The nuns' fellowship program grew out of the grantswomanship of Rev. Mother M. Regina, R.S.M., who in 1961 set up a separate legal corporation, the Sister Formation Graduate Study and Research Foundation, to procure and distribute the funds. By 1963 it had obtained grants from the Ford Foundation, the Grace Foundation, and the Raskob Foundation and had plans for two thousand fellowships.[13]

In addition there were increasing numbers of federal fellowships available, which benefited women biologists mostly. The oldest program by far was that of the National Cancer Institute of the U.S. Public Health Service (USPHS), whose program had started back in 1938 as part of the National Cancer Act. In 1958 the USPHS published a twenty-year report and a list of its 924 past and 245 current (or "active") fellows. Because it is hard to determine the sex from many of the fellows' often foreign names, one can say only that between 126 and 152 of the past fellows were women (or 13.64–16.45 percent). This list would include such now famous names as Jewel Cobb, Thelma Dunn, Helen Dyer, Charlotte Friend, Dorothy Horstmann, Berta Scharrer, and Elizabeth Weisburger. Among the estimated 56 women of its 245 "active" fellows (22.86 percent) was the young biologist Eloise E. Clark, later a high official at the NSF and president of the American Association for the Advancement of Science (AAAS). Although no data were given on the numbers or sex of the applicants to this program (and the time period did include the war years, when few men would have been applying), these proportions were among the highest in any governmental fellowship program, primarily because most of the awards were in biological fields.[14]

After the war the new Atomic Energy Commission (AEC) started its own fellowship program in the physical and biological sciences to increase quickly the number of trained professionals in radiation fields. Before its termination in 1953 (chiefly because of a panic and scandal over one male fellow's admitting to being a Communist), it had made 765 predoctoral and 155 postdoctoral awards. Of these an estimated 53, or just 5.76 percent, went to women. Thirty of the 40 women predoctoral fellows and 8 of the 13 women postdoctoral fellows were in the biological sciences. These early AEC fellows included Helen Whiteley, Justine Garvey, Betty Twarog, and Barbara Bachmann, who created a certain stir when, after holding the fellowship for two years, she publicly resigned it in September 1950 rather than sign the newly required loyalty oath.[15]

In 1952 the new NSF made its first fellowship awards, 569 predoctoral and 55 postdoctoral ones. The former included 36 women (6.33 percent), and the latter none, though physicist Hertha Sponer-Franck of Duke University declined one offered to her. Among these young winners were Barbara Bachmann, Carolyn Cohen, and Maxine Frank (later Singer). In 1953, women received only 23 of the 556 awards, or just 4.14 percent, but received 93 of the 1,274 "honorable mentions" (7.30 percent). When Bernice Brown Cronkhite of Radcliffe read this, she wrote to the NSF to ask for more data on the sex of the applicants. Several months later she was informed that 8 percent of the applicants in both the pre- and postdoctoral competitions had been women. (The proportion had varied widely among the fields, however, ranging from none in engineering to 13 percent in mathematics, 15 percent in the biological sciences, and 16 percent of the candidates in anthropology and psychology.)[16] Evidently, even if all the female "honorable mentions" had received fellowships, women would still have made up a lower percentage of winners than they did of applicants.

Many years later the NRC did a comprehensive statistical study of the NSF fellows. It revealed that between 1952–53 and 1972, when the program ended, it had awarded fellowships to 13,278 persons, including 1,592 women (11.99 percent) (see tables 4.1 and 4.2). The largest numbers overall and for women were in the biosciences, with 2,505, of whom 677 (27.03 percent) were women, and these were almost half (42.53 percent) of all the women fellows. Women's highest proportion (33.00 percent) was in psychology, the smallest field here, and their lowest proportions were in physics (2.21 percent) and engineering (1.02 percent), which is in fact a rather good showing, since they were usually received fewer than 1 percent of the baccalaureate or doctoral degrees in engineering). As the women's percentage rose steadily over time from an average of 8.55 percent in 1952–61 to 19.48 percent in 1972, their increased share of the total was owing in large part to a change in the extent to which different fields, especially the social sciences, were supported. In fact the NSF awarded 113 fellowships in the social sciences in fiscal year 1972 (with 30, or 26.55 percent, to women), figures comparable to those for the entire decade 1952–61 (126 awards, of which 28, or 22.22 percent, went to women social scientists).[17]

Meanwhile, the dropout rate for these highly selected NSF fellows was high, especially among the women. By 1974 only 69.7 percent of the 13,278 fellows—72.8 percent of the men but only 46.8 percent of the 1,592 women—

Table 4.1.

NSF (1952–1972) and NDEA Title IV Fellowships (1963–1973), by Field and Sex

	NSF			NDEA		
	Total	Women	Women as % of Total	Total	Women	Women as % of Total
Humanities and professions				14,152	4,423	31.25
Education				3,154	1,001	31.74
Psychology	603	199	33.00	1,849	569	30.77
Biosciences	2,505	677	27.03	4,624	1,049	22.69
Social sciences, other	1,254	298	23.76	6,009	1,117	19.59
Chemistry	2,002	158	7.89	2,001	300	14.99
Mathematics	2,174	160	7.36	1,789	290	16.21
Geosciences	620	32	5.10	899	91	22.54
Physics	2,168	48	2.21	1,785	96	5.38
Engineering	1,952	20	1.02	3,904	59	1.51
Total/average	13,278	1,592	11.99	40,166	9,055	22.54

Sources: Lindsey R. Harmon, *Career Achievements of NSF Graduate Fellows: The Awardees of 1952–1972* (Washington, D.C.: NAS, 1977), 8; idem, *Career Achievements of the National Defense Education Act (Title IV) Fellows of 1959–1973: A Report to the U.S. Office of Education* (Washington, D.C.: NAS, 1977), 10.

had actually completed the doctorate. This was understandable for the most recent fellows, those awarded in 1972, of whom only 5.1 percent had yet completed the degree, but for the earlier fellows this low rate was surprising. But sex differences, wide as they were, were not the only ones reported. Again certain differences by field loomed large. The completion rate for both men and women was far higher in some fields (e.g., chemistry and physics) than in others (see table 4.3). Whether this was because of the level of financial support, postdoctoral opportunities, the efficiency of the graduate training, or something else was not explained.[18]

It was not at all clear then, nor is it clear now, just which women—or men—were dropping out when or why. Certainly for many women the reason was less marriage itself than the financial burden many quickly assumed in "putting hubby through," as the phrase went. (Cornell University even formalized this practice in the late 1950s by having the dean of the graduate school and his wife award the "Ph.T." degree to many wives who stopped their own studies to take up what could hardly have been lucrative jobs near campus so that their husbands could finish their degrees.)[19] Presumably the wives would complete their studies later. Had there been fellowships for both, or had the husbands' stipends been large enough to pay for daycare, the wives might have completed their degrees and been ready to look for jobs elsewhere when their husbands were.[20]

Table 4.2.
NSF Graduate Fellows, by Field and Sex, 1952–1972

Field	1952–1961				1962–1966				1967–1971				1972			
	Total	% of All Fields	Women	Women as % of Total	Total	% of All Fields	Women	Women as % of Total	Total	% of All Fields	Women	Women as % of Total	Total	% of All Fields	Women	Women as % of Total
Psychology	141	2.92	43	30.50	193	4.75	50	25.91	222	5.76	90	40.54	47	8.80	16	34.04
Biosciences	995	20.06	207	20.80	670	16.50	201	30.00	717	18.60	225	31.38	123	23.03	44	35.77
Social sciences, other	126	2.61	28	22.22	392	9.65	63	16.07	623	16.17	177	28.41	113	21.16	30	26.55
Chemistry	979	20.27	72	7.35	547	13.47	37	6.76	437	11.34	44	10.07	39	7.30	5	12.82
Mathematics	550	11.39	34	6.18	798	19.65	55	6.89	761	19.75	70	9.20	65	12.17	1	1.54
Geosciences	283	5.86	9	3.18	168	4.14	9	5.36	147	3.81	10	6.80	22	4.12	4	18.18
Physics	1,006	20.83	19	1.89	657	16.18	14	2.13	468	12.14	14	2.99	37	6.93	1	2.70
Engineering	749	15.51	1	0.13	636	15.66	9	1.42	479	12.43	7	1.46	88	16.48	3	3.41
Total	4,829	99.99	413	8.55	4,061	100.00	438	10.79	3,884	100.00	637	16.53	534	99.99	104	19.48

Source: Harmon, *Career Achievements of NSF Graduate Fellows,* 8.

Table 4.3.

Percentage of NSF (1952–1972) and NDEA Title IV (1963–1973)
Fellows Completing Doctorates, by Field and Sex

Field	NSF Men and Women	Men	Women	NDEA Men and Women	Men	Women
Chemistry	33.9	85.9	60.1	57.6	62.0	32.7
Physics	78.6	79.2	52.1	49.7	51.3	21.9
Geosciences	72.9	75.2	31.3	43.3	46.3	16.5
Biosciences	70.1	77.6	49.9	51.1	57.9	27.8
Mathematics	65.9	67.4	46.9	44.3	50.0	15.2
Psychology	63.2	71.3	46.7	46.5	53.4	31.1
Engineering	63.2	63.6	30.0	45.6	46.0	16.9
Social sciences, other	43.3	53.9	34.6	34.8	39.3	16.4
Education	—	—	—	55.1	63.0	38.2
Humanities and pro- fessions	—	—	—	32.4	38.7	18.6
Nonscience	—	—	—	36.6	46.9	22.2
Total/average	69.7	72.8	46.8	41.4	46.9	22.7

Sources: Harmon, *Career Achievements of NSF Graduate Fellows,* 8; idem, *Career Achievements of the National Defense Education Act (Title IV) Fellows of 1959–1973,* 10.

By the time Congress discontinued the NDEA fellowship program in 1973, it had broadened it in various ways and awarded 45,829 three-year fellowships in the sciences, the social sciences, the humanities, and education. Women received 19,875 of these (21.55 percent), or almost twice their cumulative percentage at the NSF (11.99 percent), but still well below the one-third predicted (and feared) earlier by the *Wall Street Journal.* This relatively high proportion of women in the NDEA program was only partly owing to its inclusion of those fields in which women were numerous (e.g., humanities, education, and psychology). Table 4.1 breaks down by field and by gender data on the 40,166 NDEA fellowships awarded in 1963 and after (when the fellows were required to specify a field of study). Although women were more likely than men to be in the humanities, education, and psychology, they also constituted far higher percentages of such fields as physics, chemistry, mathematics, engineering, and geosciences than in the corresponding NSF programs.[21] Why there should have been such differences is not clear. With the lack of any data on the number and distribution of women's applications, one can only speculate. The NDEA program sought to train future college teachers and was administered by the USOE. Perhaps there were more women on the NDEA selection panels than there were on the NSF panels. Maybe scientists felt that the NSF fellowships, which had started earlier, carried more prestige and thus should go to their male protégés. Possibly the academic community encouraged more female college seniors to apply for NDEA awards.

But once the NDEA fellows were in graduate school, their attrition rate was even higher, as the *Wall Street Journal* had feared, than that of the NSF fellows. Of the 45,829 NDEA fellows given awards between 1959 and 1973, only 19,998 (43.6 percent) had completed their doctorates by 1974. This average was greatly lowered by the bulk of fellowships awarded in the late 1960s and early 1970s to persons who had not yet finished their degree. Nevertheless, there were strikingly large differences by sex and by field. Among the men 49.2 percent had completed their degree by 1974, compared with just 23.5 percent of the women. Among the 40,166 fellows between 1963 and 1973 only, the men's highest completion rate (63.0 percent) was in education, followed closely by chemistry (62.0 percent) and biosciences (57.9 percent); the lowest rates were in the social sciences (39.3 percent) and the humanities and professions at 38.7 percent. For women the highest completion rates were also in education (38.2 percent) and chemistry (32.7 percent), followed by psychology (31.1 percent), which had been fourth for men, and then the biosciences (27.8 percent). The women's lowest completion rates were in social sciences (16.4 percent) and mathematics (15.2 percent) (see table 4.3).

What had gone wrong? Why were so many of these graduate students with financial aid not finishing their doctorates? Maybe the limitation of the fellowship to three years of support and the ready availability of jobs for sub-Ph.D.'s in the rapidly expanding academia of the time may have prolonged or prevented the completion of the dissertation for many. Maybe they married and left the area. Maybe they were overwhelmed by childcare. Or maybe, like the unhappy women in the Cronkhite dormitory, they faced an inhospitable setting. To a large extent graduate training was and is a small-group experience, and there were numerous (later) reports of discouraging situations. Certainly women graduate students of the fifties had fewer options available to them than did their male counterparts.

To judge from the data available on the women who did complete their degrees, compared with the men, they were accepted at fewer institutions, fewer fields were open to them, and fewer professors were willing to accept them. In fact finding a successful match, while choosing among fewer options, was one of the major difficulties facing female graduate students in these years. This is evident in the data in *Earned Degrees Conferred by Higher Educational Institutions,* published by the USOE starting in 1947–48 and continuing under the National Center for Educational Statistics in various modified forms into the late 1960s and early 1970s, when economies forced it to revert to summaries.[22] The data indicate that while the number of women receiving doctorates in forty-three sciences, engineering, and the social sciences more than doubled in these years, from 215 in 1947–48 to 459 in 1960–61, the overall totals for both men and women almost tripled. Thus, the women's percentage dropped from a post–World War II high of 9.35 percent to a low of 6.16 percent in 1954 and then partially rebounded to 7.18 percent in 1961. Yet, as we shall see, there was a tremendous range in the number of women earning degrees in each field and at each institution.

When the forty-three specialties are grouped into subject clusters, which are then ranked by overall size, the totals for both sexes show that the fields in which the most doctorates were awarded were chemistry (13,480), the combined biologies (8,519), engineering (8,450), and psychology (7,208), which together account for well over half of all the science doctorates given by U.S. universities between 1948 and 1961 (see table 4.4). At the other end of the spectrum were the smallest separate sciences, including anthropology (603), home economics (416), and astronomy (208). Yet the field awarding the greatest number of doctorates to women was not chemistry but psychology (1,049), followed by the combined biologies (926), chemistry (637), and home economics (331). Thus there was a wide range in the percentages of women earning doctorates in each field and subfield (see table 4.4), from a high of 79.57 percent in home economics (itself a decline from the near 100 percent of earlier years, with the men chiefly in the new specialty of "child development") down to 17.74 percent in anthropology, 14.55 percent in psychology, 14.15 percent in sociology, and less than 1 percent in several fields and even zero in one ("earth sciences, all other"). One might have expected some greater receptivity in the newer fields, such as statistics, which had just seven women Ph.D.'s (3.15 percent), even though there had been prominent women in the field since the thirties and a few even held academic posts in the fifties.

But in several other fields the low proportion of women doctorates was the result of certain entrenched attitudes and institutional practices. Joanne Simpson, the first woman Ph.D. in meteorology (University of Chicago 1949), has recounted in several places the numerous difficulties she encountered, including skeptical professors and a lack of financial aid, and how she outwaited and outmaneuvered them all, though at considerable psychic cost.[23] In oceanography only one woman earned a doctorate in this entire period, and even this was remarkable, since women were strictly banned from oceangoing research vessels, a practice that dated back to the field's early roots in naval facilities and equipment. The prohibition was so strictly enforced that in the mid-fifties, when an intrepid female graduate student who had stowed away in hopes of being taken along surreptitiously was discovered by the ship's captain, she was immediately returned to land, and the professor in charge chastised her for wasting everyone's time and research money.[24] Similarly, in forestry, although Syracuse University awarded a doctorate to a woman in 1958–59, the other major graduate programs in the field continued their strict ban on women's admission through the late sixties.[25]

Table 4.4.
Science and Engineering Doctorates Awarded by U.S.
Universities, by Field and Sex, 1947–48 to 1960–61

Field	Total	Women	Women as % of Total
Home economics	416	331	79.57
Social sciences[a]	3,030	442	14.59
Sociology	1,979	280	14.15
Anthropology	603	107	17.74
Other social sciences	448	55	12.28
Psychology	7,208	1,049	14.55
Bacteriology	1,641	220	13.41
Medical sciences	1,773	208	11.73
Anatomy	394	69	17.51
Public health	402	67	16.67
Clinical sciences	559	36	6.44
Pharmacology	242	19	7.85
Biophysics	102	5	4.90
Pathology	63	8	12.70
Embryology	6	1	16.67
Cytology	5	3	60.00
Biochemistry	1,820	212	11.65
Biosciences	8,519	926	10.87
Zoology	2,128	260	12.22
General biology	1,751	282	16.11
Botany	1,715	173	10.09
Physiology	882	108	12.24
Biological sciences	504	53	10.52
Genetics	259	20	7.72
Entomology	886	17	1.92
Plant pathology	321	10	3.12
Plant physiology	73	3	4.11
Mathematical sciences	3,182	176	5.53
Mathematics	2,960	169	5.71
Statistics	222	7	3.15
Physical sciences	6,722	159	2.37
Physics	6,144	115	1.87
Meteorology	146	3	2.05
Astronomy	208	23	11.06
Other physical sciences	274	18	6.57
Chemistry	13,480	637	4.73
Economic sciences	3,540	158	4.46
Economics	3,079	154	5.00
Agricultural economics	461	4	0.87
Geosciences	2,675	96	3.59
Geology	1,876	42	2.24

(*continued*)

Table 4.4. (*Continued*)

Field	Total	Women	Women as % of Total
Geography[a]	626	52	8.31
Geophysics	72	1	1.39
Oceanography	58	1	1.72
Other geosciences	43	0	0.00
Forestry, agricultural and veterinary sciences	5,585	88	0.58
Agricultural sciences	5,052	83	1.64
Forestry	385	1	0.26
Clinical veterinary sciences	148	4	2.70
Engineering sciences	8,450	25	0.30
Engineering	8,090	24	0.30
Metallurgy	360	1	0.28
Total/average	68,091	4,727	6.94

Source: USOE, *Earned Degrees Conferred by Higher Educational Institutions,* 1947–48 through 1960–61, published annually as part of the USOE Circular series.

Note: Categories given are those reported by the institutions.

[a]Geography, here counted as a part of the geosciences, was becoming a social science.

Although a few black women had earned doctorates before World War II, in the 1940s (and even later) several others became the first in their field to earn a Ph.D. (see table 4.5). In 1940 Roger Arliner Young was at the University of Pennsylvania, the first black woman to earn a doctorate in biology. She was quickly followed by Mary Logan Reddick and neuroembryologist Geraldine Pittman Woods at Radcliffe in 1944 and 1945, respectively. Most of these pioneers spent their entire career teaching at black colleges, where they worked hard to improve the instruction. Marjorie Browne, the second black woman doctorate in mathematics (University of Michigan 1950), for example, was later very proud to have written the grant proposal that brought North Carolina Central University its first computer in 1960.[26]

The general pattern in Ph.D. "production" in these years was for a relatively few large university to award most of the degrees. Thus, although the USOE data show that 173 universities awarded a total of 83,113 doctorates in science and engineering between 1947–48 and 1962–63, just 25 of them accounted for 54,265 of the degrees. Or to put it more starkly, just 14.45 percent of the nation's universities awarded 65.29 percent of the science and engineering doctorates in this period. Furthermore, only a few departments (or "schools" or even "colleges") accounted for a disproportionately large number of an institution's and

Table 4.5.
First Black Women Doctorates, by Field, 1933–1973

Year	Name	Field	University
1933	Ruth E. Moore	Bacteriology	Ohio State
1934	Ruth Howard Beckham	Psychology	Minnesota
1935	Flemmie Kittrell	Nutrition	Cornell
	Jessie J. Mark[a]	Botany	Iowa State
1940	Roger Arliner Young	Zoology	Pennsylvania
1941	Ruth Smith Lloyd	Anatomy	Western Reserve
1942	Marguerite Thomas Williams	Geology	Catholic University
1948	Marie Maynard Daly	Chemistry	Columbia
	Phyllis Wallace	Economics	Yale
1949	Evelyn Boyd Granville	Mathematics	Yale
1955	Dolores Cooper Schockley	Pharmacology	Purdue
1973	Shirley Ann Jackson	Physics	MIT

Source: Vivian Ovelton Sammons, *Blacks in Science and Medicine* (New York: Hemisphere 1990), 274.

[a]First black of either sex.

the nation's total. Twenty-two of these "super-productive" units (six in chemistry and five in engineering) each awarded more than 500 hundred doctorates. (If the cutoff point had been 400 doctorates, the list would have been twice as long.) The largest number were from MIT, which awarded 1,185 doctorates in engineering, followed by the University of Illinois with 918 in chemistry and the University of California's combined total (UCLA plus Berkeley) of 792 doctorates in chemistry. Yet for reasons that are far from clear, very few of these twenty-two mammoth units (probably but not necessarily departments) awarded many degrees to women—just 50 at the University of California and 46 at the University of Illinois, all in chemistry. Perhaps there was something about the atmosphere or culture of these mass-producing departments or colleges—impersonality, hostility, restrictive policies, reluctance to make exceptions, inflexibility, or other factors—that overwhelmed the few women who were admitted.[27]

The striking fact about the institutional distribution of the women doctorates in science and engineering in these years is that 35 (or 20 percent) of the universities that awarded doctorates to men did not give *any* to women in any of the sciences covered here. These included such sizable and important institutions as Princeton University, Texas A&M, the Georgia Institute of Technology, and the combined Institute for Paper Chemistry and Lawrence College. Of the remaining 138 institutions that awarded at least one doctorate to a woman, many did so in just a few fields. This pattern was even stronger within specific disciplines, where as many as three-fourths or even more of the institutions awarding a degree in the field did not given even one to a woman. In certain notorious fields, such as metallurgy and oceanography, the proportion of such "zero-departments" was of course very high: twenty of twenty-one departments in metallurgy and ten of eleven in oceanography. In biophysics and agricultural economics the proportion was three-fourths (nineteen of twenty-five, or 76.0 percent, and twenty of twenty-seven, or 74.07 percent), in pharmacology, forty-three out of sixty-two (69.35 percent), and in entomology twenty-seven out of thirty-nine (69.23 percent). Even in physics forty-five of the ninety-four university units (47.87 percent) awarding doctorates did not give *any* to women in the period 1947–48 to 1962–63. Likewise, in economics, where women averaged 5.14 percent of the Ph.D.'s in these same years, they were totally absent from half (thirty-eight of seventy-six) of the departments awarding doctorates but constituted more than 50 percent of those receiving the doctorate at some generally small programs (three of four, or 75.00 percent, at Bryn Mawr College and three of five, or 60.00 percent, at Fordham University) and more than 10 percent at a few middle-sized ones.

Whether this skewed pattern was the result of women's not applying to these "zero-programs" (perhaps because of advice that they were unsuitable for women), quotas on the number of women admitted, or a high rate of attrition among the few intrepid women who did enroll is not clear. Certainly in some departments it was known that certain professors did not want, and would not accept, women graduate students. In fact psychologist Anne Roe found that many of the sixty-four eminent male scientists she interviewed for her book *The Making of a Scientist* (1952) refused to have women students for a variety of reasons. Some said the time would be wasted, since the women would get married and drop out (forever); others did not think women would perpetuate their name and influence as much as men students would; and still others, including one in the highly feminized field of developmental psychology, were deliberately recruiting men in order to upgrade the field, despite the fact that "some of the women are brilliant."[28] Thus, for whatever reasons—institutional or personal, written or unwritten—despite all the talk about "womanpower," many individuals, departments, and whole universities did not contribute to the training of women scientists in the period 1947–63.

Individual experiences on this issue abound, although a recurring theme is the women's astonishment at the skepticism that greeted them. Unbeknownst to most of them, who naively expected that because they had been admitted they must be welcome, they were under close scrutiny regarding just how serious they were about their work. This often extended to their social life, and at the slightest hint that they were even thinking of getting married, they could be accused of "lacking serious purpose." Evidently there was a thin line between being so serious that one was "maladjusted" and not being serious enough; graduate women were to be faulted either way.

Leona Marshall Libby, for example, decided to work with someone other than the University of Chicago's great Nobelist James E. Franck after he informed her that since she was a woman, she would never amount to much in science, and that like many Jews, she would starve to death! Archaeologist Emily Vermeule left Radcliffe Graduate School for Bryn Mawr after one year because six or seven of the thirty-five or forty faculty members in her department at Harvard did not talk to the women students and the library was very hard to use. When marine biologist Ruth Dixon Turner's professors at Cornell were so overwhelmed by veterans in the late 1940s that they had no time for her, she transferred to Radcliffe, partly because she already worked in a museum at Harvard. But even Radcliffe proved difficult, since, as she later recalled, Ernst Mayr was on her committee, and he could barely bring himself to speak to a woman. Marian Boykan Pour-El also recalled years later that for a very long time after she arrived at Harvard's mathematics department in the late fifties no one would talk to her or sit near her. Cellular immunologist Diane M. Jacobs later recalled that at the Harvard Graduate School in the sixties "the message really was that I should marry a scientist or doctor, not *be* one."[29] In the Purdue University chemistry department so few professors would take on a woman student that any who enrolled had to choose their specialties on the basis of which professors would give them lab space. When Harvard's social psychologist Gordon Allport prepared a report on the completion rates of the 421 graduate students who entered the Department of Social Relations between 1946 and 1955 that revealed that 47 percent of the women had earned doctorates compared with 65 percent of the men, he said, "I feel certain that marriage and family complications account for the discrepancy." Yet he seemed to do nothing practical to help them complete their degrees. He routinely warned his women students that they were bad risks who were in danger of dropping out, but when they tried to reenter the field after they had in fact married, had children, and reduced their professional activity for a time, he provided no help.[30] A 1967 survey at Yale of the attitudes held by departmental directors of graduate studies revealed that "a large majority" held a variety of lower professional expectations for their women students than for the men.[31]

By contrast, in this period of high mobility (for male faculty), a professor could on occasion use his leverage to force open a door at a recalcitrant institution for a woman graduate student. When, for example, chemist John Roberts of MIT accepted an offer from Cal Tech in 1952, it was on the condition that he be allowed to bring along most of his graduate students, including Dorothy Semenow, a Mount Holyoke graduate summa cum laude. In her case, a special vote was required of the Cal Tech faculty, which had rejected a similar request a few years before. But this time it voted to change its policy, the trustees concurred, and she was admitted. In 1955 she became the first in Cal Tech's trickle of female Ph.D.'s.[32]

Table 4.6.

Institutions Awarding the Most Science Doctorates to Women, by Field, 1947–48 to 1962–63

Institution	Total	Psychology	Biological Sciences	Chemistry	Home Economics and Nutrition	Sociology	Bacteriology	Biochemistry
1. Columbia	376	87	56	41	19	23	6	12
2. California—Berkeley and UCLA	371	71	84	50	15	9	19	20
3. Radcliffe/Harvard	301	45	58	21	—	32	8	11
4. NYU	269	96	55	13	12	20	3	5
5. Chicago	264	57	43	13	19	38	8	8
6. Wisconsin	257	12	70	29	31	3	20	38
7. Michigan	219	79	49	18	—	9	14	6
8. Cornell	207	16	32	14	69	12	5	17
9. Ohio State	177	51	14	26	39	15	5	5
10. Illinois	162	20	27	46	1	3	11	3
11. Yale	141	32	31	24	—	8	7	12
12. Minnesota	133	45	15	11	11	8	9	2
13. Catholic University	114	10	27	17	—	28	3	—
14. Penn State	111	21	4	18	39	6	7	5
15. Pennsylvania	108	20	29	14	—	9	9	4
16. Iowa	106	34	27	4	14	4	3	5
17. Purdue	105	34	21	16	16	3	0	5
18. Northwestern	97	33	11	20	—	1	3	4
19. Stanford	86	26	20	15	0	3	9	0
20. Michigan State	75	3	6	16	12	7	14	1
21. Iowa State	73	—	14	16	32	0	1	0
22. Fordham	72	25	23	11	—	1	1	1
23. Johns Hopkins	71	7	12	17	—	0	4	3
24. Pittsburgh	71	32	10	5	—	8	6	2
25. Indiana	70	20	20	7	—	1	1	1
113 Other institutions	1,820	466 from 86	338 from 66	300 from 66	80 from 9	86 from 24	104 from 19	95 from 37
Total	5,856	1,342	1,107	782	409	337	280	265
% of Total women doctorates	99.98	22.92	18.90	13.35	6.98	5.75	4.78	4.53

Source: USOE, *Earned Degrees Conferred,* 1947–48 to 1962–63.

Table 4.6. (*Continued*)

Medical Sciences	Mathematics and Statistics	Economics and Agricultural Economics	Anthropology	Physics and Meteorology	Geosciences	Agricultural Sciences	Other Social Sciences	Engineering and Metallurgy	Astronomy	Other Physical Sciences
13	18	21	41	10	8	—	4	2	2	13
13	19	13	21	16	9	5	0	2	5	0
44	10	32	14	9	4	0	4	2	7	0
6	22	7	3	8	0	—	18	0	—	1
8	3	5	18	8	11	—	23	—	2	—
9	8	11	1	4	4	16	—	1	0	0
9	9	8	6	2	4	0	0	3	3	0
9	2	2	8	6	1	10	2	2	—	0
2	0	5	1	0	5	5	0	3	—	1
5	16	8	—	3	7	11	—	1	0	—
6	4	6	5	2	2	0	0	0	2	0
8	7	3	0	2	1	9	1	0	—	1
1	11	9	2	6	—	—	—	0	—	—
7	4	4	2	2	—	—	2	1	0	1
5	2	2	—	2	2	—	2	0	—	—
0	2	1	—	0	—	5	—	2	—	—
4	1	0	5	1	3	—	—	0	0	—
2	4	0	1	2	1	—	0	3	—	—
2	1	1	0	4	1	3	2	2	—	—
—	3	4	—	1	0	2	—	0	—	—
4	—	3	—	3	—	—	—	—	—	—
17	1	5	0	1	2	—	0	1	—	1
4	0	3	—	0	1	—	—	0	—	—
0	1	3	—	2	2	2	0	1	—	1
0	4	3	9	1	2	—	—	—	1	—
84	84	37	5	45	42	30	4	14	3	3
from	from	from	from	from	from	from	from	from	from	from
35	57	20	4	28	14	12	3	8	2	3
262	236	196	142	140	112	98	62	40	25	21
4.47	4.03	3.35	2.42	2.39	1.91	1.67	1.06	0.68	0.43	0.36

If most of the men were earning their degrees from a relatively few large departments at 25 graduate schools (among 173 universities) in these years, there was a similar concentration effect of women, though over a smaller domain. Table 4.6 lists the 25 universities (from a total of 138) that awarded the largest number of science and engineering doctorates to women in the period 1947–48 to 1962–63. These 25 universities (18.12 percent of the total) were producing 4,036 of 5,856 doctorates (or 68.92 percent). Or to be closer to the percentages given earlier on the concentration of the total doctorate population (14.45 percent of the institutions gave 65.29 percent of the degrees), for women the comparable statistics would be that the top 20 institutions (14.49 percent) accounted for 3,677 of the degrees (62.79 percent), a strikingly similar set of proportions.

Yet if the twenty-five largest graduate schools are ranked for both sexes, men only, and women only, as is done in table 4.7, one can see that they are not identical. Twenty-one of the same universities appear on all three lists, but not in the same order. The largest producer of women doctorates was Columbia University, followed closely by the University of California, Radcliffe (the women's branch of Harvard's graduate school before they merged in 1963), and New York University, largely because of its psychology, biology, and mathematical programs. (Cathleen Morawetz later credited her professors there with not letting her quit, despite marriage, miscarriage, and motherhood.)[33] Other institutions that ranked higher in awarding degrees to women than they did overall were the mostly (but not entirely) smaller (and often private) universities—Chicago, Northwestern, Yale, Iowa, Pennsylvania, and Penn State. (Yet this was not true, of course, for all departments at these institutions; the mathematics department at the University of Chicago, for example, which had awarded 51 doctorates to women between 1908 and 1946, or 18.89 percent, changed drastically when Marshall Stone became chairman, awarding only 4 women doctorates—of 114, or just 3.51 percent—before 1960, when many faculty started to leave.)[34] Ranking about the same on both lists and thus approaching an even or sex-neutral stance were Radcliffe-Harvard, Johns Hopkins, Cornell, Berkeley-UCLA, Minnesota, and Michigan. Ranking far lower for women than for both sexes were the largely (but again not exclusively) land-grant behemoths Illinois, Purdue, and Iowa State, with Ohio State, Michigan State, Stanford, and Wisconsin slightly above them.[35]

The very largest producers of women doctorates generally held first place in a few fields and had a slightly lower standing in several others (see table 4.8). Columbia University, for example, ranked first only in anthropology (despite Ruth Benedict's death in 1948, the department's 41 women doctorates dominated the field and constituted 28.88 percent of the nation's total) and the mysterious "other physical sciences" (13 of 21 doctorates, or the very high 61.90 percent), but it also was well represented in most other fields, where it ranked

Table 4.7.
Rankings of Institutions Awarding Science Doctorates,
by Sex, 1947–48 to 1962–63

Rank	Men and Women	Men Only	Women Only
1	California	California	Columbia
2	Wisconsin	Wisconsin	California
3	Illinois	Illinois	Radcliffe
4	Harvard-Radcliffe	MIT	NYU
5	MIT	Harvard	Chicago
6	Michigan	Ohio State	Wisconsin
7	Ohio State	Cornell	Michigan
8	Cornell	Purdue	Cornell
9	Columbia	Michigan	Ohio State
10	Purdue	Minnesota	Illinois
11	Minnesota	Columbia	Yale
12	Chicago	Chicago	Minnesota
13	NYU	Iowa State	Catholic University
14	Iowa State	NYU	Penn State
15	Yale	Yale	Pennsylvania
16	Stanford	Stanford	Iowa
17	Michigan State	Michigan State	Purdue
18	Penn State	Penn State	Northwestern
19	Pennsylvania	Texas	Stanford
20	Texas	Princeton	Michigan State
21	Princeton	Pennsylvania	Iowa State
22	Iowa	Cal Tech	Fordham
23	Cal Tech	Iowa	Johns Hopkins
24	Johns Hopkins	Johns Hopkins	Pittsburgh
25	Northwestern	Northwestern	Indiana

Source: USOE, *Earned Degrees Conferred,* 1947–48 to 1962–63.

from second to fourth. Berkeley-UCLA ranked first in the biological sciences, chemistry, and physics (16 of the nation's total of 140, or 11.43 percent) and was second or third in most fields. At Radcliffe-Harvard the firsts were in the medical sciences (led by its School of Public Health, which awarded 33 of the 75 doctorates given to women in this field, or 44 percent), economics (32 of 196, or 16.33 percent), and astronomy, where its seemingly few Ph.D.'s (7) constituted 28 percent of the nation's women doctorates. (This strong showing in astronomy may have been aided by the presence on the Harvard faculty in the fifties of the famed astrophysicist Cecilia Payne-Gaposchkin and, until 1957, of her left-leaning, possibly feminist colleague Bart Bok.)[36] The University of Chicago was first in sociology and in the "other social sciences," while Wisconsin excelled in the smaller fields of bacteriology (led by E. B. Fred), biochemistry, and the agricultural sciences in general. Below this level (sixth) few universities had

Table 4.8.

Institutions Awarding the Highest Percentage of Science Doctorates
to Women, 1947–48 to 1962–63

Institution	Total	Women	Women as % of Total
Texas Woman's University	40	40	100.00
Bucknell	1	1	100.00
Southern Mississippi	1	1	100.00
Smith	4	3	75.00
Bryn Mawr	52	35	67.31
Adelphi	90	25	27.78
St. John's University	30	8	26.67
Arizona State	4	1	25.00
New School	118	28	23.73
Catholic Univerisity	485	114	23.51
Tufts	44	10	22.73
Fordham	351	72	20.51
Marquette	5	1	20.00
University of Portland (Oregon)	32	6	18.75
Baylor	62	11	17.74
SUNY—Upstate	6	1	16.67
George Washington	228	38	16.67
St. Bonaventure	6	1	16.67
St. Louis University	326	19	15.95
Clark	251	40	15.94
Emory	122	19	15.57
University of Houston	104	16	15.38
Alabama	66	10	15.15
SUNY—Downstate	14	2	14.29
Columbia	2,634	376	14.27

Source: USOE, *Earned Degrees Conferred,* 1947–48 to 1962–63.

programs ranking first or second in any field, except for home economics at Cornell (which ranked first, with 69 women doctorates) and medical sciences at Johns Hopkins (whose 17 were all in public health, a small field that Johns Hopkins and Radcliffe-Harvard dominated). The Catholic University of America, Fordham University, University of Pittsburgh, and Indiana University did not award degrees to women in all areas, but had numerical strength in a few. The Catholic University, for example, was prominent only in sociology, where it ranked third for women in the nation, and biology; the other three had large programs only in psychology and biology, with Fordham's strong showing in psychology probably the result of Dorothea McCarthy's and Anne Anastasi's presence on the faculty throughout this period. McCarthy, who earned her doctorate under Florence Goodenough and Karl Lashley at the

University of Minnesota in 1928, came to Fordham in 1932 and retired in 1971 as an expert in child development and language disorders. Anastasi, a Columbia University Ph.D. in 1930, came to Fordham from Queens College in 1947 and become a world-renowned expert on differential psychology. Her long and distinguished career at Fordham was capped in 1971 with her election as the first woman president of the APA in fifty years, since Margaret Washburn in 1921.[37]

The inclusion among the women's top twenty-five graduate institutions of two Roman Catholic universities and the relatively high rank (thirteenth) of the Catholic University also reflects the rising number of religious women completing doctorates in these years. Although some young nuns (and some not so young ones) attended secular universities, such as the University of Chicago, where they lived in the somewhat sheltered International House, most bishops preferred that they attend Catholic universities, especially the Catholic University of America in Washington, D.C., where many orders had their own residences and possibly reduced tuition.[38]

If one shifts one's view of the "production" of women doctorates in science and engineering away from the largest graduate schools, where the women were often marginal, to those other universities that, regardless of size, awarded the highest percentage of their doctorates to women, one comes up with a very different list (see table 4.8). There a few women's colleges excelled, the Catholic universities were more numerous than ever (up to seven), and even the relatively liberal Columbia University was at the bottom, ranking only twenty-fifth. At the top of the chart was the only all-woman graduate school in the nation after 1963, Texas Woman's University (formerly Texas State College for Women), which awarded its first doctorate in 1953 and changed its name in 1957. All 40 of its doctorates were in home economics, especially textiles and clothing, its main specialty after the arrival of the famed Pauline Beery Mack from Pennsylvania State College in the early 1950s.[39] At Smith College, which awarded 4 doctorates in these years, 3 were in botany, including 1 to a man, possibly a veteran, in 1948–49.[40] Similarly, Bryn Mawr College's famed graduate school, which went back to the founding days of M. Carey Thomas in the 1880s, became coeducational in the 1930s and awarded about one-third of its doctorates to men in the years under consideration here. Almost half of its 35 women doctorates were chemists (15 of 19 in that field), followed by biologists (5 of 7) and geologists (5 of 10). Geology was still such a rare field for women doctorates that if one compares Bryn Mawr's total of 5 with the numbers for much larger departments (omitting geography and the other geosciences), it tied Illinois for second place behind Columbia, with its 6 doctorates, a rather remarkable achievement.

Most of the institutions on this list lacked the large engineering, chemical, and physical departments that so dominated the earlier tables of overall totals; instead they were particularly strong in the social sciences. For example, the

New School for Social Research in New York City awarded doctorates in only three fields: psychology (17 of 38, or 44.74 percent, to women),[41] sociology (8 of 37, or 21.62 percent, to women) and economics (only 3 of 43, or 6.98 percent, to women). Similarly, at Clark University the doctorate was awarded in only four sciences or social sciences, with women in chiefly two areas: 22 of the 98 degrees in geography (22.45 percent), making Clark by far the nation's largest producer of geographers of either sex, and 17 of 81 in psychology (20.99 percent) but just 1 of 23 in economics (4.35 percent) and none among the 49 in chemistry. At the University of Portland (Oregon) all 32 doctorates, of which 6 (18.75 percent) went to women, were in psychology, and 95 of the 104 doctorates at the University of Houston were in psychology, 13 of them awarded to women (13.68 percent), as well as 3 of the five in chemistry (60.00 percent). All in all, for reasons that are not very clear, perhaps because these institutions and departments are rarely studied, some graduate schools did award many, a few even most, of their doctorates to women. These tended to be smaller, more often Catholic, occasionally women's colleges that had special strengths in the social sciences. Yet of these only Columbia University would have ranked very high on any of the many prestige rankings of the time.

In fact some of these smaller and less prestigious schools had more flexible graduate programs. When a 1964 editorial in *Science* asked once again why there were so few women in science, Allyn Rule of Newton, Massachusetts, wrote a letter to the editor in which she praised Boston University, where she was enrolled in a graduate program in psychology, as one of the few institutions that even tried to meet her need for part-time (two-thirds) study and a flexible schedule. She concluded rhetorically, "The very universities which will (or already do) need more teachers are denying education opportunities to the segment of the population that is best qualified to fill this need."[42] For example, back in 1948 Harvard professor Theodore Hunt had told Janet Guernsey, who had five children and was teaching a full load at Wellesley, that if she wanted to continue beyond her master's degree, she would need to become a full-time student. Recognizing the infeasibility of that workload, she went instead to nearby MIT, where the physics department let her be a part-time "special student" for several years. By working on her dissertation at night, she finally finished in 1955, just in time to get tenure at Wellesley.[43]

By the mid-1960s certain signs of change could be detected. In 1963–64 the Danforth Foundation of St. Louis opened its Danforth Graduate Fellowships, established in 1950, to women and also started a new fellowship program for reentry women planning careers in secondary or college teaching. Pilot-tested at Wellesley and a few other women's colleges, it even permitted part-time study.[44] Also in 1965, the editor of the journal of the Association of Graduate Schools published Mary Bunting's plea for "a more experimental approach" to the needs of part-time students, particularly student wives and mothers, about whose motivation there was no longer any doubt. Such women were closer geographically to opportunities in higher education than they might ever be again and yet for financial and other reasons were often hampered from undertaking study toward a degree.[45]

When the Ford Foundation awarded a total of $41.5 million to ten prestigious graduate schools in 1967 to reduce attrition, it tied the money to the elusive goal of completing the Ph.D. within four years. In a later evaluation a former Ford Foundation employee admitted quite frankly that the money "had little effect on attrition" and "did not have a significant effect on the ways major universities conduct their Ph.D. training programs." For this he blamed such circumstances as the Vietnam War, the "PhD glut," and a certain subversion of the grant's purposes at the departmental level.[46] Yet one suspects that the money might have been much better spent at those perhaps less prestigious graduate schools that were already showing signs of flexibility. What, after all, was wrong with eight years of part-time study if the point was, as the deans kept saying, to finish the degree?

Thus, despite all the governmental officials' talk about "womanpower" and all the deans' talk about the need to keep up the value of the doctorate by requiring full-time study, prestigious graduate programs, supported by the major philanthropies, especially the Ford Foundation, continued their traditionally inflexible and exclusionary ways. If the "womanpower" experts in and out of government had really thought that by simply increasing the fellowship funds available to women they would be meeting the faculty shortage or beating the Russians in research, they had grossly underestimated the resistance of graduate departments and academic traditions and prejudices. In fact graduate school officials at the most prestigious schools, which dominated this level of instruction, were proud of how well they had resisted any tampering or lowering of standards that might have made it easier for students to complete the Ph.D. degree. Even when the students were supported with thousands of fellowships, their attrition rate approached carnage proportions. Neither female philanthropy nor federal fellowships had been able to overcome the entrenched attitudes and practices of departments and professors. Officially "encouraged" to earn degrees in order to be ready to meet pressing national shortages, many bright women students were unable to fulfill their academic and scientific potential. Institutional attitudes and practices discouraged all but the most determined and persistent at even this lowest level of access to the academic world.

Nevertheless, record numbers of women were obtaining doctorates in the sciences in the 1950s and 1960s, with or without fellowships. They tended to be concentrated in some fields more than in others and to have completed degrees at a smaller range of institutions than had the men, partly because, except for those in biology and education, they tended to be in the fields where the money was not (e.g., psychology and home economics rather than physics or engi-neering). Yet even these highly trained women, in fact especially these women, would have a tough time in the job market. Although it was unlikely that they would fare very well in academia, where the very faculty who had expected so little of them as graduate students loomed even larger as employers, that was where the largest growth was expected to be. Just as during World War II, they would continue to be underemployed everywhere, all too often filling in around the men. And in a kind of fitting revenge, many of the "dropouts" would find that, contrary to most of the advice of the time, many of the jobs that were in practice open to women, such as those in laboratories, libraries, customer service, or governmental agencies, did not require doctorates af-ter all.

5

Growth, Segregation, and Statistically "Other"

By any count there was a tremendous expansion in the size of the scientific labor force in the United States in the 1950 and 1960s. Most sources indicate that the number of American scientists, however defined, nearly tripled between 1950 and 1970. The few manpower statistics of the time that were broken down by gender agree that the numbers increased rapidly for both sexes, though they differ on the actual rates for each. These data, like the educational statistics in the last chapter, also document certain patterns of gender differences in employment: women were more numerous in some fields and subfields than in others, prevalent at certain degree levels, and more often unemployed; when they were working, they were often concentrated in certain types of scientific employment, where their salaries were generally lower than the men's.

Yet the statistical picture is badly incomplete, for even the best data of the time on women scientists (and this was a period of rapidly improving information on scientific manpower) had many major gaps and failed to probe very deeply or meaningfully into certain other aspects of the women's situation. Little was asked, and less was published, for example, on marital status, number and age of children, or even race. Moreover, the way the manpower analysts presented their scant data on women also reveals their ambivalent attitudes and their either feigned or genuine ignorance about women's place in science. For example, their data failed to include certain fields (e.g., home economics) or to separate out those types of employer (e.g., women's colleges) or types of work or job (e.g., editing or research associate) that would have mattered more to women in science than to men. All too often the women were clustered in certain residual categories, such as "other" or "miscellaneous," which were then subdivided no further. Similarly, because data on women were never disaggregated as fully as data on men or on totals for both sexes, few comparisons between the sexes were possible. Rarely were percentages calculated, and the data on women, which were usually in a separate section that was not compared or correlated with anything else, made up an embarrassing few pages that puzzled analysts because they did not fit the overall patterns for scientists. No one ever raised questions about equity or, what is even more surprising, about degree of "utilization" or "underutilization," which, to judge from the rhetoric of the time, was presumably a pressing national problem.

This failure to disaggregate the data on women could lead to simplistic and dangerous conclusions. For example, although the overall data on salaries, that is, for both sexes, were often broken down in many ways, such as by the level of the highest degree or years in the work force or the type of work function (e.g., management), the data on women's salaries were rarely subdivided further than by field. Thus it was more likely that the women would be blamed for their low salaries; that is, rather than being viewed as the result of their relative youth, their concentration in lower-paying fields and activities, or widespread salary discrimination, their low salaries could instead be interpreted as still more evidence, even proof, of their weak performance and slow advancement, thus reinforcing their prevailing image as "bad risks."

Yet the available statistics are sufficient to show that there were tens of thousands of women scientists in the United States in the 1950s and 1960s. From what data were collected and published one can calculate a certain amount of detailed information, generally about wide differences among particular fields and specialties. But beyond this, like the official policy of the time, the manpower data were out of touch with the women's reality—insufficient in many respects, revealing in others, and politically repressive overall. Rather than trying to pinpoint the problem areas in hopes of correcting them, most governmental statisticians ignored them or, what was worse, created categories and concepts that covered them up.

The best source of overall data on women in the sciences (but not engineering) was the National Register of Scientific and Technical Personnel, which grew out of the National Roster of World War II and subsequent Cold War interest in "preparedness."[1] The National Register started to collect data in 1954 but published few before 1959, when Congress, alarmed about the Soviet Sputniks, insisted that the NSF do more about scientific manpower statistics and increased its budget accordingly. Then the National Register started its nearly biennial series *American Science Manpower,* which continued until 1970, when, after President Richard Nixon cut the rate of increase in the federal science budget, it was combined with the *Science Indicator* series.

Yet even this most comprehensive source of data on American scientists had serious limitations, since it was designed to be, not a full "census" of all American scientists and engineers (the Engineers Joint Council started its own separate National Engineers Register in 1954) or a full placement service, which in any case would duplicate parts of the U.S. Employment Service in the Department of Labor, but just a "register," or file of current information on a major portion of the nation's scientists. Partly because few additional funds were allocated initially to the NSF for this new data collection, it started small and worked through existing groups, such as those relevant professional societies that already had current addresses for their thousands of members. Even though it was realized that the data collection would not be complete, that any questionnaire that was sent out would reach only those professionals who were members of those societies being surveyed (and not all of them would return it), this was considered the least expensive and most feasible approach. The societies, with their invaluable current membership and address lists, seem to have been only too willing to be patriotic at a time of "national emergency" (but also of McCarthyism) and to make known and available the skills of their members. Accordingly, when the NSF sent out its first questionnaires in 1954 and 1955, they were distributed, not through the large AAAS, but via eight scientific associations—the ACS, the American Geological Institute, the American Institute of Biological Sciences, the American Institute of Physics, the American Mathematical Society, the American Meteorological Society, the APA, and the Federation of American Societies for Experimental Biology—which, because some were large "institutes" or "federations" of smaller groups, represented altogether two hundred scientific societies. When duplicate questionnaires (from persons who belonged to two or more groups) were discarded, the 1954–55 survey yielded usable responses from 127,000 scientists and engineers, which the NSF considered "more than half of all American scientists" and "roughly three-fourths of all scientists with a doctorate."[2]

When the NSF repeated the exercise in 1956–58, with the USPHS contributing data on sanitary engineers, the questionnaires led to 170,000 usable returns, again more than half of the estimated 300,000 scientists of the time. When the staff compared its counts by field with those given in a recent survey of doctorates awarded by American universities between 1936 and 1956 (supported by the Office of Naval Research [ONR]), it found a wide range in the degree of overlap: just 29 percent of the Ph.D.'s in engineering and 35 percent of those in the medical sciences were listed in the National Register. By contrast, 96 percent of the biologists and 105 percent of the psychologists (many of whom, it was explained, had earned their degrees abroad or held degrees in education but were members of the APA or were employed as psychologists) were listed.[3] In the 1960s the NSF's manpower statistics blossomed. More effort went into increasing the response rate, expanding the fields covered (adding social sciences in 1964 and 1966), and making the Register more useful. Thus, over time its *American Science Manpower* series contained, despite its continuing lack of completeness, the most detailed and most comprehensive statistics available on the American scientific labor force, some of which were on women in science.[4]

The National Register's surveys from 1954 to 1970 yielded the total numbers of scientists (and some engineers) presented in table 5.1. The overall total more

Table 5.1.
Scientists and Engineers, by Sex, 1954–1970

Year(s)	Total		Women		Women as % of Total
	Number	% Increase	Number	% Increase	
1954–55	115,775[a]	—	7,712	—	6.67
1956–58	166,616	43.91	12,027[b]	55.77	7.22
1960	201,292	20.81	13,551	12.67	6.73
1962	214,940	6.78	14,578	7.58	6.78
1964	223,854	4.15	17,104	17.33	7.64
1966	242,763	8.45	20,164	17.89	8.31
1968	297,942	22.73	27,833	38.03	9.34
1970	312,644	4.93	29,293	5.25	9.37

Sources: NSF, *American Science Manpower, Employment and Other Characteristics, 1954–55, Based on the National Register of Scientific and Technical Personnel, 1954–55* (Washington, DC: GPO, 1959); idem, *American Science Manpower, 1956–58: A Report of the National Register of Scientific and Technical Personnel* (1961), *1960* (1962), *1962* (1964), *1964* (1966), *1966* (1968), *1968* (1969), *1970* (1971).

[a] Does not include 4,000 on active military duty, 2,000 retirees, 700 unemployed, or 4,200 student respondents.

[b] Includes 3,288 women "other than full-time employed."

than doubled, and the number of women almost quadrupled. (The big jump between 1966 and 1968 in the number of women reported will be referred to in chapter 15 as a possible root of women's liberation in science.) Thus, according to the NSF data, the women's proportion increased from 6.67 percent in 1954–55 to 9.37 percent in 1970, a roughly 40 percent increase. Yet this seeming precision may be illusory, for the scientists may have been systematically under-counted in all years, since, as explained above, these numbers were based upon membership in certain scientific groups. In addition, there may have been a sex differential, for women may have been less likely than men to join these groups, as they would have been less likely to travel to the meetings or be elected to office, and if their husband was already a member, one copy of the journal may have been enough, as many women wrote to the ACS in the 1950s.[5]

Certainly one reason for this seeming increase in the women's percentage of the NSF total in the 1960s was artificial, a result of the previously mentioned shift in the National Register's coverage and definition of "scientists." Nineteen sixty was the last year that it included as many as 25,000 engineers, very few of whom were women; by 1962 its total included just 5,000 sanitary engineers, and thereafter it dropped them all, even the many chemical engineers who as members of the ACS had formerly been included. (The Register had never made any attempt at complete coverage of either engineers or medical scientists.) Then starting in 1964 the Register's coverage of social scientists, which—aside from the psychologists, who had been included from the beginning—formerly had been part of an entry called "all other fields," shot up to include 15,000 and more persons. Since women made up a sizable proportion in several of these newly added social sciences (linguistics, sociology, and anthropology in particular but not economics), their inclusion and the engineers' systematic exclusion could be a major reason for the increase in the women's proportion of the total. When the 1954–55 and 1970 totals are presented by field, as in table 5.2, and when one subtracts from the former all the engineers and from the latter all the social scientists (except the psychologists, who were present both times) to get two more fully comparable sets of scientists, one finds that this was indeed partly the case. Although the women's proportion of the total still rose (from an adjusted 7.46 percent to 9.02 percent), the increase was only about 57 percent of what it had been earlier without this factor. From this one can tentatively conclude that despite the tremendous expansion in science and higher education in these years, women's proportion of the nation's scientific and technical manpower (or that portion thereof in these data or at least in these core fields) rose largely because official definitions of "scientist" had shifted in their direction.

Closer inspection of table 5.2 also reveals that the increase in the women's percentage was greater in some fields than in others. The largest increases were in the biological sciences (from 10.74 percent in 1954–55 to 12.92 percent in 1970), chemistry (from 5.82 percent to 7.13 percent), and earth sciences, calculated for geography and other earth sciences combined (from 2.29 percent to 3.59 percent). Yet in one field (agricultural sciences), although the number of women rose from 34 to 57, this was not enough to offset a decrease in the proportion of women from 0.42 percent to 0.36 percent.

Yet even these figures for particular fields are in turn merely averages over many subfields within the discipline. An examination of the data on the number and proportion of women in each subfield, as in table 5.3, which is taken from the 1956–58 Register, the only one to have such gender breakdowns by subfield, reveals that in none of the ninety-six subfields was the proportion of women equal to the overall average of 6.66 percent, or 8,739 of 131,126. Instead there was a very wide range of proportions, from 45.59 percent in developmental psychology on down to zero in radiological and nuclear engineering. Thus, as the data on doctoral degrees in chapter 4 revealed, most of the women were in relatively few fields: one-third (2,995, or 34.27 percent) were in just nine of the ninety-six subfields listed (9.38 percent), which included four branches of psychology, three of biology (including nutrition and metabolism, and microbiology), and one of mathematics. Beyond this top level the concentration persisted: half of the women (4,347, or 49.74 percent) were in just seventeen subfields (17.71 percent), again especially psychological, biological, and, to a lesser extent, mathematical specialties. Further, three-fourths of the women (6,544, or 74.88 percent) were in the top thirty-three fields, or about one-third

Table 5.2.
Number of Scientists and Engineers and Growth Rates, by Field and Sex, 1954–55 and 1970

Field	1954–55 Total	Women	Women as % of Total	% of Women Scientists	1970 Total	Women	Women as % of Total	% of Women Scientists	% Change Total	% Change Women
Chemistry	32,452	1,890	5.82	24.48	86,980	6,201	7.13	21.17	168.03	228.10
Chemical engineering	8,203	14	0.17	0.18						
Other engineering	4,611	24	0.52	0.31						
Computer sciences					11,324	1,263	11.15	4.31		
Mathematics	8,670	910	10.50	11.79	24,400	2,790	11.43	9.52	181.43	206.59
Statistics					2,953	337	11.41	1.15		
Astronomy and physics	11,452	348	3.04	4.51	36,336	1,354	3.73	4.62	217.29	289.08
Meteorology	1,838	26	1.41	0.34	6,637[a]	102	1.54	0.35	261.10	292.31
Geography	522	59	11.30	0.76						
Other earth sciences	13,829	270	1.95	3.50	23,756[b]	852	3.59	2.91	71.78	215.56
Biological sciences	15,612	1,677	10.74	21.72	47,493	6,134	12.92	20.94	204.21	265.77
Agricultural sciences	8,126	34	0.42	0.44	15,730	57	0.36	0.19	93.58	67.65
Medical sciences	891	59	6.62	0.76						
Psychology	10,163	2,252	22.16	29.17	26,271	6,327	24.08	21.60	158.50	180.95
Economics					13,386	812	6.07	2.77		
Sociology					7,658	1,729	22.58	5.90		
Political science					6,493	631	9.72	2.15		
Linguistics					1,902	447	23.50	1.53		
Anthropology					1,325	257	19.40	0.88		
Other	1,244	158	12.70	2.05						
Total/average	115,775	7,721	6.67	100.01	312,644	29,293	9.37	99.99	170.04	279.39

Sources: See table 5.1.

[a] Atmospheric and space sciences.

[b] Earth and marine sciences.

Table 5.3.
Scientific Subfields, Ranked by Percentage of Women Employed Full-Time, 1956–1958

Rank	Subfield	Women as % of Subfield	Number	Cumulative Total
1	Developmental psychology	45.59	124	
2	Other biological specialties	35.10	221	
3	Educational and school psychology	29.39	323	
4	Nutrition and metabolism	23.60	265	
5	Clinical psychology	22.81	887	
6	Other psychology	20.91	60	
7	Other specialties	20.34	236	
8	Microbiology	18.63	676	
9	Other mathematical specialties	18.51	203	2,995 (34.27%)
10	Psychometrics	17.89	105	
11	Geometry	16.98	72	
12	Counseling and guidance	16.92	253	
13	Education	16.67	156	
14	Biochemistry	16.25	444	
15	Algebra and number theory	15.89	139	
16	Personality	15.33	44	
17	Anatomy	14.59	139	4,347 (49.74%)
18	Botany	14.30	218	
19	Theory and practice of computation	13.31	323	
20	Zoology	12.33	190	
21	Pharmacology	12.04	104	
22	Physiology	11.18	207	
23	Pathology	11.10	96	
24	Statistics	10.84	22	
25	Social psychology	10.63	59	
26	Genetics	10.12	69	
27	Other chemical specialties	9.64	89	
28	Experimental, comparative, and physiological psychology	9.59	111	
29	Astronomy	9.12	33	
30	Analytical chemistry	8.36	484	
31	Geography	8.26	80	
32	Biophysics	7.92	29	
33	Probability and statistics	7.64	83	6,544 (74.88%)
34	Analysis and topological algebraic structure	7.01	109	
35	Ecology	6.87	31	
36	Medical sciences	6.20	119	
37	Industrial and personnel psychology	6.16	74	
38	Topology	5.86	16	
39	Pharmaceutical chemistry	5.51	57	
40	Atomic and nuclear physics	5.29	40	

(*continued*)

Table 5.3. (*Continued*)

Rank	Subfield	Women as % of Subfield	Number	Cumulative Total
41	Social sciences (esp. economics)	4.78	20	
42	Organic syntheses	4.73	103	
43	General organic chemistry	4.59	10	
44	Hydrobiology	4.40	11	
45	Agricultural and food chemistry	4.39	83	
46	Other physics specialties	3.98	78	
47	Oceanography	3.75	9	
48	Inorganic chemistry	3.71	93	
49	Logic	3.66	6	
50	Physical chemistry	3.33	134	
51	Other organic chemical specialties	3.30	224	
52	Optics	3.04	44	
53	Actuarial mathematics	3.00	37	
54	Climatology	2.94	3	
55	Mathematics of resource use	2.91	26	
56	Human engineering	2.89	7	
57	Phytopathology	2.71	24	
58	Physical or dynamical meteorology	2.40	4	
59	Geochemistry	2.37	5	
60	Other meteorological specialties	2.36	16	
61	Solid state	2.34	45	
62	Aeronautical engineering	2.33	19	
63	Ceramic engineering	2.26	4	
64	Geology	2.16	207	
65	Nuclear physics	2.14	56	
66	Synoptic meteorology	2.10	25	
67	Mechanics and heat	2.01	17	
68	Resin and plastics	1.99	50	
69	Petroleum products	1.94	29	
70	Entomology	1.82	34	
71	Metallurgical engineering	1.58	9	
72	Electronics	1.13	28	
73	Theoretical physics	1.12	8	
74	Surveying, mapping, and photometry	1.09	15	
75	Horticulture	1.08	9	
76	Hydrology	1.03	8	
77	Engineering mechanics	1.02	5	
78	Electricity and magnetism	1.02	16	
79	Rubber chemistry	0.94	15	
80	Other engineering	0.89	9	
81	Geophysics	0.89	20	
82	Industrial engineering	0.87	7	

(*continued*)

Table 5.3.   (*Continued*)

Rank	Subfield	Women as % of Subfield	Number	Cumulative Total
83	Electrical engineering	0.81	7	
84	Acoustics	0.81	7	
85	Mechanical engineering	0.75	4	
86	Animal husbandry	0.70	7	
87	Agronomy	0.66	6	
88	Sanitary engineering	0.32	11	
89	Civil engineering	0.30	8	
90	Chemical engineering	0.26	13	
91	Forestry and range science	0.24	12	
92	Fish and wildlife	0.21	2	
93	Mining engineering	0.14	2	
94	Soil science	0.11	1	
95	Radiological and health engineering	0.00	0	
96	Nuclear engineering	0.00	0	
	Total		8,739	

Source: NSF, *American Science Manpower, 1956–58,* 80, 82.

(34.38 percent) of the ninety-six subfields listed here. The remaining quarter (2,195) of the women were scattered thinly across sixty-one other subfields, and two had no women at all. This range of women's distribution at the subfield level indicates that one cannot rely on overall averages for a true picture of women's proportion in "science." The overall averages imply that women were present uniformly in all subfields, when in fact most of the women were in a few subfields, while there were only a few women in most of them.[6]

Interestingly the names of four of the most highly feminized subfields in table 5.3 begin with the adjective *other,* including number seven, the residual, essentially miscellaneous, "other specialties," for those persons who did not fit any of the other ninety-five categories. This phenomenon is a particularly poignant example of just how marginal women's experience in science was in 1956–58. Rather than being distributed as men were in the "major" parts of a field, the women were most highly clustered in those "other" corners of specialties and disciplines that were so far outside the usual categories that they constituted mere afterthoughts of statisticians. Taken together, these patterns of distribution in fields and subfields represent just the first of several statistical differences in the National Register data that show that women's map of "science" was not only not as large as men's but less varied and more often marginal.

Still other differences both overall and by field are present in the men's and women's patterns of degree holding. An abundance of these statistics taken from the 1966 Register, the first to include anthropology, are summarized in

Table 5.4.

Highest Degree Earned by Scientists, by Field and Sex, 1966

Field	Women as % of Field	Doctorate[a]		Master's		Bachelor's		Less than Bachelor's	
		% Women	% Men	% Women	% Men	% Women	% Men	% Women	% Men
Psychology	22.25	51.10	70.47	47.13	27.58	1.56	1.85	0.02	0.03
Linguistics	21.04	44.57	62.97	38.20	24.55	16.10	9.38	0.37	0.00
Anthropology	18.61	87.13	91.44	7.60	5.35	4.09	2.54	0.00	0.27
Sociology	15.96	59.04	78.98	38.90	18.11	2.07	2.26	0.00	0.23
Biological sciences	11.29	51.33	73.77	27.76	15.81	19.36	9.40	0.54	0.41
Mathematics	10.50	13.90	25.27	53.03	42.38	30.48	28.39	1.00	1.71
Statistics	10.09	20.85	31.26	46.91	40.66	26.71	24.83	2.61	1.32
Other	8.17	12.81	17.74	48.01	33.83	38.37	46.59	0.34	0.63
Chemistry	7.58	24.00	38.02	25.24	18.31	49.21	41.30	0.40	0.86
Economics	4.34	38.18	42.75	45.18	34.98	14.71	20.48	0.70	0.59
Physics	3.37	24.97	41.30	42.00	32.07	32.52	25.70	0.10	0.26
Other earth sciences	3.31	21.25	21.95	44.65	31.84	33.18	44.24	0.31	1.28
Meteorology	2.05	10.08	10.67	32.56	22.13	43.41	47.45	6.20	11.68
Agricultural sciences	0.50	16.00	23.14	42.00	25.79	28.00	49.61	14.00	0.90
Total/average	8.31	34.23	40.36	38.07	26.54	26.31	30.75	0.49	1.05

Source: NSF, American Science Manpower, 1966, 67–69 (table A-8), 200 (table A-60).

[a]Ph.D.'s as well as professional and medical degrees (M.D., D.V.M., etc.).

table 5.4. These data indicate, first of all, that fields differed widely in the proportion of scientists (of either sex) who held doctorates. In some disciplines, such as anthropology, sociology, and biology, so many persons held doctorates (over 70 percent of the total) that they could be termed "high-doctorate" fields. (This reflects the fact that in several of these fields the doctorate was required for membership in the very professional society that was distributing the government's questionnaires.)[7] At the other end of the spectrum were certain "low-doctorate" fields, such as agricultural sciences, earth and marine sciences, other, and meteorology, where fewer than 24 percent of the scientists held doctorates.

In addition there were also certain gender differences in these high- and low-doctorate patterns of degree holding. These became evident starting in 1962, when the National Register began to provide more data on women, including data on the highest degree held. Data for 1966, given in table 5.4, indicate that only 34.23 percent of the women listed in the Register, compared with 40.36 percent of the men, had received a Ph.D. or other doctoral degree. This discrepancy held true in every field, with differences ranging from less than one percentage point in meteorology and "other earth sciences" on up to 19 or more in psychology, sociology, and the biological sciences. In table 5.4 the fields are ranked according to the percentage of women in them, from highs of more than 20 percent in psychology and linguistics to lows of fewer than 3 percent in meteorology and agricultural sciences. Thus, it does not seem coincidental that the high-doctorate fields had the highest percentages of women, while the low-doctorate fields, except for mathematics and "other," had the lowest percentages of women. From this correlation one might conclude either that women were welcome in certain fields only when they held a doctorate or that women completed doctorates and persisted professionally disproportionately in fields that required the degree.

Since so much of what is known about women in science is limited to those at the doctoral level,[8] the Register data on women with master's degrees are especially interesting. At the level of the master's degree, women predominated not only on the average (38.07 percent compared with 26.54 percent for men) but also in every field (see table 5.4). The differences range from a mere 2.25 percentage points in anthropology to 20.79 percentage points in sociology and 19.55 in psychology. In one field, mathematics, women with master's degrees were more than half (53.03 percent) of all women in the field as counted by the Register, a pattern followed closely by the women in "other" (48.01 percent), psychology (47.13 percent), statistics (46.91 percent), and economics (45.18 percent). In fact, if the data on doctoral and master's degrees were combined into one category, "those with higher degrees," the women would predominate, with 72.30 percent, compared with 66.90 percent for men. Apparently, many of the women who did not complete doctorates maintained their membership in their scientific societies (perhaps as "associate members" where such sub-Ph.D.'s were permitted) and sent in their Register questionnaires. Although it is not clear from the Register's breakdowns just what positions such persons held,[9] they would have been among the editors, librarians, community college teachers, or museum and governmental workers, a large proportion of the largely invisible and highly feminized infrastructure behind the doctoral scientists.

At the level of the bachelor's degree men predominated slightly (30.75 percent, compared with 26.31 percent for the women), as they did among those with less than a bachelor's degree (1.05 percent compared with 0.49 percent). These persons were predominantly in the "low-doctorate" fields, such as chemistry, earth sciences, meteorology, "other," and agricultural sciences, in which more than 40 percent of the scientists held only a bachelor's degree or less. Evidently, in these outdoor or industrial fields men needed no higher degree to be considered active scientists or members of the major professional society; they obtained a lot of their training and experience on the job. Yet any women who might have held these same jobs may have needed a master's degree to be considered a scientist.

The NSF Register also has an abundance of data on which sectors of the economy employed scientists full-time. Some of these data are summarized in table 5.5, which provides an overview of full-time employment in 1956–58, the first year for which the Register's totals were broken down by gender, and 1970, the last year of the Register and the end of the period considered statistically in this chapter. Table 5.5 shows once again that this was a period of tremendous growth overall and in all sectors but especially in educational institutions and the small area labeled "other." So great was the increase in the employment of scientists by educational institutions that that sector's share of the total surged ahead from 27.76 percent in 1956–58 to 44.94 percent in 1970. This jump caused a major drop in the proportion employed by industry (including those in self-employment), which fell from nearly half (49.02 percent) in 1956–58 to just 35.80 percent in 1970, a trend increased by the Register's omission of engineers by 1970. Similarly, the proportion of scientists employed by governmental and nonprofit organizations also dropped, though to a lesser extent.

At first glance, women scientists seem to have benefited from this great expansion in employment, for their numbers and percentages increased in all sectors between 1956–58 and 1970. In particular, the proportion of women scientists who were working in educational institutions, which had always been high (47.35 percent in 1956–58), increased even further to well over half (61.87 percent) in 1970. Yet there was a danger signal in this heavy concentration in educational institutions, for it meant that the proportion of women who were working in the other, possibly more lucrative sectors fell sharply. In fact they were grossly underrepresented in industry in 1956–58 (24.2 percent of

Table 5.5.
Sectors Employing Scientists Full-Time, by Sex, 1956–58 and 1970

Sector	Total[a]	% of Scientists Counted	Women	Women as % of Total	% of Women
			1956–58		
Educational institutions	38,714	27.76	4,098	10.59	47.35
Industry[b]	68,366	49.02	2,098	3.07	24.24
Self-employed	—	—	—	—	—
Federal government	19,174	13.75	1,084	5.65	12.53
Nonprofit institutions	5,453	3.91	690	12.65	7.97
State/local governments	7,217	5.17	637	8.83	7.36
Other	546	0.39	47	8.61	0.54
Total/average	139,470	100.00	8,654	6.20	99.99
			1970		
Educational institutions	130,389	236.80 44.94	15,844	12.15	61.87
Industry	97,542	42.68 33.62	3,674	3.77	14.35
Self-employed	6,314	51.91 2.18	695	11.01	2.71
Federal government	31,118	62.29 10.73	1,896	6.09	7.40
Nonprofit institutions	10,911	100.09 3.76	1,841	16.87	7.19
Other government	11,741	62.69 4.05	1,381	11.76	5.39
Other	2,116	287.55 0.73	278	13.14	1.09
Total/average	290,131	108.02 100.01	25,609	8.83	100.00

Source: NSF, *American Science Manpower, 1956–58,* 80–81, and *1970,* 197 (table A-49A), 238 (table A-61).

[a]Totals for 1970 omit 7,285 in military.

[b]Includes self-employed.

the women compared with 49.02 percent of the men), and this just got worse by 1970 (17.06 percent compared 35.80 percent), even when the somewhat feminized (11.01 percent) area of "self-employment" was included. With so many industrial doors closed to women, they took what they could get elsewhere, especially in the educational institutions, the nonprofit organizations, where their proportion was twice that of men, and the nonfederal governments (heavily populated by school and other psychologists). Others created positions for themselves in self-employment (primarily as clinical psychologists) and "other" work, where their diversity and versatility were at a premium, like, perhaps, Cynthia Westcott and her horticultural company for home gardeners.[10]

Once again, however, these overall patterns based on averages give way to a plethora of subpatterns when the Register's data are broken down by field as well as by gender and sector. In the data for 1970 (which are not presented separately here), one can detect that even though women were, on the average, substantially more likely than men to be working in educational institutions (54.09 percent compared with 40.43 percent), the opposite was true in fully seven of the fifteen fields, primarily social sciences: psychology, linguistics, sociology, anthropology, statistics, political science, and economics. In industry, where women continued to be heavily underrepresented (12.54 percent of women scientists compared with 33.13 percent of the men), the sex differences were particularly great in the fields of earth and marine sciences (just 9.98 percent of the women compared with 43.17 percent of the men), chemistry (30.03 percent compared with 60.69 percent), mathematics (7.35 percent compared with 28.50 percent) and physics (9.01 percent compared with 27.72 percent). As for the federal government, women scientists were somewhat less likely than men to be working there (6.47 percent compared with 10.31 percent) chiefly because of the large discrepancy in the agricultural sciences: 15.79 percent (or 9) of the field's few women worked there compared with 36.47 percent (or 5,716) of its men. From all this diversity among fields one can conclude that in 1970, despite the passage by then of certain equal-employment-opportunity legislation, women continued to be underrepresented in industry, in some fields within the federal government, and in statistics and the social sciences in academia.

In addition to working in a somewhat different mix of institutional settings, men and women scientists also tended to perform somewhat different roles and tasks within them. From its very beginning in 1954–55 the National Register also collected data on the types of scientific and technical work done by its respondents. These data were broken down in various ways over the years, chiefly into six categories: teaching, management, research and development, consulting (which also included clinical practice), a miscellaneous "exploration, forecasting and reporting," production and inspection, and of course "other." Two categories were further subdivided: "research and development" had a subcategory of applied research (which included, for example, the administration and interpretation of psychological tests), and "management" one for managers of research and development (R&D). Since the breakdowns of these data can be calculated separately for men and or women starting in 1956–58, they can give us a better indication of the types of work male and female scientists performed from the 1950s through 1970. Table 5.6 contains numerous percentages and numbers published in the National Register for 1970. The percentages are based on men and women who were employed and did report their type of work activity; that is, the table omits the unemployed and those who did not answer this part of the questionnaire.

Overall almost half of the scientists in table 5.6 (46.79 percent of its 25,250 women and 48.57 percent of its 266,222 men) were busy with just two tasks, teaching and management. But their proportions in each group differed great-

Table 5.6.
Type of Work of Scientists, by Sex, 1970

Type of Work	Total	Women	Women as % of Total
Consulting	14,544	2,281	15.68
Teaching	72,408	9,625	13.29
Other	8,113	970	11.96
Research	97,078	8,377	8.63
Exploration, forecasting, and reporting	14,535	1,139	7.84
Production/inspection	16,086	668	4.15
Management	68,708	2,190	3.19
Total/average	291,472	25,250	8.66

Source: NSF, *American Science Manpower, 1970,* 79, 239 (tables A-12, A-62).

ly. Women were far more likely than men to be teaching (38.12 percent compared with 23.58 percent), though again this was not true in every field. (Data on particular fields are not presented here.) This disproportionately high concentration of women scientists in teaching was undoubtedly the result of serious barriers to their employment in other types of activity areas.

If women were overrepresented in teaching, management was "men's work," occupying 24.99 percent of the men but just 8.67 percent of the women in 1970. The men's preponderance was even greater if one considers the actual numbers involved (66,518 men, or 96.81 percent of the total, but only 2,190 women). Men dominated of this job role in all fields. Among research managers women were even rarer: even though women held roughly the same proportion of research positions as men (see below), only 2.82 percent of the women (compared with 10.90 percent of the far more numerous men) worked in such managerial positions in 1970. (The actual numbers—713 women and 29,101 men—present an even bleaker statistic: women were just 2.40 percent of the nation's R&D managers.) For women this role was most common in statistics (where they constituted 8.25 percent of the R&D managers), economics (7.42 percent), agricultural sciences (6.67 percent), and sociology (6.06 percent). If one opines that the salaries for managers were higher than those for teachers or researchers (or perhaps anyone else in science, as will be discussed more fully shortly), one can begin to sense in this disproportionately high percentage of male managers the enormous gap behind all these statistics between the financial rewards going to women in science and those going to men.

The third most common kind of work activity in the NSF Register after teaching and management was that of full-time researcher, which included research workers at all levels, from laboratory technicians to postdoctoral fellows to senior scientists. In this category, for a change, there was almost parity in the proportion of women and men scientists in 1970 (33.18 percent of the women compared with 33.32 percent of the men), though once again there were sharp differences between the sexes in certain fields, reflecting the greater range of other opportunities open to men in some fields. These in turn reflect the great variety of settings in which research, basic or applied, was being done at that time—in universities and nonprofit institutions (including hospitals and federal laboratories run by nonprofit organizations), in government at all levels, and in industry, often on federal contracts.

If five-sixths of scientists in the 1970 National Register were grouped in just three categories—teaching, management, and R&D—the remaining, more diverse sixth was a combination of five highly heterogeneous work activities: consulting; exploration, forecasting, and reporting; production and inspection; and, of course, the ubiquitous "other" and "no report." Women were almost twice as likely as men to be doing "consulting" (9.03 percent of women compared with 4.61 percent of men), largely because so many women psychologists (almost 1,700) were working in clinical or counseling practice. The area called "exploration, forecasting, and reporting" represents such a mixture of unrelated tasks (editing, technical writing, information services, and regional and urban planning, as well as actual field exploration) that it is hard to know what the roughly even averages—4.51 percent of women compared with 5.03 percent of men—really mean. The men predominated in the outdoorsy earth and marine sciences, whereas the women were more likely than men to work in statistics, computer sciences, agricultural sciences, and economics. "Production and inspection," which also includes maintenance, testing, and sales, was heavily dominated by chemists of both sexes. "Other," the miscellaneous residual category for those who did not fit elsewhere, once again included women more frequently than it did men (3.84 percent compared with 2.68 percent), although this overall pattern did not exist in all fields.

What all these percentages seem to indicate is that men tended to obtain the management positions (whether in government, academia, industry, or elsewhere), and women tended to remain in teaching (except in certain fields that offered other opportunities outside academia). Besides these two major sexual divisions of labor, one-third of the scientists of each sex did full-time research. Beyond this, women tended to pick up whatever other positions they could— as clinical psychologists, technical report writers, and editors—while men dominated exploration work. Greater familiarity with the employment opportunities in each field would be necessary to identify more precisely than is possible here the usual sexual patterns and thus any notable exceptions to them. One important indicator of the politics or dynamics of the marketplace, however, remains to be examined, since salary data show that not all these tasks and fields were valued equally.

Table 5.7.
Median Annual Salaries of Scientists, by Field and Sex, 1970

Field	Overall	Men's	Women's	Ratio of Women's to Men's
Statistics	16,900	17,100	14,000	81.87
Economics	16,300	16,500	13,400	81.21
Computer sciences	16,500	16,900	13,200	78.11
Atmospheric and space sciences	15,200	15,200	13,000	85.53
Psychology	15,000	15,500	13,000	83.87
Anthropology	14,700	15,000	12,300	82.00
Physics	15,900	16,000	12,000	75.00
Linguistics	12,500	13,000	11,300	86.92
Political science	13,000	13,500	11,000	81.48
Sociology	13,000	13,500	11,000	81.48
Biological sciences	15,000	15,500	11,000	70.97
Earth and marine sciences	14,900	15,000	10,500	70.00
Chemistry	15,300	15,600	10,500	67.31
Mathematics	14,300	15,000	10,000	66.67
Agricultural sciences	12,800	12,800	9,400	73.44
Median/average	15,000	15,200	11,600	76.32

Source: NSF, *American Science Manpower, 1970,* 238 (table A-59).

Because the National Register rarely subdivided its data on women scientists' salaries by sex and field, the data in its report for 1970 are worth examining. Table 5.7 shows that though the median annual salary in 1970 for more than 312,000 responding scientists was $15,000, it varied greatly by field (from $12,500 in linguistics to $16,900 in statistics) and by sex (with women earning on average 76.32 percent of the men's median, a rate that could be compared favorably with the roughly 67 percent of men's salaries that, according to the U.S. Census, women in professional positions earned in general in 1970).[11] Overall the fields with the highest median salaries for women were statistics ($14,000, or 81.87 percent of men's) and economics ($13,400, or 81.21 percent of men's). Earning significantly more than their field's female median were women in certain possibly overlapping subgroups: economists who managed research and development groups ($21,500); economists with twenty-five to twenty-nine years' experience ($19,700); those (generally biologists) with medical degrees ($18,500); economists in the federal government ($17,400); and self-employed psychologists ($17,000).[12] Although these were the best salaries going to women, they were well below those of men with comparable characteristics.

Yet consciousness about salary discrimination was very low in 1968, as demonstrated in a lengthy report that year on economists' salaries (based on the data in the 1966 Register) by Arnold Tolles and Emanuel Melichar, two leading men in the field. Although this was one of the better-paying fields for women, Tolles and Melichar admitted that there was a significant discrepancy between men's and women's salaries that was even a bit unjust. But then, in two short sentences in a 153-page report, they quickly dismissed it as of minor importance to anyone but the victims themselves. Moreover, they did this on the incriminating grounds that the field had attracted too few women for it to matter much: "Being a woman tended to reduce salary by 19 per cent, a significant effect both statistically and for the economists affected. The pronounced difference was relatively unimportant in explaining total variation, however, because only 3.6 per cent of the registrants were women."[13]

Nor were chemists any more concerned that women chemists were near the bottom of the salary scale in 1970 (see table 5.7, where, however, the figures are not controlled for length of service, degree level, number of publications, etc.), and for a while in the 1950s their starting salaries were actually dropping. Yet one would not have noticed such an unusual occurrence by reading the weekly *C&EN,* the main source of such extratechnical information in the field. On the contrary, its very few articles on women's salaries did not compare them with men's at all. Instead they stressed the positive news, as reported in the 1959 article "Women Chemists Best Paid Grads"—summarizing a recent report on women college graduates by the WB, which had not surveyed any men—that women chemistry majors were upon graduation paid more than women nurses, teachers, or stenographers![14] When the *C&EN* reported biennially on the most recent starting salaries, the fact that women's totals, medians, and averages were always substantially lower than men's was considered of such little consequence to the profession that it was omitted from the reports' titles and buried far down in the text as an anomaly in an otherwise rosy picture. For example, a 1958 article entitled "Starting Salaries Still Rising" indicated that there had been increases for most chemists, but not for women: "The one group [at the bachelor of science level] with a lower 1958 median was women chemists. Their monthly starting salary dropped by 6% to $374 [the lowest of any group counted], about the same as 1956." No reason was given—part-time work was counted elsewhere—and no concern was voiced. The same rosy non sequiturs appeared in a subsequent article about the ACS's placement service at a recent Chicago meeting,

Women chemists seem to be doing well in locating employment. In Chicago, the seven unemployed women [registered at the job service] represented only 9% of the total unemployed registrants, the lowest figure on record for all fall meetings 1951–1958. The average of unemployed women for the same period is 21%. A previous article reported that 1958 starting salaries for women in chemistry

dropped to 1956 levels. . . . One conclusion is that women are having little difficulty finding or retaining jobs but are doing so at reduced salaries.[15]

Not until 1970, when Madeleine Polinger, "*C&EN*'s woman-chemist-turned-
journalist," as an editorial called her, presented the unequal salary ratios in the
article "Women Chemists: Concerned over Rights," was this dismal situation,
which had existed for decades, brought forcefully to readers' attention.[16]

The chemists' situation raises the somber subject of unemployment, a factor
that loomed much larger in women's careers than in men's. For many years the
preparers of the National Register had so little interest in the unemployed that
they either merely listed them briefly in the report's preface as one (N = 700) of
the several groups excluded from further consideration in the text, as they did
in 1954–55, or included a residual category, "not employed," among the sectors
of employment, as they did from 1964 on. This category was broken down by
field and by gender but presumably still contained both the retired and housewives as well as other unemployed persons. The Register's figures for 1970
showed that although relatively few scientists were unemployed at that time,
women made up a disproportionately large share of them. At a time when the
Register recorded an overall unemployment rate of 4.13 percent (12,908 of
312,644), 11.18 percent of women, or 2.71 times the overall figure, were unemployed. Put another way, at a time when women were 9.37 percent of all
scientists in the Register, they constituted a substantial 25.36 percent of the
unemployed. Yet as so many times before, this overall average varied greatly by
field—from 8.10 percent for biologists to 20.01 percent for physicists and 21.05
percent for agricultural scientists.[17]

But even these figures may present a distorted or partial picture, especially
since it is likely that a great many other unemployed scientists were not included at all. Because the National Register relied on scientists' returning
questionnaires sent to them by their professional societies, it could have seriously undercounted the long-term unemployed, who might not bother to
complete and send in the questionnaire, which could well be a painful reminder and admission of professional failure, if indeed they were still paying
their dues and their scientific society had their current address. Yet even if the
National Register systematically missed many of the unemployed, why there
should be such a range of unemployment rates among those men and women
who did respond remains puzzling.

Although it would be helpful to have immigration data on the "brain drain"
into the United States for this whole period broken down by gender, such
statistics were not available until data for 1965 were published in the late 1960s.
At that time women were shown to be 8 percent of the scientists and engineers
(32 percent of the social scientists, 17 percent of the "natural scientists," and 2

percent of the engineers) and to be more likely than men to come from Eastern Europe, South America, and Asia.[18] No reasons were given, but these trained women may have been underemployed in their native land, or perhaps they had married Americans. Also, in 1965 Congress enacted important changes in the American immigration laws, the first since 1952 and in some ways the most significant since the 1920s. One major modification was to drop the former preference for persons from northwestern Europe. This quickly led to an upsurge of Asian, including Indian, immigrants, as well as of groups from other regions, including Eastern Europe and South America. A second change was to institute a preference for certain occupations and for persons trained in certain "critical" fields. For a few years, science and engineering were among these fields, but increasing unemployment in the United States around 1970, especially among engineers, caused these fields to be dropped from the preferred list by 1972. A third change, which may well have benefited women in science, was to create a preference for the spouses and relatives of Americans, thus easing immigration for those foreign women who had married American men.[19]

Because the National Register omitted any information on the marital status of its men and women scientists and engineers, one must turn for information on this matter to other sources and studies. Generally the marriage rate varied by field, date, and degree level. In 1969 Helen Astin reported that in her study of the women doctorates (in many professional fields) of 1957 54.7 percent were currently married or had been married (59 percent if one left out nuns). An average of 44.6 percent were "married, living with husband," and 10.1 percent were either divorced (6.4 percent), widowed (2.7 percent), or separated (1.0 percent). The percentages varied widely by field: 42.1 percent of those in education were or had been married, compared with 77.8 percent of those in psychology, who were also the most likely to be divorced.[20] When Deborah S. David examined a 1962 postcensus study of scientists and engineers, she found that 58 percent of the women doctorates had been married, as had 51 percent of the women with a master's degree, and 65 percent of those with just a bachelor's degree. The rates varied greatly by field.[21] Similarly, when Louise Dolnick Solomon obtained information on 774 women chemists listed in the eleventh edition of the AMS (1965–67) and supplements (to 1969), she found that 73 were nuns and that 393 of the remaining 701, or 56.06 percent, had married. But she found large differences over the years: although only 34 percent of the women Ph.D.'s of 1925–29 had ever married, 60 percent of those who earned degrees between 1950 and 1959 had.[22] When Barbara Reskin collected data on the marital status of 177 women Ph.D.'s in chemistry, she found that 42.4 percent of the women (compared with 63.3 percent of a control group of men) were married at the time of their degree (between 1955 and 1961); another 40 (22 percent) had married by her cut-off date of 1970, and a few others married after this.[23]

Table 5.8.
Notable Scientist Couples, by Field, 1950–1972

Anthropology
  Ann Fischer and John
  Ruth S. Freed and Stanley
  Frances Shapiro Herskovits and Melville
  Madeline Appleton Kidder and A. V.
  Dorothea Cross Leighton and Alexander
 * Margaret Mead and Gregory Bateson
  Betty J. Meggers and Clifford Evans
  Harriet Klinge Pomerance and Leon
  Margaret Park Redfield and Robert
  Felicia Harbern Trager and George (linguistics)
  Beatrice Blyth Whiting and John
  Michelle Rosaldo Zimbalist and Renato
Astronomy
  Priscilla Bok and Bart
  E. Margaret Burbridge and Geoffrey
  Antoinette de Vaucouleurs and Gérard
  Helen Sawyer Hogg and Frank
  Margaret Walton Mayall and Nicholas
  Cecilia Payne-Gapsochkin and Sergei
  Vera Rubin and Robert (physics)
  Mary H. Shane and C. D.
  Charlotte Moore Sitterly and Bancroft
  Emma Williams Vyssotsky and Alexander
 * Constance Sawyer Warwick and James
Biochemistry
  Evelyn Anderson and Webb Haymaker (neuropathology)
  Molly Bernheim and Frederick
  Anne Briscoe and William A. (physiology)
  Essie White Cohn and Byron (physics)
  Mildred Cohn and Henry Primakoff (physics)
  Gerty Cori and Carl
  Helen Tredway Graham and Evarts (surgery)
  Mary Ellen Jones and Paul Munson (pharmacology)
  Beatrice Kassell and Harris Friedman (pharmacology)
  Elizabeth Miller and James A.
  Ethel Ronzoni and George Bishop (physiology)
  Jane Russell and Alfred Wilhelmi
  Rosemary Schraer and Harald (biophysics)
  Sofia Simonds and Joseph Fruton
  Olive Watkins Smith and George (gynecology)
 * Elizabeth K. Weisburger and John H.
Biology
  Meridian Ball and Gordon

(*continued*)

Table 5.8 (*Continued*)

Jean Clark and Katsuma Dan
Laura Hunter Colwin and Arthur
Ingrid Deyrup-Olsen and Sigurd
Louise Johnson Eagleson and Oran
Marie Poland Fish and Charles (oceanography)
Mary-Helen Martin Goldsmith and Timothy H.
Frances Hamerstrom and Frederick
Ethel Browne Harvey and E. Newton
Ruth Hubbard and George Wald
Laura Hubbs and Carl
Sally Hughes-Schrader and Franz
*Wanda S. Hunter and E. W.
Mary Juhn and Richard Fraps
Elsie Klots and Alexander
Helen Martin and Fred
Margaret Murie and Olaus
Margaret Morse Nice and Leonard
Marian Irwin Osterhout and W. J. V.
Phyllis Parkins and William
Ruth Patrick and Charles Hodge IV (entomology)
*Grace Pickford and G. Evelyn Hutchinson
Dorothy Pitelka and Frank
Virginia Ferris Rogers and John F.
*Elizabeth Shull Russell and William L.
Liane Russell and William L.
Berta Scharrer and Ernst
*Bodil Schmidt-Nielsen and Knut
Elizabeth R. Schwartz and Charles
Dorothy Skinner and John Cook
Clara Szego and Sidney Roberts (biochemistry)
Marvalee Wake and David
Norma Ford Walker and Edmund (entomology)
Barbara Webster and Grady
Anna R. Whiting and P. W.
Anna Allen Wright and Albert
Botany
Edith S. Clements and Frederic
*Mary Clutter and Ian Sussex
*Margaret Bryan Davis and Rowland
Isa Degener and Otto
Catherine Gross Duncan and Robert E.
*Sylvia Earle and Jack Taylor (zoology)
Mildred Mathias and Gerald Hassler (geophysics)
*Sylvia Earle Mead and Giles (ichthyology)
Earlene Rupert and Joseph (agronomy)

(*continued*)

Table 5.8 (*Continued*)

Miriam Carpenter Strong and Forrest C.
Alice Tryon and Rolla (biology)
Chemistry
  Leonora Neuffer Bilger and Earl
  Elizabeth A. Bridgeman and Oscar
  Marjorie Caserio and Fred
  Chicita Culberson and William L. (botany)
  Helen Miles Davis and Watson
  Mary Fieser and Louis
  Mary Good and Bill (physics)
  Lilli Schwenk Hornig and Donald
  Nancy Kolodny and Gerald
  Mary Rising and Julius Stieglitz
  Marjorie Vold and R. D.
Crystallography
  Gabrielle Hamburger Donnay and J. D. H.
  Isabella Karle and Jerome
  Rose Mooney and John Slater (physics)
  Clara Brink Shoemaker and David
Economics
  Juanita Kreps and Clifford H. (banking)
Engineering
  Thelma Estrin and Gerald
  Irmgard Flügge-Lotz and Wilhelm
  Lillian Gilbreth and Frank
  Margaret Hutchinson and W. C. Rousseau
  Sheila Widnall and William
Geology
  Esther Applin and Paul
  Margaret Fuller Boos and C. Maynard
  Anna I. Jonas and George Stose
  Eleanora Bliss Knopf and Adolph
  Helen Plummer and Frederick B.
  Virginia Franklin Ross and John (chemistry)
  Della Roy and Rustum
  H. Catherine Skinner and Brian T.
  Dorothy Tilley and Aubra
  Judith Weiss and Clifford Frondel
Mathematics
  Sophia Levy McDonald and John H.
  Cathleen Synge Morawetz and Herbert (chemistry)
  Klara von Neumann and John
  Klara von Neumann Eckart and Carl (physics)
  Julia Robinson and Raphael M.
  Georgia Caldwell Smith and Barnett (biology)

(*continued*)

Table 5.8 (*Continued*)

Olga Taussky Todd and John
Hilda Geiringer von Mises and Richard (applied mathematics)
Medical sciences
  Grete Bibring and Edward
 *Jennifer Buchwald and Nathaniel
  Calista Eliot and O. Causey
  Margaret Gey and G. O.
  Ingegerd Hellstrom and Karl
  Mabel Hokin and Lowell
 *Virginia Johnson and William Masters
  Margaret Lewis and Warren H.
  Elizabeth Cavert Miller and James A.
  Faith Stone Miller and James A.
  Maud DeWitt Pearl and Raymond
  Gertrude Rand and Clarence Ferree
  Mila Scheibel and Arnold
  Margaret H. D. Smith and Morris F. Shaffer (microbiology)
  Sophie Spitz and Arthur Allen (physician)
  Caroline Bedell Thomas and Henry M.
Meteorology
 *Joanne Starr Malkus and Willem (geophysics)
  Joanne Simpson and Robert H.
Microbiology
  Lucia Lewis Anderson and Allan G. (mathematics)
  Mary I. Bunting and Henry (medicine)
  Mary A. Eccles and John (physiology)
  Gertrude Henle and Werner
  Rebecca Lancefield and Donald
 *Esther Lederberg and Joshua (genetics)
  Margaret H. Sawyer and Wilber A. (international health)
  Beatrice Carrier Seegal and David (medicine)
  Marjorie Roloff Stetten and DeWitt (medicine)
 *Barbara Wright and Herman Kalckar
Nutrition
  Helen Cherington Farnsworth and Paul (psychology)
Other social sciences
  Demography
  Irene Taeuber and Conrad
  Housing policy
  Catherine Bauer Wurster and William (architecture)
Physics
  Fay Ajzenberg-Selove and Walter
  Betsy Ancker-Johson and Harold Johnson (mathematics)
  Ann Chamberlain Birge and Robert W.

(*continued*)

Table 5.8 (*Continued*)

Mary Boas and Ralph (mathematics)
Jenny Rosenthal Bramley and Arthur
Mildred Dresselhaus and Gene
Phyllis Freier and George
Gertrude Scharff Goldhaber and Maurice
Sulamith Goldhaber and Gerson G.
Lucy Hayner and Bernhard Kurrelmeyer
Evans Hayward and Raymond
Margaret Kivelson and Daniel
Luiet Lee-Franzini and Paolo Franzini
Leona Marshall Libby and Willard
*Leona Marshall and John
Claire Marton and Ladislaus
Maria Goeppart Mayer and Joseph (chemistry)
Edith H. Quimby and Shirley
Laura Roth and Willard (biology)
Hertha Sponer and James Franck (chemistry)
Mary Wheeler Wigner and Eugene
C. S. Wu and Luke Yuan
Psychology
Augusta Fox Bronner and William Healey
Mamie Phipps Clark and Kenneth
Else Frenkel-Brunswik and Egon
Eleanor Jack Gibson and James J.
Marcia Guttentag and Paul Secord
Margaret Kuenne Harlow and Harry
Josephine Hilgard and Ernest
Ruth Howard and Albert Beckham
Dorothea Jameson and Leo Hurvich
Margaret Hubbard Jones and F. Nowell
Mary Cover Jones and Harold
Florence Kluckhohn and Clyde (anthropology)
Beatrice Lacey and John I.
Catherine Cox Miles and Walter
Lois Murphy and Gardner
Louisa E. Rhine and J. B.
Anne Roe and George Gaylord Simpson (paleontology)
Pauline Snedden Sears and Robert
Georgene Seward and John
Carolyn Wood Sherif and Mazur
Janet Taylor Spence and Kenneth
Emily Stogdill and Ralph (business psychology)
Lois H. Meek Stolz and Herbert
Thelma Guinn Thurstone and L. L.
Ruth Tolman and Richard (physics)

(*continued*)

Table 5.8   (Continued)

Public health
    Leona Baumgartner and Nathaniel Elias (chemical
       engineering)
    Katherine Rotan Drinker and Cecil
    Dorothy Nyswander and George Palmer
Sociology
    Jessie Bernard and Luther
    Eleanor Glueck and Sheldon (criminology)
    Helen MacGill Hughes and Everett
 * Marie Jahoda and Paul Lazarsfeld
    Patricia Kendall and Paul Lazarsfeld
    Helen Lynd and Robert
    Alice Rossi and Peter
    Dorothy Swaine Thomas and W. I.
Statistics
    Hilda Freeman Silverman and A. Clement (public health)
    Aryness Joy Wickens and David (economics)

---

*Subsequently divorced.

The Register had no data on a subgroup of special note here, the scientist couples, those women scientists and engineers married to other scientists (not necessarily collaborators). Although they are presumed to be a rather rare group statistically, table 5.8, which does not claim to be comprehensive, lists almost two hundred of the better known scientist couples of the period 1950–1970.

Our examination of several sources of statistics on men and women scientists of the 1950s and 1960s, chiefly the National Register of Scientific and Technical Personnel, produced rather reluctantly by the NSF, has revealed that there was such a great increase in the number of women scientists that by 1970 there were, by the best count, about twenty-nine thousand (ten thousand with doctorates, despite the graduate schools). Although there were women in all fields and almost all subfields by then, most were concentrated in relatively few, including some so marginal that their name began with the word *other*. Most of the employed women worked at educational institutions (to which we turn next), although not all of these were teaching. As far as we can tell from the salary data here, all were underpaid, and far too many were unemployed. Yet actual patterns varied so much by field that these aggregates tell little about the experience of any individual woman.

Despite all these data, the manpower analysts missed many aspects of women's lives in science. Even the NSF's National Register, the most authoritative source on scientific personnel, stressed fields, degree levels, and salaries but failed to ask about the barriers and obstacles faced by those women, information that might have been important to anyone interested in recruiting more of them. Instead the Register took a detached stance, as if it did not want to come to grips with certain realities and preferred to minimize and ignore prominent social factors. Thus, it omitted race, marital status, and any hint of children, though such realities of daily life (and their social and institutional consequences, e.g., segregation and antinepotism rules) might loom large to the many young wives and mothers presumably included in the counting. Nor, since the Register's statisticians did not consider home economics a science or specify at what types of educational institution women were employed, was there any way to measure what remained of such former bastions for women in science as the women's colleges and schools of home economics. Nor was there any attempt to gauge such institutional phenomena as advancement (or the lack of it), marginality, ghettoization, stratification, discrimination, and underemployment. The picture presented (and thus, in a sense, the official view) was instead one of bland, neutral, sanitized aggregates. Such reports (and all others based upon them) then perpetrated to the public the view that science was fair, harmonious, and impersonal, with some women on the sidelines, earning less and in the less important "other" fields, for reasons that were not very clear but that were probably inevitable for women. Although these reports persisted until 1970, they are politically light years away from the "status of women" reports that proliferated explosively around 1970 and that, looking at largely the same population and evidence, saw instead patterns of widespread inequities and discrimination. But before we consider those developments (in ch. 16), we will focus on the period to 1968.

6

Faculty at Major Universities

The Antinepotism Rules
and the Grateful Few

The period from the mid-1950s to the late 1960s was one of tremendous growth and prosperity for science at American educational institutions. In these golden years almost every campus built (often with the help of federal funds) new science laboratories and libraries and obtained new equipment, including its first computers and electron microscopes. At the same time, the numbers of both faculty and nonfaculty employed at such institutions also jumped dramatically, as was well documented in the increasingly complete governmental manpower statistics. Yet the women scientists employed in academia did not fare as well, either proportionally or hierarchically, as all the talk of shortages and the rhetoric of the manpower experts might have led them to expect. In fact as far as rank and marginality can be measured, academic women scientists were in some ways worse off in 1970 than they had been in 1945. Overall growth and prosperity did not necessarily benefit scientists of both sexes equally; in fact increased resources could and did unleash ambitions and provide the means to undercut the women's former niches and masculinize academia to record levels.

This curtailment of women's influence was the result of the convergence of several economic, demographic, and psychological factors in the 1950s and 1960s. For example, most large and/or prestigious campuses clung tenaciously to their so-called antinepotism rules for faculty employment despite the rising proportion of married women scientists and a presumably pressing national shortage of faculty members. At the same time, the tremendous expansion of research work on campus necessitated the institutionalization of large numbers of long-term positions for research associates. But overriding everything was the new affluence and ambition at countless colleges and universities, whose men and women trustees and presidents sought (and for the first time in many years could afford) to expand their faculties and upgrade their institutions' prestige. Aiding this were the record numbers of American men, educated to the Ph.D. and near-Ph.D. level on the GI Bill, and foreign men, either trained in the United States or recruited from abroad, that were available for and intent upon desirable faculty positions. The result of these several forces was that even though by 1970 there were many more women scientists employed in academia than ever before, they were even less often than earlier in faculty or administrative positions and more often research assistants, librarians, or other non-faculty personnel on someone else's grant. Thus this "golden age" for science was for women in academia the very dark age that the many "status of women" reports of 1969–70 and after would so angrily discover, depict, and deplore.

Let us turn first to the position of women scientists at the prestigious universities, where they were often trained but had never been employed in prominent or numerous positions. This exclusionary pattern of finding very few single or married women who measured up to their arbitrary standards persisted in the 1950s and 1960s. Some of these universities had simply never hired women faculty, but others, especially public institutions, codified the practice in so-called antinepotism rules, which prohibited hiring relatives, especially the spouse, of a faculty member. The result was that despite all the expansion, faculty women remained very rare at prestigious institutions. In the late 1960s some departments were rather proud never that they had never had any female faculty members, and others were systematically not replacing those that they had had in such footholds as "social economics" in the past.[1] Thus the government's "womanpower" message did not penetrate the faculties of the major universities, despite the so-called pressing national shortages, even though they were the very groups and institutions that were accepting and spending federal funds on research.

Some of the affected women were angry but felt powerless to do much either for themselves or for the group as a whole. A few had protested in the late 1940s, and occasionally one committed suicide, but most went quietly. Several, including a number of leading researchers, left for the proliferating nonprofit institutions and the expanding federal laboratories (discussed in chs. 11 and 13); one, unsuccessful at her institution, spurred the AAUW to publish a study of the phenomenon in 1960; and still another got the University of Kansas to change its rules in 1963. Yet when in 1964 a female sociologist—one of just two in the nation who were full professors at major universities—who thought herself an expert on women faculty wrote a book entitled *Academic Women*, she denied seemingly authoritatively that there was any discrimination going on. We might, therefore, typify those who, like her, did get onto the faculty as the grateful few.

By the 1950s most major universities had antinepotism rules. Designed to protect institutions from being forced to employ the less able relatives of persons in power, they could be so broadly worded or interpreted as to prohibit the employment of any two members of the same family by the same department or institution or even system of institutions. Nor was this practice by any means restricted to state-run institutions. When the College and University Personnel Association surveyed college employment practices four times between 1949 and 1958 as part of its goal of standardizing procedures at the increasingly bureaucratized universities, it found that 54–62 percent of the growing numbers of colleges and universities responding to the surveys (whose total more than tripled from 42 in 1949 to 143 in 1958) reported a wide range of practices, ranging from a total ban on the employment of any relatives or a prohibition on hiring them permanently to a limitation on specific relatives (e.g., immediate family members or married couples) in specific jobs. Some simply required that one relative should not be the supervisor of another. The most common restrictions were on members of the same immediate family working in the same department. But some interpreted existing antinepotism rules much more broadly. In Illinois, for example, the state's boards interpreted the rules as prohibiting the employment of relatives down to the level of second cousins, and the regents of the new California State College system, with its eighteen campuses, would not permit couples to teach at the same college, let alone in the same department. Some said that exceptions could be made in cases only of "recruiting difficulty," while others could make an exception only if the board of trustees approved, a daunting requirement at a large multicampus university.[2] In general there was such a wide range of practices nationwide that a couple contemplating a move to a new university would have been well advised to look at these reports (e.g., that for 1953–54, which itemized which practices prevailed at which universities) rather than simply believing the assurances of department chairmen eager to hire the husband that, of course, some effort would be made to find a position for the wife. All too often couples arrived only to learn that local rules prevented any such easy solution to the wife's continuing her career. But of course the "rules" were merely a pretext; for a variety of petty reasons that could loom large in a small college town (such as the fear that a junior, two-income couple could live visibly better than their senior colleagues), local officials were reluctant to upset traditional hierarchies as well as college finances by treating the new spouse equally (and thus differently).[3]

The true extent of those affected is difficult to determine, but it may have been great, especially at the major universities. A 1970 survey of 336 faculty members at Berkeley found that 58 spouses had been affected by such university restrictions. These included 23 spouses with Ph.D.'s, 22 of whom were in the same field as their husband or a closely related one.[4] Notable among these would have been mathematician Julia Robinson, economist Margaret Gordon, and psychologists Mary Cover Jones, Ravenna Helson, Susan Ervin-Tripp, and, earlier, Else Frenkel-Brunswik.[5] In a 1972 study of married women chemists with Ph.D.'s, 42 percent reported that they had been adversely affected by antinepotism rules at some point during their career (60 percent of the husbands were chemists and 35 percent other scientists).[6] But the true figures may never be known, for others were so sure that marriage to a fellow faculty member would end a professional woman's career that, like Madeline Kneberg and Thomas Lewis of the University of Tennessee, a famous team in southeastern archaeology, they chose not to marry until they both retired.[7]

In 1953–54 one woman found a way to protest. Although she was unsuccessful in the short run in keeping her (tenured) job, her call for outside attention was of great significance in abolishing the whole practice in the long run. In April 1953 Josephine Mitchell, who had been a student of Hilda Geiringer at Bryn Mawr Graduate School in the late 1940s and was in her sixth year at the University of Illinois mathematics department, was awarded tenure. This was unusual for any woman at Illinois, but Mitchell was one of the department members who had held an NSF research grant. Then in August of that year, perhaps feeling that tenure protected her, she married an assistant professor in the same department. (He was several—possibly seven—years younger than she and had been previously married, factors that may have made the match all the more unconventional.) Thereupon she was informed that despite her tenure, she would not be reappointed at the end of the year because of Illinois's antinepotism rule. (Apparently it did not occur to anyone that her husband, who was still untenured, should be the faculty member not renewed.) Being mathematicians well versed in logical proofs, both Schoenfelds battled the ruling with forceful memorandums. Among other points, they argued that such antinepotism rules conflicted with other university statutes that claimed that only merit was involved in making faculty appointments and that they had not been warned when they were hired initially. But even though the department and the local chapter of the AAUP voted to support them, the Illinois provost, as Mitchell later described their meeting, politely "suggested that if we could not live happily under this policy, we could find positions elsewhere," as he did not intend to make any exceptions for them.[8]

Fortunately the episode did not end there, for Mitchell, who had expected this kind of outcome, had applied to the AAUW for its Marion Talbot Postdoctoral Fellowship for 1954–55. Suspecting that the circumstances of her case might interest the AAUW's national staff, Mitchell enclosed a copy of her forceful "Statement of the Nepotism Policy of the University of Illinois" with her acceptance of the fellowship. AAUW staff member Eleanor Dolan held onto the document, and five years later, when she was running an AAUW project to encourage women doctorates to take up academic careers, she used Mitchell's 1954 memo as the basis for a national survey of the current status of antinepotism rules, one of the first published signs, though muted, of smoldering discontent.[9]

But by then Josephine Mitchell Schoenfeld and her husband had left Urbana for a series of short-term industrial and academic positions before settling down in 1958 for what turned into eleven years at Pennsylvania State College (later University). The mathematics department there was not as prestigious as that at Illinois (which ranked twelfth in the nation in both 1957 and 1964), but hiring the Schoenfelds was part of the university's effort to upgrade itself.[10] Moreover, Penn State was one of the very few universities known at the time for its willingness to hire academic couples. In fact its College of Liberal Arts seems to have been the only institution in the nation that used the expansion of the postwar years as an opportunity to upgrade itself by hiring *more* women. It managed to add faculty despite its less than competitive salary scale by hiring couples, some of whom came after searching for a year or more for any university that would hire both of them. Among these recruits were the sociologists Jessie and Luther Bernard, bioscientists Rosemarie and Harald Schraer, mathematicians Lowell Schoenfeld and Josephine Mitchell, psychologists Carolyn and Muzafer Sherif, and mathematician Marian Pour-El (but not her husband biochemist Akiva). One measurable result of these new hires was the rapid rise in the reputation of the Penn State mathematics department, which in 1957 had not been ranked in a national survey of graduate programs but by 1964 had moved up to "adequate plus" or between thirty-third and forty-seventh in the nation. Although most of these women held ranks junior to their husbands', none became a chairman or administrator, and many moved on after a few years, most felt grateful at the time to the college and its dean, Ben Euwema, for this relative oasis in the otherwise unwelcoming academic world.[11]

Meanwhile, in the 1950s, while Schoenfeld's complaint was sitting at the AAUW, interest there and elsewhere was focusing more on protesting the antinepotism rules. They came in for criticism in March 1958 at the annual meeting of the Advisory Board on Education of the NRC. At a section meeting on the topic "Education in Science," devoted to a discussion of what might be done to improve the teaching of high school and college science, the fifty participants felt strongly enough to pass the following resolution:

> It is in the national interest to encourage gifted women to prepare for careers in science, and we should take full advantage of this trained woman-power. Women should be employed without prejudice. Some advantages would accrue from changing restrictive employment policies, for example, restrictions against the employment of a husband and wife by the same institution.[12]

This suggestion was an indication that college and university antinepotism rules had already been identified as major obstacles to the better "utilization" of women that was so generally advocated and that they would be an early target for emerging protest groups. But this idea was so controversial in 1958 that the passage of the resolution was not even mentioned in the NRC's annual report.[13]

Then in 1959–60 Eleanor Dolan of the AAUW and her assistant, Margaret P. Davis, wishing to encourage women to reenter academia and earn higher degrees, began a study of the antinepotism rules to show that they were not the severe detriment to women's careers that many thought. Because the AAUW was chiefly concerned with those schools responsible for the training of women, Dolan and Davis omitted from their survey all men's colleges (as too unlikely ever to hire wives to be worth considering)[14] and included many small private ones, especially Roman Catholic and Lutheran colleges. Of the 363 institutions they contacted, 285 replied (78.51 percent), and of these 158 (55.44 percent) reported no restrictions on the employment of spouses (in fact several of the 158 said that they benefited greatly from the restrictions at other nearby institutions [especially public universities], because they could never have attracted so many highly qualified faculty otherwise). Although Dolan and Davis found that 44.56 percent of their respondents had such regulations, this was far fewer than in the earlier CUPA studies. The discrepancy was the result of the AAUW's using a larger and a more sympathetic group of colleges and universities.

Nevertheless, after seizing on this ray of optimism, the AAUW surveyors also had to report that seventy-five institutions, or 26.32 percent of their respondents, would not hire spouses at all. This was particularly disturbing because in this group were many large public universities and junior colleges that were slated for rapid expansion in the near future. (The study made no attempt to count how many actual or potential jobs were involved, just the number of institutions and their overall policies, a stance that made its conclusions rosier than they might have been otherwise.) Dolan and Davis also found fifty-two respondents (18.25 percent) in a third, more ambiguous category. These had no fixed policy or written rules on the subject but instead an abundance of various labor practices, which they described (and the authors quoted) at length: lower salaries for wives despite their credentials or accomplishments, no permanency or tenure, slower advancement, and the like. The respondents felt that these practices were justified, for they did not seem particularly embarrassed to describe them matter-of-factly to the AAUW, as ways in which a chairman or other administrator could prove that he was not showing undue favoritism to the wife he had hired. Several respondents predicted that they would be using these practices less in the future, when they anticipated there would be a faculty shortage and they might have more competition from other schools. Thus the report, which remains to this day the classic study of what was a very serious yet elusive problem, has the tone of rather forced cheerfulness—the coming expansion would force institutions to change their ways—while the extensive detail of the actual unfair practices (they talked of "barriers," not discrimination) would have provided most readers with chill-ingly vivid evidence for a darker conclusion concerning the devious ways of exploitative deans.[15]

Because the extensive governmental statistical surveys of educational insti-tutions and scientific manpower in the 1950s and 1960s did not break their data down in a way that would allow one to focus on women scientists at prestigious universities, actual data on where the women were at leading universities seemed unavailable. Yet in 1960 John Parrish, a professor of labor economics at the University of Illinois and a researcher on women's "underutilization" in science and research, simply collected two sets of very relevant numbers. After reading in an unspecified NSF report that 90 percent of the faculty doing scientific research were employed at just 143 institutions (and 60 percent were concentrated at just 67 colleges and universities), Parrish wrote to each of twenty major universities (the ten private universities with the largest endow-ments and the ten public ones with the largest enrollments, which collectively awarded about one-quarter of all doctorates) requesting the number of its faculty members broken down by department, rank, and gender. Perhaps sur-prisingly, all responded.[16] Parrish's unpublished data, summarized here in table 6.1, present quite a different manpower picture from the bland aggregates of the NSF manpower reports. Not only were women just 6.72 percent of the total science faculty (474 of 7,057 at the level of assistant professor or above) at these twenty leading institutions but their distribution varied greatly by field, university, and rank. These numbers demonstrate how totally exclusive the closed recruitment described in Theodore Caplow and Reece McGee's famous ethnography, *The Academic Marketplace* (1958), was in practice.[17]

Parrish's data demonstrated strikingly the persistence in 1960 of both "terri-torial" and "hierarchical" segregation at those very leading doctoral univer-sities at which most of the nation's science and engineering was being done.[18] The "territorial" ghettoization is evident at the top of table 6.1, where is is revealed that almost two-thirds (303, or 63.92 percent) of the 474 total women faculty in twenty-two science and social science disciplines were in just one very highly feminized field, home economics. The remaining 171 women faculty were in twenty-one other fields, where their distribution ranged from highs of 15.38 percent of the geneticists and 12.12 percent of the anatomists down to fewer than 1 percent of the geologists, engineers, and entomologists. Thus, although women might constitute 10.59 percent of the scientists employed at educational institutions by the late 1950s (as shown in the National Register data cited in table 5.5), the academic hiring practices at the richest and largest universities kept the proportion of women faculty there far below this average in most scientific fields.

Besides this "territorial" concentration of women faculty in home econom-ics, there was also a "hierarchical" segregation in the lower faculty ranks. Thus,

Table 6.1.

Science Faculty (Assistant Professors and Above) at Twenty Leading Universities, by Sex, Field, and Top Rank, 1960

Field	Total	Women	Women as % of Total	Full Professors Total	Full Professors Women	Full Professors Women as % of Total
Home economics	331	303	91.54	98	86	87.76
Genetics	13	2	15.38	6	0	0
Anatomy	99	12	12.12	30	3	10.00
Psychology	416	30	7.21	184	5	2.72
Physiology	143	10	6.99	58	0	0
Bacteriology	102	6	5.88	46	1	2.17
Astronomy	57	3	5.26	29	1	3.45
Mathematics	521	27	5.18	229	1	0.44
Biochemistry	142	7	4.93	56	1	1.79
Botany	165	8	4.85	69	0	0
Sociology	259	12	4.63	108	2	1.85
Economics	435	16	3.68	205	4	1.95
Anthropology	94	3	3.19	46	1	2.17
Pathology	71	2	2.82	23	0	0
Zoology	255	5	1.96	132	0	0
Chemistry	487	9	1.85	238	4	1.68
Political science	336	6	1.79	161	2	1.24
Physics	545	6	1.10	228	2	0.88
Geography	97	1	1.03	44	0	0
Geology	216	2	0.93	113	0	0
Engineering	2,207	4	0.18	901	2	0.22
Entomology	66	0	0	35	0	0
Total/average	7,057	474	6.72	3,039	115	3.78

Source: Unpublished data in John B. Parrish Papers, SLRC.

as also shown in table 6.1, at Parrish's twenty leading universities in 1960 very few women were full professors in these twenty-two sciences and social sciences—only 115 of the 3,039 persons at this rank, or 3.78 percent—and 86 of these were in home economics. This left only 29 women full professors in twenty-one fields at twenty major universities, or about 1 per university and 1 per science in 1960. Yet even this would be too high for some fields, for there were no women full professors at *any* of these twenty schools in eight sciences, including the substantial fields of zoology (with 132 full professors), geology (113), botany (69), physiology (58), geography (44), and entomology (35). When president E. B. Fred of the University of Wisconsin suggested adding 1 woman faculty member per department, he must have seemed liberal and utopian, for even that tokenism would have been quite an advance from the norm at prestigious universities at the time.[19]

The situation was only slightly better for the women at the levels of associate and assistant professor in Parrish's data (not in table 6.1). There were 170 women associate professors, or 8.66 percent of the 1,963 associate professors at these institutions in 1960, but although this was higher than women's proportion of full professors, it was still particularly low, for the rank of associate professor was usually the highest and terminal rank reached by women after decades of service teaching heavy loads of introductory courses. (Often they retired as "associate professor emerita.") Thus, all too often a rank that for men was a steppingstone to a near-automatic full professorship was not such a temporary rank for women. Since the 1950s were also a period of increasingly rapid rates of promotion, women's stagnation at the level of associate professor stigmatized them even further. Once again 103 of these 170 were in home economics, leaving 67 in all other sciences, or just 3.61 percent of the total associate professors in the twenty-one other sciences, with actual zeroes in four (geology, pathology, genetics, and entomology).

Similarly, of the 2,051 assistant professors in Parrish's data, 187 were women (a slightly higher 9.11 percent), of whom 114 were in home economics and 73 were in the other sciences, or just 3.79 percent of the total assistant professors in those twenty-one sciences. Yet even this distribution was uneven, for in five (genetics, astronomy, anthropology, geography, and again entomology) there were no women assistant professors at all. Evidently, in the previous eight years (i.e., almost the entire decade of the fifties) either no young women had been hired by any of these twenty leading departments, none of those who had been hired had stayed very long, or those women that were hired were given the lower rank of instructor. For example, the first woman faculty member at Cornell's College of Arts and Sciences was Martha E. Stahr, a Wellesley graduate and Berkeley Ph.D., appointed an assistant professor of astronomy in 1947; but she married another Cornell faculty member in 1951 and thus in 1954 was appointed a research associate rather than promoted to associate professor.[20] Psychologist Frances Clayton encountered several of these practices at the University of Minnesota when she was about to complete her doctorate there. After offering her a mere instructorship the chairman informed her, lest she think she might have a future there, that the department "had an explicit policy of not hiring women for more than one year because there was no possibility of their staying permanently and they felt it would be unfair to hire them for longer and then let them go!"[21] Thus, although the late 1950s were years of increasing academic expansion, wherever the record numbers of women Ph.D.'s were going, they were not being hired or retained by Parrish's leading universities.

Table 6.2 may help to demonstrate how thin and spotty was the distribution of John Parrish's 474 women science faculty members (all three ranks) in 1960. Broken down by university and by field, it shows that in some fields all the women or a good proportion of them were in just a few departments. The largest number of non–home economists were in the rather large field of psychology (416 professorships [table 6.1]): 30 women, including 6 at New York University. It was followed by mathematics (27, with 8 at the University of Illinois[22] and 7 at Penn State) and economics (16 at ten universities). Thus, the leading institutions for women science faculty not in home economics were Penn State, Illinois, and Columbia with 18 each, followed by New York University with 17 and Ohio State with 15. The worst of these leading twenty universities were Princeton and MIT, with none. Similarly, 17 of the 29 women full professors were at just four institutions: Chicago and Columbia each had 5, Penn State had 4 (including Jessie Bernard, the only woman full professor in sociology in Parrish's data), and Michigan had 3.

Table 6.3 supplements Parrish's data for March 1960 with actual names for more than one hundred of these prominently placed women scientists, identified by field and rank at nineteen universities. (Table 6.3 omits six of Parrish's universities—Northwestern, College of the City of New York, Ohio State, Michigan State, New York University, and Penn State—but includes five other institutions of particular interest to scientists and engineers or with high prestige rankings at the time—Wisconsin, UCLA, Pennsylvania, California Institute of Technology, and the Rockefeller Institute, soon to be Rockefeller University.) Selecting from university catalogs for 1960, primarily those of the colleges of arts and sciences, is not a foolproof way to find all the women science faculty, because women were often deliberately camouflaged behind unexpected job titles and in distant administrative units.[23] Thus, table 6.3 includes a column for women scientists who have come to light in "other" units or rubrics at these institutions. Table 6.3 also includes only seventeen fields, though some of the omitted fields (e.g., physiology and genetics) are included under "other."[24]

Some of the faculty women listed in table 6.3 were highly acclaimed and internationally renowned, such as the physicists Maria Goeppart Mayer at the University of Chicago and C. S. Wu at Columbia.[25] Others were master grantswomen of their time, holding numerous NSF, ONR, or National Institutes of Health (NIH) grants over the years, such as biochemist Barbara Low, first at Harvard Medical School but after 1956 at Columbia's College of Physicians and Surgeons, zoologist Grace Pickford at Yale, anthropologists Ruth Benedict at Columbia and Cora DuBois at Harvard, and psychologists Helen Peak at Michigan and Jean W. Macfarlane at Berkeley, who struggled for years to maintain support for the longitudinal studies at its Institute for Human Development.[26] DuBois and Peak were, however, full professors at these top universities only because of two cases of "creative philanthropy" started years before (see ch. 2).[27] In fact during the 1950s it became clear that the Zemurray Stone–Radcliffe Professorship at Harvard had in fact opened a "small wedge,"

Table 6.2.
Women Science Faculty (Assistant Professors and Above) at Twenty Leading Universities, by Field, 1960

Institution	Total	Home Economics	Psychology	Mathematics	Economics	Sociology	Anatomy	Physiology	Chemistry	Botany	Biochemistry
1. Cornell	87	83	1	—	—	1	—	—	—	—	1
2. Ohio State	68	53	3	1	2	2	1	3	1	2	—
3. Penn State	62	44	2	7	—	2	—	—	3	—	—
4. Illinois	56	38	4	8	2	1	2	—	1	2	1
5. Michigan State	56	45	2	1	—	—	2	1	—	1	—
6. Minnesota	29	22	1	1	1	—	2	—	3	1	2
7. Columbia	18	—	—	1	2	—	2	—	—	—	2
8. NYU	17	—	6	3	2	2	—	2	—	—	—
9. Indiana	14	9	—	—	2	—	1	1	—	1	—
10. California	13	7	1	1	1	3	1	—	—	—	1
11. Chicago	13	—	1	—	2	3	1	—	—	1	—
12. Stanford	11	—	3	1	1	—	1	1	—	—	—
13. Harvard	8	—	—	3	—	1	2	—	—	—	—
14. Northwestern	6	2	1	—	1	—	—	—	—	—	—
15. Michigan	6	—	2	—	—	—	—	2	—	—	2
16. Yale	4	—	1	—	—	—	—	—	—	—	—
17. Johns Hopkins	3	—	1	—	—	—	—	—	—	—	—
18. City College of New York	3	—	—	—	—	—	—	—	1	—	—
19. Princeton	0	—	—	—	—	—	—	—	—	—	—
20. MIT	0	—	—	—	—	—	—	—	—	—	—
Total	474	303	30	27	16	12	12	10	9	8	7

Institution	Bacteriology	Physics	Political Science	Zoology	Engineering	Astronomy	Anthropology	Genetics	Geology	Pathology	Geography	Entomology
1. Cornell		1										
2. Ohio State												
3. Penn State	1		1	1							1	
4. Illinois						1			1			
5. Michigan State	2				2					1		
6. Minnesota												
7. Columbia	1	2										
8. NYU		1	1	2								
9. Indiana			1					1				
10. California		1						1				
11. Chicago	2		2							1		
12. Stanford			1		1				1			
13. Harvard												
14. Northwestern							3					
15. Michigan				1								
16. Yale				1		2						
17. Johns Hopkins												
18. City College of New York												
19. Princeton		1			1							
20. MIT												
Total	6	6	6	5	4	3	3	2	2	2	1	0

Source: See table 6.1.

133

Table 6.3.

Women Science Faculty at Nineteen Leading Universities, by Field, 1960

Institution	Total	Mathematics and Statstics	Psychology	Biochemistry	Anatomy	Microbiology	Economics	Zoology	Sociology
Columbia	13			Abell (C) Low (C) Mandl (C)	Murray (A) Johnson (C)	Seegal (A) Knox (C) Silva (C)			Sitgreaves (C)
Illinois	13	Armstrong (C) Bateman (C) Chandler (C) Halton (C) Pepper (C) Reiner (C) Schubert (C)	Jonietz (B)				Beckett (B) Weston (B)		Chandler (B)
Chicago	11		Koch (A) Stock (C)	Eichelberger (A) Vennesland (A)	Rhines (B)		Reid (A)	Price (A)	
California	11	Fix (B) Scott (B)	MacFarlane (A)		Simpson (A)		Huntington (A)	St. Lawrence (C, genetics)	
Stanford	10		Dowley (B) Maccoby (B) Rau (C)				Peffer (A)		
Minnesota	9	Carlson (B) Yamabe (B)			Sundberg (B) Padykula (C)				
Harvard	6	Worcester (B)							
Cornell	5		Smith (B)	Daniel (A)					Goldsen (B)
Wisconsin	5					McCoy (A) Ball (B)		Bilstad (B) Szego (B)	
UCLA	5		Ruhlman (B)						
Indiana	5			[Cohn (B)]	Strong (B)		Crawford (B)		
Pennsylvania	4								Thomas (A) Boll (C)
Rockefeller	4			Perlmann (B)		Lancefield (A) Gottschall (C)		Lee (C)	
Michigan[b]	3		Peak (A)		Crosby (A)				
Yale	2			Simmonds (B)				Pickford (B)	
Johns Hopkins	2		Ainsworth (B)						
Cal Tech	0								
MIT	0								
Princeton	0								

Institution	Physics	Botany	Political Science	Engineering	Astronomy	Anthropology	Nutrition[a]	Other
Columbia	Wu (A) Hayner (B)							Biophysics Quimby (A) Library science Bryan (A)
Illinois				Wilson (C)				Chemistry Bartow (B)
Chicago	Mayer (Vol. A) Marshall (Vol. C)	Ralser (B)						Human development Neugarten (C)
California						Gayton (A)		Education Jones (A) Linguistics Haas (A) Physiology Timras (A) Speech Ervin-Tripp (C)
Stanford	Waggoner (C)		Harris (C)	Flügge-Lotz (A)				Education Sears (B) Geology Keen (C)
Minnesota		Dodsall (C)						Food econonomics Farnsworth (A) Pharmacology Cranston (C) Biophysics Herrick (A) Plant pathology Hart (A) Child development Fuller (A) Templin (B)

(continued)

Table 6.3 (Continued)

Institution	Physics	Botany	Political Science	Engineering	Astronomy	Anthropology	Nutrition[a]	Other
Harvard			Shklar (C)		Payne-Gaposchkin (A)	DuBois (A)	Trulson (B, public health)	
Cornell							Young (A)	Labor relations McKelvey (A)
Wisconsin		Fisk (B)	Penniman (B)					Veterinary science Wipf (C)
UCLA		Mathias (B)						Education Seagoe (A)
Indiana		Shalucha (C)						Physiology Wertenberger (B) Education Mueller (A)
Pennsylvania								Human relations Epstein (C)
Rockefeller								
Michigan[b]					Dodson (A)			
Yale								
Johns Hopkins								Physiological optics Sloan (B)

Sources: University catalogs, 1959–60.

Note: A = professor; B = associate professor; C = assistant professor; [ ] = should have been in catalog.

[a]Other than home economics.

[b]Catalog omits first names, so list may omit some.

Table 6.4.
Women Full Professors at Harvard, 1948–1963

Year	Name	Field	School
1948	*Maud Cam	English history	Arts and sciences
1954	Sirarpie Der Nersessian	Byzantine art	Arts and sciences (Dumbarton Oaks)
	*Cora DuBois	Anthropology	Arts and sciences
1956	Cecilia Payne-Gapsochkin	Astronomy	Arts and sciences
1957	Martha May Eliot	Maternal and child health	Public health
1959	Henrietta Larson	Business history	Business
1961	Grete Bibring	Clinical psychiatry	Medical school
1962	Jane Worcester	Biostatistics	Public health
1963	Anne Roe Simpson	Psychology	Education

*Zemurray Stone–Radcliffe Professor.

as astronomer Cecilia Payne-Gaposchkin later put it, for a slow trickle of other women, including herself in 1956, to rise to full professorships on the university faculties (see table 6.4).[28] Although several of the other women in table 6.3 were top-notch in various ways, the best women researchers in academia around 1960 were not necessarily on university faculties at all; they were more often among the research associates there, and some of the very best women researchers of the time, such as geneticist Barbara McClintock or zoologist Libbie Hyman, had long since left academia for more welcoming nonprofit institutions.

Nor were the few tenured women faculty at prestigious universities necessarily very comfortable there, as far as can be determined. In 1954 Grace Stewart, a professor of geology at Ohio State, retired nine years early (at sixty-one) because she felt "discouraged." Gertrude Perlmann, long an associate professor at the Rockefeller Institute/University, was evidently not in full charge of her own laboratory and had great difficulty in finding out from Detlev Bronk and others what her future prospects there were.[29] Harvard's Cora DuBois later complained that she had never been a part of her department's politics and that she had found faculty meetings a waste of time but said that fortunately she was often away.[30] At Cornell, associate professor Rose Goldsen never fit in and was on leave as much as possible, often working for the Ford Foundation in Latin America, where she preferred to be.[31]

Of those few women who were full professors in 1960 several were in the final years of long academic careers dating back to the 1930s or even the 1920s. At Berkeley, when women who had been hired in the social sciences back in the twenties finally retired in the sixties, they were not replaced by other women. The last woman the psychology department hired on the tenure track was Jean Walker Macfarlane, in the twenties, and the last in economics was Emily Huntington, who had come in 1928 to carry on Berkeley's feminized tradition in "social economics," started by her mentor Jessica Peixotto. For decades Huntington worked on such consumer issues as the cost of living, the minimum wage, unemployment insurance, health insurance, and Medicare, becoming a national expert in these areas. During World War II the University of California started a new School of Social Welfare on the Berkeley campus and reallocated most instruction in these subjects to that unit. Feeling increasingly uncomfortable in the economics department, which was moving rapidly into mathematical modeling and econometrics, and bored by the professional journals, which were so narrow and specialized as to be almost beyond her comprehension, Huntington retired two years early (at sixty-five) and was not replaced with another "social economist." Instead this subject, once an important part of "women's work" in the social sciences, was excised from the "mainstream" and increasingly theoretical field of economics and reassigned to a less prestigious part of the campus.[32]

Meanwhile, a few institutions designed ways around the antinepotism rules for exceptional couples. At the University of Chicago, for example, problems arose when the spouse of a full professor wished to be employed. Thus the addition of Maria Goeppart Mayer, an award-winning physicist but also the wife of Joseph Mayer, Eisendrath Professor of Chemistry, occasioned the creation of a new category, that of the "volunteer professor" of physics. This ingenious status was also used for other scientist-spouses. Leona Marshall (later Libby), then the wife of physics professor John Marshall, was designated "volunteer assistant professor of physics" at the university's Enrico Fermi Institute for Nuclear Studies rather than in the physics department itself.[33] In these cases the words "volunteer" and "voluntary" would have meant unpaid, lest someone think that the wife of a well-paid academic man was also earning a full salary. (Mayer was employed half-time that and other years at the Argonne National Laboratory, as many of her colleagues may have been as well.) As far as placement in the catalog and what that reveals, a "voluntary" professor was, as shown in figure 3, right up there with the other full professors, suffering relatively little loss of status, as Mayer always insisted when queried about this quaint practice by skeptics years later.[34] Evidently, her physics colleagues were so proud of her many accomplishments, rather than jealous, and so eager to have her share their high-status visibly that they had invented for her this new title, awkward in one way but humane in a sense too. By 1960, of course, several colleagues would have been involved in her election to the National Academy of Sciences (NAS) in 1956 (only the fifth woman). Perhaps they were in their own way trying to embarrass their administration for its humiliating exclusionary rules.

Other university administrators, seeking to build up new programs, were making a few exceptions to their own antinepotism rules even in the 1950s.

DEPARTMENT OF PHYSICS
Officers of Instruction

WILLIAM HOULDER ZACHARIASEN, PH.D., SC.D., Chairman of the Department of Physics and Professor of Physics.

HAROLD ROBINSON VOORHEES, PH.D., Assistant to the Chairman, Departmental Adviser, and Associate Professor of Physics.

ROBERT SANDERSON MULLIKEN, PH.D., SC.D., Ernest DeWitt Burton Distinguished Service Professor of Physics.

SAMUEL KING ALLISON, PH.D., Professor of Physics.

MARCEL SCHEIN, PH.D., Professor of Physics.

GREGOR WENTZEL, PH.D., Professor of Physics.

ANDREW WERNER LAWSON, PH.D., Professor of Physics.

HERBERT LAWRENCE ANDERSON, PH.D., Professor of Physics.

MARIA GOEPPERT MAYER, PH.D., Volunteer Professor of Physics.

MARK G. INGHRAM, PH.D., Professor of Physics.

JOHN ALEXANDER SIMPSON, PH.D., Professor of Physics.

S. CHANDRASEKHAR, PH.D., SC.D., Morton D. Hull Distinguished Service Professor of Theoretical Astrophysics.

JOHN MARSHALL, PH.D., Professor of Physics.

NICHOLAS C. METROPOLIS, PH.D., Professor of Physics.

JOHN RADER PLATT, PH.D., Professor of Physics.

RICHARD H. DALITZ, PH.D., Professor of Physics.

YOICHIRO NAMBU, PH.D., Professor of Physics.

VALENTINE L. TELEGDI, PH.D., Professor of Physics.

ROGER H. HILDEBRAND, PH.D., Associate Professor of Physics.

CLEMENS C. J. ROOTHAAN, PH.D., Associate Professor of Physics.

MORREL HERMAN COHEN, PH.D., Associate Professor of Physics.

S. COURTENAY WRIGHT, PH.D., Associate Professor of Physics.

CARL M. YORK, PH.D., Assistant Professor of Physics.

FREDERICK REIF, PH.D., Assistant Professor of Physics.

ULRICH E. KRUSE, PH.D., Assistant Professor of Physics.

ALBERT VICTOR CREWE, PH.D., Assistant Professor of Physics.

PETER MEYER, PH.D., Assistant Professor of Physics.

RUSSELL DONNELLY, PH.D., Assistant Professor of Physics.

HELLMUT FRITZSCHE, PH.D., Assistant Professor of Physics.

EUGENE N. PARKER, PH.D., Assistant Professor of Physics.

DEREK ALBERT TIDMAN, PH.D., Assistant Professor of Physics.

RICARDO LEVI-SETTI, PH.D., Assistant Professor of Physics.

REINHARD OEHME, PH.D., Assistant Professor of Physics.

JOHN ROBERT SCHRIEFFER, PH.D., Assistant Professor of Physics.

WILLIAM L. LICHTEN, PH.D., Assistant Professor of Physics.

JUN JOHN SAKURAI, PH.D., Assistant Professor of Physics.

CLAYTON F. GIESE, PH.D., Instructor in Physics.

GIOVANNI DEMARIA, Ph.D., Research Associate in Mass Spectroscopy.

JEAN DROWART, PH.D., Research Associate in Mass Spectroscopy.

BERNARD J. RANSIL, PH.D., Research Associate in Molecular Structure.

MARTIN W. TEUCHER, PH.D., Research Associate in Cosmic Rays.

PHILIP G. WILKINSON, PH.D., Research Associate in Spectroscopy.

HARVEY BRACE LEMON, PH.D., Professor Emeritus of Physics.

GEORGE SPENCER MONK, PH.D., Associate Professor Emeritus of Physics.

Fig. 3. Faculty roster of the Department of Physics at the University of Chicago, 1959–1960, publicly listing Maria Goeppart Mayer as its volunteer professor. Reprinted with permission of the Department of Special Collections, University of Chicago Library, from *University of Chicago, Announcements: Graduate Programs in the Divisions, Sessions of 1959–1960*, 59, no. 2 (1958): 185–86.

They might hire the wife, but often her rank was lower than she merited and her subsequent advancement would lag years, even a decade or more, behind her husband's, even discounting age differences. At Yale University "Miss" Sofia Simmonds, an associate professor of biochemistry and microbiology, had been the wife since 1936 of Joseph Fruton, the Eugene Higgins Professor of Biochemistry, department chairman, and director of the science division.[35] Stanford had several faculty scientist couples, perhaps because it used a second faculty position as an effective bargaining chip in attracting eminent couples from the East. These included psychologists Robert and Pauline Sears in psychology and education, who came from Harvard in 1953, and Helen and Paul Farnsworth in food economics and psychology, respectively.[36] At the University of Wisconsin Medical School, despite the antinepotism rules that had forced some off the faculty in the late 1940s, Elizabeth and James Miller were a famous team of cancer researchers, he a full professor of oncology since 1952 and she an associate professor from 1959 to 1969 and then a full professor, long after they had both been winning national awards.[37]

Meanwhile, the opening of the private Brandeis University, one of the miracles of mid-twentieth-century philanthropy, required the appointment of a whole new faculty. Because the founders wished to assure high academic standing for their fledgling enterprise, the president often sought out proven senior people from already prestigious places. One inducement for such persons to trade the known for the unknown might be a faculty or even just a research positions for their spouse. Thus a few of Brandeis's early appointments included couples, such as Lawrence Levine and Helen Van Vunakis, both of the New York State Department of Health, to a full and an associate professorship, respectively, in the well-funded biochemistry department and Maurice and Raquel Rotman Sussman, both of Northwestern University, to a faculty and a research position in the biology department. A few years later the anthropology department at Brandeis hired David Aberle and his wife, Kathleen Gough, as chairman and assistant professor, respectively; however, as she was politically outspoken, their stay was brief.[38]

More couples were hired even at public institutions in the 1960s. When, for example, meteorologists Joanne Malkus (later Simpson) and her husband of the time negotiated with UCLA about faculty appointments, they became the first couple appointed to professorships in the same department there in 1960.[39] After its experience with the Schoenfelds in the mid-fifties, the administration at Illinois may have felt it was taking a big step when in 1963 it hired both entomologist Judith Willis and her husband John. Despite their similar Harvard doctorates in biology, however, they were not only put in different departments but given different ranks and salaries: she was hired as an "emergency" instructor, while he became an assistant professor, with a 20 percent higher salary.[40] In 1968 the still new branch of the State University of New York at Stony Brook found itself able to offer a distinguished professorship and a full professorship to sociologists Lewis and Rose Coser, respectively, something that his employer, Brandeis University, would not.[41]

Occasionally some precipitating event necessitated the long overdue promotion. In some cases the wife stoically professed not to notice this status discrepancy but worked tirelessly until there was such an overwhelming accumulation of outside honors that the institution was finally embarrassed into promoting her to full professor. The classic example, because it was so outrageous, was the case of biochemist Gerty Cori of Washington University in St. Louis. A longtime research associate, she was not promoted to a full professorship there until 1947, the same year that she shared the Nobel prize in medicine with her husband Carl Cori and an Argentine.[42] Similarly, Stanford lecturer and spouse Irmgard Flügge-Lotz, after more than a decade of outstanding work on discontinuous automatic control systems, was finally promoted to full professor of engineering mechanics, aeronautics, and astronautics, the first full professorship for a woman in Stanford's College of Engineering, in 1960. The administration did not apologize for its years of humiliation but simply claimed that her admittedly long overdue promotion had reportedly been prompted by her invitation from Moscow to be the only American woman delegate to an international congress there on automatic controls.[43]

More dramatically, at the University of Kansas the rules were deliberately liberalized in 1963 after one outrageous case finally convinced a senior administrator to have the trustees change them. As the university expanded its offerings in the 1950s, under the prevailing antinepotism rules it hired persons from elsewhere rather than local wives, regardless of how eminent they might be. This approach led to the absurdity in the late 1950s that the faculty in the psychology department were offering a course on persons with disabilities, while Beatrice Wright, an internationally renowned expert on the subject, was barred from teaching it simply because she was married to psychology professor Herbert Wright. When she appealed to then chancellor Franklin Murphy, he decided in January 1959 not to try to change the rule, which he saw as unjust in some cases but in the best interests of the university as a whole. Then, as if to demonstrate the political value of an overt precipitating event, her textbook *Physical Disability: A Psychological Approach* came out in 1960 and was used by the very class that she could not teach. Apprised of this, the new chancellor, W. Clarke Wescoe, decided that the time had come to propose a change to the board of regents, who approved his recommendation.[44]

Even the very few women who made it onto the faculties of the major universities faced the symbolic and real indignity of their continued exclusion from the campus faculty club and the several private dining groups that met there and off-campus as well. Although some university faculty clubs began to admit women in the 1950s—at the University of North Carolina in 1953 (as associate members only) and at Duke University in 1956 (with the perhaps unexpected result that Frances Campbell Brown, professor of chemistry at the Duke Woman's College, became its first woman president seven years later)— on the whole, women continued to be excluded more or less totally from these clubs.[45] This was the pattern at Cornell University, where 1,517 male charter members (staff as well as faculty) started the Statler Club in 1950, deliberately excluding, however, the substantial number of women faculty at the College of Home Economics.[46] Elsewhere women were still excluded, even when the group meeting there had invited a woman to speak (as happened to chemist Icie Macy Hoobler several times at the University of Michigan Faculty Club) and even when official faculty meetings, which the senior women faculty were entitled to attend, were held there.[47] Elizabeth Scott, a professor of statistics at Berkeley, complained that she had not been admitted to the Berkeley Men's Faculty Club for faculty meeting in the 1960s even though she was head of an important faculty committee; when the petite and resourceful Berkeley anthropologist Laura Nader was similarly rebuffed at the doorway, she first compliantly walked away but then, when no was looking, nimbly climbed through a window in the low-set building![48] At some point anthropologist Cora DuBois became the first woman to eat in the main dining room of the Harvard Faculty Club, an event that was worth mentioning in her obituary years later. But in 1969 her successor Professor Emily Vermeule still had to enter by the north door of the dining room.[49]

Among the other campus groups that continued to exclude women faculty as well as staff in these years were the untold numbers of dining clubs and private "research" or social coteries that existed on various campuses, such as the Biosystematists and the Geneticists Associated among the biologists at Berkeley, as well as the more formal but less specialized Research Clubs, first of the University of Michigan and later at Cornell. The last, started in 1918, was still going strong, with more than two hundred members, in 1965.[50] These clubs maintained their men-only status even when they met at each other's homes. This meant that the wife of the host, herself banned from a faculty position by antinepotism rules and from club membership by prejudice, was expected to provide the dinner or refreshments even if she was a scientist herself and so a potential but ineligible member of the group. She was expected to put aside her anger and play the smiling gracious wife who withdrew from the men's important conversation. Most must have gritted their teeth and done as expected, for it was considered quite a rare event, even years later, when Lauramay Dempster of Berkeley, refused—a rather daring but probably misinterpreted act of protest in a small, ingrown college town in the 1950s.[51]

At some campuses, such as Berkeley and Columbia (with its Teachers College and its College of Physicians and Surgeons), there had been enough women faculty as far back as the 1920s to justify separate men's and women's faculty clubs, and by the fifties and sixties some women were proud to have been members for several decades. Because maintaining a separate dining room or clubhouse could be expensive for the much smaller group eligible for membership in a women's faculty club, the facilities were often less lavish and extensive, and the qualifications for membership were often very broad, including, for example, research staff, librarians, the registrar, even wives and graduate students. Thus, at the University of Pennsylvania in 1948, when several women recently added to the lower echelons of the faculty (because of the growing undergraduate women's college there) desired to start their own separate women's faculty club, they invited women staff members at nearby (and in some sense at least quasi-affiliated) research units to join.[52] Similarly at Yale in 1961, when officials decided that the new graduate women's dormitory, Helen Hadley Hall, completed in 1958 (see ch. 4 above), should, by analogy with other Yale "colleges," have a group of senior "fellows" who on occasion dined and perhaps intermingled with some of the students, this honor was deemed suitable for the assorted women staff and faculty dispersed among the departments, museums, and other research units at Yale, as well as for wives of important Yale men. Although there was no difficulty in keeping the group suitably small, the result was far from analogous to that at the far more lavish undergraduate colleges. In effect the group had to huddle around a small snack bar in a dimly lit corner of the dormitory's living room, an even greater reflection of the peripheral status of women at Yale than anyone had intended.[53]

Yet even such broad membership eligibility did not assure the continued viability of some of these separate women's faculty clubs. Rising costs, such as increased wages resulting from unionization and mandatory social security payments starting in the early fifties, led to a cut in the size of the staff and thus greatly eroded many of the amenities of club life. At Columbia University, although the Men's Faculty Club had obtained a liquor license in 1942, the Women's Faculty Club never did, thus undercutting the number of mixed groups that wanted to dine there. By the late sixties several of the older women's faculty clubs were facing dwindling membership levels and financial difficulties. They were ripe for mergers with other groups on campus.[54]

At some universities the women fought their exclusion by forming women-only clubs that did not try to follow the "faculty club" model. Among these was the Women's Research Club (WRC) of the University of Michigan, which had started as early as 1902, when Lydia DeWitt and eight other women were outraged at being excluded from the men-only Research Club. DeWitt's group is still in existence, with the simple format of listening monthly throughout the academic year to women speakers discuss their research. When at some point it found itself with an excess of funds, the WRC formed a small loan fund for women graduate students or other impoverished researchers that has served its own useful purpose well over the years.[55] On other campuses, such as Cornell's, there were and are active chapters of women's fraternities, such as Sigma Delta Epsilon, limited to scientists, that serve, among other purposes, much the same function.[56]

Although such separatism, or a club of one's own, was one answer to overt exclusion from men's groups, it was only a partial solution, as many members were fully aware. Although such clubs made it easier for women to meet each other, discuss scientific or other matters, and perhaps form political coalitions on university issues, the main social events for faculty, such as the reception for newcomers, annual banquets, dinners for visitors, and the like, continued to take place at the men's faculty club. What was perhaps even more important, a great deal of informal politicking or decision making took place there over meals or drinks. Thus the continuing exclusion of women from such clubs was a sign of their low level of participation in the whole administrative side of campus life. They were totally out of the running for any position higher than dean of women (or where that no longer existed, assistant dean of students).[57]

In fact women were rather rare at even the lowest step of university administration, the chairmanship of a department, in major universities before 1970. Cecelia Payne-Gaposchkin was one of the very few woman scientists to head a department at a prestigious university when she served briefly as chair of Harvard's astronomy in the 1950s, and political scientist Clara Penniman was probably the first woman to chair her department at Wisconsin, from 1963 to 1966. Other inroads into administration were made by astronomy professor Helen Dodson Prince at Michigan, who served as associate director of the McMath-Hurlbert Observatory for several years after 1957, and by statistician Elizabeth Scott of the University of California at Berkeley, who was an assistant dean of the College of Letters and Science from 1965 to 1967 and then chair of her department from 1968 to 1973. Meanwhile, in the mid-sixties at the University of Wisconsin, economist Barbara Newell became a rising star in even higher academic administration, serving as assistant to the chancellor (Robben Fleming) from 1965 to 1967. When he became president of the University of Michigan in 1967, she followed as, successively, assistant to the president, 1967–68; acting vice president for student affairs in the difficult years 1968–70; and special assistant to the president, 1970–71.[58]

If such a nearly total exclusionary system was the norm at most major American universities in the fifties and sixties, it was thus somewhat unfortunate that it was one of the faculty women at the atypical Penn State who undertook in the early sixties an authoritative book on [all of] academic women. At first Jessie Bernard had planned a somewhat autobiographical venture, but over time it grew longer and became more of a general treatise. The result was a curious amalgam because she found it hard to see discrimination and feel angry at injustice when personally she had been relatively well treated at Penn State for over a decade.[59] Ironically, Bernard's interest in explaining her own experiences as an academic woman and exploring those of the group as a whole had been triggered by her very unhappy year as the first female visiting professor of sociology at Princeton University in 1959–60. There she was treated so rudely by the students (men only then), who hissed or booed her and even refused to come to class, that, stalwartly refusing to take any of their hostility personally, she tried instead to understand it sociologically. How did the sex of the professor matter to a class of men used to male faculty? Why should they feel so cheated just because she was a woman? Could a woman ever be, or be taken seriously as, a "man of knowledge"? Or could women be only "teachers" (as was common in elementary language instruction, perhaps even at Princeton), merely passing on noncontroversial tools and skills to an uncritical assemblage?[60] By 1961, after discussions with some women faculty at Penn State, including Margaret Matson, also a wife and professor of sociology there, Bernard had determined to write a more general survey of the whole emerging but still fragmentary subject of academic women. After all, as she mentioned in the preface, she had been an academic woman all her life—graduate student, faculty wife, mother, collaborator, professor, sociologist—participating, observing, and now analyzing the multiple roles involved.[61]

What did capture immediate and lasting attention was Bernard's cautious political stance and triumphantly righteous tone in the book's preface and elsewhere. There she claimed (falsely) that because she herself had never been discriminated against, she could be an objective analyst, able and willing to examine all sides of the issue, rather than a mere special pleader for her own downtrodden group:

My own biases, I suppose, are clear enough not to require specific articulation. I object to the current approach to high-level personnel in terms of "manpower utilization" or "conservation" or of "using" or "losing" brainpower. I "believe" in women, and I applaud every attempt to institutionalise ways to help them achieve what they want and are able to achieve, but I do not believe that their contribution as academic women is necessarily any better or more socially useful in every case than their contribution as wives, mothers and community leaders. . . .

I write, finally, as a woman—not, however, as a militant feminist, to vindicate the rights of women. I have never, so far as I know, experienced professional discrimination from my colleagues because of my sex (although I was once chased out of the sacred precincts of a faculty club at a great university when I inadvertently stepped over the invisible line). I have, rather, been treated with extraordinary chivalry by administrators, colleagues, and confrères who have made allowances for the demands of maternity—three times—sustained me in bereavement, and rewarded me with many official honors, including the vice-presidency of the American Sociological Society, the presidency of the Eastern Sociological Society and of the Society for the Study of Social Problems, and the secretary-treasurership of the last-named society, as well as membership on im-

portant committees, including the research committee of the International Sociological Association.

 I write, therefore, without bitterness or rancor, knowing full well that a bland
approach is less likely to be appreciated than a rapier thrust. I have enormous
pride in my profession; the people who staff our colleges and universities, women
as well as men, for all their fairly obvious defects . . . rank high in my opinion . . .
and I concur in the popular polls which have assigned them great prestige.[62]

 As if these protestations were not enough, Bernard also fortified the book
with a foreword and an introduction by two prominent men—Ben Euwema,
her longtime dean at Penn State, and David Riesman, the Harvard
sociologist—who would vouch further for her credibility. But this strategy
backfired, for they both undercut her position by proving feminist enough to
say outright what she could or dared not: that there was a great deal of
discrimination against women in academia, that they had seen it, and that
Bernard should have been more of a "crusader" for academic women. Thus,
even before the reader got to the actual text of the book, he or she would
be aware that it might be more cautious and beholden to the powers that
be than actual conditions warranted. Yet, since it was the only book-length
analysis of the subject, the only comprehensive survey by a competent sociologist, the book was read by many young women contemplating an academic
career.[63]

 But rather than proving an inspirational guide to the challenging world of
academia, Bernard blamed the women themselves for not participating more.
She explained the decline in the percentage of women on academic faculties
since the 1930s (omitting World War II) as the result of a shortage of trained
women. First, she interpreted contemporary reports on financial aid for graduate students to mean that women tended, for whatever reasons, to drop out of
graduate school voluntarily;[64] and second, she hypothesized that the declining
proportion of women on the faculty of the women's colleges was the result of
the colleges' new insistence on doctorates, as well as increased opportunities
for women doctorates outside the women's colleges.[65] With hindsight one can
see that she was too eager to conclude that "certainly there is no reason to
suppose that the decrease reflects a discrimination against women."[66]

 From this scant evidence Bernard extended her view to all colleges and
universities: the reason for the overall decline was a shortage of available wom-
en, that is, unless there was a desire to upgrade the prestige of the school by
hiring men preferentially. That was a real possibility—and one that Mirra
Komarovsky, a sociologist at Barnard College, saw as being very likely the
case[67]—but one that Bernard had immediately dismissed, for as she reasoned,
it was "functional": "That men might be thought to bring prestige to a profession is in itself an interesting sociological phenomenon but one that cannot be

pursued farther here except to note that if it is true that men do upgrade a faculty, then their selection is functional."[68]

Evidently, when preferences were "functional" they were justifiable, acceptable, and inevitable, or at least unchallengeable by sociologists of the time. Thus, despite her many experiences as an academic woman, as a trained sociologist she was unable in 1964 to find and charge that active discrimination was in effect rampant in academia. Although the evidence was there—she mentioned several times that though the women were brighter than the men, they taught more, published less, were paid less, and were honored less than the men—she continued to believe that it was all justified: for unknown reasons, the women chose their inferior status and gravitated toward heavy teaching loads and low-prestige subjects.[69] Bernard evidently believed that because those women who did persevere to find academic positions adapted their behavior in order to avoid most outright rejection, discrimination was not happening. Bernard thus remains an interesting example of a conflicted participant, a little too grateful for her atypical situation to follow her evidence where it might have led.

Yet it was typical of the mid-sixties that these somewhat political difficulties with Bernard's interpretation were barely noticed or ever emphasized. The book won two prizes, one from the Pennsylvania State University Press and another, of fifteen hundred dollars, from Delta Kappa Gamma, a women's educational fraternity, which sums, along with the royalties, Jessie Bernard donated to the AAUW. Those who might have argued with her interpretation (e.g., perhaps, Ruth Eckert, Alice Rossi, or Betty Friedan) did not review the book, and those who did (a far less well known group) chose the more cautious path of applauding the appearance of any book on the subject as a sign of its evident importance and praising her cautious and objective tone (a great relief, Komarovsky said in *Social Forces,* from the "shrillness" of other recent analyses, perhaps a reference to Betty Friedan). Those few who noticed some interpretational problems cut short any adverse criticisms, lest even this undercut the women's cause still further. Evidently, the women's movement was still so weak and unformed at the time that there was not yet any clear, consistent viewpoint on academic or professional women with which to confront Bernard's eclectic, muddled interpretation.

Thus women faculty at the prestigious schools continued to be few and isolated and at best marginal in status and power. As sociologists Caplow and McGee had depicted, in the fifties they were easily dismissed as mere "exceptions." The whole "womanpower" message of the federal government was remote and failed to penetrate in any meaningful way. Although faculty women at leading universities did not complain publicly, a few of the captive spouses were working and plotting to improve their own status in particular and the system in general. Their strategies tended toward maximizing outside recognition as a way of bringing pressure upon the local institution to negotiate a private exception or, as in Josephine Mitchell's or Beatrice Wright's case, to try to change the system. Yet this was difficult, since, as Edwin Boring had put it in 1951, the whole culture was involved. A lot of attitudes would have to change before academic practices could be coerced into something closer to fairness. This was something on which some of the unhappy research associates to whom we turn next were already working.

When in the late 1940s Smithsonian herpetologist Doris Mable Cochran, pictured here among her collections, learned that recently hired men were being paid the same salary as she, who had been on the staff since 1919, she began a series of protests to her superiors.

Photograph courtesy of Smithsonian Institution Archives, RU 95, Photographs.

After a front-page article in the *Wall Street Journal* in 1958 complained that women were bad risks for federal fellowships because they would marry and drop out of graduate school, a letter to the editor reminded readers of a counterexample: applied mathematician Dr. Cathleen Morawetz, married with four children, pictured here, who was still at work at the Courant Institute of New York University.

Photograph courtesy of Cathleen Morawetz.

Although some professors refused to have any women graduate students, a few, like mathematician Lipman Bers, trained several, including the three pictured here (left to right, Lesley Sibner, Linda Keen, and Tillie Milnor).

Photograph from Donald J. Albers et al., More Mathematical People: Contemporary Conversations *(Boston: Harcourt Brace Jovanovich, 1990), 12.*

Although physicist Ann Chamberlain Birge, Radcliffe Ph.D. 1951, may *seem* to be recording cosmic ray data for her husband Robert's dissertation, this was a deliberate distortion introduced by the photographer, for it was her dissertation, and she had built the apparatus. An additional irony was that the photograph was later published in a 1956 book purporting to encourage women to take up graduate study.

Photograph reprinted by permission of the Radcliffe College Archives. Information on the story behind the picture is from Monica Healea to Mrs. Dunkelbarger, January 17, 1955, Alumnae in Science File, Special Collections, Vassar College Library.

Marguerite Thomas Williams, the first black woman Ph.D. in geology (Catholic University of America, 1947), shows off her new doctoral gown on her family's back steps in Washington, D.C. *Photograph from Women's Bureau,* The Outlook for Women in Science, *WB Bulletin 223-1 (1949), 59.*

Biochemists Gerty and Carl Cori, co-winners of the 1947 Nobel prize for medicine or physiology, worked together for decades and trained many other women and men in their laboratory at the Washington University School of Medicine in St. Louis.

Photograph reprinted by permission of the Washington University School of Medicine Library and Biomedical Communications Center.

Opposite, top: All of the five women at the NSF-sponsored summer institute for college chemistry teachers at Rensselaer Polytechnic Institute in 1965 were nuns. Pictured, from left to right, are Sr. M. Gertrude of Sacred Heart College, Wichita, Kansas; Sr. M. Angelique, Mount St. Mary's College, Hookset, New Hampshire; Sr. Lean, Seton Hill College, Greensburg, Pennsylvania; and Sr. M. Emily, Regis College, Weston, Massachusetts. The fifth, Sr. M. Phileman of Salve Regina College, Newport, Rhode Island, was not present.

Photograph reprinted by permission of the Regis College Archives, Weston, Massachusetts.

Opposite, bottom: Geneticist Barbara McClintock, long at the Carnegie Institution of Washington's Cold Spring Harbor Laboratory on Long Island, pictured here at work in 1947, was the most honored American woman scientist of the twentieth century.

Photograph from Women's Bureau, The Outlook for Women in the Biological Sciences, *WB 223-3 (1949), 4.*

APRIL 9
1951

After Katharine Burr Blodgett of General Electric, known best for her work on nonreflecting films, won the American Chemical Society's Garvan Medal for women, she appeared on the cover of the *Chemical & Engineering News.*

Reprinted with permission from Chemical & Engineering News, *April 9, 1951, 29(15).* © *1995 American Chemical Society.*

7

Resentful Research Associates

Marriage and Marginality

Because burgeoning research budgets in the 1950s and 1960s in biological, social, and other fields enabled faculty members to hire trained assistants, for women scientists the expansion at major universities was in the swelling ranks of the research staffs. Bright women, including married ones and even spouses, with or without doctorates, were hired for these positions and at times they were preferred, which helped to make such positions in many respects a new kind of "women's work." Although one might have thought that employers as solicitous of employee welfare as universities presumably were, with their protective antinepotism rules, might also have prevented one spouse from working directly under another, lest favoritism or abuses arise, in practice these research positions were generally private fiefdoms. Thus a research staff member could be employed within her husband's department or outside all departments, in the growing sector of affiliated research units, which operated on funds from outside grants or other "soft money." This two-tier labor practice meant that certain secondary parts of the university might, in much the same way as small colleges next to large universities with antinepotism rules, be able to take advantage of captive spouses in the area and employ better-trained, more talented persons at lower salaries than they might have otherwise. But because these employees' status was subordinate and marginal and initially assumed to be temporary, the positions also presented certain disadvantages. Such workers were essential, yet they were also liable to a certain exploitation, especially over time. These jobs were designed to be invisible and to gain the worker no professional recognition. Thus, even though such a worker's efforts added to the reputation of the professor, the project, the university, and even the funding agency, his or her own status remained precarious, for even after many years or decades the university usually was not obligated to offer the staff member any tenure or long-term security. In fact, all the new federal and other "soft money" only increased the prevailing sex-segregation in academic science (with men on the faculty and women as their assistants). The opening in the fifties and sixties of some new universities and medical schools released the mounting pressure a bit by allocating a few faculty positions for women as well as men, but the structural problem remained.

Many women in the fifties were glad and even proud to be a part of a prestigious university in any way at all, even if they had a low or marginal personal status within it.[1] Others took such positions as the best ones available locally. Yet over time some research associates grew resentful, and a handful managed to salvage their own careers and reputations by mustering certain leverage and negotiating private concessions that put them on the faculty. One angry sociologist did research on the issue in the mid-sixties and found that dissatisfied research associates in fact published more than the uncomplaining ones. Alice Rossi, jolted by direct exploitation, in the early sixties became a founder of the women's liberation movement when she started to question and rethink the whole deployment of men's and women's roles in the university. By the end of the sixties, "consciousness" was rising among the rapidly growing ranks of disgruntled lecturers and research associates that something was seriously wrong with the entire system. Several of them would be among the leaders working to change it.

Because the basic causes for the expansion in the number of such temporary and especially research positions were the ever-increasing pressure for faculty to publish and the spiraling amounts of federal and other money being spent on research in the fifties and sixties, most of the lecturers and research associates tended to be concentrated at a relatively few large research institutions. In fact, a 1953–54 NSF survey of college and university scientists found that these ranks were so rare beyond the 190 doctorate-producing universities (of which 180 responded) that it did not need to include the more than 900 other colleges and universities in its "comprehensive" survey.[2] Another NSF survey, in 1961, found that just twenty leading universities accounted for 50 percent of the research staff (in full-time equivalents) at 306 institutions.[3] Moreover, certain parts of the university—the medical school, the oceanographic institute, the opinion research center, the agricultural experiment station, and various other "research units"—employed most of the scientific staff. The problem of how to incorporate this growing underclass into the university structure had been one of the postwar issues facing the major universities.

Although one might suspect that special personnel practices for the non-faculty research staff evolved slowly over decades perhaps as the staff members stayed on and some were felt to be worthy of honors and rewards, according to Carlos Kruytbosch, a sociologist at the social sciences project of the Space Science Laboratory at Berkeley and head of Berkeley's Academic Research and Professional Association in the 1960s, this underclass developed suddenly there in the early 1950s. Apparently there had been in those years such a flood of outside, especially military, funding for high-priority research projects and such a lack of doctoral staff able to perform such work that the contractors had been able to offer salaries far in excess of what Berkeley was paying its own junior faculty. The result was a rash of defections to the higher-paying research staff. In an effort to exert its control over the operations of such research projects, the Berkeley central administration in 1952 set up specific job titles, fixed salary scales, and introduced other restrictions on the research staff that ensured that these positions were, and would remain, clearly subordinate to the faculty, especially those who were members of the academic senate. Once one campus had worked out this defense against the encroaching demands of well-funded contractors, the practice spread, and by the 1960s this division of labor, status, power, and security was almost universally accepted at American research universities, as evidenced in the ever larger personnel manuals spelling out faculty and staff rights and privileges.[4]

There are, however, scant data on this elusive class of research workers. When John Parrish wrote to 460 departments in March 1960 in search of data on faculty rank and sex, 22 departments in the physical and biological sciences volunteered additional data about their research associates, most of whom held doctorates. Of the 155 persons reported on, 36, or 23.2 percent, were women.[5] A 1966 survey at Berkeley found that 23.92 percent of its research staff were women.[6] A 1970 report on Berkeley added that compared with male research staff, most of the women were ranked lower and paid less, even though they supervised 5.1 graduate students (compared with 3.2 supervised by men).[7] Data in the NSF's National Register indicate that most of the research associates who answered NSF questionnaires in the 1960s were in the fields of physics (including astronomy), biological sciences, and chemistry (see table 7.1).[8] Their numbers also grew sharply in the short period between 1966 and 1968 and then stabilized in 1970, as workers seeking faculty positions were increasingly stockpiled as temporary research staff.

There is also evidence that the salaries were kept as low as possible in these positions. A contemporary study showed that $2,450, or 27.6 percent, of the "direct costs" of the typical NSF grant ($13,741 in fiscal year 1956) went to the research associate's salary. Such an assistant must have been either employed part-time or supported by a second grant, for in 1962, the first year the National Register's biennial manpower report included salaries by academic rank, it reported that the median annual salary for research associates was $8,000.[9] The 1966 survey at Berkeley found that 80 percent of the single women but only 41 percent of the married women were employed full-time.[10] Such low and partial salaries could also mean that the research associates were being deprived of fringe benefits, for at many institutions one had to be employed at least half-time to get them. Thus, years later the former research associate, perhaps

Table 7.1.
Research Associates, by Field, 1966–1970

Field	1964	1966	1968	1970
Physics	56	63	198	221
Biological sciences	93	51	212	183
Chemistry	46	35	194	183
Psychology	5	2	39	33
Mathematics	9	32	23	27
Earth and marine sciences	3	7	32	27
Agricultural sciences	3	3	6	20
Economics	6	6	16	19
Sociology	3	3	27	14
Meteorology / atmospheric and space sciences	2	5	10	13
Computer science	—	—	3	12
Political science	—	—	10	9
Statistics	—	3	7	4
Linguistics	3	1	1	4
Anthropology	—	4	2	3
Other	6	4	—	—
Total	235	219	780	772

Source: NSF, *American Science Manpower, 1964,* 106–7; *1966,* 171–72; *1968,* 202–3; *1970,* 189–90.

divorced by then, might have a very small pension, if any, to live on. Sociologist Helen Hughes reported how her monthly pension check of $19.58 always reminded her of her seventeen years of part-time work for the *American Journal of Sociology* at the University of Chicago in the 1940s and 1950s.[11]

In a few cases the practice of hiring spouses for research and lecturer positions extended to the husbands of women faculty, such as Sergei Gaposchkin, who was long a research astronomer at the Harvard College Observatory, and Gerald Hassler, a researcher at UCLA, where his wife Mildred Mathias was climbing the faculty ranks in botany.[12] Far more commonly the spouse who held the invisible staff or subfaculty position was the wife, who was usually younger and all too well socialized for such a subordinate role. Some professors even preferred women staff. In fact, if one can believe some of the growing lore of the time on how to run a laboratory efficiently, a strong motivation for getting a grant in the first place was to be able to hire a highly feminized staff of assistants, who would run the experiments, do the library work, handle the paperwork, and supervise the staff for the increasingly harassed faculty "researcher." One particularly explicit example of this revealing genre was E. C. Pollard's "How to Remain in the Laboratory Though Head of a Department," which appeared in *Science* in 1964. Almost a caricature of hints on how to remain scientifically productive while spending ever more time on budgets, committees, and trips, though deadly serious at the time, this article offered such "pieces of wisdom" as the following: "You must have a laboratory assistant, preferably female. The purpose of this assistant is to require daily instruction about what to do. Thus you are inescapably forced to plan for the operation of another pair of hands. A female is better because she will not operate quite so readily on her own, and this is exactly what you want. Any head of a department who lets his laboratory assistant go has, by that action, resigned himself from future laboratory work."[13] A postdoctoral fellow (presumed to be male), by comparison, could be quite disruptive in the laboratory, for he was there full-time, thought for himself, and could be quite independent until the time came to seek his next position elsewhere. All in all, Pollard, chairman of the department of biophysics at Penn State and former head (1954–60) of the department at Yale, felt that the boss's status was more secure with bright and obedient women assistants.[14]

At most research universities there were by the 1960s several ranks for research personnel. Usually these were threefold: *research fellow,* for the recent doctorate, especially the increasing numbers of postdoctoral fellows, who would visit for a year or two and then move on; *research associate,* for longer-term personnel tied to a particular project or faculty member; and *senior research associate,* with, one hopes, somewhat better benefits and privileges, for those who stayed for many years, perhaps ten or more, and worked with distinction. But few universities had any provision for any other advancement. Among those that did, after a fashion, was the University of Chicago, which appended to its research associates' titles a kind of parafaculty rank. Thus the catalog included such titles as "research associate," "research associate (assistant professor)," "research associate (associate professor)," and, very rarely, "research associate (professor)."[15] Since some of the research associates were married to well-paid faculty members (and so presumably unpaid), their titles would combine the "voluntary" and the dual rank, such as the title "voluntary research associate in microbiology (associate professor)" for Lucy Graves Taliaferro, wife of the department chairman. Other universities, like as the University of Maryland, achieved the same effect by calling such women as Mary Juhn in poultry science and Mary Shorb in poultry husbandry simply "research professor."[16]

As the woman research associate grew older, she might take on a bit of authority and move into a kind of "shadow" or vicarious (but still subordinate) management role, actually running the laboratory without seeming to. Ruth Hubbard, herself a longtime research associate at Harvard and second wife of the eminent biologist George Wald there, later described the scientific laboratory and its resident staff as a "patriarchal household."

In this household, as in others, roles are not evenly shared between women and men. The males, by and large, are (like sons) there to be trained in order to leave for the world in which they will establish their own households. The women are there largely to help the men—as secretaries, dishwashers, technicians, assistants. If they fit in well, it is anticipated that they will remain at least until they marry and have children. Many laboratories also have a wife-mother figure, usually an older secretary or technician whose loyalty to the professor and the laboratory is what keeps both going. Occasionally she is a scientist who, after surveying the scene and seeing how few desireable and autonomous positions there are for women, has made a semipermanent niche for herself within the patriarchal home. Sometimes she is the professor's wife, who combines a nurturing, low-level managerial role at home with one in the laboratory. In European laboratories, the wifely role is sometimes relegated to a male of lower social class who has no chance to advance in the larger society and makes the professor's laboratory his permanent home.[17]

Such helpful subordinates were not always portrayed so favorably. In his 1964 reflections on the scientific career, an older biomedical scientist, attempting to be witty and clever, provided a distinctly negative view of competent female assistants. After outlining the various types of personalities (all men) he had met in long years of laboratory life in Europe, the United States, and Canada, physiologist Hans Selye of the University of Montreal described in a chapter presumably to be read by the young, entitled "Who Should Do Research?" one female type:

> The desiccated-laboratory female. She is the bitter, hostile, bossy and unimaginative female counterpart of the Cold Fish [previously described]. Usually a technician, she rarely gets past the B.Sc., or at most, the M.Sc. degree, but she may be a Ph.D., less commonly an M.D. In any event, she assumes a dominant position in her own group, has very little understanding of human frailties among her subordinates and almost invariably falls in love with her immediate boss. She may be very useful in performing exacting, dull jobs herself and in enforcing discipline upon others, but tends to create more tension and dissatisfaction than the results warrant. Some women make excellent scientists, but this type never does.[18]

Here a rather capable scientist is being blamed for wanting a little more authority and credit than her boss or the system would allow her. She sounds like the highly talented but underemployed and therefore frustrated scientist one might expect to find in a field that had as few women in faculty ranks as John Parrish had found earlier among physiologists: even the best women in the field were blamed for presumed personality defects.

Although the research associateship offered scientists with or without a doctorate a chance to participate in cutting-edge research without a lot of the tensions and uncertainties of a tenure-track position, it did not offer the protections of one either. A research associate who had been hired because of a personal relationship with her employer, might find herself on her own if such bonds broke. If there were a divorce, for example, as happened in the case of Esther and Joshua Lederberg at Stanford University in the late 1960s, she could suddenly find herself in mid-career looking for a new lab or a new employer, possibly on short notice.[19] When the faculty member grew older and faced retirement, the research associate would have to find another job at midlife. Astrophysicist Charlotte Moore Sitterly, who had worked for Princeton professor Henry Norris Russell for twenty-five years, luckily was offered a suitable job at the NBS in 1945.[20] Usually such older professors knew in advance that they would have to close down their laboratory, but it could take years, as immunochemist Dan Campbell of Cal Tech found in the late 1960s, to locate suitable positions for his longtime staff.[21] Rarely, the PI, or principal investigator, died suddenly, leaving the research associate to clear away everything as well as find herself a new location. But regardless of how well qualified she was or of how closely identified with the professor's work she was, such a research associate would never inherit the faculty position.

Because the research associates and lecturers were concentrated at a relatively few research universities, one can list many, though certainly not all, of the nation's important women in these ranks in sixteen sciences by collecting the names of those included in the catalogs of the colleges of arts and sciences of the twenty most prestigious universities for the academic year 1959–60 (see table 7.2). These women certainly constitute an eminent group, with several Guggenheim fellows, other prizewinners, and future presidents of professional societies among them. They were by far most numerous in the biomedical fields, especially biology, followed by sociological and "other" specialties, including "human development" at the University of Chicago and oceanography at Scripps Institution of Oceanography (SIO), which in 1959–60 was still a part of UCLA. Two of Princeton's three women research associates were connected with its world-famous Office of Population Research, established in 1936 with assistance from the Rockefeller Foundation and the Milbank Memorial Fund. From there and from her home in Hyattsville, Maryland, Irene B. Taeuber published the *Population Index,* the official journal of the Population Association of America, as well as other works, and strove to make the field of demography statistically rigorous around the world.[22] Cal Tech's several women in biology and chemistry contributed importantly behind the scenes to the winning of several Nobel prizes by its illustrious faculty members in the 1950s, 1960s, and 1970s. Perhaps the most eminent collectively (and so the most underemployed) were the women at Harvard, several of whom achieved fame and an independent reputation for distinguished research. Among these were vertebrate paleontologist Tilly Edinger of the Museum of Comparative Zoology (MCZ), a refugee from Nazi Germany, who worked out the evolution of

Table 7.2.
Notable Women Science Subfaculty and Research Staff at Thirteen Leading Universities, by Field, 1960

Institution	Zoology	Oceanography	Anthropology[a]	Other Earth Science	Sociology	Chemistry	Astronomy
Harvard	Deichmann (C) Hubbard (RA) Lawrence (C) Turner (RA)		Whiting (L) Gimbutas (RF) Stone (RA) Proskouriakoff (RF)	Paleontology Edinger (RA)	Kluckhohn (L)	Dory (RA) Fieser (RF)	Wright (L) Dieter (RA)
Columbia	Barth (RA) Hughes-Schrader (RA) Sager (RA) McClintock (RA)	Tharp (RA)	Bunzel (L) Mead (Adj)	Geology Hamilton (RA)	Kendall (RA)		
Chicago	Overton (RA)			Meteorology Bradbury (RA) Geography Larimore (RA)	Star (RA)		Burbridge (RA)
Yale	Pickford (RA)	Deevey (RA)					Hoffleit (RA)
UCLA/Scripps		Orton (ARS) Sweeney (ARS) Robinson (ARS)		Paleontology Loeblisch (L)	Sheldon (RA)		
UC—Berkeley							
Pennsylvania					Hottel (L)	Jouille (A)	
Cal Tech	Vogt (SRF)					Caserio (SRA) Garvey (SRA)	
MIT				Meteorology Austin (RA)			
Cornell			Leighton (SRA)				
Other		Washington Henry (RA) Rhode Island Fish (RS)					Carpenter (RA)

Institution	Mathematics and Statistics	Psychology	Botany	Biochemistry	Anatomy	Other
Harvard	von Mises (RF)			Cohen (RA)		Criminology Glueck (RA) Education Roe (L)
Columbia				Briscoe (A)		
Chicago	Lange (L)					Human development Kitazawa (A)
Yale					Van Wagenen (L)	Microbiology Vishniac (L/RA)
UCLA/Scripps						
UC—Berkeley		Honzik (L) Jones (L)	Cave (RA)			
Pennsylvania					Nelson (L)	Political science Brownlee (A) Physics Ralph (A)
Cal Tech	Todd (RA)					
MIT						Engineering Brazier (A)
Cornell						
Other		Minnesota Senders (L)	Michigan Kanouse (C)			Princeton/demography Taeuber (RA)

Sources: University catalogs; *AMWS*, various editions.

Note: A = associate; Adj = adjunct faculty; ARS = assistant research scientist; C = curator; L = lecturer; RA = research associate or assistant; RF = research fellow; RS = research scientist; SRA = senior research associate.

[a]Includes Archaeology.

horse brains in several definitive works;[23] Russian émigré Tatiana Pros-kouriakoff of Harvard's Peabody Museum of Ethnology and Archaeology, who, with funding for a time from the Carnegie Institution of Washington, decoded the pictorial language of the Mayan Indians;[24] Eleanor Glueck, who with her husband, a member of the Law School faculty, published numerous works on juvenile delinquency and the life patterns of criminals;[25] biochemist Ruth Hubbard, who worked on vision independently and with her husband George Wald;[26] malacologist Ruth Turner, also at the MCZ, an expert on wood-boring mollusks;[27] and psychologist Anne Roe, to be discussed shortly.

The field of oceanography saw tremendous developments in the twenty-five years between the late 1940s and the early 1970s largely because of considerable funding by the Navy. Although women were nearly totally banned from the ships that collected the data, two women were able to play an important part in making sense of them back on shore. On the East Coast research associate Marie Tharp worked with geophysicist (initially a graduate student) Bruce Heezen of Columbia University's Lamont-Doherty Geological Observatory on mapping the floor of the Atlantic Ocean. When in 1957 their preliminary map showed a continuous north-south ridge and rift valley down the middle of the ocean, it provided pivotal evidence for the emerging theory of "plate tectonics." Thereafter they continued to work together in a kind of inspired collaboration until his death at age fifty-three in 1977.[28] On the West Coast, at the SIO, near San Diego, long a part of the University of California, former schoolteacher Margaret K. Robinson spent twenty-six years recording data on the world's oceans. In 1946, at age forty, she was hired at Scripps as a "senior clerk" to do paperwork. In particular she was to prepare reports and maps from the immense quantity of data collected at sea by a new scientific instrument (the "bathythermograph") invented in 1938 to record ocean temperatures at various depths. Robinson earned a master's degree in oceanography, despite discouragement from the institution's director, who felt women had no need for such a degree, and then spent more then two decades at the SIO in a progression of staff positions ("senior engineering aide," "assistant oceanogra-pher," "associate specialist," and finally "specialist"), assisted by a staff of other women, in an arrangement reminiscent of Edward C. Pickering's "harem" of women "computers" and astronomers at Harvard at the turn of the century. Before her retirement in 1973, she and "her girls" had prepared atlases of temperature distribution for the North Pacific, the Gulf of Mexico, the Carib-bean, the Red Sea, the Black Sea, and the Mediterranean. In 1961 she spent six months in Bangkok teaching the staff of the Thai Hydrographic Office how to chart their waters.[29]

In the social sciences the founding of the National Institute of Mental Health (NIMH) in the late 1940s opened the door to funding of various short- and long-term projects. One that relied directly on such grants was the Institute of Human Development at the University of California at Berkeley. It grew out of three earlier projects started in the late twenties and early thirties to collect over a number of years a variety of data on the growth and adjustment of several hundred essentially normal local children. Since such a longitudinal study did not generate enough publications in specific disciplines to interest rising young faculty, whose tenure lines were in departments, the institute had to rely on staff psychologists, including many women, such as Mary Cover Jones, Nancy Bayley, and Dorothy Eichorn, as well as professor Jean Mac-farlane.[30]

Similarly, the whole field of opinion polling, or "survey research," grew up in the postwar era, as various loosely organized and funded units had by 1950 affiliated in some way with a major research university, as the Institute for Social Research at the University of Michigan, the Bureau of Applied Social Research at Columbia University, and the National Opinion Research Center (NORC) at the University of Chicago. Although they were "independent" in the sense that they had their own separate boards of directors, such organizations were often subsidized by the universities (e.g., given rent-free space or the services of a librarian) and hoped to capitalize on joint appointments in particular academic departments and the energies of bright graduate students and wives. The new jobs created and supported by outside contracts were often those of the middle managers or "study directors" (or even "associate study directors"). This was where the women were found, such as Shirley A. Star, first at NORC and later at the Bureau of Applied Social Research at Columbia University, frantically helping to hold the place together, hiring people, arranging for machine time, and seeing projects negotiated or initiated by others through to completion.[31] But even success in such work, challenging and essential as it was, did not lead to promotion to the top levels. Instead one remained a hired hand or staff member in a professional kind of limbo and part of the research underclass at the prestigious university. Patricia Kendall (Lazarsfeld) was therefore fortunate when in 1965, after twenty-three years as a survey researcher at Columbia's Bureau, she became a full professor of sociology at Queens College.[32]

Yet indispensable as some of these research associates might have seemed at the time, and as they were occasionally acknowledged to be, all too often their share in the public credit was less than one might have expected. (Unfortunately, there does not seem to be a way to compare the relative fates of male and female research associates in this matter.) Marguerite Vogt, for example, was a close collaborator of Renato Dulbecco of Cal Tech and the Salk Institute from 1953 until 1972. When in the early 1950s the U.S. Employment Service required evidence that, though she was an alien, her services were urgently needed in America, Cal Tech officials anxiously assured governmental officials that there was "no one else in the United States who is available for this important project on poliomyelitis sponsored by the National Foundation" and that the work would "be of substantial benefit prospectively to the welfare of the United States."[33] Yet word of her importance apparently did not get to Stockholm, for when Dulbecco won the Nobel prize in 1975, he shared it with two former male students but not with Vogt. (This lack of Nobel recognition will be explored more fully in ch. 14.)

One kind of restriction that rankled research staff a great deal, even more, according to a 1957 survey at Berkeley, than the lack of sabbaticals, laboratory space, graduate students, and even parking privileges, was the ban against applying for outside research grants in one's own name.[34] From a staff member's point of view, this prohibition perpetrated the notion that one was merely a "hired hand," however indispensable, on someone else's project. This was tolerable until, as happened with bright researchers, one got ideas of one's own and wished to pursue them. Then one had just two possibilities: to find an agreeable "ghost" or "shadow" PI who would sign the necessary forms (and expect to be paid a modest sum for his nominal supervision) or to induce someone who already had a grant for a related project to let one bootleg some time for his or her own project.[35] Both alternatives involved difficulties, but a few women managed to find a way around them and become, despite their "research associate" status, PIs and in time even faculty members.

A very small number of determined women managed to obtain outside grants in their own name despite university restrictions. Dorrit Hoffleit, an astronomer at the Harvard College Observatory, in fiscal year 1954 received a two-year NSF grant of $5,500 for a study of "Variable Stars in the Milky Way."[36] And Elizabeth Ralph, a research associate in the department of physics and the University Museum of the University of Pennsylvania, in fiscal year 1957 received a two-year grant of $20,000 from the earth sciences program for work on the "Half-life of Carbon-14."[37] When marine biologist Ruth Turner of Harvard's MCZ discovered how little funding it had for her work on mollusks, she was able to convince the ONR to support it for twenty years. (Years later she still very grateful to Dixy Lee Ray, rather than anyone at Harvard, for introducing her to the Navy program officers.)[38]

Even more unusual than the cases of these three research associates turned PIs was the case of Hilda Geiringer von Mises, a highly accomplished applied mathematician who had fled the Nazis in the 1930s and was by the 1950s chairman of the mathematics department at Wheaton College, a woman's college about forty miles south of Boston, and the wife and (on weekends and in the summer) co-worker of Richard von Mises, a professor of aerodynamics at Harvard University. When he died in the mid-fifties leaving much important work undone, the ONR, which had been funding his researches, wanted someone to finish them. Hilda Geiringer von Mises was uniquely qualified, but it had been a Harvard-based project. The unusual solution was for Harvard to hire her as a "research associate," despite her full-time position elsewhere, and for her to hold the new grant in her own name.[39] It seems unlikely that without her outside independent position she would have fared so well with the Harvard grants office; nor, one suspects, would she have received as many other awards late in life as she did if she had acquiesced all along to what would for most women would have been the far easier path of quietly living in Cambridge as the brilliant but invisible wifely assistant thanked in the preface and forgotten thereafter.

One notable research associate who did not fare even this well and who exemplifies the indignity and humiliation that could come with this form of academic exploitation was another widow, psychologist Else Frenkel-Brunswik of the University of California at Berkeley. Trained at the University of Vienna and on the faculty there, she had fled the Nazis in 1938 and joined a former colleague, Egon Brunswik, in Berkeley. There they were married, and he became a faculty member, while she accepted a research associateship at its Institute for Child Welfare. But her work outstripped that designation, and among other projects, she co-authored *The Authoritarian Personality* (1950), one of the landmarks of twentieth-century social science. Meanwhile, her husband had grown despondent over ill-health. His condition exacerbated by the depressive side effects of some of the early anti-hypertensive medication, he took his own life in July 1955. Two and a half years later, in December 1957, the psychology department, realizing that the antinepotism rules no longer applied, finally voted her a faculty position. Whether she knew this or not is unclear, but in March 1958 she committed suicide as well.[40]

A few others managed to escape successfully from their research associates to a faculty position. Among these was sociologist Rose Goldsen of Cornell, who apparently did this in 1958 without the help of a powerful husband or the lure of a big new grant. As she later explained to Jessie Bernard, after doing social surveys on soft money for almost a decade,

> I confronted the usual discrimination. . . . At that time I was a senior research associate at Cornell. John Dean resigned from the department, and there was an associate professorship up for grabs. My department went through the usual motions of circularizing leading academic centers looking for a replacement for John Dean and never even asked me to prepare a folder. It was at that point that I dug in my heels and demanded to be considered, a procedure guaranteed not to make one well beloved. That's how I got the original appointment at Cornell.[41]

When Jewish philanthropy created a new university and a new medical school on the East Coast in the 1950s, several former research associates and lecturers were among its newly appointed faculty. Brandeis University hired such research associates as biochemist Mary Ellen Jones, who came with her own American Cancer Society support, and others.[42] When Albert Einstein Medical School in New York City, a part of Yeshiva University, opened in 1955, Ernst Scharrer, the new chairman of its anatomy department, made it a point to hire such accomplished female research associates as his wife Berta and geneticist Salome Glueckson Waelsch, long with L. C. Dunn at Columbia. (Both were elected to the NAS in the 1970s.)[43]

Similarly, when the University of California opened several new campuses in the sixties, the new deans hired an occasional woman faculty member. Among these were several highly qualified voluntary faculty. Two of the first to accept full professorships at the new San Diego campus, opened in the early sixties, were Maria Goeppart Mayer, the long unpaid "voluntary professor" at the University of Chicago, in physics, and her husband Joseph, in chemistry.[44] In 1961 Mary A. B. Brazier, an English neurophysiologist and research associate for thirty years (1930–60) at various institutions, including MIT in electrical engineering from 1953 to 1960, became a full professor of anatomy and physiology at UCLA's new Brain Research Institute, where she simulated brain functions with computers.[45] In 1965 the university's new Irvine campus hired Marjorie Caserio, for nine years a research associate at Cal Tech, as an assistant professor of chemistry, and in 1966 UCLA's geology department took the highly unusual step of promoting its own senior lecturer and former associate research geologist Helen Loeblich, also the mother of four, to a full professorship.[46] Other female research associates who moved to faculty positions at the University of California's other new campuses were the ecologist Jean Langenheim, a 1953 Ph.D. and research associate at Harvard for several years, whose recent divorce freed her to move west to the biology department at Santa Cruz in 1966, and botanist Beatrice Sweeney, long a research associate at the SIO and then at Yale, as well as the mother of four, who was hired by Santa Barbara in 1967. Although well beyond their doctorates, both came untenured and spent their whole careers in their new departments, bringing them national distinction.[47] Similarly, the Irish physicist Anne Kernan, who joined the faculty at the Riverside campus (though as an associate professor) in 1967, after four years of teaching at University College, Dublin, and five years on the research staff of the Lawrence Berkeley Laboratory and the Stanford Linear Accelerator, has put it on the international map. Yet old patterns persisted at the University of California as well. For example, when astrophysicists E. Margaret Burbridge and her husband Geoffrey went to San Diego in 1962, he was hired as an associate professor of physics, and she, who had already spent the 1950s in assorted research capacities at Yerkes Observatory and Cal Tech, was at first hired as a research associate in chemistry to get around the antinepotism rules; however, in 1964 she became a full professor of astronomy.[48]

Two other research associates, both psychologists, remained at their husbands' institutions and were later appointed full faculty members only after they obtained large outside grants. After several years as a lecturer and research associate at Harvard's Graduate School of Education, Anne Roe, author of the classic work of the 1950s *The Making of a Scientist,* an intensive study of the family and personal backgrounds of sixty-four eminent male scientists, and wife of paleontologist George Gaylord Simpson, received, as her husband put it, such "an enormous grant" ($443,056) from the NIMH to head her own Center for Research in Careers for 1963–67 that her dean became convinced that a full professorship was more suitable. When the governing boards agreed, she became the first woman to hold such rank at Harvard's Graduate School of Education, the ninth in the university (see table 6.4), and she and her husband became the first couple on the Harvard faculty.[49]

In the late 1940s the psychologist Eleanor Jack Gibson followed her husband, also a psychologist, to Ithaca, New York, when Cornell University offered him (but not her) a faculty position. She later described her sixteen years of second-class citizenship in a section entitled "Life as a Research Associate" in the prestigious series *History of Psychology in Autobiography,* where she was one of the very few women psychologists included.[50] Although, as she said diplomatically, "Cornell did not give me the opportunity to seek my own outside support at first," "in desperation" she worked on someone else's animal project at one of the university's farms. Before long, NSF had started supporting research in psychobiology, and she had such original ideas that she was able (with a professor) to obtain a "generous" two-year grant of $13,000 from NSF in 1955 to study the "development of visual perception," and she received others from other agencies later on. Yet she was allowed to apply in her own name only after her husband threatened to leave for another university.[51] Such confrontational success and her subsequently more visible and independent stature undoubtedly played an important role in the considerable later professional recognition that Gibson enjoyed. Starting in 1965, she became a full professor of psychology at Cornell, won many medals and honorary degrees, and was elected to national academies. From that happy vantage point she could conclude her autobiography by saying, "I think I can truly say that I have never felt any real bitterness about my inferior status during those sixteen years that I did research and paid the university a large overhead on my grants and contracts. I do lift my eyebrows, however, when people tell me how lucky I was to have all that time to do nothing but research. The people who tell me that, of course, are never women."[52] Thus, for research associates like Hoffleit, von Mises, Roe, Gibson, and others, the successful struggle to become PIs was an important step in forging a sufficiently independent research identity to make them eligible for those scientific honors that were rarely awarded to mere research staff.

Another strategy of research and revenge was undertaken by sociologist Rita Simon in the sixties. After marrying University of Illinois professor of economic marketing Julian Simon in 1961, she spent two years trying to get a suitable faculty position in the sociology department at Urbana. By the time she did she had designed a study of recent women doctorates that demonstrated in part that those who complained of antinepotism restrictions did in fact publish significantly more than those who did not. If, she concluded, the whole point of such rules had been to protect the university from unqualified employees, in practice their effect was just the opposite: they immediately disqualified some highly suitable people.[53]

Others just left. Some took (or created) positions in nonprofit institutions where a woman could be her own boss, get her own grants, if necessary, and be visible enough for outside recognition. Some (to be mentioned in ch. 13) found jobs in the federal government, which hired couples and was expanding in several areas in which former research associates often had special talents, such as atomic energy, physics, medical research, grants administration, and data collection. Others may have found faculty positions in the area of home economics, which faced a shortage of doctorates in several biological, biochemical, and social fields. In time, some research associates received certain honors, such as the presidency of their scientific society or an AAUW achievement award, but not membership in the NAS or, as Vogt showed, the Nobel prize. For that, one needed to have published in her own name and formed her own research identity.

But leaving one's research associateship behind did not correct the system or make it any easier for anyone else to follow. In fact joining the faculty was tantamount to joining the system. In the case of research associate Alice Rossi in sociology at the University of Chicago, for example, it was her shabby exploitation, even betrayal, by her "shadow PI" that precipitated a drastic rethinking, with wide-ranging consequences, of where women fit or did not fit in the whole burgeoning world of science and careers (see ch. 16). Had Chicago's sociology department created the title "voluntary professor" for her, as its physics department had done for Maria Goeppart Mayer, or even just "voluntary associate professor," Rossi might have joined the grateful few there rather than leading the resentful many. Unlike Gibson, quoted above, who minimized the problems of research associateship after she left it behind, Rossi in the late 1960s made it one of her first recommendations to change the system that allowed any new faculty member to apply for grants in his or her own name but denied this to research associates even of very long standing.[54]

Meanwhile, as Parrish's data in chapter 6 show so vividly, in these years most of the women scientists at prestigious institutions who were faculty members continued to be those in the female ghetto of home economics. To them we turn next, for their whole field was under siege in the 1950s and 1960s, primarily because it was so highly feminized. The deans of home economics faced constant pressure in these decades to raise the prestige level of their colleges. The easiest way to do this was to hire doctorates, who were all too often young men.

8

Protecting
Home Economics,
the Women's Field

By far the most numerous and the highest-ranking women scientists employed on the faculties of the nation's top universities in the 1950s and 1960s were those concentrated in the highly but not totally feminized area of home economics.[1] Yet data specifically on this field indicate that few of its largest employers, such as the land-grant colleges of home economics, which were most likely to employ scientists and to produce research, grew substantially in the period 1947–63. Instead, all numbers and the numerous writings about home economics in these years indicate that the field was under attack the whole time from several directions and for seemingly contradictory reasons. Chief among its on-campus difficulties were continuing communication problems with the central administration (uniformly male at the time), which held skeptical and hostile attitudes about home economics while at the same time admitting unabashed ignorance about what the field was and what it was trying to do.[2] These behavioral patterns were symptoms of a growing sense of embarrassment, usually at prestige-conscious universities, about the field's strong vocationalism or explicit links to teacher education (which were usually required by state law) and the low proportion of doctorates among its faculty (usually slightly higher than in engineering). But chiefly the field's high proportion of women (usually 90–100 percent), especially older and single ones, was mentioned with increasing openness in the sixties as further evidence that the field was out-of-date. Such a high ratio was intolerable and would need some rectification. One can also sense a rising frustration and even anger in these years among the field's leaders, who had not only to fight losing battles for personnel and resources but also to define and defend the field's intellectual rigor and social importance. They had to do this repeatedly, not only on campus but also at the national level, where through numerous committees and commissions they attempted to convince skeptics of the field's need for greater research support. Although the USDA supported research in home economics, the NSF, the NIH, and the NIMH, the chief supporters of academic research in these years, did not recognize "home economics" as a field of science, though they did support various aspects of it, such as nutrition and developmental psychology.

A practical and increasingly frequent remedy for this embattled position, and one pursued by several of the remaining women deans themselves, was to exhort the women students to earn more doctorates (see ch. 15) but then to hire men preferentially, especially those with doctorates in specialties such as child development, who might attract new research funds and hence prestige. Only in this way could a dean hope to preserve her job and her college and salvage some vestige of this subject, formerly a respectable field, especially at the land-grant universities. Nevertheless, despite these women's best efforts, many of the larger programs actually shrank in these years of record academic expansion, and several major programs were discontinued, removed, or dismembered in some way. Then, ironically, by the mid- to late sixties, when the Great Society provided substantially more funds for these subjects and several of the women deans were in ill health or near retirement age, the predatory men, sensing both a big opportunity in an area they had formerly ridiculed and a moment of strategic weakness, seized the chance to take over. By 1968 the old college of home economics, built up over the decades by a succession of devoted deans, was being forcibly, often brutally, renamed, restaffed, and reconstituted. To a large extent these "reforms" were symptomatic of the sexism, ageism, and misogyny of the age.

Very useful for examining the state of home economics, unique in fact, was the USOE's specialized series of data "Home Economics in Degree-Granting Institutions," which appeared biennially from 1939 until it was discontinued in 1963 (another sign of the decline of the field). Home economics long merited this closer federal interest, because ever since the passage of the Smith-Hughes Act of 1917 the federal government had been subsidizing the training of home economics teachers. This had led to the creation of a home economics education branch within the USOE, which, long headed by Edna Amidon (see ch. 13), collected and published extensive data on enrollments, degrees, and faculty at the many colleges and universities (not just those in the land-grant system) that awarded degrees in home economics.[3] According to the USOE's data, the total number of such institutions increased modestly (4.6 percent), from 388 in 1947–48 to 406 in 1962–63, and the total number of faculty in terms of "full-time equivalents," or FTE, rose even more (7.1 percent), from 2,574 to 2,756.7. Yet the growth was uneven. Of the fifteen largest programs, only four grew in size—those at the University of Minnesota, Pennsylvania State University, Texas Tech University, and, most spectacularly of all, the private Brigham Young University, which exploded from 5.5 FTE in 1947–48 to 38.3 in 1962–63 under an expansionist president who saw home economics (or "family living," as he called BYU's department) as a way to promote traditional female roles among his thousands of Mormon women students.[4]

But eleven others suffered losses in these fifteen years, ranging from four FTE at Purdue to 51 FTE at Cornell. At Cornell the loss was due primarily to the secession in 1954 of its male-dominated school of hotel administration to become an administratively separate entity.[5] When interviewed in 1964, E. Lee Vincent, Cornell's dean of home economics at the time, made light of the loss, insisting that she had wanted it to happen because the hoteliers relied on private (especially Statler) money rather than state funds. Although she felt that it was important for women to hold deanships, lest they all go to men, she stepped down from her post a year after the secession, at the age of fifty-seven, because, as she later explained, she had wanted to get back to teaching and lucrative textbook writing.[6] Overall the largest programs were shrinking downward, from an average of 51 FTE in 1947–48 to just 39.2 in 1962–63, or a loss of about 23.1 percent.

It is important to note that this rather striking decrease in the size of the faculty in most of the nation's largest home economics programs between 1947–48 and 1962–63 was not accompanied by (and perhaps therefore partially explained by) a decrease in the number of undergraduate (or graduate) degrees awarded. On the contrary, other data from these same reports indicate that most were turning out *more* graduates in 1962–63 than fifteen years earlier, when they had had far more faculty: the number of graduates increased moderately at these fifteen schools, from 1,628 in 1947–48 to 1,916 in 1962–63, or an increase of 17.7 percent. Although some men had heeded the call in the late 1940s and after to take courses in home economics, fewer than 2 percent stayed to major in it.

But even this rising number of home economics graduates could be a bit worrisome to the deans of the time, for it was not increasing enough for them to hold onto their "market share" of the student body. Although the number of women graduating from college nationwide almost tripled (from 96,165 graduates in 1947–48 to 277,116 twenty years later), the number majoring in home economics fields did not even double (increasing from 7,204 to only 12,455). Put another way, whereas 7.49 percent of all women graduates in 1947–48 majored in home economics, by 1967–68 their percentage was just 4.49 percent, a substantial (40 percent) drop.[7]

To a certain extent this slippage was the inevitable result of a breaking down of the internal segregation long practiced at many of the land-grant colleges. Although officially they had been coeducational for decades, several still had few women students anywhere except in the college of home economics. At Iowa State University the trend was particularly noticeable: 64.5 percent of the women graduates of 1947–48 had majored in home economics compared with only 31.8 percent in 1969–70. This was partly due to the university's approval in

1960 of a new major in women's physical education, still taught by the College of Home Economics. When in 1968 the university rather belatedly created a separate College of Education, 630 of its first 821 enrollees came from the College of Home Economics. Thus, increasingly in the sixties, as the land-grant institutions decentralized and diversified their curriculums, women students were moving into other parts of the university, and the formerly monolithic women's world of home economics was splintering, even at Iowa State.[8]

Another trend among academic home economists in these years was a new stress on graduate degrees, particularly doctorates. The total number of such degrees awarded tripled between the late 1940s, when about 14 were given annually, and the mid-1950s, when it leveled off at about 50 until it started to rise again in the early 1960s. By far the largest doctorate-granting program in the country was at Cornell University, which awarded about one-fifth of all such degrees nationwide in these years. For a time it was followed by substantial programs at Texas Woman's University, Pennsylvania State, Teachers College, and Ohio State (see table 8.1).[9] One reason why the number of doctorates awarded remained roughly constant for about a decade was the near total lack of fellowships for graduate students in the field. A 1954 NSF survey of financial aid for graduate students revealed that there were just 32 fellowships in home economics in the whole nation, none of which were from the federal government, whereas in other fields, especially the biological sciences, hundreds of fellowships were awarded by the government, private sources, and the universities themselves.[10]

Yet despite the increasing rhetoric about the need for more doctorates in the field, not even these fifteen traditional producers of doctorates in home economics persisted. In fact in these years of expanding graduate fields in other sciences there were some notable terminations and retrenchments, starting with the distinguished program at the University of Chicago in the mid-1950s. Founded as a department of household administration in 1904 under Marion Talbot, a close friend of Ellen Swallow Richards, founder of the field of home economics, the program had later been relegated to the university's School of Education. Under Katharine Blunt in the 1920s, however, it had moved back to the colleges and graduate schools of arts, literature, and sciences and had increased its faculty to seventeen members. But fed up with the lack of adequate facilities and especially the cancellation of a promised new building, Blunt left in 1929 to head the new Connecticut College for Women. Her protégée Lydia Roberts, an outstanding nutritionist and researcher who served on the first Food and Nutrition Board during World War II, succeeded her and trained many important women. Upon her retirement in 1944 the university appointed one of her students, Thelma Porter, then at Michigan State College, to head the program, which was notable for the high proportion of doctorates on its faculty at a time when this was rare. Among them were Hazel Kyrk, re-

Table 8.1.

Institutions Producing the Largest Number of Doctorates in Home
Economics, by Sex, 1947–48 to 1962–63

Institution	Total	Men	Men as % of Total
Cornell	112	30	26.79
Texas Woman's University	52	0	0.00
Penn State	48	3	6.25
Teachers College	45	17	37.78
Ohio State	42	0	0.00
Iowa State	37	0	0.00
Florida State	33	11	33.33
Wisconsin	33	0	0.00
Iowa[a]	28	14	50.00
California	24	9	37.50
NYU	19	1	5.26
Michigan State	18	4	22.22
Chicago	17	0	0.00
Purdue	16	0	0.00
Minnesota	15	3	20.00
Other institutions	46	12	26.09
Total	585	104	17.78

Source: USOE, *Earned Degrees Conferred,* 1947–48 to 1962–63.
[a]In child development and family relationships only.

nowned in the area of consumer economics, Helen Oldham, and Margaret
Davis Doyle.[11]

But even the faculty's distinguished past and strong credentials were not
enough to save the department, for by the late 1940s president Robert Maynard
Hutchins and other male administrators could find neither funds nor a place
for it at the University of Chicago. At a reorganization meeting in 1949 it was
decided to split home economics among three separate faculties: biological
sciences, social sciences, and humanities. In 1950 a new dean of biological
sciences refused to administer his portion of the old home economics, and in
1952 the department was reduced to the level of a committee, which could not
make faculty appointments. This and new tenure rules adopted about the same
time encouraged faculty to leave or retire. In 1956, when Porter, one of only
three remaining faculty members, resigned to accept the deanship at Michigan
State University, the Chicago chancellor terminated the program, attributing
his decision, as he stated in a letter to the program's eight hundred alumnae, to
the university's inability to compete with the land-grant universities in this
field. This stirred Marie Dye, recently retired from Michigan State, to write a
history of the defunct program. In it she laid the blame squarely on the shoul-

ders of the (male) administrators for skimping on funds for decades and for not making any place for this eminent program in the university's otherwise expanding postwar future.[12] She also praised Porter, who had "withstood the pressures brought to bear against the Department with dignity and maintained her scholarly attitude and standards in spite of the temptation to find an easy way out." Perhaps Porter, so long thwarted at the University of Chicago, got at least some satisfaction at Michigan State, where the College of Home Economics grew rapidly under her leadership. But if such a distinguished program as Chicago's, with so many doctorates on its faculty, could be so vulnerable, it remained to be seen just how strong the competing land-grant colleges of home economics might be, for the leaders at many of those institutions were themselves in the 1950s and 1960s showing signs of wanting to emulate the private, prestigious universities—like Chicago! Although the pattern of public stoicism followed by a very different historical account was not at all effective, it would recur all too often elsewhere.

A less dramatic restructuring, more of a retrenchment, took place in the late forties and fifties at Columbia University's Teachers College. Although it had been a strong force in the field of foods and nutrition before the deaths of Mary Swartz Rose in 1941 and her successor and protégée Grace MacLeod in 1944, the Teachers College's effort had subsequently centered on the large department of household science and art, run by Helen Judy Bond. A series of reorganizations of the college relegated the nutrition program, however, to a subordinate status within the department of science education, where the highest-ranking of Rose's last followers was a mere "lecturer." In later years the program moved into the area of "nutrition education" within a new division of health services, a compromise between the basic research in nutrition done in Rose's day and the new curricular needs at Teachers College.[13]

Meanwhile, out at Berkeley the chancellor, Clark Kerr, was wrestling with the new master plan for all of higher education in California, including several new campuses to the east and south. In 1954 Agnes Fay Morgan retired after devoting more than forty years to strengthening the home economics program at Berkeley and to a lesser extent statewide, and a long-delayed, expensive new building was dedicated in her honor at a ceremony attended by Kerr and a bevy of male agricultural administrators. Just a year later, however, the Berkeley academic senate voted to drop household science entirely and send it to the Davis and Santa Barbara campuses. Most of the male administrators saw home economics as an embarrassment to the prestige of the great university that they were trying to build in the West. Only the concerted protest of many hundreds of California home economists and AAUW members, whose strength administrators later admitted they had underestimated, brought about an eventual compromise: the graduate program in nutritional sciences, the only area of home economics in which Berkeley had trained many doctorates, would remain there (in Morgan Hall with its superb research facilities), while the rest of the "household sciences" were dispersed elsewhere.

One result was that the program at Davis flourished (under Gladys Everson and Lucille Hurley, one of Morgan's star pupils at Berkeley). Another was that the Berkeley administrators, faced with having to retain a graduate program in nutrition after all and assuming that another Agnes Fay Morgan did not exist, considered only male researchers, some of whom declined to come for unspecified reasons. Some participants suspected that they had an aversion to working in a "home economics" unit. In the end George Briggs, a biochemist and former chief of the nutrition unit at the NIH, became the dean, and two years later, in 1962, the phrase "home economics" was dropped from the unit's title, leaving it with the more gender-neutral term "nutritional sciences." Although at the time the move was seen as a great blow to Morgan, she for her part kept a circumspect public silence, and even when she completed a history of her unit years later, she did not mention the incident.[14]

Also in the fifties, more and more men were earning doctorates in home economics (see table 8.1). Their proportion of the total awarded reached its peak in 1957–58, when they earned 20 of 34, or 58.8 percent. Numerically the men were concentrated in a few programs, chiefly those at Cornell, Teachers College, Florida State (the former Florida State College for Women), Iowa, and California, but there were no men at all at several others. At Cornell this welcome to men was greatly aided, according to a history of the program, by a special W. T. Grant Foundation fellowship for them. This was the only fellowship at the New York State College of Home Economics, as the program was officially known, until it got a General Foods Fund in 1956 and it started an NIMH traineeship program in 1958. Interestingly, the W. T. Grant Foundation fellowship, which was awarded twenty times between 1951 and 1962, was restricted to men, preferably married, who already held a master's degree in psychology, sociology, education, or some other field related to "family life"; accordingly, the fellowship provided the "substantial amount" a family man would need.[15]

The men's greatest inroads into the field of home economics were in certain specialties, as shown in the USOE's data after 1955–56, when data began to be broken down by subfield as well as by institution and by gender. Table 8.2 summarizes the annual data through 1962–63. It reveals that fully two-thirds of the men (42 of 63) earning Ph.D.'s in home economics in these years were in just one area, child development and family relationships, where they received nearly half (49.41 percent) of the total doctorates awarded.[16] The rest were spread far more thinly over four additional fields, and there were none at all in general home economics or institutional management.

Certainly the field of foods and nutrition, in many ways the bedrock of home economics, remained almost totally feminized. The largest graduate

Table 8.2.

Male Doctorates in Home Economics, by Specialty, 1955–56 to 1962–63

Specialty	Total	Men	Men as % of Total
Child development and family relationships	85	42	49.41
Home economics education	83	9	10.84
Foods and nutrition	77	8	10.39
Clothing and textiles	31	2	6.45
Other home economics	35	2	5.71
General home economics	53	0	0.00
Institutional management	5	0	0.00
Total	369	63	17.07

Source: USOE, *Earned Degrees Conferred,* 1955–56 to 1962–63.

program in the nation in this field (with 13 doctorates, none to men between 1955–56 and 1962–63) was at Iowa State, where Pearl Swanson, Ercel Eppright (two of the last of the long line of Yale doctorates under Lafayette B. Mendel), Wilma Brewer, and others carried on a research tradition that went back to the 1920s. In fact, by so long outlasting the Teachers College program in nutrition, the Iowa State department of foods and nutrition was the program that continued the Mendel protégée chain into at least the 1960s and, through their students, even beyond.[17]

One pleasant challenge and a rather exotic diversion for many home economics faculty in these years involved much travel to foreign nations, especially what were beginning to be called "underdeveloped" nations, for purposes of research, extension, and administration. Perhaps the first to do this was Flemmie Kittrell, then of the Hampton Institute in Virginia, a black woman who had received her doctorate in nutrition from Cornell in 1935. As a consultant to the U.S. State Department, Kittrell conducted a "nutritional survey" of Liberia and five other African nations in 1947–48.[18] Then in 1950–51, on one of the first Fulbright grants, she went to Baroda, India, to study the local diet. She returned there in 1953–55 on another State Department assignment to start up a college of home economics, and she received a departmental citation for her work there. Cornell professor Hazel Hauck went to Thailand in 1952–53 as part of a university survey team and then to Nigeria in 1959–60 on a Fulbright. She returned so sick, however, that she was barely able to write up her results on the Nigerian diet before passing away in 1964.[19] One other sign of the growing international flavor of American home economics in the fifties was that the (ninth) International Congress on Home Economics was for the first time held in the United States, in 1958. Hosted by the American Home Economics Association (AHEA), it was held at the College Park campus of the University of Maryland and attracted 1,041 registrants. Hardly any, however, were from "underdeveloped" nations: nearly half were from the United States (489), followed by large contingents from the United Kingdom (123), Scandinavia (111), Canada (88), and France (80).[20]

In the mid-fifties and increasingly in the sixties, more and more home economics administrators became involved in the foreign exchange programs, as the U.S. Agency for International Development (AID) and the Ford Foundation set up long-term institutional development projects abroad, especially in India. Between 1952 and 1973, for example, under an AID contract, six land-grant colleges had an exchange program with several agricultural colleges in India, part of which involved several American home economists' (quaintly called "home scientists") spending from a few months to seven years, as in the case of Fanchon Warfield of Ohio State, in India. In return many Indian women spent varying lengths of time at American land-grant colleges.[21] The American home economists' greatest impact abroad, however, must have been at the college Flemmie Kittrell had started at Baroda, India, in the mid-fifties, for in 1960 Iowa State signed a ten-year agreement with the Ford Foundation to exchange many of its faculty with Baroda. As a result Dean Helen LeBaron, already very busy with her large college in Ames, made seven trips to India between 1960 and 1970. Additionally, as part of the agreement, many Iowa State faculty went to Baroda, where they taught and interacted with faculty (some of whom were from other colleges in India), while more than thirty Indian graduate students attended American universities; ten of these earned Iowa State doctorates by 1971.[22]

Besides all this outreach to foreign lands, many home economics faculty members performed scientific research on their home campuses in these years. The food and nutrition scientists were particularly innovative and well funded, especially in what was known as "cooperative" research, that is, research funded by the USDA and performed jointly by two or more agricultural colleges and/or experiment stations.[23] Congressional passage of the Research and Marketing Act of 1946 provided even more funds for this work. Advised by a national advisory "committee of nine," which by statute included one home economist, the USDA increased the number of research projects relating to home economics fivefold by 1949.[24] One of the largest of these projects was a nationwide nutritional survey that involved fifty experiment stations. When the survey was completed in 1958, its more than two hundred contributing nutritionists had examined the state of health of about twelve thousand persons ranging in age from five to over eighty. Summarizing the findings of the project's 178 publications in one master bulletin, *Nutritional Status, USA*, Agnes Fay Morgan, who was by then retired but still very active, concluded complacently that Americans were quite well nourished, the best ever in fact— a finding that was later challenged by journalists and muckraking reporters of widespread poverty in America.[25]

This modest expansion of federal funding for home economics research starting in the late forties was never enough, however, to satisfy the relentless demand in these years at many campuses for ever more research. At some, long-awaited new buildings with freshly equipped laboratories and new classrooms were finally completed and energized researchers whom administrators had been exhorting to greater efforts for years. In addition, women researchers at a few campuses, particularly Iowa State, Texas Woman's, and Cornell, were able to supplement USDA project funds with grants from the chemical and food industries, private foundations, the NIH, and eventually even NASA, though there was also some frustration that even these research funds were short-term and often had restrictions on their use. This was all part of the growing realization by leaders in the field of home economics that published research was necessary and important not only for the well-being of the nation's families in an increasingly technological age but also for the prestige and very preservation in academia of the field of home economics itself.[26]

In 1955 the AHEA, jolted by an attempt within the USDA, which had fostered the field of home economics since its infancy, to reduce its budget for textile research, and possibly spurred by the termination of the graduate program at the University of Chicago, created its own committee to consider ways to increase future federal aid for home economics research. This Committee on Federal Research Related to Home Economics, which existed from 1955 until 1962, consumed much of the time of several of the leading researchers and research administrators in home economics, whose names are repeatedly mentioned here—Agnes Fay Morgan, Pearl Swanson, Marie Dye, Helen LeBaron, Hazel Kyrk, Cornell dean Helen Canoyer, and Penn State dean Grace Henderson—in producing a series of reports, speeches, and grant proposals over seven years.[27] In 1957 the group even formulated an ambitious proposal for a new federal agency, either a foundation like the NSF or an institute, perhaps like the NIH, that would provide scholarships and fellowships to train much-needed new personnel and award grants to researchers at universities and elsewhere. Such funds were needed, they felt, to help combat such continuing problems of the American family as juvenile delinquency (then much in the news), unwise consumer spending, divorce, misuse of leisure, and bad diet (despite their own optimistic report!). The problems were so pervasive in American life, and the government's response was so scattered (across thirty-five agencies, according to a Library of Congress study), that they felt one new central agency was needed to address the problems properly.[28]

The committee members received a certain amount of encouragement, especially in the first few years, including even a public endorsement (as it must have seemed) from none other than Arthur S. Flemming, currently president of Ohio Wesleyan University but formerly director of the ODM and soon to be Eisenhower's powerful Secretary of Health, Education and Welfare. In a major address to the closing session of the AHEA's annual meeting in June 1957, he offered his opinion and his help: "I have been very much interested in the proposal for a Research Foundation for the American Home. Such a Foundation, it seems to me, could render a tremendously significant service, and I have been delighted to have the opportunity of talking with some of the officers of your Association about the prospects for such a Foundation. I am interested, and I am certainly willing to do anything I can to help you achieve your objective."[29] But as the years wore on, and the AHEA women made the rounds of governmental officials, they were told repeatedly that other projects— military preparedness, ICBM missiles, and a balanced economy—held higher priority in the federal budget. By 1962 the discouraged group recommended that its committee be dissolved and a new one be appointed. A new committee was appointed, but it merely recommended that in the light of the near impossibility of getting such an agency funded the goal become a long-term one for the association, and indeed resolutions recommending action continued to be passed at the annual meeting for several more years.[30]

It is interesting to speculate why the proposal failed, for presumably the late fifties were years of rapidly increasing federal funds for research, with both the NSF and the NIH (and in early sixties, NASA) reporting sizable gains. First, it does not seem that the AHEA committee went about selling its project as well as it might have. A long list (two pages, single-spaced) of persons contacted, probably dated August 1961, included no congressmen or their staff members, but instead many administrative officials, including the formerly enthusiastic Secretary Flemming and the heads of biological and social science research at the NSF, who might have been considered rivals for any new funds in this area.[31] Such naiveté seems a little strange, for the AHEA, although not officially a lobbyist group (lest this endanger its tax-exempt status), did have a legislative committee and staff person who regularly monitored activities on Capitol Hill of interest to the association. Second, the political stance of the women may have been a little too bland or ladylike to dramatize the urgency of the problems sufficiently. Middle-aged or older, Protestant, white, largely from the Midwest, they tended to see a society that was affluent, increasingly suburban, and generally well fed with perhaps some juvenile delinquency, mental problems, divorce, and racial prejudice but generally enjoying the best of times. These women wanted substantial funds to work on those problems that were not yet sufficiently pressing or dramatic to convince officials to allocate much additional federal funding. When just a few years later (ca. 1967) shocking and sensational reports of hunger, poverty, and racism appeared, opening the path to the the civil-rights movement, the Great Society, the War on Poverty, and Head Start, these "consciousness-raising" accounts were written by others.[32] The home economists, especially the deans and directors of research, were too decorous and ladylike to speak in such strident terms and start such a social and political revolution. Nevertheless, it does seem strange that no foundation—not Kellogg, their first choice, Ford, or Russell Sage, among those listed as contacted by 1961—wanted to pursue the proposal.

Just two years later, however, the political tide was beginning to turn, and the field did get some new research funds starting with the passage of the Vocational Education Act of 1963. This windfall grew out of the appointment in 1961 of Iowa State's very effective Helen LeBaron to a White House panel on vocational education. LeBaron made sure that home economics research was specifically included, and by 1965 up to $12 million, and by fiscal year 1967, $22.5 million, was provided for this purpose.[33] But ironically this sudden federal interest in funding home and family projects was to backfire on the women home economists, for rather than finally helping them to do research and to earn prestige, it was a signal to powerful men in academia that however trivial "home economics" had once seemed, it was now a potentially lucrative field, worth taking over as soon as possible.

Meanwhile, things had begun to get rather nasty within the American Association of Land-Grant and State Colleges and Universities (AALGSCU), whose leaders were designing a reduced, less visible status for home economics. The association's division of home economics, composed of the deans and other administrators of home economics units (both instruction and extension) at the nation's fifty-two land-grant universities, had long been deeply involved in the struggle to find new resources for research. Over the years, especially since the passage of the Research and Marketing Act in 1946, the division, many of whose members were also on the AHEA's committee on federal support, had commissioned authors to summarize the status of research in their specialties and to present their reports at workshops for male and female administrators. Discussion at the workshops generated a recommendation for still another attempt to define home economics in such a way as to convince others that the field merited more support.[34]

This was the pattern in the spring of 1959, when Dean Helen Canoyer of Cornell, the division's new representative to the association's imposing executive committee (half university presidents and half representatives of other groups, including home economics), presented the most recent definition (or defense) of her field. She had expected the men to be pleased with this latest effort, since for years they had been badgering her for still another one, but instead they received it scornfully. One university president (reportedly Eric Walker of Penn State) threw his unread copy on the table, denounced it as still inadequate, and moved that the association "request a study of the present status, objectives, and future of Home Economics in Land-Grant Institutions." Startled and aghast at what might be happening, Canoyer stammered that any such request should first be approved by the home economics division. In 1960 that division agreed, but with misgivings, that such a report by an outside "objective" group might misrepresent the field in an unsympathetic and even damaging way. For several years nothing happened, for it was not until November 1964 that the association was able to raise the two hundred thousand dollars needed for such an ambitious study, and then only after what Canoyer recalled as a "confrontation meeting" of association officials with Florence Anderson of the Carnegie Corporation, who barraged them with questions. Perhaps one factor in the final Carnegie decision in 1964 may have been its new president, John Gardner, along with its growing interest in the status of women in America.[35]

Two addresses at AALGSCU annual meetings—by Paul Miller, provost of Michigan State University, in 1960, and Reuben Gustavson, professor of chemistry at the University of Arizona, in 1961—openly recommending that the colleges of home economics hire more men were indicators of increasingly direct hostility in the upper levels of land-grant education toward women home economists.[36] Then in 1965, as the result of a reorganization of the association masterminded by Frederick L. Hovde, president of Purdue University, the division of home economics ceased to exist. Although its members felt that it was serving a useful purpose, Hovde said that the new organization would be able to compete more forcefully with other proliferating associations of "state universities" if it deemphasized its former land-grant, agricultural, even "cow college" emphasis. Although the new format, eventually arrived at through compromise, permitted the existence of a "commission for home economics," it eliminated the women's seat, even the uncomfortable one Canoyer had filled, on the executive committee. Effectively excluded, the home economists immediately formed their own separate but closely affiliated Association of Administrators of Home Economics, which met annually with the National Association of State Universities and Land-Grant Colleges (NASULGC), formerly the AALGSCU, and occasionally on its own.[37]

The Carnegie grant was to pay Earl J. McGrath, a former U.S. commissioner of education and now director of the Institute of Higher Education of Teachers College, Columbia University (which had done similar studies of other areas of professional education in the past, including one on liberal arts and home economics), and an associate to complete the study in two years. This included processing questionnaires from the 75 institutions (of the 101 members of the newly expanded NASULGC) and the thousands of home economists on their faculties (1,672 responded), collecting other degree data, interviewing as many home economists and administrators as necessary, and preparing a final report.[38]

The final 121-page report, *The Changing Mission of Home Economics,* published in 1968, was a curious blend of ideas and tone. It began by claiming that although the field had broadened its emphasis in recent years, from the home to the family, it was still too heavily dominated, the authors felt, by the classic areas of foods and nutrition; home economics was out of touch with recent work in the social sciences and was thus "increasingly outside the mainstream of American undergraduate education" (27). The report summarized much of its data and said directly and explicitly what no one inside the field would have said, namely, that 90.2 percent of the faculty in the field were women, more than half of whom were single and 45 percent of whom were over age forty-five (29); that only 27.8 percent of the women held doctorates, compared with 61.7 percent of the men (32); and that most of the doctorates were from relatively few universities, just ten universities accounting for three-fourths of the doctorates (31, 34). Beyond this, relatively little research was being done (just 533 projects), and most of that was concentrated at nine top universities (58). Nor did the report's accusatory tone change when it came to discussing the touchy topic of sex-differentials in salaries. Although just 695 home economists reported their salaries, the women earned much less than did the far fewer men in the same rank, a galling inequity that might be especially demoralizing to the women, who were generally older than the men. McGrath and Johnson immediately minimized this striking differential, which they had bothered to raise and document, by comparing it with other data that showed that women faculty in other fields at the university were even worse off.

After collecting and presenting these "objective" or at least quantitative data, McGrath presented conclusions and recommendations that were not only anticlimactic and relatively muted but almost positive and thus a relief to the worried women. In fact, as McGrath admitted in an address to the AHEA's annual meeting in Minneapolis in June 1968 (with seven thousand attendees, the largest ever by more than a thousand), he had changed his impressions of the field in the course of doing the study. He had assumed at first that the report's recommendation would be to discontinue or dismember an outmoded field,[39] but in the course of the two years spent on the project he and his associate had become increasingly impressed with the strong corporate demand for home economics graduates and the predictions of even greater employment prospects in the the Great Society and abroad in the years to come. In fact his staff had come to feel that home economics was so useful in so many ways to so many employers that training in this area should not be terminated but instead should be increased in a variety of ways, at community colleges, through "general education" courses for nonmajors in colleges, and even at the graduate level. Then, though earlier in their report McGrath and Johnson had deplored the small number of universities offering doctoral degrees in home economics as evidence of the field's narrowness and inbreeding, in their conclusion they stated almost the opposite: that because doctoral training was so expensive, the ten major existing programs were sufficient, though they and others should expand their enrollments. This proved to be the report's most controversial recommendation. Critics felt that if there was such a shortage of doctorates in the field, new programs were justified, and departments needed graduate programs in order to attract top-quality staff and produce research.[40]

In addition, McGrath and Johnson recommended that home economics programs have more contact with other disciplines on campus, especially those in the social sciences, and do more collaborative research with them. Whatever reorganizations did or did not take place, however, it was essential that home economics programs no longer be tied to the college of agriculture. Its experiment station funds were important for research (and more funds were needed from federal and industrial sources), but the future of home economics was in the cities and foreign nations, not in rural America. McGrath also felt that some university should step forward and start the long-awaited home economics research journal (which Iowa State did in 1971).[41]

All in all, the deans of home economics that McGrath interviewed had turned around his thinking on most of the key issues within a few years. McGrath, unlike many other men at the time, did *not* recommend that the field drop the name "home economics" and take on something else; he did not think that this would accomplish much.[42] Nor did he recommend that men be hired preferentially for faculty or administrative positions, even though the detailed data on gender, age, and marital status in the text of the report seemed to indicate that he would press for that as part of the overall "mainstreaming" of the field. Moreover, by 1968 McGrath was disillusioned with and even critical of some of the land-grant administrators who were intent on upgrading their institutions' image and prestige. He deplored the trend of "a growing number of faculty members and administrators . . . to believe that institutional respectability can be gained only by training analysts and critics rather than active practitioners" (98). "This exaltation of the scholar and disdain for the practitioner has long been endemic in some prestigious academic institutions, but until recently the land-grant colleges have kept it in proper and socially necessary balance." It made the faculty members in the applied programs in professional schools "feel increasingly insecure and defensive about their role" and so undercut the likelihood of interdisciplinary teaching and research (98). Although one of the historic strengths of the land-grant institutions had been to combine theory and practice, McGrath feared in 1968 that this was soon to be lost.

Yet by the time the long-dreaded McGrath report appeared in 1968, some of the very changes that the women home economists had feared most had already occurred or were then taking place on certain campuses. Often accompanying the retirement of a longtime dean of home economics, these changes could be quite brutal and humiliating, involving such deliberate ruptures with the past as name changes, appointment of male deans without qualifications in the field, reorganizations, termination of some departments (especially home economics education), masculinization of the faculty and to a lesser extent the student body, and a dramatic increase in the level of funding. The purpose of this forcible change was evidently to oust powerful women and to demonstrate that men were now in charge, holding all the prominent places in this newly prestigious university.

The first university since Berkeley to change the name of its home economics program and to hire a male head was West Virginia University, which in 1962 appointed as its new president the notorious Paul A. Miller, provost at Michigan State University, who in 1960 had had the effrontery to suggest in an address to the assembled deans of home economics that they hire more male faculty in the future. Once at West Virginia, where the department of home economics was part of a combined college of agriculture, forestry resources, and home economics, Miller created a separate college of education and human resources with a new division of "home and family studies," with William Marshall its male head.[43]

At Pennsylvania State University, which had, as mentioned earlier, one of the largest and most rapidly growing colleges of home economics in the country, the change seems to have been even more brutal. Everyone there knew that before long the ailing Grace Henderson, director and dean since 1946, would be retiring. Thus it was an indicator of things to come when in 1963 Penn State's president Eric A. Walker appointed a committee on the university's future that included no one from Henderson's college. The sense of betrayal increased when one of this committee's recommendations was for a new college of "human development," which would contain portions of the colleges of physical education and home economics. After Henderson's retirement, a one-year term as acting dean by food chemist Dorothy Houghton, and many other retirements, the college's name was officially changed to the College of Human Development, its home economics education program was transferred to the college of education, and a psychologist, Donald H. Ford, was appointed dean. By the spring of 1968, Ford had dissolved seven departments, added programs in nursing and criminal justice, and created four "problem-oriented divisions" within the new college. In addition, President Walker had provided funds to add many new faculty members. As shown in table 8.3, the size of the faculty more than doubled, from 78 in 1965 to 169 in 1972. The undergraduate student body expanded fivefold, from 882 in 1965 to 4,075 in 1972, and the number of graduate students increased from 135 to 153. Significantly, the proportion of women dropped strikingly at all levels in these years. With the new dean, the new mandate, and the new money, the percentage of women on the faculty dropped from 78.2 percent in 1965 to a bare majority (50.3 percent) in 1972. The percentage of women graduate students dropped substantially, from 91.9 percent to 62.1 percent, while that of undergraduates decreased from 76.4 percent to 58.0 percent. If one counts up the winners and the losers in such a reorganization, the name change, reorganization, and hiring of a male dean

Table 8.3.
Men and Women at Two One-Time Colleges of Home Economics, 1965 and 1972

	Total	Men	Women	Women as % of Total
			1965	
Penn State				
Faculty	78	17	61	78.2
Graduate students	135	11	124	91.9
Undergraduates	882	208	674	76.4
Cornell				
Faculty	108	19	89	82.4
Graduate students	145	17	128	88.3
Undergraduates	738	0	738	100.0
			1972	
Penn State				
Faculty	169	84	85	50.3
Graduate students	153	58	95	62.1
Undergraduates	4,075	1,712	2,363	58.0
Cornell				
Faculty	139	53	86	61.9
Graduate students	219	45	174	79.5
Undergraduates	1,162	63	1,099	94.6

Source: Theodore R. Vallance, "Home Economics and the Development of New Forms of Human Service Education," in *Land-Grant Universities and Their Continuing Challenge,* ed. G. Lester Anderson (East Lansing: Michigan State Univ. Press, 1976), 88, 101.

had been only the most visible signs of the drastic dismemberment and systematic masculinization of home economics in just seven years. The only public reaction of Henderson, who had long complained of inadequate resources and worked to expand her college, was that she wished she could have had ten more years to build it up herself. It must have been galling to see all the resources long denied her handed over so readily to her male successor.[44]

A few years later another kind of takeover, an academic mutiny by the male faculty, abetted by the president, unfolded at Cornell University's New York State College of Home Economics. Once again, everyone knew that Helen Canoyer, dean since 1954, would reach the required age for retirement in 1968. Thus, in 1965 President James Perkins, formerly head of the Carnegie Corporation, appointed a committee to study the future of the college. Although it was chaired by Sara Blackwell of the college's department of home economics education, the committee had only two other women (one from cooperative extension and Patricia Smith of the psychology department, who went on leave midway into the project). The majority of the committee members were men, including three from the college of home economics, one from agriculture, and two from arts and sciences. The group met numerous times, collected some data (including the unflattering assertions that the women students in the college of home economics were weaker academically than those in the arts and agricultural colleges), and then in its final report, dated December 1966, made certain recommendations. They recommended that the name of the college be changed to reflect its now broad focus, that men be admitted as undergraduates, that graduate work be stressed even more, that in recruitment and research the college be more closely tied to social science disciplines, and that the college be reorganized into five departments rather than the current seven (with the termination, apparently, of textiles and clothing and Blackwell's own home economics education).[45] The stage was thus set for a series of dramatic changes in 1968. In May the long-awaited new wing of Martha Van Rensselaer Hall was dedicated; in June Helen Canoyer retired (to become dean at the college of home economics at the University of Massachusetts at Amherst); in July David C. Knapp, formerly an administrator at the University of New Hampshire, but with no previous experience in home economics, became the new dean; and in November the remaining faculty (for there had been many retirements) voted unanimously to change the name to the College of Human Ecology.[46]

The results of Knapp's first four years in office can be seen in table 8.3. Like Ford at Penn State, Knapp was given the resources to increase the size of the college, though not as dramatically as at Penn State: the number of faculty members increased from 108 in 1965 to 139 in 1972, the number of undergraduates increased from 738 to 1,162, and the number of graduate students, from 145 to 219. But again, at every level the proportion of women was down: on the faculty dramatically, from 82.4 percent to 61.9 percent (with a drop in the actual number from 89 to 86), but less so among the undergraduates (from 100 percent to 94.6 percent) and the graduate students (from 88.3 percent to 79.5 percent). (As shown in table 8.1, Cornell by this time already had a tradition of training men in home economics at the graduate level.) Interestingly, the titles of the college's administrators were also upgraded. Whereas earlier administrators, such as Catherine Personius, had been content to be called, or even insisted that they be called, "coordinators" (e.g., of research), their replacements were given the more impressive title of associate dean, to enhance their status before their fellow administrators at Cornell and elsewhere. Thus within just seven years the influx of men who had joined the Cornell home economics faculty since the 1940s and still made up less than 18 percent of the total in 1965 had reshaped the school to fit their image. Although the Cornell committee report had only barely mentioned any underlying gender issues, the results indicate a spurt of preferential recruitment of males. Canoyer, who knew what was happening and approved it, was praised at her retirement for having graciously acceded to the inevitable. (Her Cornell obituary notice also applauded her enthusiastic encouragement of young men in the field over the years.)[47]

Similar changes took place at several other colleges of home economics in the late sixties. At the University of Wisconsin, whose School of Home Economics had been long headed by Frances Zuill despite her lack of a doctorate, agriculture dean Glenn Pound later recalled the steps he had taken in the late sixties to dismember it. First he moved the heart of the school, its Department of Foods and Nutrition, to the agricultural college, where some faculty were assigned to its Food Science Department, others to a new Department of Nutritional Sciences. Then he gave the latter additional funds with which to recruit new (male) scientists, often with a joint appointment in biochemistry, animal nutrition, poultry science, or pediatrics. (Several of these departments previously had not permitted joint appointments with home economics.) Then the dean grouped the remaining departments into a smaller School of Family Resources and Consumer Sciences, with William Marshall, the male dean from West Virginia University, as its head.[48] In 1970 the program at Michigan State University changed its name to the College of Human Ecology.[49]

Yet this changeover did not occur everywhere or even at most institutions. A 1974 article in the *Journal of Home Economics (JHE)* by, among others, Marjorie East of Penn State, AHEA president in 1972–73, on the name and organizational changes of home economics programs between 1962 and 1972 reported that though such alterations had seemed to be widespread and much in the news, even epidemic, in the late sixties, only 10 percent (22 of the 214 institutions reporting) had actually made a change. Of those that had, half were small programs (with fewer than seventy-five majors) and half were among the ninety-five largest, most visible, generally land-grant institutions. But 90 percent, including the large established programs at Iowa State, Kansas State, Purdue, and Minnesota, had made no change at all.[50]

The key factor in bringing on the brutal, forced break with tradition that occurred on a few campuses in the 1960s was the presence of an ambitious president (urged on by an aggressive board of trustees) who was anxious to remake overnight what had been perceived as a "cow college" into a prestigious university. Among the better known of those trying to follow this path, though not always with success, in the fifties and sixties were Clark Kerr at Berkeley, Milton Eisenhower at Kansas State and Penn State (and then Johns Hopkins, which had no home economics nor, at that time, any women undergraduates), John Hannah at Michigan State, Frederick Hovde at Purdue, James Perkins at Cornell, Eric Walker at Penn State, and Fred Harrington at Wisconsin.[51] In the highly laudatory, even adulatory, biographies that recount many triumphs as their campuses generally grew and prospered in the 1950s and 1960s, home economics is generally treated anecdotally, rather than analytically, when it is mentioned at all. Generally these titans of academe found the field baffling and a bit of an embarrassment. Their instinct was to appoint a young male aide as dean to straighten the college out, but unless they choreographed things very carefully, for example, with university-wide committees or commissions on the future, which deliberately left out the women home economists, their plans could be thwarted in various ways. Among these tactics might be the opposition of a strong dean (such as Margaret Justin at Kansas State, who long battled Milton Eisenhower),[52] a protest by the alumnae (such as at Berkeley), or disapproval by the few women on the board of trustees (as was the case at Purdue in 1952 and again in 1962 when Frederick Hovde pondered appointing a male dean).[53] Thus, these men who were able to work wonders in so many other parts of the university were sometimes stymied and forced to endure a woman dean a little longer after all. The woman dean, in turn, feeling the hostility and given new resources only to hire men, determined to stay on as long as possible in hopes of outlasting the president. Thus her long-postponed retirement date became the focal point for a mutiny by the male faculty and the male president. Accordingly, when Helen LeBaron, who succeeded so well at Iowa State for more than two decades and under three presidents, wrote in 1971 that a lot of the college's success had depended on having the right president, she was not making a modest understatement. By then it was all too evident what stress and turmoil the wrong president could and had caused her colleagues elsewhere. (Interestingly, the year before she had married former president James Hilton.)[54]

From this vantage point, then, the crux of the home economists' problems in the 1950s and 1960s was not so much the numerical data on enrollments or lack of faculty holding Ph.D.'s, as was so often claimed, but the hostility and related lack of communication between a new breed of ageist, sexist, and misogynist university presidents who wanted to get rid of home economics, whatever it was,[55] and the many women deans who tried repeatedly and futilely to define the field for their unwilling ears. At the same time, they also sought to improve and expand their programs; however, they found little outside support, since most foundations, public and private, did not consider home economics a sufficiently legitimate science to merit much money. For a while the home economists exhorted each other to train more Ph.D.'s, and they worked very hard to seek out more research funds. But by the mid-sixties, when such support was suddenly forthcoming for the very topics they worked on, several of even the strongest women deans were too old, in ill heath, or otherwise slated for retirement. Long-thwarted male administrators on several major campuses seized upon the opportunity to reshape this formerly female bastion into the somewhat more gender-neutral subjects of "nutritional sciences," "human development," or "human ecology."

Thus, home economics, once a female subject, taught for decades by women faculty to women students on all land-grant and many other colleges and universities in America, faced difficulties in the 1950s and 1960s. Its well-established tradition of "women's work" or a form of "territorial segregation" had by the early sixties become unacceptable to many—less attractive to the women students and intolerable to many of the male faculty and administrators. This was not because of any great influx of male students, though there had been some. Rather, the male faculty, though generally making up less than 20 percent of the total and clustered within the subfields of child development and family studies, found their female deans and their schools' reputation embarrassing. With the help of a sympathetic male president, they could replace the female deans and much of the older female faculty with younger men and thus revert to the more traditional and comfortable "hierarchical segregation," with the remaining women in the lower ranks.

The ousted women rarely protested publicly. Some took to preparing scrapbooks or writing the history of their program as they had known it, extolling the pioneers and concluding with an ambivalent last chapter on the recent changes.[56] Even as late as 1968, when sociologists and others elsewhere within academia were fueling the National Organization for Women, or NOW, hardly any of the home economists looked to the burgeoning women's movement for help in publicizing or protesting their imminent takeover.[57] Politically and stylistically the female home economists of the late sixties may have been (although more work needs to be done on this) fairly far removed from most such groups. In fact some of the men in home economics may have been more liberal than they were, for when feminists at Cornell University offered its first course in "female studies," in 1970, it was sponsored by the College of Human Ecology and listed under Human Development and Family Studies 390, Professor Harold Feldman's course on the evolution of the female personality.[58] But before moving that far ahead, it is necessary to consider the unfortunately all too similar set of events that was equally quietly reconstructing those other parts of academia where women had once dominated, the teachers colleges and the women's colleges.

9

Surviving in "Siberia"

In addition to the major universities, which had never had many women on their faculties except, perhaps, in home economics, there were hundreds of colleges and universities that had at one time employed many women as faculty members. Among such former female bastions were the nation's numerous teachers colleges and junior colleges, where nearly half the faculty were women in the 1940s. Various forces, among them the baby boom, local boosterism, foundation support, regional accreditation, and pension plans, inspired and enabled trustees and officials at hundreds of institutions across the nation not only to expand to meet the new demand but also to upgrade what had been small and poor teachers colleges into bigger, richer, more diversified and prestigious "state colleges" and eventually "state universities." The visible steps of this transformation were larger appropriations from the state legislature, a geographically and taxonomically broader name (for example, New Haven Teachers College became Southern Connecticut State College), many new buildings, more male students, and a much larger faculty. The new faculty members would also earn higher salaries, hold more doctorates, teach fewer courses, and do more research than their predecessors. If these budding researchers needed laboratory space and equipment, the college (or its new "research foundation") would supply it, perhaps from a governmental grant for new facilities specifically at such "developing" institutions. Then if these new faculty members were tempted to leave, early tenure might retain them. Yet all too often these new and pampered faculty members were men. When self-congratulatory histories were written at the end of this golden age of expansion at the former teachers colleges, much was recalled nostalgically and anecdotally about building the new campus—so much disruption, all that mud, the daily fight for a parking space—but there was general silence about the forced departure of the old-time women faculty who had served the institutions for decades.[1]

The most informative data on women's deteriorating position in higher education in the 1950s were collected and presented, not by such timorous governmental agencies as the USOE or the NSF, but by the outspoken staffs at the Bureau of Labor Statistics (BLS) and the research division of the nongovernmental NEA. Between 1954 and 1965 the NEA's Research Division conducted an annual statistical survey of several hundred "degree-granting" institutions on various timely and long-range issues. For example, its 1954–55 survey of 673 such institutions of higher education revealed certain distribution patterns. Its responses yielded data on 63,751 faculty members, including 27,083 faculty members in thirty-four science or related fields (see table 9.1). Because the NEA included home economics among its fields and teachers colleges among its institutions, 3,641 respondents (13.4 percent) were women, the highest percentage in any survey of women science faculty of the time. (The overall percentage, as shown at the bottom of table 9.1, would have been just 8.4 percent without home economics.) By field the largest group of women science faculty in this NEA survey was the 1,488 home economists, followed by 559 in all the biologies, 490 in several social sciences, 365 in mathematics, 309 in the physical sciences (204 of these in chemistry), and 212 in psychology. The rest were distributed among other specialties, and four had no women at all. The women's proportion of the total was highest in home economics (96.4 percent) and physiology (28.0 percent), which was often taught at women's colleges or in conjunction with women's physical education and hygiene elsewhere. Their proportion was above 15 percent in only four other specialties: "all other social sciences," 22.6 percent; general biology, 17.6 percent; geography, 15.8 percent; and bacteriology, 15.4 percent.[2]

But what makes the NEA survey different from all the others of the time is that it deliberately broke its results down by type of institution as well as by gender and field. It reveals that a wide range of institutions employed these more than thirty-six hundred women scientists in 1954–55. The largest proportions of women faculty were at the poorest and least prestigious institutions, and the lowest proportions were at the most prestigious colleges and universities. Thus, women constituted a high of 24.9 percent of the science faculty at the many small (and probably poor) private colleges but a mere 7.2 percent at the fifty-six large private (occasionally quite wealthy) universities that bothered to answer the NEA's questionnaire. The same pattern held for particular sciences. In the field of geography, for example, women's substantial proportion of the faculty at the teachers colleges (30.6 percent) was five times their proportion at the public universities (6.2 percent). Since geography was still being taught in the elementary and high schools in the fifties, the subject had to be a part of the curriculum at teachers colleges. But at the more prestige-oriented larger universities, where, ironically, many of the students might also be going on to jobs as schoolteachers, women geographers on the faculty were very rare.[3] In other fields as well, such as the biological science, mathematics, and psychology, the women were three times as prevalent on the faculties of teachers colleges and small private institutions as at the larger, wealthier public and private universities (see table 9.1).

Table 9.1.

Women Science Faculty at 673 Degree-Granting Institutions, by Field and Type of Institution, 1954–55

Field	Total	Women	Women as % of Total	176 Private Colleges with Less Than 500 Students	94 Teachers Colleges	103 State Colleges	112 Private Colleges with 500–999 Students	50 Private Colleges with More than 1,000 Students	82 Public Universities and Land-Grant Colleges	56 Private Universities
Home economics	1,544	1,488	96.37	100.0	92.7	97.8	100.0	89.6	95.9	100.0
Geography	499	79	15.83	21.4	30.6	17.2	18.2	29.2	6.2	9.4
Biological sciences	3,760	559	14.87	30.1	20.1	15.0	18.5	26.7	10.7	12.1
Physiology	182	51	28.02							
Biology	992	175	17.64							
Bacteriology	228	35	15.35							
All other[a]	811	115	14.18							
Zoology	618	86	13.92							
Botany	488	68	13.93							
Anatomy	149	20	13.42							
Biochemistry	141	8	5.67							
Entomology	151	1	0.66							
Mathematics	2,567	365	14.22	23.6	23.0	17.7	12.7	10.3	13.6	7.4
Psychology	1,586	212	13.37	25.4	19.5	20.0	16.5	14.0	8.1	10.9
Social sciences	4,722	490	10.38	17.4	13.9	13.0	11.5	10.2	9.4	7.4
All other[ab]	829	187	22.56							
Anthropology	139	20	14.39							

Field									
Sociology	1,112	141	12.68						3.7
Political science	1,066	65	6.10						
Economics	1,576	77	4.89						
Physical sciences	5,168	309	5.98	16.9	7.0	5.6	8.4	10.3	3.8
Astronomy	68	10	14.71						
All other[a]	318	30	9.43						
Chemistry	2,445	204	8.34						
Physics	1,657	50	3.02						
Geology	538	15	2.79						
Metallurgy	84	0	0.00						
Meteorology	58	0	0.00						
Agriculture	2,797	116	4.15	0.0	0.0	0.0	0.0	5.0	4.8
All other[a]	2,097	112	5.34						
Forestry	266	2	0.75						
Animal husbandry	434	2	0.46						
Engineering	4,440	23	0.52	3.9	0.0	0.6	0.0	0.4	0.5
All other[a]	1,399	13	0.93						
Chemical	329	3	0.91						
Electrical	835	4	0.48						
Mechanical	994	3	0.30						
Civil	752	0	0.00						
Aeronautical	131	0	0.00						
Total/average	27,083	3,641	13.44	24.9	23.5	18.2	15.6	14.2	11.0
(Less home economics)	25,539	2,153	8.43	20.1	16.3	10.2	12.0	11.4	6.3

Additional right-hand column values:
- Agriculture: 0.0
- Engineering: 0.4
- Total/average: 7.2
- (Less home economics): 5.5

Source: Calculated from data published in "Teacher Supply and Demand in Degree-Granting Institutions, 1954–55," *National Education Association Research Bulletin* 33, no. 4 (1955): 134, 136 (tables 7 and 9).

[a] Includes those faculty members in the field whose specialty was unspecified.

[b] Omits those listed under history and international relations.

Had this pattern of institutional sexual stratification continued unchanged into the 1950s and 1960s, there would at least have been opportunities for women scientists at the less prestigious colleges, either as these institutions expanded or as the older women faculty retired after decades of service. But this easy substitution did not take place, and the proportion of women on college faculties began to decline, despite the so-called pressing national shortages. Thereupon the staff of the NEA's research division, which included several women, began to insert into their biennial reports a section asking rhetorically "Can Women Be Used in Larger Numbers?" or "Will More Women Enter the Teaching Ranks?" They even polled the heads of all the institutions in their survey (all those colleges and universities in the USOE's *Education Directory* of the previous year) concerning whether women might be employed there in larger numbers in the future. The response was mixed. In 1957–58, for example, of the 913 institutions answering this question only 260 (28.48 percent) said they would employ women in larger numbers in "all fields," a substantial 210 (23.00 percent) said they would not employ more women in any field, and the remaining 443 specified certain fields, especially social sciences (130), English (115), and mathematics (110) but ominously few in physical sciences (77), biological sciences (28), psychology (21), and geography (1).[4] Adding women faculty evidently was not a priority for many administrators, and the percentage of women at those institutions (as counted biennially by the USOE) continued to decline, from 30.99 percent in 1945–46 to just 21.97 percent in 1961–62; at the teachers colleges down from 47.58 percent in the fall of 1947 to just 33.00 percent in the fall of 1963; and at the junior colleges from 48.63 percent women in 1945 to just 28.5 percent in the fall of 1963.[5] Most of the persons retiring or leaving for other reasons must have been women, and most of the persons being hired must have been men, perhaps those being educated at public expense under the GI Bill.

In fact administrators and trustees at most modest institutions were intent on shedding the "normal school" image and upgrading their school and faculty to full university status. There was more to this than merely insisting uniformly on doctoral degrees. Often the trustees of an institution intent on an expanded role and enhanced status hired as top administrators venturesome persons from institutions that had higher prestige than their own who would attempt to make the current institution more like the ones they had just left. They would appoint deans preferably from research universities, who then in turn recruited chairmen with instructions to transform their departments as quickly as possible. To ensure their effectiveness, such men were often allowed to bring with them enough colleagues and former students to outvote the current faculty members. Before long the increasing numbers and visibility of the new men on campus (as well as the college's new inflated title of "state university") made it easier to recruit and, with the adoption of the tenure system endorsed by the American Association of University Professors (AAUP) in 1940, to tenure still more downwardly mobile men. Thus within a few years, a decade at most, departments, whole campuses, and even new university "systems," such as the State University of New York (SUNY), the California State University, or the Wisconsin State University System, were so totally remade that even the memory of the women once there was dim indeed.[6] It is staggering to contemplate the number of jobs at these expanding institutions that were *not* going to women.

Despite a public silence on the transformations under way, we can assume that such enforced changeover was quite brutal and traumatic in many places, as the one study of the phenomenon admitted.[7] Although officials knew that this practice was upsetting to many current faculty members, especially the older women, they felt that allowing such malcontents to linger would demoralize and undermine the enterprise. The best many such women could do was to retire as soon as possible, hurt, confused, and angry that they were being so hastily and unceremoniously cast aside, their life's work denigrated as worthless. It must have been difficult for these women, many of whom were quite old, to survive financially on what were probably totally inadequate pensions, often hastily introduced in order to expedite their departures.[8]

Those women too young to retire who tried to leave for a better job found it difficult, since so many other institutions were undergoing the same process, did not value their long years of teaching rather than publishing research, and in any case did not advertise their openings. In such a situation one might resort to the job advertisements in *Science* and other journals or the placement service at a professional convention, but this was unlikely to lead to much improvement within academia, for many of the advertisements specified the sex desired (e.g., "men only" or "woman preferred"). Frances Clayton later described her experience at the American Psychological Association (APA) meetings in the early 1950s, where thirty-seven of about fifty job vacancies posted explicitly stated that only men need apply. Of the thirteen others for which she could and did apply, only one would talk to her. In fact having to resort to such a source of information was a sure sign of one's outsider status in the clubby academic world of the fifties and sixties; only those institutions outside the private male network of graduate professors, such as the smaller Catholic colleges or the rapidly expanding former teachers colleges, for instance, found it necessary to advertise in professional journals or, worse, through commercial employment agencies. Oftentimes advertisers were so embarrassed about having to resort to these necessities that they used blind or anonymous box numbers, such as at *Science* magazine, for the return address, making it impossible then or now to discover their identity.[9] In fact a major reason for changing an institution's name to include the title "university" was to impress and recruit a higher level of new faculty.

Thus rather than escape, the more common response for women faculty employed full-time in what many persons pejoratively termed "academic Siberia" seems to have been stoic acceptance, devoting themselves to the teaching and to the students perhaps to the point of exhaustion, enduring the inequities and indignities of the new regime as long as they could. Unfortunately, the rewards were private, personal, and nonmonetary, for even decades of solid performance brought almost no outside recognition and, as a 1959 survey by Ruth Eckert of 706 men and women faculty in the state of Minnesota revealed, surprisingly little even inside.[10]

Although the tendency in histories of individual institutions is to assume rather simplistically that what came later was a major improvement over what had gone before, this view minimizes and understates the often heroic efforts by the early women who had very few resources, such as zoologist Vesta Holt, who taught for more than thirty years (1926–57) at what later became California State University at Chico. An effective and beloved teacher, Holt formed a student (then alumni) club for biology majors, ran popular fieldtrips, and led a successful movement to establish a biological station at nearby Eagle Lake, the second largest freshwater lake in California.[11] Accordingly one can have only mixed reactions to this consciously sex-linked upgrading of the teachers colleges and junior colleges. On the one hand, it was surely an improvement in an increasingly technological society finally to have more money spent on such institutions of higher learning—more buildings, more faculty, and more subjects, including more science. All these undoubtedly benefited the students, who would graduate better and more broadly trained than their predecessors and thus more able to take on a wider variety of positions in society. As for the faculty, hiring Ph.D.-holders was advantageous, for on the whole they would have had better and more extensive training than their forerunners. But why had the proportion of women not remained the same? Such colleges had long been employers of large numbers of women faculty, and more women were obtaining doctorates in the sciences in the 1950s and after than ever before (despite their low percentage of the total, as discussed in ch. 4). One would have expected such women to be succeeding the Vesta Holts of earlier years. Instead she herself had hired numerous men, including several Berkeley, Stanford, Colorado, and other doctorates. When she retired in 1957, she reported that the department, which had had two faculty members in 1926, now had nine full-time and four part-time persons, of which one full-time and one part-time member were female.[12]

Because raising salaries was one relatively rapid and direct way to make a school competitive, the topic received a lot of attention in the fifties and sixties. In fact the pressing need to raise faculty salaries was such a truism that some private foundations, especially the then new Ford Foundation, spearheaded a national campaign. In 1955 Beardsley Ruml and a collaborator published a report, commissioned by the Ford Foundation's daughter Fund for the Advancement of Education (FAE), in which they showed that, as expected, faculty salaries and earning power had not kept pace with those of other professionals, such as physicians, dentists, and school principals, over the previous fifty years. Because it was one of the Ford Foundation's goals to "influence public policy," the report was printed as a pamphlet and circulated to newspaper editors nationwide, who then amplified its message in editorials.[13] Having thus set the ground politically, the Ford Foundation, bursting with funds from the recent forced sale of more than $200 million in Ford stock, made even more news in 1955 by donating more than $260 million, a lot of money then, to all 630 private, four-year accredited colleges and universities in the United States for the very purpose of improving faculty salaries. When these benefactions were followed by what later came to be called 4, 3, or 2 to 1 "challenge grants," they unleashed prodigious fund-raising efforts at each of these schools. Although some already had alumni "development" offices by the mid-fifties, these big grants created a continuing industry, for once the Ford "challenge" had been met, almost every private college or university could find other needs for still more funds.[14] In addition, the public colleges and universities were able to use all the talk about higher salaries at the private schools to justify legislative help in raising theirs as well.

Yet the whole area was rife with gender politics. Although very few data on faculty salaries were broken down by gender in these years, a 1952 BLS survey of fourteen fields in the social sciences and humanities showed that in every field the median salary for academic women was substantially less than that for academic men, despite the fact that in every field their median age was higher. Interestingly, in 1954, a time when the ICWP was too afraid to survey its members about their status, BLS author Cora Taylor did not hide this inequity in a footnote or appendix or omit it altogether but instead highlighted it by providing extensive data on salaries, broken down by field, gender, and type of employer (i.e., colleges or federal government), and even went so far as to feature them in a special chart (reprinted here as fig. 4), a rather feisty thing for a governmental agency to do in 1954.[15] But college administrators seem not to have used the new money to equalize the women faculty's lagging salaries; it is more likely that they used it to recruit still more men to those parts of the academic profession, including teachers colleges, colleges of home economics, and women's colleges, that had formerly paid so poorly that they attracted only women. Thus, rather than decreasing the substantial salary differential between the sexes, the new money probably (though proof of this is hard to pinpoint) increased it still further by raising the valued men's salaries, starting or otherwise, to new highs and adding on the new "fringe benefits."[16] Some colleges even subsidized or built houses for their new faculty members.

Part of the reason given at the time for the women's lower salaries was that

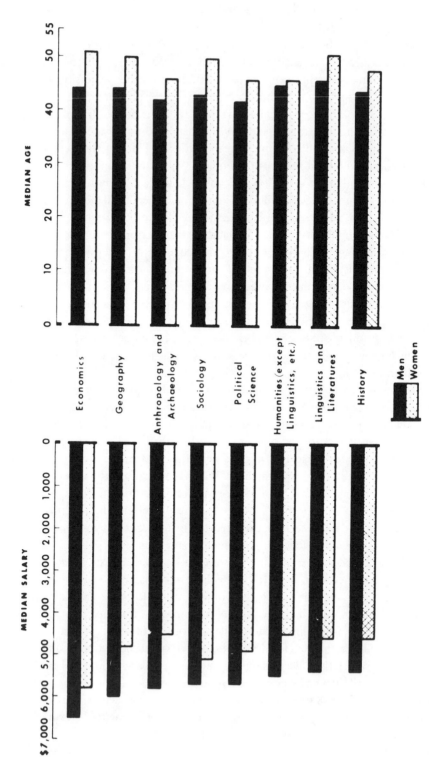

Fig. 4. Median annual salaries of Ph.D.'s employed by colleges or universities in the social sciences and humanities, by age, field, and sex, 1952. Reprinted from *Personnel Resources in the Social Sciences and Humanities: A Survey of the Characteristics and Economic Status of Professional Workers in 14 Fields of Specialization*, BLS Bulletin 1169 (Washington, D.C., 1954), 35 and table A-29.

women held fewer doctorates. Most had been hired in an age when a Ph.D. was not required at teachers colleges and other schools. Yet even though the system now seemed to favor, rhetorically at least, those job applicants with doctorates, the little evidence available indicates that even women with such degrees were rarely hired by academic institutions, except in the field of home economics. In fact despite all the talk in the 1950s of the need to recruit Ph.D.'s in order to achieve full accreditation, data collected by the Research Division of the NEA, the lone voice on this seemingly important issue, indicated that all too often institutions hired men (and women) without a doctorate instead of women who already had them. The NEA's ten large annual surveys between 1955–56 and 1964–65 of "teacher supply and demand in universities, colleges, and junior colleges" constitute the most ambitious and innovative attempt of the time to come to grips with the issues of what types of persons (broken down by field, gender, and especially degree level) were being hired to fill vacancies. Over time the surveys grew, with the help of funding from the FAE, to include about 1,080 colleges and universities, which were collectively hiring up to 13,000 new faculty members per year. But what was of particular concern to the NEA staff was that only about 25 percent of these newly hired persons held doctorates, compared with the 40.5 percent of current faculties, as shown in its own earlier 1954–55 report. Thus by the mid-fifties the rapid expansion in higher education was already causing the quality of faculties to decline rather drastically, despite denials by many in higher education and proposals by others for new "doctor of arts" degrees to speed up production even more. At Northern Illinois University the president was so eager to count faculty doctorates that for two years in the catalog he listed one faculty member as holding the coveted degree even though he had not yet completed the requirements.[17]

The NEA research staff found that although the percentage of doctorates among these newly hired faculty members varied widely by field—with the highest in the sciences in psychology (54.0 percent in 1957–58) and biology (53.6 percent) and the lowest in two fields that had not traditionally awarded the doctorate or required it for faculty positions, home economics (13.5 percent) and engineering (13.2 percent)—only 10.03 percent of those hired with doctorates, presumably for what now might be called tenure-track positions at the better institutions, were women.[18] Why was this so? The usual explanation was that because women received only about 10 percent of the doctorates awarded, no more than 10 percent were available for employment. Yet other NEA data indicate otherwise.

The NEA's reports for 1957–59 on who was earning doctoral degrees and who was being hired (both by field and by sex) reveal that of the 8,942 doctoral degrees awarded in 1958, 10.8 percent, or 964, went to women, 781 in scientific fields. Yet the number of women doctorates hired in either academic year 1957–58 or 1958–59 was not close to half of this number (see table 9.2). Had all the

Table 9.2.

Women Awarded Doctorates and Women Doctorates Hired by 936
Academic Institutions, by Field, 1957–1959

Field	Doctorates Awarded to Women 1958	Women Doctorates Hired	
		1957–58	1958–59
Education	341	27	35
Biological sciences	138	23	26
Social sciences	96	37	25
Psychology	84	11	13
Physical sciences	74	22	26
Mathematics	15	4	5
Home economics	14	21	14
Geography	9	—	1
Agriculture	6	—	1
Engineering	4		
Total	781	145	146

Source: NEA, *Teacher Supply and Demand in Universities, Colleges, and Junior Colleges, 1957–58 and 1958–59* (Washington, D.C., 1959), 24, 59–60 (tables C and D).

women awarded doctorates been hired by these nearly 1,000 colleges or universities, there might have been valid claims that there were too few to go around and more were needed in these fields. But instead the opposite was happening: only about 20 percent of even these very highly trained and qualified women were taking up academic positions at any of these colleges and universities. Evidently educational institutions were not hiring the many women doctorates who were trained and available in the 1950s (except in home economics, where they were in great demand), but, in many cases, men and women with lesser credentials.

Unfortunately, although the NEA came closer than any other of the several organizations collecting data on academic women in the 1950s to showing just how "underutilized" they were and how rarely academic units hired the best-qualified women, its reports were rarely cited by foundation officials and academic spokesmen, except when they wanted reasons *not* to provide help to women's programs. For example, one of the few times that the NEA reports were used was when Katherine G. Clark prepared a confidential report for the Ford Foundation, "Women as a Potential Resource for Relieving Teaching Shortages in College Science and Mathematics Departments," in the late 1950s. Citing the recent NEA poll of department chairmen about their reluctance to hire women faculty (as well as a recent NSF-supported survey of AAUW members, part of which showed that most were unwilling to become science teachers immediately), she could not say that support of retraining programs for reentry women would necessarily lead to any breakthroughs. Evidently it was not worth funding any attempts to train women for jobs they might never hold. Thus the climate was so negative that the NEA's strategy of documenting the extent of the problem had backfired: providing evidence of the pattern and its intractability had in fact helped reluctant funding agencies rationalize why they should not waste valuable resources on trying to solve the problem.[19]

One of the few groups that attempted to protest sex discrimination in higher education in the 1950s was the AAUW, which was most famous for its graduate fellowships and other awards but was also something of a watchdog for college women. Its membership policy, from the time of its founding in the 1880s, stipulated that only graduates of approved institutions could become members. This did not include normal schools or teachers colleges, but when hundreds of former teachers colleges and other new branches of existing colleges were upgrading themselves to full collegiate status in the 1950s and early 1960s, their alumnae wished to be treated as equal to those of other colleges. This provided a great deal of work for the AAUW's Committee on Standards and Recognition, which over time had evolved a rather cumbersome, two-step approval process that could take years. For an institution to be approved, its president had to submit first a preliminary and then a final application to the AAUW providing information about the institution's trustees, faculty, staff, facilities, pay scales, and curriculum. The AAUW committee liked to see (but did not actually require) a woman on the board of trustees; a woman, preferably with a doctorate, in the upper ranks of the faculty; some women on the administrative staff, including, it was hoped, a dean of women, in charge of the female students' extracurricular life; and about 50 percent of the degree requirements in "liberal education," or the humanities broadly defined. This last criterion long excluded many schools and certain vocational curriculums, such as those in music education, medical technology, and home economics. Thus, until 1959, some graduates of an institution might be eligible for AAUW membership, while others might not, and even this could change over time. (For example, Cornell's College of Home Economics had a hard time getting its graduates approved for AAUW membership until 1947, when Dean Sarah Gibson Blanding finally convinced the AAUW committee that many of the courses taken by her students that sounded technical—because home economics was usually trying to look "scientific"—were in fact quite humanistic.)[20]

To a certain extent, the hard-working members of the committee, which over the years included several scientists (usually from the Northwest) and deans of women, almost enjoyed prolonging the process and preserving the AAUW's high standards. Usually there were several rounds of correspondence, one or more site visits and reports by an AAUW member, and then requests for further clarifications and improvements before an eventual decision. This could lead to friction. For example, the president of Bowling Green State

University in Ohio withdrew in disgust in 1951. But others wrote long letters pleading their intention of improving conditions for women if they were first given AAUW approval. In March 1956, at perhaps its highest point, the committee, long headed by Eunice Roberts, assistant dean of faculties and director of women's educational programs at Indiana University, voted to revise its guidelines to require "at least one woman" on the faculty, for certainly some of the larger institutions applying should be held to this higher standard.[21] Her committee was further encouraged in August, when the Middle States Association, a regional accrediting group with which the AAUW shared evaluations and paperwork, issued a statement favoring some consideration of female faculty at coeducational institutions, part of which was published in the *AAUWJ:*

> Higher education for women, and career opportunities for women in higher education, are established concepts in our society. Each institution must decide for itself in the light of its own objectives and policies whether to admit women as students and whether to use them as teachers and administrators.
>
> Clearly it is desirable that women should be prominently represented among the faculty, administration, and trustees of coeducational and women's colleges, but an institution's first duty is to find the ablest people it can for every appointment. To select women simply because they are women unduly limits the range of choice but arbitrarily eliminating half the human race from consideration does too, and makes it more difficult to ensure the invigoration, breadth of view, intellectual stimulation, and normal social attitudes a cosmopolitan academic community inspires.
>
> Putting the institution's welfare first and appointing the best qualified candidate whoever he or she may be is the more defensible policy.[22]

For its time this was a landmark statement, a kind of trophy on the wall for the AAUW and its Committee on Standards and Recognition. Although both associations were later proud, it is debatable how widely known it was at the time, for outside of publication in the *AAUWJ,* hardly any copies exist.

Within a few years the committee's attempt to enforce its new standard led to its own demise. When in 1959 the committee (renamed the Committee on Higher Education) rejected Clark University, which had admitted women undergraduates in the 1940s, on the grounds that it had no women trustees and too few on the faculty (psychologist Thelma Alper had left for Wellesley in 1952), officials of two local AAUW branches protested strongly. The committee, confident that Clark's continued exclusion was in the best interests of college women generally, disregarded their action. Thereupon the critics, capitalizing upon a general rift between the (housewifely) branches and the (professorial) national committees, took their complaint to the AAUW national convention, where in June 1963 they succeeded in passing resolutions that overturned the relevant bylaws. Henceforth any institution approved by any one of the six regional accrediting groups would be automatically approved for AAUW membership! (Ironically, two months earlier, in April 1963, the committee had approved Clark University, which had appointed a woman trustee and given one of the two women on the faculty tenure.) Accordingly, at its fall 1963 meeting the committee recommended a record 275 institutions, including Bowling Green State University, for immediate approval.[23] Thus by the early sixties even this relatively weak form of leverage over the few remaining unapproved universities had been discontinued.

Having realized by the early 1950s that the AAUW's approval procedure was not particularly effective at increasing opportunities for women on college faculties—that all too often college presidents would rationalize to the Committee on Standards and Recognition that their lack of female faculty was due to the lack of women applicants—and that the AAUW had had modest success during and after World War II with various rosters of women qualified for public office, Eleanor Dolan of the AAUW staff began to consider the idea of a roster of academic women. Yet when the idea was presented to the AAUW Committee on the Status of Women in November 1953, the committee, led by Marjorie Child Husted of Minneapolis (the former "Betty Crocker" at General Mills) and Lt. Col. Mary-Agnes Brown of the VA, said that since this was not likely to be an effective way to fight sex discrimination in academia, it should not receive AAUW funding.

Finally, in 1958, after the AAUW had started its own separate foundation, Dolan got the first of several outside grants to make the roster functional. These grants were from the U.S. Steel Foundation, the FAE, and the Alfred Sloan Foundation for the AAUW Roster of Earned Doctorates, expanded and renamed the Roster of Advanced Degrees in December 1964, when women with master's degrees were added. The goal of the project was to show an institution seeking to hire qualified faculty or administrators that many such women were indeed available. To accomplish this, the AAUW Roster got a list of eleven thousand women with doctorates from that NRC's Earned Doctorates File and sent each a postcard requesting that she submit her credentials if she was interested in being listed. From the more than one thousand replies, the Roster's staff could, using punched cards, create specialized sublists, organized by field (seventy-six specialties), location, position sought, age, salary level, and preference for part-time work. Nevertheless, despite this vision of the roster as a means of breaking through what would later be called the old-boy network, it never did succeed in becoming an efficient placement service and was discontinued in August 1965, just as, Dolan felt, it was on the verge of finding its niche and becoming self-supporting. Yet in its last and best year the Roster succeeded in placing only ten women, not in itself a bad showing but, a seemingly paltry outcome after the boasting about the thousands of women contacted.

One problem was the fees that the Roster had to charge. Because its own funds were so limited, the AAUW Foundation had to charge unpopular fees: one dollar to the individual woman, which upset those who were already AAUW members, and fifteen dollars for an institution, which caused officials there to balk, because lists from other placement groups were free. One critic summed up the problem succinctly as follows: faculties were already so reluctant to hire women that to charge a fee for a list of them did not make it any easier to get some considered. Those relatively few who had used the Roster, however—generally small colleges, including many Catholic and women's institutions, perhaps the most inclined to hire women anyway—were enthusiastic and wanted the Roster to continue. One man wrote Dolan from Bucknell that because his university was a coeducational institution (and perhaps pushed by the AAUW approval process or the statement by the Middle States Association), it had decided to hire a woman for a faculty position, but all of the names sent to them by various individuals and groups in that field had been men's. Without her list from the AAUW's Roster, they would not have had any women candidates at all. This was just what Dolan wanted to hear, and yet in 1965 she could find no more outside support and was particularly disheartened by the refusal of the board of directors of the AAUW Educational Foundation to find funds to tide it over. By early 1967 Dolan, who had been a mainstay of the AAUW national staff since 1950, had left for a job at the USOE.[24]

Yet despite their being unwelcome and their many difficulties, several of the women scientists who were hired by these "developing" institutions in the 1940s through the 1960s built the places up considerably. For example, Cynthia Irwin-Williams, a Harvard doctorate in archaeology who had made newsworthy discoveries while a graduate student, found in 1964 that the only job offered to her that was anywhere near her husband's at Sandia Laboratories in Albuquerque was at the recently renamed Eastern New Mexico State University at Portales, formerly a junior college, 240 miles away. (The much more convenient University of New Mexico at Albuquerque was, as one account put it, "in the process of trying to lay off or to retire the only women anthropologists on its faculty.")[25] Discouraged and yet determined to continue her career, Irwin-Williams, with the help of her mother, who was herself a contributor to archaeology,[26] commuted weekly between the two cities and over the years built up the department, capitalizing on the unusual local fossils and ruins to direct expeditions and publish research.

In another case, botanist Lora Maneum Shields, a mother and possibly a widow, arrived at New Mexico Highlands University in 1947 with her brand new doctorate from the University of Iowa. Four years later she became chairman of its fledgling biology department, and over the next twenty-seven years she proved to be a master grantswoman, obtaining support for her research on plant nitrogen, lipids, and the impact of radioactivity on local Navajo tribespeople from the NSF, the NIH, Sigma Xi, Squibb and Company, the Searle Corporation, and the AEC. Her accomplishments led to both regional and national recognition: in 1959–61 she was president of the Rocky Mountain and Southwest Region of the AAAS (and thus a vice-president of the national group), and in 1962–63 she served as vice-president of the Ecological Society of America. Clearly she was one woman who found a way to put her work, herself, and her school on the map scientifically.[27]

Similarly, bacteriologist Cora Downs, hired in 1919 at the University of Kansas, where her grandmother had been the first woman trustee in 1882, did such interesting and important work on biological warfare at Camp Detrick during World War II that afterwards she continued her work on tularemia (Rocky Mountain fever) and fluorescent methods of microbe identification with the help of numerous NIH, ONR, Army, Air Force, and other grants until her retirement in 1963. One listing in her faculty file at the university's archives of the monetary amounts of her many outside grants gives a total of $270,000 as of 1959, to which another $100,000 or so could be added from later awards. She also trained twenty-two master's degree candidates and thirteen doctoral students. In her last year at Kansas Downs achieved the rare honor of holding a "distinguished professorship."[28]

In the mid-1960s, as even more money became available for research, especially in the areas of vocational guidance and childhood education, at least one woman psychologist quickly eclipsed even Downs's grant totals and achieved the entrepreneurial goal of heading a center of her own. Dorothy Adkins (Wood), who moved to the University of Hawaii late in her career, became head of a Center for Research in Early Childhood Education, serving from 1966 to 1973. It was one of her chief duties to write the grant applications that yielded ten successive awards, totaling $1.3 million, from the U.S. Office of Equal Economic Opportunity.[29]

In 1966 another "developing" institution, Ohio's Bowling Green State University, which was undergoing a spurt of growth and had ambitions to be nationally recognized in some fields, did the unthinkable and hired a couple— Patricia Cain Smith, the first woman full professor at Cornell University's endowed arts college, and her husband, Olin Smith, a research associate there. (Bowling Green may also have been responding to pressure from the AAUW to hire more women.) Although the Smiths were strongly attached to Ithaca— Olin had been born there, and they both had been trained at Cornell—by the mid-sixties they were growing dissatisfied with the limitations of the psychology department there and especially with the annual uncertainty about Olin's funding. When they let it be known that they might be willing to move, the chairman of the psychology department at Bowling Green was able to offer two professorships, including a research one for Olin. The Smiths jumped at the chance and spent the rest of their careers there building up Bowling Green's new department. In retrospect, one would have expected more up-and-coming departments or institutions to take similar steps and to find ways around their own antinepotism rules in order to attract such couples and give their programs instant visibility. The Smiths' readiness to leave a prestigious university is partly explained by the circumstance that here it was the husband, and not the wife, who had the soft-money position. It does show what moves might have taken place if "Siberia" had responded more readily or opportunistically to the needs of some scientist couples.[30]

As for department chairmanships, the lowest level of academic administration, women scientists still so rarely held these positions at coeducational institutions in the forties, fifties, and mid-sixties that one can almost count these exceptions on two hands. In chemistry there were a few female chairs, starting with the controversial Leonora Bilger at the University of Hawaii from 1943 to 1954;[31] Clara deMilt at Tulane from 1946 to 1949, when the administration there was starting to combine the Newcomb and Tulane science departments at the graduate level;[32] and Jean Simmons, with a Ph.D. from the University of Chicago in 1938, who left her chairmanship of the department at Barat College, Lake Forest, Illinois, in 1958 to follow her husband, Glen Simmons, to Princeton, New Jersey. Shortly thereafter she became a professor at Upsala College in East Orange, where in 1965 she chaired the chemistry department and in 1966 the whole natural sciences division.[33] In physics, again at Tulane, Rose Mooney, who had worked on the Manhattan Project during World War II, was appointed chairman of the joint physics department in 1948, making her perhaps the first woman physicist ever to head a coeducational department. As with chemist deMilt, this was one of the very few cases in which a woman faculty member was promoted to a higher position as a result of a merger between coordinate colleges.[34] Among the several biologists to chair departments were Vesta Holt, mentioned above, chairman at what later became the California State College at Chico from 1931 until her retirement in 1957; Cornelia Marschall Smith at Baylor University, 1943–67; Lora Shields at New Mexico Highlands University, 1951–78; Harriet George Barclay, at the University of Tulsa, 1953–58, after the death of her husband, who had been the botany department's first chairman; and Winona Welch, chair at DePauw University in Indiana from 1956 until her retirement in 1961.[35] Among anthropologists, Erna Gunther (Spier) chaired the department at the University of Washington from 1937 to 1966 and then upon retirement at age seventy took up the chairmanship at the University of Alaska.[36] Two of the few women to head a coeducational psychology department in these years were Thelma Hunt of George Washington University from 1938 to 1963, best known for her work in developing the Medical College Admission Tests (MCATs), and Dorothy Adkins at the University of North Carolina from 1950 to 1961, where she was the only woman department chairman, having been favored for her position as the protégée and appointee of the senior professors there, L. L. and Thelma Thurstone.[37]

Even more notable than these few grantswomen and department chairs were two women at opposite ends of the country who made long-term contributions as fund-raisers and builders. The first was Erna Gunther of the University of Washington, who was not only the chair of its anthropology department for three decades but also longtime director of its on-campus anthropology museum, the Washington State Museum, from 1929 to 1961. Long housed in a dilapidated old building left over from an earlier world's fair in Seattle, and dependent largely on volunteer help, Gunther's museum finally got approval for a new, more suitable structure in the late 1950s. It was largely through her efforts that the complicated three-way funding (from the NSF, the state, and a sizable private bequest) necessary for the building's completion in 1962 was finally agreed upon. But according to a history of the museum, because Gunther had made so many enemies in the course of the long struggle to have the building built almost the way she wanted it, the university administration retaliated by forcing her to retire from the museum directorship in 1961, when she turned age sixty-five, rather than making an exception to its own rules and letting her stay on to open the new building. Instead, she saw her successor (Walter Fairservis, Jr., Harvard Ph.D., 1958) installed in time to open the building, but she also had the satisfaction of seeing him resign six years later for lack of local cooperation in running it. Meanwhile, and somewhat revealingly, Gunther refused to acknowledge her forced retirement from the directorship and continued to list herself in the *AMS* as holding the museum job until 1967, when she left for Alaska.[38]

An even more successful and more important woman builder, who did get much of the credit for her accomplishments, was Gertrude Cox, chair of the department of experimental statistics at North Carolina State University during World War II (1940–1944); then, as head of the statewide institute of statistics of North Carolina's consolidated university, she oversaw its explosive growth in 1945–60. Cox had come highly recommended by her professors and co-workers at Iowa State University at a time when hardly anyone worked in the area of experimental statistics. In later years she became widely known for co-authoring with W. G. Cochran the famous textbook *Experimental Designs* (1950; rev. ed., 1957). Besides training many of this and other nations' future faculty in this field, Cox was also instrumental in building up the Research Triangle in North Carolina, where she helped many agencies in the sixties to get funding for large-scale statistical projects and evaluations, including the development of "standard statistical package," since used by social scientists everywhere. Cox was thus one woman scientist who, like Cora Downs, got a good start during World War II and just kept on growing and building as her field broadened and the federal and other funding blossomed. She not only managed to ride the wave of Big Science in the 1950s and 1960s but to be enough ahead of it to help shape the form it took and the impact it had on her university, field, and region. As she grew older, rather than being ousted as Gunther was, she was accorded many professional honors, not only national ones, such as being elected president of the American Statistical Association and even a member of the NAS, but local ones as well, such as having a new building named for her and a scholarship fund set up in her honor. If this was the norm for many male scientists in the postwar era of big growth, it has been all too rare among women, and Cox must stand, until further research disputes it, unique in this regard.[39]

Women deans anywhere but in a school or college of home economics (see ch. 8) or a woman's college (see ch. 10) were rare. Possibly the nation's first woman dean of a graduate school (other than the Radcliffe or Bryn Mawr graduate schools, for women only) was applied mathematician Mina Rees, who served as the first dean of graduate studies at the reorganized City University of New York from 1961 to 1968.[40] Psychologist Leona Tyler served as graduate dean from 1965 to 1971 at the University of Oregon, where she spent most of her long career. Known for her several textbooks as well as for her research on interest development and counseling, she was then elected the fourth woman president of the APA, for 1972–73.[41] The first woman dean of a school of public health was Grace A. Goldsmith of Tulane, who graduated at the top of her medical school class there in 1932, served an internship at the famed Mayo Clinic in Minnesota, and then in 1943 was promoted to associate professor at the Tulane School of Medicine, at which point her department chairman told her that this was as far as she could hope to rise. Yet following hard work in the clinical nutrition of the B and C vitamins, many grants, long years of service on many important committees (including chairing the Food and Nutrition Board of the NRC from 1958 to 1968), the reception of many awards, presidencies of two (and later a third) major professional societies, by 1967 times had changed enough in the South for the trustees of Tulane University to approve this talented woman's appointment as the first dean of their new school of public health.[42]

Above this level, at a time of proliferating layers of new associate provosts, provosts, vice chancellors, chancellors, and vice-presidents of several sorts, there was a dearth of women scientists even at the institutions where women once had been so prevalent and still had considerable seniority. When one begins to imagine the hundreds of institutions of higher education with ever more administrators, so many that even a chairmanship or a deanship no longer meant reporting to the president, one begins to sense just how disadvantaged in comparison with their predecessors even those women were who were firmly on the faculty.[43]

Between 1950 and 1970 most parts of academia were transformed. Overall there was not only a tremendous expansion in faculty and physical plant but also a process of "upgrading" that did not affect the two sexes equally. In fact there are signs that the newly affluent academic institutions frequently enacted policies and practices that pushed out many of the women, who had once been numerous in some fields and at some kinds of institutions. It is probably safe to assume that many of them were aware that all this "progress" was undercutting rather than enhancing their position. Certainly the researchers at the BLS and the NEA documented many sexual differences in academic employment and in the 1950s tried repeatedly to bring the issues to public attention, only to have their reports used as evidence of how hopeless it was to expect academic institutions to hire women. Similarly, the members of the AAUW Committee on Standards and Recognition, some of the few remaining feminists of the time, tried until they were prohibited in 1963 to embarrass those institutions that had few women faculty or no women trustees. Yet several of the women scientists who did work at such institutions accepted what was available and strove for decades to build up those institutions. By 1970 they had made notable accomplishments.

Majors, Money, and Men
at the Women's Colleges

The women's colleges, which historically had trained and employed many women scientists, had a golden age of their own between 1945 and 1968. They prospered financially and continued to attract top-level students; about twenty new colleges for women were established in the early 1960s, when the number of degrees awarded peaked as well. Philanthropy was important at these usually privately controlled institutions, where many presidents and trustees ran successful fund drives for new dormitories, classroom buildings, libraries, and laboratories. The faculty benefited from higher salaries, federal fellowships and summer institutes, and occasional governmental grants for their research. But as elsewhere in academia, prosperity here brought other important changes as well. In particular the women's colleges hired more male faculty at higher salaries than ever before. The trustees evidently desired this and chose presidents (both women and men) who would hire men, often preferentially, to "normalize" the faculty, which was thought to have more than its share of older single women. Rarely did the remaining women faculty protest. But the women's colleges had hardly made over their faculties when the period ended cataclysmically with a loss of their traditional clientele and angry attacks by the alumnae. Starting in 1968, many men's colleges voted to admit women, and in 1969 Vassar College became the first major women's college to accept men. In addition, some former coordinate colleges were absorbed into fully coeducational institutions, often with adverse consequences for the women faculty, if there had been any, on the women's side. The numerous Roman Catholic colleges faced additional problems in the late 1960s, including the dwindling number of nuns and legal requirements that colleges be owned and run, not by religious orders, but by lay boards of trustees. When the Catholic women's colleges voted to admit men and even non-Catholic students, their formerly clear identity became doubly blurred. At the same time, some former alumnae angrily accused women's colleges of never having been very feminist. In 1968 NOW, founded by Betty Friedan (Smith 1942), had a subcommittee on education that produced the pamphlet *Token Learning,* written by Kate Millett, an angry critique of how some of the women's colleges had failed womankind. The year 1968–69 thus marked a turbulent end to a period that had started out quite differently.

The usual starting place for data on women's colleges is the USOE's annual count of the number of such institutions, although its totals are undercounted due to its omission of several of the "coordinate" colleges for women, including such large ones as Barnard and Douglass, which it counted as parts of "coeducational" institutions. Nevertheless, the number of separate four-year colleges and universities for women remained roughly constant, averaging 183 over the period 1948–59. This number included five institutions that awarded doctoral degrees (Ph.D., D.S.W, or M.D.): Bryn Mawr College, Radcliffe College, Smith College, Texas Woman's University, and the Woman's Medical College of Pennsylvania. Then in 1960 the number of women's colleges began to inch upward, reaching an all-time high of 204 in the fall of 1965. (The number of men-only colleges peaked a bit earlier, in the fall of 1961, at 198.)[1]

One large group comprised the many nondenominational private women's colleges, such as, to name alphabetically just a few of the better-known ones of the time, Goucher, Mills, Mount Holyoke, Smith, Vassar, and Wellesley. There were two private colleges for Negro women in the South (Spelman in Georgia and Bennett in North Carolina) after 1954, when the trustees of a third (Barber-Scotia in North Carolina) voted to admit men and other races. There were also in the South seven public "state colleges for [white] women," in Georgia, Mississippi, North Carolina, Oklahoma, Virginia, Texas, and South Carolina (Winthrop College). (The Florida State College for Women had been officially transformed into the coeducational Florida State University after the campus was overwhelmed by hundreds of veteran-students in the late 1940s.)

Almost 60 percent of these women's colleges were affiliated with the Roman Catholic Church. Although very little has been written on the history of the Catholic colleges for women in the fifties and sixties, their story can be told in terms of energetic builders, especially those mothers superior who, having obtained some land or a suburban estate from a wealthy benefactor, devoted their lives to raising funds, erecting new buildings, and strengthening the faculty and curriculum to attain regional accreditation. Among this group would be Sister Margaret Burke of Barat College of Chicago, Mother Eleanor O'Bryne of Manhattanville, Sister Mary Emil of Marygrove, Sister Madaleva Wolff of St. Mary's of Indiana, and Sister M. Jaqueline Grennan of Webster College in Missouri. A 1961 statistical survey of 4,700 faculty members at 93 Catholic women's colleges indicated that 75 percent of the science faculty were sisters, mostly in chemistry and biology. Of these Sister M. Emily, who taught chemistry at Regis College for decades, and Sister Florence Marie Scott, a biology professor at Seton Hill College, who spent numerous summers at the

Marine Biological Laboratory at Woods Hole, were outstanding examples (see also table 10.1). Besides increasing the number of nuns with doctorates, a major concern in these years was the hiring, promotion, and possibly tenuring of lay men and women. Prominent among the last would have been the marine biologist Mary Dora Rogick of the College of New Rochelle from 1935 to 1964, the anthropologist Ruth Sawtell Wallis at Annhurst College in Connecticut, 1956–74, and the chemist Lilli S. Hornig at Trinity College in Washington, D.C., from 1964 to 1969.[2]

Through another set of data from the USOE, that on earned degrees conferred, it is possible to identify the women's colleges, including the "coordinate" ones, and the fields (usually the same as a department) that were graduating the most women. Among the majors listed were fifteen sciences and social sciences, including such related fields as home economics and engineering. Thus, one can discover which colleges and even which departments were training the most women science majors in a particular year.[3] Since any year in the early to mid-sixties would be suitable, let us look at 1962–63, the last year in which the very large University of North Carolina's Woman's College at Greensboro, with an enrollment of 3,677 in the fall of 1962, was counted as a woman's college. (In 1963 it became coeducational.)

Table 10.2 gives an overview of the thirteen separate and coordinate colleges for women that graduated more than one hundred science or social science majors in 1962–63. Five would be best counted as "coordinate colleges for women," four were run by orders of Roman Catholic nuns, and four others were the major independents, the ubiquitous Mount Holyoke, Wellesley, Smith, and Vassar. Topping the list, with 223 science or social science majors, 200 of them in seven fields, was the rapidly expanding Douglass College, followed by Barnard, with 191 in ten areas, including 8 in physics (the most in that field in the nation). Vassar and Mount Holyoke had majors in the most fields (twelve and eleven, respectively). But these figures are deceptive, because at some schools the smaller fields were combined into joint departments. Wellesley, for example, which offered almost every conceivable major, including classical archaeology, reported no anthropology majors, partly because anthropology courses were offered as part of the sociology department.[4] Also, the biological sciences were also undergoing a consolidation on many campuses in these years. Thus Radcliffe reported a separate biochemistry major (with eleven graduates in 1962–63), and the Mississippi State College for Women (not in the table) had ten majors in bacteriology that year,[5] but most others would simply have counted such students under the biological sciences. Physiology, which had had a long history at many women's colleges, was also rarely listed separately by 1962–63; in fact Mount Holyoke's ten majors and Vassar's five were the only fifteen majors in the whole country that year.

Almost as interesting to look at as the majors drawing large numbers, which

Table 10.1.

Notable Nuns in Science, by Field and College, 1950–1970

Field	Name(s)	College
Chemistry	Sr. Emily (Martha Cahill), C.S.J.	Regis College, Weston, Mass.
	Sr. St. John Nepomucene (Elizabeth Fennessey)	Trinity College, Washington, D.C.
	Sr. Martinette (Hagan), C.D.S.	Mundelein College, Chicago, Ill.
	Sr. Claire Markham, R.S.M.	St. Joseph College, Hartford, Conn.
	Sr. Amata (Rosalie McGlynn)	St. Mary-of-the-Woods, Ind.
	Sr. Claire McNamara, C.S.J.	Regis College, Weston, Mass.
	Sr. Mary Ellen Murphy, R.S.M.	St. Joseph College, Hartford, Conn.
	Sr. Marian José Smith	College of St. Elizabeth, Convent Station, N.J.
Biology	Sr. Francis Solano Geisler, S.S.J.	Nazareth College, Rochester, N.Y.
	Sr. Elizabeth McDonald Seton	College of Mt. St. Joseph on the Ohio, Ohio
	Sr. Cecilia Agnes (Virginia Mulrennan), C.S.J.	Regis College, Weston, Mass.
	Sr. Anna Lawrence Roche, C.S.J.	Regis College, Weston, Mass.
	Sr. M. Rosari Schmeer, O.P.	College of St. Mary of the Springs, Ohio
	Sr. Florence Marie Scott	Seton Hill College, Greensburg, Penn.
Physics	Sr. M. Ignatia Frye, I.H.M.	Marygrove College, Detroit, Mich.
	Mother Marie Kernaghan	Maryville College, St. Louis, Mo.
	Sr. Aloysius Marie Metz	Trinity College, Washington, D.C.
	Sr. Mary Therese, B.V.M.	Mundelein College, Chicago, Ill.
Mathematics	Sr. Elizabeth Markham, R.S.M.	St. Joseph College, Hartford, Conn.
Anthropology	Sr. M. Inez Hilger	College of St. Benedict, St. Joseph, Minn., and elsewhere
Psychology	Sr. Mary Viterbo McCarthy, C.S.J.	Regis College, Weston, Mass.

Sources: AMS, various editions; *WWAW,* various editions; Sr. Mary Oates, C.S.J., Regis College, Weston, Mass., to author, Apr. 18, 1991.

Table 10.2.

Baccalaureate Degrees Awarded in Science at Thirteen Women's Colleges, 1962–63

Institution	Total Baccalaureates	Science Baccalaureates	Science Baccalaureates as % of Total	Political Science	Sociology	Psychology	Biology	Mathematics	Chemistry	Economics	Physics	Anthropology	Geology and Geography	Physiology	Astronomy	Engineering
Douglass	645	223	34.6	23	34	52	41	34	23	16	—	—	—	—	—	—
Barnard	796	191	24.0	40	26	38	22	19	11	20	8	6	1	—	—	—
Mount Holyoke	356	161	45.2	37	15	20	23	16	17	14	3	—	4	10	2	—
Wellesley	392	155	39.5	36	12	18	32	8	10	26	5	—	7	—	1	—
Mary Washington	288	147	51.0	4	41	43	16	20	17	6	—	—	—	—	—	—
Smith	468	140	29.9	50	14	14	15	13	8	16	7	—	2	—	1	—
Pennsylvania—Woman's College	444	114	25.7	25	21	10	16	14	14	4	—	9	1	—	—	—
Vassar	333	113	33.9	27	12	11	8	21	6	8	5	4	4	5	2	—
Regis (Weston, Mass.)	188	104	55.3	13	15	31	5	14	13	13	—	—	—	—	—	—
Mundelein	292	103	35.3	—	31	12	15	16	23	5	1	—	—	—	—	—
Trinity (D.C.)	210	102	48.6	25	10	13	12	7	11	16	8	—	—	—	—	—
Emmanuel	349	101	28.9	—	26	13	29	17	14	—	2	—	—	—	—	—
Radcliffe	293	101	34.5	17	25	5	22	10	5	3	5	9	—	—	—	—
Total	5,054	1,755	34.7	297	282	280	256	209	172	147	44	28	19	15	6	0

Source: USOE, *Earned Degrees Conferred, 1962–63.*

indicate institutional strengths and diversity, are some of the majors that drew smaller numbers, which indicate areas in which faculty and administrators were facing constraints and choosing among the various fields. This was the situation at the many colleges listed in table 10.2, which, being poorer than Wellesley and some other colleges, did not offer majors in ten or more sciences. Because of a lack of resources, to a certain extent these institutions had to go along with the prevailing stereotypes and concentrate their efforts on those fields that the women students were most likely to major in. As shown in table 10.2, most of the students majored in just seven sciences or social sciences: sociology, psychology, political science, biological sciences, mathematics (including statistics), chemistry, and economics. Partly because relatively few women's colleges ventured beyond this cluster to offer separate and thus countable majors in the fields of physics, anthropology, geology, geography, physiology, or astronomy, the totals in these fields remained small and the majors almost esoteric.

In fact one of the persistent curriculum issues at the women's colleges in the fifties and sixties was whether to add (or drop) the physics major. Although 572 colleges and universities awarded baccalaureate degrees in physics in 1962–63, according to the USOE, only one-quarter of them (145) granted *any* to women. Of these, only 24 were women's colleges (compared with 147 women's colleges granting degrees in mathematics, 136 in chemistry, and even 54 in economics), largely because hardly any of the many Roman Catholic women's colleges offered a physics major. (Trinity and Emmanuel were exceptional in this regard.) Often the decision about a change in the curriculum was tied to (or rationalized in terms of) facilities, such as equipment or laboratory space, though in fact it often also had to do with personnel, which in the long run may have been more costly. Thus at Douglass, when the longtime professor Wilfred Jackson died in 1959, Dean Mary Bunting decided that because it would be too expensive to modernize his laboratory properly for the few likely majors, she would not replace him but would have the undergraduates needing elementary instruction in physics cross-register and take the courses on the Rutgers campus.[6] That same year, however, the faculty at Mary Washington College of the University of Virginia made the opposite decision: now that its new science building was finally completed, it was time to add a major in physics,[7] though as table 10.2 reveals, Mary Washington's physics department still had few majors in 1962–63. Nor did Bryn Mawr's, despite the presence of Walter Michels, an active participant and leader in many physics education and teaching projects in the fifties and sixties.[8] By contrast, Trinity College of Washington, D.C., had 8 physics majors among its 210 graduates in 1962–63; and the even smaller Sweet Briar College had 4 in its graduating class of just 117 in the same year, thus comparing well with the immense University of California at Berkeley, whose total of 5 women physics majors made it the coeducational institution with the largest number that year. Meanwhile, Smith College, with 7 majors in 1962–63, had been proud in 1959 to be chosen as the first women's college to have a student section (29 members) of the American Institute of Physics.[9]

Similar tales could be told of the earth sciences, astronomy, and physiology majors at the women's colleges in these years.[10] A few new majors were offered (e.g., geology and geography at Mary Washington College in 1964), and a few departments were discontinued (e.g., geography at Wellesley in 1965, even though it had contributed six of Wellesley's seven in earth sciences in 1962–63).[11] But although they were rare (there were none at any of the Catholic women's colleges), these departments played a very important role nationally in the careers of women in these fields, for they were among the largest sources, and often were *the* source, of women majors in those fields in the country. Although larger universities routinely offered majors in these fields to students of both sexes, in practice very few of the women undergraduates at those institutions completed majors in these subjects (unlike their female classmates in the sociology or psychology departments). Thus, those astronomy majors or geology departments that did exist were more important for future women in science than their small numbers might at first indicate. Accordingly, behind the numbers in table 10.2, even behind the dashes or zeroes, lies a tale of the choices made at the women's colleges in the fifties and sixties not only by the undergraduates deciding which field to major in but also by the faculties and administrators determining which fields they could afford to offer. Only the relatively wealthy women's colleges, partially listed in table 10.3, could afford to defy stereotypes and to mount and maintain a major in a field that required expensive equipment and laboratories but, for whatever reasons, might graduate relatively few women per year.

But public appreciation for these achievements was even rarer than the actual research, as was shown repeatedly in the case of Mount Holyoke College. By almost any count (e.g., the number of graduates who earned doctorates or the number of articles published by the faculty), Mount Holyoke continued to rank as one of the "most productive" chemistry departments among liberal arts colleges in the United States.[12] Yet this anomalous status bewildered and confused many chemists, even those who were concerned with research at small colleges. For instance, when in 1959 chemists surveyed liberal arts colleges and then held a meeting to encourage them to do research (and to encourage the NSF and other foundations to support more of it), they were astonished to discover that several women's colleges were already quite active in this regard. In fact so many (sixteen) responded to their questionnaire that they asked Anna Harrison of Mount Holyoke to analyze the data. She did, and since she attended their meeting, the final report included her in the group picture, even though the women's colleges were all omitted from the final report!

Table 10.3.
Endowments at Fourteen Women's Colleges, 1963

Institution	Endowment	
	Market Value	Average per Student
Wellesley	$83,634,000	$48,038
Vassar	53,151,000	34,469
Smith	42,406,000	17,855
Bryn Mawr	30,523,000	29,577
Mount Holyoke	27,208,000	16,550
Radcliffe	24,594,000	19,946
Scripps	10,041,000	31,575
Chatham	9,232,000	16,575
Goucher	8,085,000	9,478
MacMurray	4,437,000	4,570
College of Notre Dame (Md.)	911,000	887
College of Mt. St. Joseph-on-the-Ohio	840,000	855
St. Joseph (Conn.)	570,000	688
Caldwell College for Women	357,000	400

Source: D. Kent Halstead, *College and University Endowment, Status, and Management*, USOE, Bureau of Educational Research and Development (Washington, D.C.: GPO, 1965), 75, 77.

Entitled *Research and Teaching in the Liberal Arts College*, it did not include any data or discussion on chemistry at the women's colleges but instead blithely limited itself to coeducational and men-only institutions. It thereby added to, rather than lessened, the misimpression that chemical research was something done only by men.[13]

Yet this rather deliberate underrecognition was mild compared with the systematic omission of the women's colleges from the major study of the time concerning the undergraduate institutions that had trained many of currently active scientists. When Robert H. Knapp and H. B. Goodrich of Wesleyan University sought to determine the collegiate background of those scientists in the classes of 1924 through 1934 who were later listed in the *AMS*, there were many women in their institutional totals or raw data, since despite its title, the directory included them. But when the authors learned (erroneously) that women were just 2 percent of the science Ph.D.'s, they dropped them from further calculations and the text:

A further important decision was to compute the index for male graduates only. Our raw data included graduates from men's, women's, and coeducational institutions. Since women relatively rarely enter careers in science (less than 2 per cent of the PhD's in science [physics?] are women), it is obviously unfair to make direct comparisons based on the total number of graduates from these three types of institutions. Various statistical adjustments to make allowance for the percent-

age of women graduates might have been utilized. It was finally decided, however, to use the simple device of eliminating women altogether from our primary
data. Therefore women's colleges were excluded from our lists, and only male
graduates and male scientists were included in our study.[14]

Thus when readers and, later, social policy analysts looked in *Origins of
American Scientists* to see where American scientists had gone to college, once
again they were led to think that only men did science and that a few private
liberal arts colleges, such as Oberlin, Swarthmore, and Reed, had trained a
high proportion of their graduates in science. The fact that the study omitted
the productive women's colleges like Mount Holyoke, Vassar, Wellesley, and
Smith was ignored. The study may also have diminished the relative standing
of such major coeducational universities as Cornell, Berkeley, or the University
of Michigan, many of whose women graduates were listed in the *AMS*. Yet by
omitting the women, this broadly titled and seemingly neutral and authoritative effort at social science perpetrated the prevailing notion that women did
not do science and denied the considerable contributions of even the strongest
women's colleges. Yet hardly anyone noticed, and no one protested at the time.
In fact Knapp and Goodrich's findings with regard to the value of small liberal
arts colleges (such as, it should be noted, their own Wesleyan, which had
supported the study) were immediately accepted as proven truths by the manpower analysts of the time.[15]

For their science classes the women's colleges also had the usual continuing
need for modern but increasingly expensive scientific equipment and apparatus. Sometimes instruments, such as Smith's new sixteen-inch reflecting telescope in 1964, were the gift of one or more individuals, perhaps in memory of
another. At other times local businesses donated chemicals or surplus electronic equipment. On occasion a nearby research university might loan indefinitely
a machine which it no longer needed, or the department or a faculty member
might purchase a new piece of equipment with the help of a small grant, for
example, from Sigma Xi, the Research Corporation, or, later, the Brown-
Hazen Fund. But increasingly the federal government was the source of new
funds for equipment. Sometimes this funding was designed to enrich ongoing
instruction, as in the case of the AEC's efforts to add radioisotopes and nuclear
science to the undergraduate curriculum in the 1950s. These efforts included
visits to campuses in a van equipped with radioactive materials and Geiger
counters (e.g., to Agnes Scott College, outside Atlanta, in 1959), as well as
grants to colleges for the purchase of supplies and equipment. Wellesley College, for instance, was proud to report in 1961 that its AEC grant had enabled
the departments of botany, zoology, and chemistry, as well as physics, to use
radioactive substances in their classes and laboratories.[16]

As federal grants grew larger in the sixties, some women's colleges used them to expand their course offerings. Goucher College, for example, with the aid of an NSF grant, in 1961 purchased an IBM 1620 computer, one of the first woman's colleges, it claimed, to do so. Dorothy Bernstein, a professor of mathematics and in 1961–67 director of Goucher's new computing center, was so successful in adding applied mathematics to the Goucher curriculum that when she needed a newer machine with a larger memory in the late 1960s, the NSF provided it.[17] Elsewhere the rationale for the new federally financed equipment was often faculty research, though governmental and college officials both hoped that the new techniques would find their way into the undergraduates' advanced courses. Such was the case at Wellesley in 1964 when it obtained the first electron microscope not only at a woman's college but at any undergraduate liberal arts institution. (Haverford and Reed were soon to follow.) Because Helen Padykula, a prominent cell biologist, was rejoining the Wellesley faculty after ten years in the lower ranks at Harvard Medical School, the NIH, which was continuing to support her researches, provided Wellesley with the new machine. Wellesley cooperated by making her the director of its new Laboratory for Electron Microscopy; her duties were to include introducing the new molecular biology to the undergraduates.[18] Finally, in the late 1960s, when grants were briefly available for even more ambitious "institutional development" projects, Bryn Mawr president Katherine McBride, an astute builder and one of the very few women to have served on many NSF committees, was able to obtain a large grant ($403,000) to continue to include biochemistry among her college's offerings.[19]

Much more could be written than is possible here on the impact of the new federal and private programs, especially those of the Ford Foundation, on undergraduate education at the women's colleges in the 1950s and 1960s. For example, eight women's colleges were included in the series of five conferences supported by the NSF in 1953–54 on research in astronomy, biology, chemistry, geology, and physics at small colleges. (The one on biology was held at Bryn Mawr in April 1954.) Similarly, when the Research Corporation made grants totaling about $6 million to one hundred colleges between 1945 and 1968, it included many women's colleges, and ten of the fifty institutions surveyed in its final report were women's colleges.[20] Suffice it to say that on many, if not most, of these campuses the pace began to quicken in science departments as the energetic, ambitious, and more research-oriented members of the faculty began increasingly to go off on stimulating summer programs (e.g., "institutes") or year-long sabbaticals (as science faculty fellows) and returned refreshed, enthusiastic, and full of new ideas. Many of the women holding these awards were nuns.

These and other faculty also held research grants for work on campus. Among the women at the women's colleges who received NSF research grants in the fifties were physicist Rosalie Hoyt of Bryn Mawr ($3,400 to study the bioelectric behavior of algae with an analog computer), astronomer Alice Farnsworth of Mount Holyoke ($1,500 for a project on the Milky Way), chemist Sister Mary Martinette of Mundelein ($1,200 to continue work on the stereochemistry of several inorganic compounds), and embryologist Dorothea Rudnick of Albertus Magnus ($3,800 to study the effect of an enzyme on chick embryos, probably at nearby Yale University, where she was also a "research guest" from 1940 to 1971).[21] In the sixties the NSF's science education division also made institutional grants to colleges specifically to pay students to help professors during the summer. Barnard had a grant in 1961–64, and Wellesley had one in 1962–65, from this Undergraduate Science Education (later renamed Research Participation) Program, which often supported research projects for upperclassmen seeking honors. In addition, once the research was completed, the college usually kept the new equipment for use in teaching. All over the country, from Wellesley to Mundelein to Spelman, science faculty and science majors were attracting new interest and new resources in the 1960s.[22]

Besides these generally modest faculty research projects and the extensive efforts of the chemistry department at Mount Holyoke, three women's colleges had four more ambitious efforts that bordered on being full-scale research institutes. The first and second were at Smith, where Albert Blakeslee, starting in 1943 and continuing until his death in 1954, headed the Smith College Genetics Experiment Station. After retiring from the Carnegie Institution's department of genetics at Cold Spring Harbor, New York, in 1942, Blakeslee had gone to Smith as its William A. Nielson Professor, a visiting position for distinguished researchers. While in Northampton Blakeslee, who was also still heavily involved in seeing *Biological Abstracts* through the war, arranged to reestablish his laboratory at Smith and move two of his longtime staff members—Sophia A. Satin, a Russian émigrée, and Amos Avery—there as well. Largely supported by the American Cancer Society, the NSF, and the Rockefeller Foundation, the team did important work on plant tumors, often in the absence of Blakeslee, who traveled around the world receiving foreign honors.[23]

Meanwhile, upon her return from war work Professor Gladys Anslow of Smith's physics department obtained a grant from the new ONR, the first person at a woman's college to do so. This grant and a renewal that lasted until 1954 enabled her and Dorothy Wrinch, an English-American mathematician with controversial ideas about protein structure, to transform a former coalbin into a basement laboratory and equip it with several spectrophotometers in order to pursue their "physical and chemical studies of biologically important molecules," including insulin and hemoglobin. When, however, their results failed to agree with those of others (notably the alpha helix of Linus Pauling), the ONR refused to support the women's project any further. This situation lasted until 1960, when some new evidence for the women's glycocol structure allowed the NSF to step in and support their controversial work a few years longer.[24]

A large research unit for a women's college in the fifties and sixties in terms of staff and equipment was the one that nutritionist and master grantswoman Pauline Beery Mack set up at Texas Woman's University in the fifties. After giving up her Ellen Richards Institute at Pennsylvania State University in 1952, Mack moved to the expanding TWU and soon, in the era of concern over fallout from atomic tests, had grants from the Surgeon General's Office and the NIH to test bone density in human feet. While pursuing this work, her group developed such strong ties to the Oak Ridge Institute of Nuclear Studies that in 1960 TWU was elected the thirty-ninth member (the first and only woman's college) of this regional consortium. By then TWU's president could claim that twenty-two members of the faculty and staff were working in the areas of radiation chemistry and biology. In 1963 Mack's group started a new series of grants and contracts with NASA to study the loss of calcium (or "bone demin-eralization") among astronauts and others subjected to long-term immobility. This work continued through the Apollo flight in 1969. Meanwhile, the univer-sity's exuberant male president announced in 1962 that TWU had $368,500 worth of federal research projects under way and that the five-year total for 1957–62 was $1.13 million. In 1964 he announced the start of doctoral programs in radiation chemistry and radiation biology.[25] Although this was obviously a little farther and faster than most women's colleges were willing to go in the 1960s, it is an indication that under certain circumstances these institutions could and did perform as modest research universities and get some public recognition for it.

The fourth and last (but not least) research project at a woman's college in these years was the antifeminist Mellon Research Program at Vassar College from 1952 to 1958. Funded by a $2 million endowment from Mary Conover Mellon (the first wife of Paul), who died in 1946, it employed a staff of five male psychologists, headed by Nevitt Sanford, later of UCLA, who studied the undergraduates in the classes of 1954 and 1958, as well as a control set of alumnae (the classes of 1929–35). Although at first this research had been expected to yield insights helpful to the current faculty, who were thought to be in some ways out of touch with their students, before long it had undercut them so badly that there was little communication between the project and the faculty. In general the researchers set up categories, typical of the antifeminist spirit of the times, that classified the compliant, conformist "under-achievers" as well adjusted to their future roles and the independent, nonconformist high achievers (apparently there were some!) as being maladjusted. As David Ries-man, a close student of the women's colleges, described the group's findings in 1964, the (male) researchers discovered in testing and talking to the women students that many were so repelled by "some of the unmarried women on the faculty [that] they may fear that such involvement would cut them off from the life of a normal average woman, and they are persuaded that it is more important to be a woman than to become some kind of specialist." Luckily, most of the project's staff had left Vassar before many of their results were published in a massive volume in 1961.[26]

Besides enriching their science curriculum, most women's colleges in the fifties and sixties erected several new buildings, including new science laboratories and classrooms. (Several of the buildings erected or remodeled in those years are listed in table 10.4.) They ranged from a sixty-five-thousand-dollar observatory donated to Agnes Scott College in 1949 (to house its newly purchased thirty-inch reflecting telescope)[27] to Smith College's multimillion-dollar Clark Science Center, completed in 1966. Officials at Smith claimed, probably accurately, that Clark was the largest to date at any woman's college in the nation. It comprised three buildings—one renovated older laboratory and two new ones (McConnell Hall and Sabin-Reed, the latter named, appropriately, for two prominent alumnae who had been close friends). Estimates of the size of the new complex varied, depending on whether one counted the renovated space, and the total cost increased from $6.6 million to $8.5 million when the furnishings and equipment were included.[28]

Fund-raising for necessarily expensive science buildings in the fifties and sixties heightened the existing dissatisfaction with the low level of corporate support going to the women's colleges generally. Several attempts were made to correct this deficiency by strengthening the bonds between industry and the women's colleges. One of the first attempts was a fund drive at Mount Holyoke in 1952 for a new chemical building. Several of the retired and soon-to-retire faculty members of its famed chemistry department—Emma Perry Carr, Mary Sherrill, and Lucy Pickett—gave speeches, ran symposia, including one in October 1952 entitled "Science, Industry and Education," and prepared appeals for support from chemical, especially petroleum, companies that included long lists of their many graduates (with master's degrees as well as bachelor's) who had gone on to careers not only in teaching, research, and medical work but also in industrial chemistry. In December 1952 the Mount Holyoke board of trustees announced the formation of an Industrial Associates for Women in Science program that promised "to pioneer in the development of new forms of co-operation" between industry and a woman's college. Specifically the college was to develop projects of interest to industry that would in turn help to "advance the educational aims of the college." But it is not clear just how much financial support this effort yielded, for a timely bequest in 1953 by Newcomb Cleveland, a former advertising executive in New York City, provided the $1 million needed for the new chemistry building. Completed in 1954, the structure was dedicated as Cleveland Hall, though its wing of ten laboratories was named for Emma Perry Carr herself. The account of the opening in the *C&EN*

Table 10.4.
New Science Buildings at Selected Women's Colleges, 1945–1970

Year	College	Building	Size	Cost (Source of Funds)
1949	Agnes Scott	Bradley Observatory (A)		$65,000 (Family?)
1951	Agnes Scott	Campbell Science Building (B, C, P)		3 million
1953–54	Mount Holyoke	Cleveland Hall (C)		1 million
Mid-1950s	Goucher	Hoffberg Science Building	Smaller than previous	
1957–58	Bryn Mawr	Science Center I (B)	30,000 sq. ft.	(Federal)
1957–59	Mary Washington	Morgan Combs Science Center (A, B, C, G + G, M, P)		675,000 (800,000 equipment)
1962–63	Bryn Mawr	Science Center II (C, G, M, P)	45,700 sq. ft.	(Federal)
1962–63	Stephens	Science Center (B, C, G, M)	One floor	
1963	Douglass	Davison Hall (HE, Ps)		(State)
1964–66	Smith	Clark Science Center (A, Ba, B, C)	220,000 sq. ft.	6.6–8.5 million
1966?	Texas Woman's University	Biology		(Federal)

Sources: Alumnae bulletins, college histories.

Note: A = astronomy; B = biology; Ba = bacteriology; C = chemistry; G = geology; G + G = geology plus geography; HE = home economics; M = mathematics; P = physics; Ps = psychology.

reported that of all the ACS-approved colleges and universities in the country, Mount Holyoke College had graduated the largest number of women majors in 1953–54. Although with all the retirements at Mount Holyoke, one era was ending, surely with the new building and strong undergraduate program a new one was beginning.[29]

In other ways too the whole scale of college philanthropy was changing in the 1950s. Starting in 1954 with the "corporate alumnus" program of the General Electric Charitable and Educational Fund, increasing numbers of business foundations were stimulating alumni giving by matching employees' gifts to their alma mater. Unfortunately, very few of General Electric's professional employees were women, and the program did not extend to the donations by employees' wives, a point raised several times in the *Wellesley Alumnae Magazine* between 1956 and 1961. In these five years, however, the number of such matching gifts to Wellesley increased from three to eighty-four.[30] Although the Ford Foundation's highly publicized 1955 gift of $260 million to 615 private colleges and universities for faculty salaries stimulated even more corporate and alumnae contributions, alumnae giving was already well under way at the major women's colleges. In fact, when Vassar's president Sarah Gibson Blanding upon her retirement in 1964 assessed the changes during her eighteen years in office, she gave as a major one the great increase in the size and organization of what had come to be called the "development office." Whereas in 1946 the comptroller's office had employed a part-time secretary to pen polite thank-you notes, by 1964 the operation had become a large annual effort employing several persons. It was generally successful in meeting its goals. In 1955, for example, Vassar had announced a ten-year plan to raise $25 million, which it accomplished a year early, just in time for Miss Blanding's retirement. Then when the Ford Foundation gave Vassar and Smith three-for-one matching grants of $2.5 million in the 1960s, their well-organized development offices were able to raise another $7.5 million to complete new projects within just a few years, though Vassar's long-awaited new science center, Olmsted Hall, was postponed until the 1970s.[31]

Alumnae giving at Smith and Wellesley was sparked by two prominent women, Elizabeth Reeve Cutter (Mrs. Dwight) Morrow, longtime chairman of the Smith board of trustees, and her good friend Florence Corliss Lamont of New York City, from the Smith class of 1893, who also gave generously to Wellesley and other "Seven Sister" colleges. If Wellesley's president Margaret Clapp could be credited with tripling the college's endowment in twenty years (1947–66), she was greatly aided by the fund-raising skills of women like Lamont and Dorothy Bridgman Atkinson Rood of Minneapolis (who also masterminded the AAUW Million Dollar Fellowship Fund). In some ways these energetic and successful women philanthropists and fund-raisers were more important to the wealthy women's colleges than were their actual presidents, about whom much more has been written.[32]

Table 10.5 lists some of the longer serving or more eminent women faculty at a few women's colleges in the period 1950–65, chosen to demonstrate a diverse as well as strong group of women's colleges. The continuation into these years of a few of the old protégée chains, particularly at Wellesley College (where botanist Harriet Creighton, physicist Janet Guernsey, chemist Eleanor Webster, psychologist Thelma Alper, and geneticist Dorothea Widmayer all succeeded their professors) and Bryn Mawr (where anthropologist Frederica de Laguna, mathematician Marguerite Lehr, and geologist Maria Luisa Crawford took up the baton), is evident. More common by this time was the hiring of graduates of other women's colleges (for example, Mount Holyoke graduates Jean Crawford and Virginia Mayo Fiske both taught at Wellesley).

Something of a silent revolution was also under way in the 1950s with regard to the marital status of women faculty members at the women's colleges. Whereas formerly all women faculty had been single and had been required to resign their positions upon marriage, during World War II and increasingly thereafter they were allowed to stay. What was perhaps even more of a break with the past, women who were already married and even had children (such as Janet Brown Guernsey, perhaps Wellesley's first assistant professor to be not only married but also the mother of five)[33] were hired for tenure-track positions. In fact, married faculty quickly became the norm at Wellesley College in the fifties, as a big black binder on college statistics kept by President Margaret Clapp demonstrates. In the fall of 1952 the proportion in the column marked "% Married" was 20 percent. Just two years later it had doubled to 39 percent; by the fall of 1962 it was up to 50 percent; and in October 1964 the last entry was 61 percent. Evidently, almost all of the new persons hired (and these figures included male and female instructors and lecturers) were married, while most, if not all, of the retirees and others departing were unmarried.[34] When Virginia Mayo Fiske, the mother of three, was asked in 1954 whether the Wellesley administration had any prejudice against married women on the faculty, she replied, "Not consciously so. . . . The administration does show a distinct lack of enthusiasm toward expectant mothers on the staff even though these women as a group are more conscientious than ever in their maintenance of high standards in their teaching [and] carrying their full responsibilities. I suppose the difficulty here is that once in fifteen years or so one of them gives up or feels she cannot carry a full program for health reasons."[35]

The women's colleges also employed a few scientist couples in these years, In some cases both partners held faculty rank: Elsa Siipola and Harold Israel were both professors in Smith College's psychology department; young Maria and William Crawford taught in Bryn Mawr's geology program;[36] and for a while Georgene and John Seward were both assistant professors of psychology at the Connecticut College for Women. In December 1944, however, the new (and, as it turned out, temporary) president of Connecticut College informed the Sewards that though they had been assured by previous administrators that they would both get tenure, she did not approve of two incomes for one family, which in any case the college could no longer afford. She gave them one month to leave! Luckily, by 1948 they both landed posts in southern California that led to tenure.[37] Years later the presidents of two other women's colleges discussed the advantages and disadvantages of such dual faculty status for couples. Katherine McBride of Bryn Mawr was more positive about this practice than was Smith's Benjamin Wright. Although both agreed that in a small department a couple could create unspecified personnel problems, McBride saw the advantage that if both were good, faculty status for both would be a great boon to the college in a time of high faculty mobility, for neither partner would be likely to leave.[38] It is interesting that neither of these presidents of major women's colleges at a time (1954) when the government was exhorting more women to go into science expressed the opinion that it was important for the students to

Table 10.5.

Notable Women Science Faculty at Eight Private Women's Colleges, by Field, 1940–1970

Field	Bryn Mawr	Duke	Mount Holyoke	Regis	Smith	Spelman	Vassar	Wellesley
Anthropology	*de Laguna						Lee Codere	Goodman
Astronomy			*Farnsworth		Williams		Makemson	Hill
Biology	*Gardiner *Oppenheimer	Hunter	*Adams *Haywood Boyd *Sprague Eschenberg Kallenbach Beeman	Lawrence Roche *Muirennan	Sampson TeWinkel Carpenter Horner Hobbs	Albro Patterson	Pierce Dewey Wright Lumb Zorzoli	*Austin Fiske Padykula *Widmayer
Botany		Addoms Philpott	Stokey Reed		*Bache-Wiig *Kemp			*Creighton
Chemistry			Sherrill Pickett Harrison	*McGarry	Burt		Ellis Hillis	Crawford *Webster
Economics		Kreps			Lowenthal		Newcomer Myers	Bell

Geology	*Crawford						Kingsley
Mathematics	Pell-Wheeler Lehr	Litzinger Bates	Burke	Kierstead Stobbe Dickinson	Caldwell Falconer McBeay	Newton *Asprey	Schafer
Microbiology	Bliss						
Physics	Hoyt Sponer-Franck	Allen		Genung Robinton *Anslow Wrinch[a]		Healea	Wilson Heyworth *Guernsey Kohn Fleming
Physiology						Conklin	
Political science		Muus Schuck					
Psychology		Banham Reichenberg-Hackett		Carter *Siipola		*Gleason *Kambouropoulou	Heidebreder *Alper *Zimmerman

Sources: Assorted histories, alumnae magazines, and obituaries; *AMWS*.

* Alumna.

[a]Lecturer and visiting professor for twenty-four years.

Table 10.6.

Faculty (Assistant Professors and Above) at Three Women's Colleges, by Sex, 1940–1971

	Smith			Vassar			Wellesley		
Year	Total	Men	Men as % of Total	Total	Men	Men as % of Total	Total	Men	Men as % of Total
1940				109	29	26.6			
1947–48	223	89	39.9						
1948–49							99	21	21.2
1949–50							116	29	25.0
1950–51				106	35	33.0			
1951–52							110	33	30.0
1956–57	200	102	51.0				84	34	40.5
1960–61				138	62	44.9			
1964							102	40	39.2
1965	230	149	64.8						
1970–71				154	97	63.0			

Sources: [Myra Sampson], "A Study of the Teaching Faculty of Smith College, 1956–1957," and idem, "Repor[[on the] Status of Women Faculty in Academic Departments in Smith College, 1955–56 vs. 1965–66," Augus[18, 1966, both in Myra Sampson Papers, Smith College Archives; Vassar course catalogs; Statistics Notebook[President's Office, 1 DB/1899–1966, Wellesley College Archives.

Note: Figures for Wellesley in 1956–57 and 1964 are for full-time faculty only.

see married women performing responsible scientific work and holding appropriate professorial status.

Besides the inclusion of married women and couples, an even more striking change in the faculties of the women's colleges in the decades after World War II was the large influx of men. This was not necessarily because the size of the faculty at the women's colleges was increasing—in fact, at some colleges it dropped first before rising in the late 1950s and after—but because of high turnover due to retirements, resignations, and other terminations. Table 10.6 provides data on the size of the faculty, both total faculty and men faculty, at the level of assistant professor and higher at three important colleges from 1940 to 1971. There was a steady rise in the proportion of men on the faculty from the 1940s through the 1960s, reaching a high of over 40 percent at Wellesley in 1956–57, 64.8 percent at Smith in 1965, and 63 percent at Vassar in 1970–71 (a year after its trustees had voted in favor of admitting male students). Evidently women were leaving (either as older women retiring or as younger women whose contracts were not renewed) and being replaced by men, who often started at higher levels than had the women. Since the process did not end with the waning of the GI Bill veterans in the early 1950s but continued at least into the early sixties, the erosion of the formerly quite segregated women's bastion was fueled or at least made possible, as elsewhere, by the rising salaries of which

Presidents Blanding, Clapp, and others were so proud. *Newsweek* reported favorably on the phenomenon in a 1956 piece on Sarah Gibson Blanding's first ten years at Vassar. Praising her for sharply increasing the number of male faculty there from 46 to 75 (out of 182, or 41.21 percent), it added that this was "a balance in a female institution much to be wished for."[39]

It became a truism of the fifties that higher salaries were necessary for the liberal arts college to maintain or increase the quality of its faculty in the face of great competition from industry and universities. Even leaders of major women's colleges, most notably Sarah Gibson Blanding of Vassar College, wrote and spoke widely on this theme. Yet what was rarely said in the fifties and sixties was that this competition involved a great deal of gender politics, for the higher salaries were only necessary if one wished to hire men with families and provide the proper fringe benefits, including at times even free faculty housing. Rarely would a woman candidate have an alternative offer from industry, a university, or even another liberal arts college. Nor was she likely to forsake academia for higher-paying fields such as medicine, law, or industrial chemistry. Thus the frequent public discussion of the need for "competitive salaries" can be seen as a kind of code phrase for the mainstreaming and consequent masculinization of the women's college faculty.[40]

The process was so blatant and offensive to some at Smith College, long something of a battlefield in the area of faculty gender politics, that a series of protests broke out in the mid- to late 1950s. An early encounter was at a faculty discussion in the spring of 1954 where recent hiring practices, including the "sex ratio," were reportedly discussed quite heatedly.[41] Then in early 1957 Smith's retired zoologist Myra Sampson (aged seventy-one, at Smith since 1909, and possibly a former suffragist) sat down and collected from old catalogs, annual reports, and directories some relevant statistics, which she then "presented at an informal meeting of a small group of women of the faculty," as she inscribed the copy she later sent to the college archives. In this she compared the number of men and women in each rank for the academic years 1947–48 and 1956–57. (She also presented data on each department and the length of service and rate of promotion for men and women.) She found that although the size of the faculty had dropped in these nine years from 223 to 200, the number of men had increased from 89 to 102 (see table 10.6). The number of women had simultaneously decreased sharply, from 134 to 98, or from 60.1 percent of the faculty in 1947 to just 49.0 percent in 1956. Not only had far fewer women than men been hired in the previous decade but they had been paid less at each rank, promoted less often, and forced to start as instructors rather than as assistant professors, as men with the same qualifications routinely did. She concluded her presentation, to which there was no official response though word of it must have spread, with some suggestions for future recruitment and appointment; these included looking beyond New England for future faculty, paying equally, encouraging more seniors to go on to graduate school, and "modify[ing] the Tenure plan."[42]

Sampson, a 1909 graduate of Pembroke, may have mentioned the need for a broader study of the topic to her friends there, possibly at her fifty-year reunion in 1959, for in 1961 Frances Clayton, an associate professor of psychology at Brown University, investigated the issue at the behest of the editor of the *Pembroke Alumna*. From a comparison of the 1940 and 1960 course catalogs of eight women's colleges she found that indeed the proportion of women faculty members was down considerably since 1940. Yet unlike Sampson, Clayton insisted that this could not be due to discrimination or preferential hiring, since these schools were the traditional employers of women faculty; instead she attributed the decline to the colleges' rising standards and their requirement that faculty have doctorates. She felt that since the proportion of women doctorates was low, the decline was justified. But then either Clayton or the editor undercut her own argument by providing a chart showing that the actual numbers of doctorates awarded to women had more than doubled between 1940 and 1957.[43]

But Professor Sampson did not give up. Because the situation for women faculty at Smith worsened in the next decade, in August 1966 she renewed her efforts and sent an even more extensive memorandum on the status of Smith's women faculty to the president, with copies to the trustees, the chairman and vice-chairman of the board of counselors, and the officers of the alumnae association. This time she compared the figures for 1955–56 with those for 1965–66. She found that the proportion of women had fallen even further in these years, from 49.0 percent (101 of 206) to just 35.2 percent (81 of 230). This defeminization occurred at all ranks except that of instructor, the only one where women were in the majority in 1965–66. The drop was particularly great at the level of associate professor, at which women had been 60.5 percent of the total in 1955–56 (23 of 38) but were only 22.9 percent in 1965–66 (11 of 48). Evidently women were being neither appointed nor promoted to the higher ranks at Smith College. Of the total number of promotions in ten years, 123 (64.7 percent) went to men, and only 67 (35.3 percent) went to women. The situation with regard to initial appointments was even worse, for of the 117 made in this decade only 22, or 18.8 percent, were women. In two departments, astronomy and botany, there were no women faculty at all, not even instructors. In six departments, including chemistry, physics, sociology, and government, there were five times as many men as women.

Nor did the courageous and outspoken Sampson see any prospect of future improvement. On the contrary, because of the compulsory retirements in the next five years (1966–70) of fourteen of the remaining twenty-seven women professors and one of the eleven associate professors, the percentage of women in these ranks and overall would be dropping to even lower levels. Obviously there was something of a crisis at Smith even if she was the only one willing to document and talk about it. Sampson concluded her report with the menacing statement that "there will be few if any Women Professors or Associate Professors in Smith College, in the future, unless the present procedure is revised."[44] There is no evidence, however, that this second, more forceful appeal led to any changes or even received an official response. But there is considerable evidence that the college trustees and administrators not only knew what they were doing but had presumably clear reasons for doing it. Hiring and preferring men was a longstanding though unwritten policy of the college, and the recent prodigious fund-raising was making it more possible in the fifties and sixties than ever before.

The policy of preferring male faculty members at Smith reportedly extended back to its first presidents, though it was Marjorie Hope Nicolson, dean of the college from 1929 to 1941 and a very close associate of President Neilson, who put it most succinctly when in her oral history memoir late in life she asserted that Smith had been the best women's college of its time because it had the most men on the faculty![45] She and others admitted that in order to attract and retain such men, certain inducements were offered, including higher ranks and salaries than those of comparably qualified women, reduced teaching loads, convenient scheduling of classes (to permit holding a second job, living in New York City, or writing for long stretches of time), two series of subsidized publications (in the humanities), and relatively rapid promotions. As a result the men on the Smith faculty became known for their many publications; and in 1947 four Smith men in the humanities won Guggenheim fellowships, compared with Harvard's three and only four others in all of New England.[46]

This subsidized hyperproductivity of Smith's male faculty was inadvertently confirmed quantitatively in the mid-1950s by a joint committee of faculty and trustees at Radcliffe College that was preparing a report on its past doctorates. Hoping to show what a productive group the Radcliffe Graduate School had trained, the committee sought to compare their publication records with those of some suitable control group. Finding none, its staff associate (Cecilia Kenyon, an assistant professor of government at Smith) chose, presumably for the convenience of the data, the Smith faculty, men and women. This was unfortunate, for her data revealed that the Smith men outproduced not only the Smith women faculty but the Radcliffe doctorates as well, rather than vice versa. Accordingly, the committee's final report included only one brief paragraph about this comparison, on rank. It stated that 69 percent of the men at the level of professor had published "extensively" (two books or twenty or more articles), compared with 56 percent of the Smith women faculty at this level; the discrepancy was even greater at the level of associate professor: 73 percent of the men but only 27 percent of the women. (The data were not broken down by field.) Several pages of the report were then devoted to hypothetical explana-

tions for the men's advantages, such as financial pressures, contacts with pub-
lishers, few domestic responsibilities, or desire for jobs elsewhere. The impression given was that the women were weak sisters—less hard-working, less
motivated, and less persistent—and that the Smith administration was neutral
rather than, as some committee members, including Kenyon, began to suspect,
indeed quietly favoring and rewarding male faculty and their publications.[47]

The assumption that the women published less, as well as a touch of what
would now be called ageism, may have been behind the statement of Smith's
sixth president, Thomas C. Mendenhall, when he wrote to his former Yale
colleague, NSF head Alan T. Waterman, in 1960. (Waterman's interest in Smith
was considerable, since he had grown up in Northampton, where his father
Frank Waterman had chaired Smith's physics department for thirty-six years,
1897–1933.) In responding to Waterman's congratulations on his new job,
Mendenhall described the current lethargy in the sciences at Smith, which he
felt was due to the considerable amount of "deadwood" on the faculty, a
reference, perhaps, to the disproportionate number of older women who,
according to Myra Sampson's second report, were, and would be, leaving and
retiring between 1955 and 1966.[48]

One consequence of these older women's retiring and their either not being
replaced or being succeeded by younger men was the end of certain long-term
protégée chains that went back several academic generations, in some cases to
the start of the college. Three examples are particularly notable. When zoologist Charlotte Haywood retired in 1961 after thirty-four years on the Mount
Holyoke faculty, she deeply regretted, according to an obituary by a male
colleague, that she had not found a protégée to succeed her. Although there
were still other women in the department, including one alumna, she felt
keenly the end of a tradition that went back via her professors Abby Turner and
Ann Haven Morgan to Cornelia Clapp in the 1870s.[49] Meanwhile, at Vassar in
1957 Maud Makemson, the Maria Mitchell Alumnae Professor of Astronomy,
had also retired. Not an alumna of the college, she was, nevertheless, the fourth
woman in a position that not only dated back to the 1860s but was named for
America's first woman science professor. Although it might have been hoped
or expected that a prominent woman would be found for such a position,
instead, after briefly considering Nancy Roman, who was about to leave the
Yerkes Observatory for NASA, the administration hired Henry Albers of
Bucknell University as an assistant professor, and during 1958–59 the physics
and astronomy departments were combined. The alumnae may not have been
aware of what was happening and apparently did not protest.[50]

A third case reflects some of the racial and civil-rights issues of the fifties and
sixties. In 1953, when the longtime president (since 1926) of Spelman College
Florence Read, previously an associate of the Rockefeller Foundation's International Health Board in New York City, finally retired, she was succeeded by

the black dean Albert Manley of North Carolina College, who held an Ed.D. from Stanford. Then in 1960, when biologist Helen T. Albro retired after twenty-nine years as chairman of her department, she was the last of a long line of white New England women (including several associated with Mount Holyoke College) who had founded the college in 1881 and served it ever since. She was replaced as chair of the biology department by Barnett Smith, a black with a doctorate from the University of Wisconsin who had been in the department since 1945. Also in 1960, Rosalyn Patterson, a Spelman alumna with a master's degree from Atlanta University, joined the department as an instructor. (By 1970, after earning a Ph.D. at nearby Emory University, she was a full professor.) Evidently, in the mid- to late 1950s the leadership of this black women's college was moving away from its white missionary and Rockefeller-dominated past to a new generation of black scholars with doctorates. Yet even in this new racial pattern, it would be the black men who were holding the top positions—president, dean of instruction, department chair—with a few black women on the junior faculty.[51]

Just why strong women were not found to replace these and other retiring giants is hard to pinpoint. The few contemporary discussions of the situation, which are not entirely convincing, sidestep the real issue and point instead to the presumed unavailability of young women for these positions. According to this view, very few women were completing doctorates in the 1950s, and of those who were, most were married and unwilling to move to small college towns far from their husbands' jobs.[52] Presumably, those who were single were also unwilling to come to such remote localities, fearing, as one journalist put it melodramatically in 1961, that they would be "entombed" far from the eligible young men. Having thereby logically rejected both the married and the single women as too immobile or too unwilling to come to their particular college towns, employers felt justified in making little or no effort to recruit any women at all. It would just be energy wasted, and all had heard tales of cases that seemed to support this negative interpretation.[53]

Yet plausible as this negative view, which blames the victims and exonerates the administrators, may seem at first glance, and pervasive as it was at the time, it has several serious deficiencies. First, and perhaps chiefly, the whole placement process was so personal and private, if not secret, in these years that it excluded many persons who did not already belong to certain privileged circles. Such matters were often handled via private correspondence, in smoke-filled rooms, or, increasingly, via long-distance telephone between professors at major graduate school and department chairmen, deans, and even presidents at the women's colleges. Usually, however, these powerful graduate professors (almost all men) recommended everyone else before their women students, even for those jobs at the women's colleges for which they had formerly been considered to be very well suited and even uniquely qualified. President Marga-

ret Clapp mentioned the professors' new interest in Wellesley for their profusion of male students in a January 1958 report to her alumnae council. She
attributed this new responsiveness to her recently improved salary scale.

> The results of good salaries in the lower ranks . . . already show forth in spotty
> ways. A telephone call from a university placement office on the west coast . . . a
> letter from the chairman of a department in a great mid-west university stating
> that he had heard somewhere that Wellesley had an opening for a young person in
> his field and asking if he could make a recommendation . . . a letter from a
> chairman of another department, this in a great eastern university, expressing
> delight in recommending his best young man for Wellesley's opening. This very
> weekend several first rate appointments are being planned.[54]

Second, the changeover was not only among professors at the graduate
schools. At the women's colleges the old protégée chains were not working as
they had earlier to recall bright young women back to their alma mater once
they had earned their doctorate. In fact the opposite was true in at least one case
at Smith College, where instead of regretting, as Mount Holyoke's Charlotte
Haywood reportedly had, that she had not found an alumna successor, Margaret Kemp, associate professor of botany and the last of five women students
appointed to the faculty by chairman William Ganong (1894–1932), worked
very hard in the late 1940s to break the pattern and hire a man to head the
department. Correspondence in the Smith College Archives indicates that
although the department had produced a few graduates who had later earned
doctorates, the only persons sought, suggested, or seriously considered in this
case were men. In the end, after several men had turned down the position for
reasons no one wanted to put into writing, Kenneth Wright of Rhode Island
College (since renamed the University of Rhode Island) accepted.

   Why the department members were so sure that they needed a male chair
remains open to speculation. Evidently none of the current faculty—two or
three older alumnae, including Kemp, with doctorates, publications, and more
than twenty years of teaching apiece—were suitable for the chairmanship. A
middle-aged woman from the outside might have had difficulty in fitting in and
exerting authority, assuming that the president would appoint her. Besides, a
man might be better able to present the department's problems to the senior
administration, which had not exactly favored the botany department in recent
years. The department's facilities had been more disrupted than some others' by
the military occupation of Smith's campus during World War II, and enrollments were reportedly lower than they once had been; an interdepartmental
course with zoology and physiology was starting up; and a merger of the
departments, the pattern on many campuses in these years, was likely to follow.
Perhaps a male chairman who got along with the president could hold off such

a merger until the department was stronger. In this last area Wright may have been successful, for the eventual merger did not take place until the building of the new Clark Science Center in the mid-1960s. But if these women botanists also thought that a male chair could or would have them promoted to full professors, they were wrong, for all of these women retired, often after thirty or forty years of teaching, with the title "associate professor emeritus," perhaps a sign of of how weak they had felt all along. In any case, such a skewed search indicates that contrary to President Benjamin Wright's private assertion in the mid-1950s that searches at Smith could not possibly be biassed, since they were generally run by women, in this case the opposite was clearly true. The women thought that they knew what was best for their department and did their utmost to accomplish it. The episode also helps to explain, if other searches followed similar lines, why one item discussed at the faculty protest meeting in the spring of 1954 was the flagrant lack of women candidates. This glimpse into the antifeminist leanings of the Smith botany department also helps to explain why its merger with the feminist zoology department (Sampson and Parshley) was so long delayed.[55]

Third, when a woman, especially a married one, turned down a job offer, officials attributed her decision to her marriage rather than to the characteristics of the two professional opportunities. Although it is generally difficult to know anything about such private matters, and although one case is far from the whole story, another example from Smith in the 1950s casts some light on this issue. In 1953 associate professor of astronomy Marjorie Williams, aged fifty-three, who had been at Smith since 1925, either left, perhaps in protest, or was let go.[56] When the college then sought a junior replacement, it chose from the seven candidates for the position a Smith alumna and recent Radcliffe Ph.D., Constance Sawyer Warwick, from the class of 1947. But when she was offered an instructorship at Smith, as was the usual practice there for women faculty, Warwick declined, because both she and her husband, also an astronomer, had gotten research positions at the Air Force's new solar observatory on Sacramento Peak in New Mexico. Yet rather than admit that Smith's offer was not competitive—that it was low in rank and probably in salary, without research facilities, and with no sign of a job for the husband—Smith's President Wright later claimed that Warwick had turned Smith down to go somewhere else with her husband. This interpretation thus served to reinforce the prevailing view that (all) married women would rather be unemployed elsewhere with their husbands than accept jobs in Northampton (which here was not the case), confirming longstanding expectations that this was standard female behavior.[57]

Ironically, at the same time that Smith's trustees (and perhaps those at other women's colleges) were intent upon masculinizing their faculties, a different group, one of the many foundation-supported committees of the time that were interested in the upcoming shortage of college faculty, recommended that colleges hire women—as faculty assistants. In 1955 the FAE formed a committee on the utilization of college teaching resources to advise on how to better use current faculty. (Composed of eighteen college and university presidents, provosts, and chancellors, its only woman was Millicent McIntosh, president of Barnard.) Although the FAE was particularly eager for colleges to take up educational television, it was also willing to think of such other teacher-extenders as part-time assistants, both students (graduate and undergraduate) and nonacademics. Of the thirty-four grants approved by this committee, totaling $502,000, the three smallest were to women's colleges: Sweet Briar ($7,000), Wellesley ($6,350), and Smith ($5,000).

Smith's project, "a study of the feasibility of using competent liberal arts graduates in the community for teaching assistance," is of particular interest, because it was aimed at women, especially college graduates, in the Northampton-Amherst area. (Perhaps the camouflaged title was needed to conceal from McGeorge Bundy and the committeemen the fact that housewives were being considered for subfaculty positions.) The rationale for the study was that in a time of national emergency, as some thought the Korean crisis would become, local housewives might, in the tradition of Gray Ladies in hospitals, pitch in to help the real faculty cope with difficult circumstances. The first step taken by Smith's faculty Committee on Educational Policy was to interview department chairmen to discern what need might exist for such assistants. Some departments, including astronomy, had such low enrollments that they had no need for any extra help. Others, including botany, chemistry, and geology, had increased enrollments and would be glad for assistance. All chairmen created conditions: some wanted any such new persons to be required to take an orientation course; others insisted on attending several lectures as well as familiarity with Smith and the particular department; many wanted the women to have taken or audited the course before they became lab assistants or graders. Then Dean Helen Randall, who eventually wrote the college's reports to the FAE, hired three faculty wives familiar with the situation to contact as many women in the area as possible. Through local women's groups (the League of Women Voters, alumnae clubs, and the Amherst College Ladies' Association), likely prospects in the phone directory (wives of professional men), and personal contacts, the energetic threesome sent out 1,165 questionnaires. Of the 1,150 that were delivered, there were 366 responses (31.83 percent). Randall's group was surprised at how many respondents held doctorates (12), master of arts or master of science degrees (64), other master's degrees (36, including master of education and master of library science degrees), and bachelor's degrees (292). Twenty-four women had taught on college faculties in the past, and 36 had been assistants in college. Four Ph.D.'s, as well as 33 holders of master of arts or master of science degrees, were not working.

Sixty-five expressed interest in further academic training, including 17 in the sciences and social sciences. Randall and her assistants estimated that 128 women, including 59 in the sciences were suited (i.e., met the chairmen's requirements) for some form of college work the next fall. Sixty-nine others, including 42 in the sciences, expressed possible later interest.

While collecting this census, Randall's committee also held a meeting with eight women who had already served as such faculty assistants. They pointed out several pitfalls: the workload would have to be standardized, for there was a wide discrepancy in how much an assistant in a particular course had to do; because the jobs involved much drudgery, there should be a lot of "public recognition and appreciation"; and wages should be high enough to cover the rising cost of childcare and domestic assistance. But rather than pushing ahead with the project, the Smith committee recommended merely that if such a roster was to be useful, it had to be kept up to date. The final report also suggested that the training program for the new faculty aides might itself blossom into some sort of general postgraduate adult education for "students at large" in the area. Thus, little seems to have come of the report or the suggestions; the group sought no great change in existing practices at Smith but merely to systematize and extend ongoing practice. Like other rosters before and since, complete coverage could be very time-consuming and costly. Chances were that the colleges in the area could do fairly well without such a formal roster. But if a very serious crisis were to occur, there were evidently a great many highly trained wives currently stockpiled in Amherst-Northampton who could be mobilized for some sort of assistant's position. It seems not to have occurred to anyone that some of the women were sufficiently well trained to merit regular faculty positions even without a crisis.[58]

One is left with some discordant impressions of the faculty hiring at the women's colleges in these years, Smith in particular (because much information on its has come to light). In a way it was a truism of the time that there was a faculty shortage: men were hard to find, and when they were finally located, they merited "competitive" salaries. Women were so few as to be unavailable and therefore hardly worth the search. And yet a simple survey of the area around one college turned up a great many qualified women right there in town. In fact by 1956–57 many such women, often with higher degrees, were already "underutilized," to use a favorite term of the time, in college towns all across America. But these women, being the wives of the men who had been recruited from afar, were deemed barely eligible even for essentially technician-level positions as faculty assistants. Nor were the major women's colleges at all eager to hire them for any jobs. Thus, at least at Smith and probably elsewhere as well, there was a two-tiered, sex-linked system wrapped in a cloak of privilege and prejudice. The trustees were anxious to replace the older, single women on the faculty who had served thirty to forty years with young, married, and family-oriented men, perhaps in tweed jackets and with pipes, who, though it was never stated, would give the school a more "normal" collegiate image. Although these men rarely had credentials in research, they were given advantages so that they would justify their higher salaries by increasing the school's prestige. Three decades of prosperity had brought the women's colleges, formerly the bastion of women faculty, to this impasse: although their purpose was still to train women for meaningful lives, because of their thinly disguised ageism, sexism, and perhaps even homophobia, they refused to hire them.

And then starting in 1968 some alumnae turned upon the women's colleges, less because of the large number of men on the faculty, whose students many of them had been, than because of their lack of support for feminism and, at some, the watered-down curriculums designed to train women to be second-rate citizens. At the same time the trustees at Bryn Mawr and elsewhere continued to elect male presidents, and those at Vassar (and increasingly elsewhere) voted to admit men. And all faced major economic and other dislocations in the early 1970s.[59] The golden age, such as it had been, was over.

Nonprofit Institutions and Self-Employment

A Second Chance

*I*n contrast to the reluctant recruitment, systematic ghettoization, and minimal advancement facing women scientists in other sectors in the 1950s and 1960s, their participation was relatively welcome and their achievements were substantial at many nonprofit institutions. Although only a tiny proportion (about 1 percent) of the nation's scientists and engineers worked at such institutions, several of the more talented and dedicated women researchers of the time, unwelcome or not advancing in academia, capitalized on the opportunities there to do good science. Many service-oriented women scientists, such as librarians, editors, abstractors, educators, and association staff members, upon whom so much of the scientific community relied but preferred to keep invisible, also worked there, as did those "institution builders" who, finding existing job opportunities closed to them (for reasons of geography, marriage, or race), created alternative organizations to fit their own as well as some larger scientific needs. Often rather small and financially precarious, these organizations were frequently characterized by short promotional ladders, flexible work styles, and admittedly often low (and in some cases nonexistent) salaries. Yet they provided titles and positions, allowed women to accept grants and hire others, and sometimes lived on beyond them. (Very few women ever headed an organization that they had not founded themselves.) Other innovators forwent such institutional trappings and simply became self-employed. Although barriers remained, especially in the top ranks and at the larger and more prestigious foundations and research institutions, and new ones arose as some institutions outgrew their humble origins, many eminent and not so eminent women of science found the diversity and flexibility of the nonprofit and self-employment sectors more responsive to their needs than academia, government, or industry, and a few even flourished there. The chance for autonomy over their own work challenged and energized them to their greatest efforts.

Table 11.1.

Scientists and Engineers Employed by Nonprofit Institutions,
by Field and Sex, 1956–58 and 1968

Field	1956–58			1968		
	Total	Women	Women as % of Total	Total	Women	Women as % of Total
Other specialties	98	29	29.59			
Statistics	8	2	25.00	132	18	13.64
Psychology	885	189	21.36	2,318	616	26.57
Geography	21	4	19.05			
Biological sciences	1,163	195	16.77	2,893	443	15.31
Chemistry	1,249	174	13.93	2,121	400	18.86
Education	17	2	11.76			
Astronomy	28	3	10.71			
Mathematics	451	40	8.87	681	43	6.31
Medical sciences	232	17	7.33			
Earth and marine sciences	93	5	5.38	249	22	8.84
Physics	559	24	4.29	882	29	3.29
Social sciences	32	1	3.13			
Linguistics				118	39	33.05
Sociology				343	84	24.49
Anthropology				25	5	20.00
Political science				200	26	13.00
Economics				505	22	4.36
Meteorology	42	1	2.38			
Engineering	402	4	0.01			
Agricultural sciences	173	0	0.00	148	2	1.35
Atmospheric and space sciences (astronomy and meteorology)				114	4	3.51
Computer sciences				475	54	11.37
Total	5,453	690	12.65	11,204	1,807	16.13

Source: NSF, *American Science Manpower, 1956–58,* 81, 83; *1968,* 56–58, 252.

Table 11.1 summarizes the only systematic data broken down by field and by gender in the 1950s and 1960s on scientists and engineers at the nonprofit organizations, which included research institutes, hospitals, clinics, museums, botanical gardens, zoos, scientific academies and societies, journals, foundations, and miscellaneous other organizations. Since the number of women there nearly tripled from 690 in 1956/58, the first year these data were available, to 1,807 in 1968, the end of our time period, while the total number of scientists and engineers reporting employment in this broad and diverse category about doubled, the women's proportion of the total jumped markedly from a calculated 12.65 percent in 1956/58 to 16.13 percent in 1968. This was their highest proportion of any of the employment sectors covered by NSF (academia, industry, or government). The chief reason for this relative feminization of this employment sector was the mix of fields situated and supported there—those portions of biology, psychology, biochemistry and those generally marginal specialties (the ubiquitous "other," statistics, geography and several social sciences) that were even in the age of "Big Science" often still dependent on private philanthropy and federal grants.[1]

One reason for the increase in the number of persons working at such nonprofit institutions was the enormous, even explosive, increase in the federal and private funding of medical research in the 1950s and 1960s. This encouraged both new organizations to start up and older ones that had been started back in the 1920s to obtain large grants, expand their staffs and facilities, and perform some world-class research, including some by women medical scientists. Although it is beyond the scope of this book to examine the practice of medicine in any systematic way, certain specialties—e.g., cardiovascular diseases, cancer research, children's specialties, and "mental health," as it was euphemistically renamed in these years—did employ many medical and other scientists. In the midst of such growth and wealth, some institutions that had once employed a high proportion of women on their scientific staffs attempted to "mainstream" themselves or join with nearby universities. The consequences of this were not always clear.

Particularly notable in the realm of cardiovascular studies was the private Cleveland Clinic of Ohio, where, starting in 1928, four surgeons devoted up to 25 percent of their total income to a small research division. Over the years it employed a series of women chemists and biochemists, including Maria Telkes, Arda Green, and Lena Lewis, an expert on arteriosclerosis and methods of detecting cholesterol in the blood. But it was the coming of the NIH, with its large budgets for long-term clinical studies in cardiology and hypertension, that enabled the Cleveland Clinic to run studies of national importance in the 1960s, such as the National Diet-Heart Study of 1960–68, headed by Helen Bennett Brown (a Yale doctorate under Lafayette B. Mendel), and some studies on the latest antihypertensive drugs by Harriet Dustan, who in 1976–77 served as the second woman president of the American Heart Association (after pediatric cardiologist Helen Taussig).[2]

Another small local institution that underwent a tremendous transformation in the 1940s and after was what later became known as the Institute for Cancer Research (ICR). Originally the Lakenau Hospital Research Institute, founded outside Philadelphia in 1927 with a gift from Rodman Wanamaker, it barely survived the Depression, and then only with the help of frequent fund-raising efforts by its "women's auxiliary." Its early scientific staff included several women, such as Irene Diller, with a Ph.D. from the University of Pennsylvania, who also served as a librarian and editor of the institute's *Journal of Growth,* and chemist Grace Medes, a pioneer in the use of carbon-13, who in 1955 won the ACS Garvan Medal for her work on tyrosinosis, a rare inborn error of human metabolism. In January 1945 the institute's leaders opportunistically changed its name in order to take advantage of new sources of funding, and in 1949 they completed a large new building at nearby Fox Chase, Pennsylvania. Among the many federal and private grants its director obtained in the sixties was a five-year grant of $5.25 million from the National Cancer Institute in 1962. A few members of its greatly expanded and better-paid staff in the sixties were women, including geneticist Beatrice Mintz, an associate professor at the University of Chicago who came in 1960, crystallographer Jenny Glusker, a student of Dorothy Crowfoot Hodgkin's in Oxford, England, and the couple Betty and A. L. "Lindo" Patterson from Bryn Mawr College, where in 1934 he had invented the "Patterson function," which transformed modern crystal studies. Patterson was also unusual at the time in that he was willing to hire young mothers as part-time research associates; he had five on the staff in 1963, for which other men kidded him. Meanwhile, in 1962 the director signed an agreement with the University of Pennsylvania under which many ICR staff members accepted "secondary appointments" there, gave lectures, and attracted graduate students to ICR laboratories. (The consequences of this academic encroachment for women members of the ICR was not at all clear.) Thus, in just two decades an impoverished local institution had transformed itself, somewhat wrenchingly in this case, into a national institution that continued to provide major research opportunities for a few top women of the time.[3]

There was also great activity and expansion at the nation's children's hospitals in these years, as three age-old scourges of childhood—polio, rheumatic fever, and leukemia—were eliminated, controlled, or even cured. Besides these dramatic breakthroughs, a whole process of subspecialization was under way at the large urban hospitals as the traditional single field of pediatrics gave way to the research specialties pediatric oncology, hematology, immunology, cardiology, nutrition, and others. One pioneer in this process was Icie Macy Hoobler, whose specialty ever since graduate school in the 1920s had been the chemistry of milk, especially mother's milk, whose composition and properties were then unknown. Macy found just the right position for sustained work in this area as director of research for the Merrill-Palmer School in Detroit. Provided with a laboratory and access to the many nursing mothers at the nearby Children's Hospital and aided by a joint appointment as director of research of the Children's Fund of Michigan, Macy and her staff were able to complete several hundred publications on this little-known subject before the Fund was completely spent, capital and all, by 1956.[4]

At the Children's Hospital of Philadelphia virologists Gertrude and Werner Henle, who had fled the Nazis in the late thirties, did important work on infectious mononucleosis and the immunology of childhood tumors for three decades.[5] Similarly, in New York City physician and microbiologist Ann Kuttner was director of research at Irvington House, a convalescent home for children with rheumatic fever, from 1936 until 1968.[6] In Boston, whose historic Children's Hospital grew in these years into a sprawling multiunit medical center, biophysicist Carolyn Cohen, later at Brandeis, did important work in 1955–56 and from 1958 until 1972 at its Children's Cancer Research Foundation, housed in the new Jimmy Fund Building, constructed in the 1950s with funds donated by fans of the Boston Red Sox.[7]

Besides the children's hospitals, there were in some large cities private hospitals for women, which were usually affiliated with a medical school and run by a board of male trustees and a staff of male physicians. At one, however, the Free Hospital for Women in Brookline, Massachusetts, Harvard-trained biochemist-turned-endocrinologist Olive Watkins Smith spent her long career starting in 1929. She and her husband George, an expert on diseases of women's reproductive organs, together published hundreds of papers dealing with the toxemias of pregnancy or the reasons why so many pregnancies ended in miscarriage. When her husband became a professor of gynecology at Harvard Medical School in 1946, she succeeded him as director of the hospital's Fearing Research Laboratory and served until 1960. (She returned in 1970 and continued until 1980.) Although this "first couple of endocrinology" helped many women to have babies successfully (including, in cooperation with Priscilla White, formerly hopeless diabetic mothers), they became most famous for one remedy that later had tragic consequences: the DES (diethylstilbestrol) pill. Endorsed wholeheartedly at the time by the Smiths and prescribed widely by physicians for women with a history of miscarriages, it was later shown to have caused vaginal cancer in the patients' daughters and other side-effects in their sons. The horror of this outcome haunted the couple until their deaths in the late 1980s.[8]

The years 1946–66 were also what Roy Menninger of the Menninger Foundation later called "golden years for mental health," as numerous wartime and postwar writings on various mental illnesses, neuroses, and psychotherapies caused an explosion of interest in clinical and social psychology. This movement culminated in the passage in 1947 of the National Mental Health Act, which created the NIMH, which immediately began to support research and training in this area. One of the best-regarded institutions in this field in these years was the Menninger Sanitarium in Topeka, Kansas, which was founded in 1919 and expanded rapidly in the mid-forties in order to be ready to train needed professionals after the war. Already affiliated with a local state hospital and after 1946 with a nearby VA hospital, in 1941 the Menningers added a foundation," and in 1946 schools of psychiatry and clinical psychology, which within a few years had hundreds of graduates ready for positions all across the nation.[9] In 1951–52, however, there was, for reasons that are unclear, an upheaval at the foundation's division of research, and Sibylle Escalona, a psychologist there during the war, one of the first to win an NIMH award, and director of research since 1949, was asked to leave. (She became an assistant professor of child study at Yale.) She was replaced by Gardner Murphy, a psychologist long at the City University of New York, whose wife Lois Murphy, also a psychologist, long on the faculty (in religion) at Sarah Lawrence College, also moved to Topeka and developed a long-term project on how normal children learn to cope with everyday life. (An NIMH program officer was initially skeptical of the theoretical importance of such a seemingly mundane topic, but she convinced him to provide some funding.) Also at Menninger in the fifties was psychologist Grace Heider, who as a faculty wife had been ousted from her teaching job at the nearby University of Kansas in Lawrence when its antinepotism rules, which had been suspended during the war, were reinstated in 1949.[10]

Besides the clinics and medically oriented laboratories, many other independent nonprofit research organizations employed women scientists of renown in the 1950s and 1960s. Varying in size and financial status, many of them endowed years earlier by the bequests of wealthy industrialists, these institutions often outgrew their limited income and became increasingly dependent on federal and other grants for new buildings, research support, and even salaries. Many of these institutions had one or two women on their research staff. Whether single, married, or divorced, they often stayed for decades, perhaps for lack of other options. Rarely promoted, probably paid a bare minimum, they rarely complained (publicly at least) and stuck to their collections and experiments long after official retirement (unlike the many in academia, industry, and government who retired early). Possibly their pensions, when they existed, were too small to permit any other choice. Years later those without a regular job or a pension might end up as homeless "bag ladies," which happened to at least one woman by the 1980s in a large midwestern city.[11]

Because the National Register for 1968 broke down its salary data on women by field and work sector, we can see that although overall the median annual salary paid to women scientists in all nonprofit organizations was exactly the same as that for women scientists in general ($10,000), there was a very wide range ($7,200–$13,800) by field (see fig. 5). Those who reportedly earned less at the nonprofit organizations than women scientists in general were the numerous women psychologists, biologists, chemists, and linguists whose median salary may possibly appear lower because of unpaid, honorary, or voluntary positions at these institutions. Those at the organizations who earned more

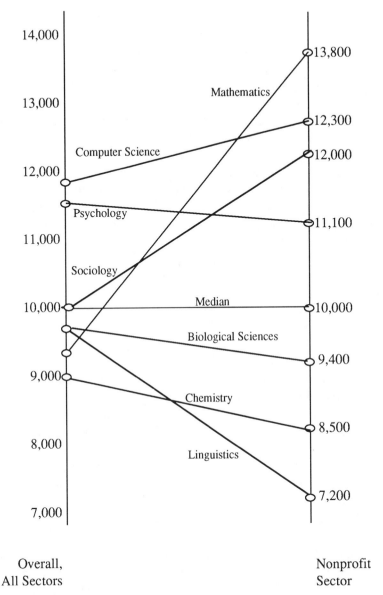

Fig. 5. Median annual salaries of women scientists overall and employed at nonprofit institutions, by field, 1968. Based on NSF, *American Science Manpower, 1968* (Washington, D.C.: GPO, 1969), 254 (table A-65).

than women scientists in general were the relatively few women working in mathematics, computer science, and sociology. These were probably the women employed by such groups as the RAND Corporation, the Battelle Memorial Institute, the Stanford Research Institute, and the Systems Development Corporation, which, to judge by governmental hearings and the often commissioned, selective and highly anecdotal chronicles that are beginning to appear, featured conspicuously high salaries. These organizations also favored former Air Force personnel with strong personal ties to well-placed industrialists and Pentagon officials from whom they could obtain highly lucrative contracts for research and development, a hiring practice that necessarily eliminated most women scientists and engineers, unless they were the needed mathematicians, librarians, or other service staff.[12]

Particularly prominent numerically and qualitatively among the independent research institutions was the large, decentralized, and international Carnegie Institution of Washington, which over the years counted many women among its distinguished staff. These included anatomist Mary Rawles and placentologist Elizabeth Ramsey at its Department of Embryology in Baltimore, crystallographer Gabrielle Donnay at its Geophysical Laboratory and astronomer Vera Rubin at its Department of Terrestrial Magnetism, both in Washington, D.C., geneticist Evelyn Witkin at its Department of Genetics at Cold Spring Harbor on Long Island, New York,[13] astronomer Henrietta Swope in California,[14] and the meticulous architect-turned-linguist Tatiana Proskouriakoff, whose brilliant paper in 1960 unlocked the key to the subsequent decipherment of the Mayan script, at its Department of Archaeology in Cambridge, Massachusetts.[15]

But, of course, the Carnegie Institution's most famous woman scientist, one of the most eminent in the nation, was geneticist Barbara McClintock, who came to Cold Spring Harbor in 1941, just before the benighted University of Missouri faculty denied her tenure. She stayed on past official retirement in 1967 (with the new title "distinguished service member") to win numerous awards, including the Nobel prize for medicine in 1983. At one point in 1945 she confided to her old friend and fellow graduate student at Cornell years earlier George "Beets" Beadle,

> After all, Beets, I am very much a wanderer and have'nt [sic] much personal ambition in the usual interpretation of this phenomenon. The goals I have aimed at can not be compared with the goals that many people are able to aim at. If I aimed at them, my life would have been a series of frustrations. The compensation lies in the fact that I am much freer to wander around and do the things I like to do without feeling that I must maintain a standard pattern of behavior. This leads me up to the question of returning to Neurospora work and where to do it.[16]

Although her five decades at Cold Spring Harbor were productive ones, there were enough frustrations, especially in the winter, with the isolation and lack of winterized facilities, that in 1953 McClintock, at a low point in her career, regretfully submitted her resignation. Feeling that neither she nor her work was being adequately supported, she, perhaps typically, blamed herself in the letter for being possibly too much of a "perfectionist" for even the Carnegie Institution. Alarmed, her superiors, Miloslav Demerec and Vannevar Bush, finally comprehended her spoken and unspoken moods and needs and countered with such resounding moral support (and probably long overdue financial support as well) that, suffused with gratitude, McClintock withdrew her letter. In addition, shortly thereafter George Beadle, now head of the biology division at Cal Tech, who was aware of the crisis, invited her to spend a few months in Pasadena. Presumably she went and returned refreshed.[17] Successful as this call for help was, McClintock was one of the very few women at these research institutes who could muster such leverage and rouse solicitous associates. But there must have been certain gaps in communication or responsiveness for her unhappiness to have reached the level of a letter of resignation in the first place.

Other eminent women biologists at other independent research laboratories in these years were geneticists Elizabeth Shull Russell and Margaret Green at the Roscoe Jackson Memorial Laboratory in Bar Harbor, Maine;[18] botanists Lela V. Barton and Norma Pfeiffer at the Boyce Thompson Institute for Plant Research, then in Yonkers, New York;[19] anatomists Margaret Reed Lewis and her husband Warren, still, in their seventies, at the Wistar Institute in Philadelphia;[20] and invertebrate zoologists Olga Hartman Petersen and Irene McCulloch, into her nineties, at the Allan Hancock Research Foundation in southern California.[21] Elsewhere oceanographers Mary Sears, Betty Bunce, and, during the 1950s, meteorologist Joanne Malkus (Simpson) were at the Woods Hole Oceanographic Institution on Cape Cod in Massachusetts;[22] distinguished crystallographer Elizabeth Armstrong Wood, a graduate of Barnard under Ida Ogilvie with a doctorate in geology from Bryn Mawr College, spent twenty-four years (1943–67) at Bell Laboratories in New Jersey;[23] and classical archaeologist Hetty Goldman, was the first, and between 1936 and 1947 the only, woman professor ("member") at the Institute for Advanced Study in Princeton, New Jersey.[24]

Besides these nonprofit institutes, several that were devoted to the quantitative social sciences had notoriously few women in visible (or any) positions (see table 11.1). Among these was the independent National Bureau of Economic Research, established in 1920 in New York City and long supported by the Rockefeller Foundation. Its staff specialized in painstaking quantitative studies of business cycles, national income, and other statistical measures of the American economy. A high-powered place, famed as a nursery for Nobel prize winners as well as presidential advisers, its staff had recommended the creation of the Council of Economic Advisers in the late 1940s. Yet women's place at the bureau was as bright, even brilliant, but nearly invisible assistants. Among these were Phyllis Wallace, a black doctorate from Yale whom the bureau hired for her first professional job in 1948, and especially Anna Schwartz, who came in the late 1930s and stayed several decades, co-authoring with Milton Friedman several books that helped *him* to win the Nobel prize in economic science in 1976.[25]

The place of women scientists in the nation's large nonprofit natural history museums, botanical gardens, and zoos was undercut in these years by continuing roles there for unpaid females, either as patrician philanthropists or volunteers and amateurs in the library or education department, where their often decades of selfless service were warmly applauded.[26] Rarely was a woman scientist a paid curator, for such jobs were usually reserved for men, and natural history museums, some of which went back to the nineteenth century, were quite strict about this segregation. The most important such private museum was the American Museum of Natural History (AMNH) in New York City. It is best known in women's history for providing the attic out of which worked anthropologist, lecturer, essayist, and social commentator Margaret Mead, hired in 1927 but not promoted to full curator until 1964, by which time she had been world-famous for decades. (Despite her fame, her museum salary remained low and led to such a small pension that she continued to lecture and write well after her official retirement.)[27] Less widely known than Mead but highly acclaimed in her own field of invertebrate zoology was Libbie Hyman, who, after inheriting and earning enough money to support herself (from royalties on two popular laboratory manuals), moved to New York City and in 1937 to the AMNH, where, unpaid, she worked daily for more than three decades to produce a series of six massive encyclopedic reference works that earned her two rare awards: election to the NAS in 1957 and the Gold Medal of the Linnean Society of London (the first woman and only the third American) in 1960.[28] Besides these two stars, one poorly paid and one unpaid, the museum also had some women associate curators: Dorothy Bliss, a Radcliffe doctorate whose work as editor in chief of a ten-volume reference book on crustaceans was long supported by the NSF; Bella Weitzner in anthropology, Peruvian textiles in particular; and Angelina Messina in micropaleontology, who, with the help of assistant curator Eleanor Salmon part of the time, prepared in thirty-four years at least ninety-five catalogs of foraminifera. One obituary writer put it well: "She spent her life in making the work of others easier and more effective."[29] Some women who worked in the museum's education department were widely known to New Yorkers, such as Farida Wiley, an expert on ferns, who for decades led early morning bird tours through Central Park; Lois Heilbrun, who provided special tours and children's books; and Evelyn Shaw, who directed its undergraduate programs.[30]

But the expected role for women in a museum that relied heavily on volunteers and philanthropy is evident in stories told about Elsie Reichenberger, a widow who became a part-time volunteer in the ornithology department in 1915. In 1917 she had the temerity to ask for a small salary and was awarded $37.50 per month. Later that year she became a full-time worker on some new museum collections and was paid the munificent sum of $100 per month, despite fears that her salary would become a burden on the museum's payroll. When she remarried in 1933 she gave up her salary but kept on with her own work on birds, eventually publishing a notable volume on Brazil. But overshadowing her contributions to scholarship was her later philanthropy, for her new husband was none other than banker Walter Naumberg, then head of the New York Stock Exchange. As prosperity returned, the Naumbergs underwrote several expeditions by the ornithology department and then in 1959 bequeathed to the department a block of IBM stock worth almost $2 million—more than a hundred times times the total salary paid to Elsie Reichenberger years before. The moral of the tale, as told in the anecdotal histories of the museum, was that recompensing impoverished society women could pay off very handsomely indeed—a rather snide minimization of her scholarly career and contributions. There seemed to be great relief that she had, after all, conformed to the expected female role of Lady Bountiful.[31]

Elsewhere across the nation there were very few women scientists on the staffs, paid or unpaid, of the major private natural history museums. The most notable among these few was ecologist Ruth Patrick, a freshwater ecologist at the Academy of Natural Sciences in Philadelphia, who served as a volunteer for ten years before raising the funds for a department of limnology, which she then headed from 1947 until 1973.[32] In Honolulu, Hawaii, at the Bernice Bishop Museum, botanist Marie Neal, initially a department secretary at Yale, and librarian Louise Titcomb, formerly at the AMNH, worked hard for several decades to build up the museum's collections on the Pacific regions.[33] Yet the situation at some of these institutions could be quite difficult, as the case of botanist-horticulturist Elizabeth McClintock at the California Academy of Sciences in San Francisco demonstrates. Hired as an associate curator in 1948, possibly to replace the long-lived Alice Eastwood, McClintock spent three decades at the academy doing what she could for its collections. But she felt underappreciated, understaffed, and underpaid the whole time and retired in 1977 on a barely adequate pension.[34] Genteel poverty could give way to anger and bitterness, especially in this case, for she was denied the title "emerita."

Of the several independent botanical gardens in America only two, to judge from the little that has been written on these institutions (and most of that focuses on their directors, who until 1972 were all men), have had many women on their staffs. Premier in this regard has been the New York Botanical Garden, where fern and moss expert Elizabeth K. Britton, wife of an early director, was long a powerful figure. Since then the New York Botanical Garden has had several other important women researchers, such as Caroline K. Allen, an expert on laurels, Annette Hervey, keeper of a large collection of fungi, Marjorie Anchel, an authority on mushrooms, and mycologist Alma Whiffen Barksdale, formerly of the Upjohn Company in Kalamazoo, Michigan, an expert on the hormones of bisexual water molds, before she was stricken by a crippling disease. More often the most prominent woman at a botanical garden was the chief librarian, as was the case with the long-lived Elizabeth C. Hall, who came in 1937, earned a reputation for near omniscience before her retirement in 1967, and then, bored elsewhere, returned as a volunteer for twenty years more.[35] Similarly, Nell Horner and Carla Lange were legendary librarians at the Missouri Botanical Garden in St. Louis.[36] At the Brooklyn Botanic Garden there were two high-ranking women of note: Frances Miner, who joined the staff in 1930, became curator of its department of instruction in 1945 and by 1965 was deputy director of the whole garden; and South African–born Elizabeth Scholtz, who arrived in 1960 to head the adult education department, did so well that in 1972 she became the first woman possibly worldwide to head an important botanic garden.[37]

At two of the nation's more prominent private zoos, a few women held challenging positions, and one woman even held a managerial position (in fact if not in name). At the New York Zoological Society, which, though it still runs the Bronx Zoo, is often considered the most research-oriented zoo in the United States, Jocelyn Crane, fresh out of Smith College in 1930, became one of naturalist William Beebe's many female assistants. Although others, including Gloria Hollister, who was very helpful during Beebe's bathyspheric diving experiments in the 1930s, left upon marriage, the single Crane stayed and even accompanied Beebe when in 1949 at age seventy-two he moved his division of tropical research to an estate in Trinidad. There, according to his biographer, he let it be known that Crane was his companion as well as his colleague. When Beebe officially retired three years later, Crane succeeded him as director of tropical research, but he stayed on in Trinidad, growing increasingly senile. She nursed him "like a daughter" through his final illness and death in 1962. But this did not ease her situation, for after Beebe's death and with the growing university interest in animal behavior studies, the Zoological Society, anxious to close down the Trinidad branch, merged it with a new program in animal behavior at the Rockefeller University. Crane then returned to New York City and, possibly without a position, in 1965 married Donald Griffin, the director of the joint program. In subsequent years Crane completed a massive monograph on her own specialty, the fiddler crabs of the world, on which she had been working with occasional NSF support since the 1950s.[38]

An even better-known zoo woman was Belle Benchley of San Diego, a former schoolteacher, cashier, and divorced mother who began at the new zoo in 1925 as a bookkeeper and stayed on for decades to run the place. One of her greatest triumphs was building up the museum's specialty of monkeys and apes, especially obtaining in 1931 the first pair of baby gorillas for an American zoo. (In 1950 she got three infant gorillas from the French Cameroons and induced psychologist Joan Morton Kelly of Penn State to test and observe their heretofore unstudied learning habits.) When Benchley, known on the West Coast as the "zoo lady," finally retired years later, she was still officially only the zoo's "executive secretary," a title she had held since 1927. But the trustees voted her the title of "director emeritus" in long overdue recognition of the real status she had held for decades on the job. This move has justified some calling her (either incorrectly or finally correctly, depending on one's faith in titles) the first woman director of an American zoo, which she long was in fact but not in name.[39]

Like museums and zoos, which had long relied on volunteer and clerical help, scientific societies were growing rapidly in these years, adding new functions, and hiring new staff members. But sometimes the functions outstripped the available staff, and volunteers continued to be needed here as well. This was often the case when the opportunity arose to host an international congress, as occurred several times in the fifties and sixties. Often sponsored by a scientific society and supported in part by fees and a grant, they required the services of someone in or close to the discipline who had about a year or two (even longer if she edited the proceedings afterword) to devote to this service work. Thus it was a temporary task at which some women scientists of the time shone. The home economists had their ninth international congress at College Park, Maryland, in 1958, and the women engineers had their first one in New York City in 1964 (see ch. 15). Particularly notable among the others were the First International Oceanographic Congress, held in New York City in 1959, at a time of great breakthroughs in the field, whose very effective chairman was Mary Sears of the Woods Hole Oceanographic Institution, and the 1961 International Astronomical Union meeting, which brought 1,140 people to Berkeley and the mountaintop Lick Observatory. Mary Shane, herself a Ph.D. in astrophysics but also wife of Donald Shane, director of the Lick Observatory in California and chairman of the organizing committee, took on the heavy task of local arrangements. Largely unassisted, because she could not get a secretary at the top of Mt. Hamilton, she spent two years full-time at the task. For her efforts, however, the grateful members of the U.S. National Committee devised a special award for her: a citation showing her on roller skates, with a pair of wings and a cap with an antenna that predated the cordless telephone.[40]

Some scientific societies grew into small publishing houses as the number of their journals proliferated. Many women contributed decades of selfless service as "editorial assistants" or "assistant editors," titles that could and did cover a wide range of duties. One example of a particularly overworked "editorial assistant" was Beulah Brewer of the Society for Research in Child Development, who, as the society's only paid employee, had duties greatly in excess of her modest title. As one account described her job, "Although her title was only editorial assistant, she managed the whole NRC office devoted to the work of the committee and the society. Not only did she edit all the copy for publication, type it or have it typed for planographing, maintain the membership and subscription lists, and mail the journals, but she also kept minutes of the work of both the committee and the society whose meetings she arranged, handled the correspondence and, mirabile dictu, preserved the records for our [later] use." Only her unexpected death in 1947 jolted the society's leaders into realizing that it would need to hire a whole staff to replace her.[41] One of her successors (as "executive officer," 1966–71) was psychologist Margaret K. Harlow, who had started out as an assistant professor at the University of Wisconsin but after her marriage to another faculty member there in 1948 had set up and managed the APA's first publications office, in 1950–51.[42] Other long-term journal editors in the fifties and sixties included Wilma Fairchild, who in 1949 succeeded the redoubtable Gladys Wrigley as editor of the *Geographical Review* and served until 1972; Marjorie Hyslop, who spent forty years at the American Society for Metals editing a series of journals (*Metal Progress, Metals Review, Metal Literature, Metals Abstracts,* and *Metals Information*) as an information explosion hit the field of metallurgy; Agnes Creagh, who served as managing editor from 1952 to 1962 (and editor from 1962 to 1964) of the many publications put out by the Geological Society of America (GSA); and reentry geologist Patricia Clabaugh, who was managing editor of the *Journal of Geological Education* in the 1960s.[43]

Some scientific associations (and some consortia of several societies, such as the American Institute of Physics [AIP]) moved in these years into the area of STINFO, or scientific and technical information and documentation, as their staffs indexed or abstracted the expanding literature in their field. In some, but not all, of these projects women were prominent, such as Irene Taeuber, long the editor of the *Population Index,* Marjorie Hyslop of *Metals Information,* and Pauline Atherton (Cochrane) and Rita J. Lerner of the AIP's documentation service.[44] But in other cases the task outstripped the traditional roles of editors plus staff members and grew into a multimillion-dollar nonprofit business in its own right, one that owned a building, met a payroll, accepted governmental grants, and might even have its own board of directors.[45] The largest by far of these abstracts services, long run by the ACS, was *Chemical Abstracts,* known in the mid-1950s as Chemical Abstracts Services, Inc., and later simply *CA.*[46] One of *CA*'s longer-serving staff members was Mary Magill, who started in 1933 and after two years at the Johnson Foundation of the University of Pennsylvania spent the rest of her career with *CA* at its Columbus, Ohio, headquarters, retiring in 1971. For many years she held one of the toughest jobs at *CA,* that of index editor for organic chemistry. This was difficult because so many newly discovered compounds had not yet been officially named; they had to be arranged in some logical "index" on the basis of their complex structures, explainable only in diagrams.[47]

By contrast, one woman did much better at the smaller and younger *Biological Abstracts,* later BIOSIS, started in Philadelphia in the 1920s with grants from the Rockefeller Foundation (but no support from any biological society). Over the years, during Depression and war, it struggled to expand its coverage. In 1953 it hired Goucher College graduate and faculty wife Phyllis Parkins as an assistant editor. She showed such interest and managerial ability that before long she was put in charge of the other staff women in the editorial department. Then when Miles Conrad, her mentor and the director of *Biological Abstracts,* died suddenly in 1964 in the midst of a financial crisis, the board of trustees took five months to search the nation and then, only then, found Parkins qualified to hold the top job. She was brimming with ideas on how to solve the financial problems. When she retired in 1975 after eleven years as director, she had brought BIOSIS out of the red and into the computer age.[48]

In addition to specialists in editing and publishing, a few of the larger scientific societies, especially those with a substantial headquarters building, also had other specialists on their staff. When these included librarians, some might be women, such as the series of map librarians at the American Geographic Society building, long in upper Manhattan, in New York City, and a similar series at the Marine Biological Laboratory at Woods Hole.[49] When the society was one devoted to a woman's field, such as the AHEA, which purchased a Washington, D.C., headquarters in the 1950s and proceeded to set up a separate foundation as well, the number of women in specialized staff positions could proliferate. The AHEA had, for example, a steady profusion of legislative analysts, field secretaries, and specialists on various temporary topics and enthusiasms.[50]

At the top of this emerging career line was the position of "secretary" or "executive officer" of the whole scientific society. Only rarely was this position held by a woman, such as Matilda White Riley, executive officer of the American Sociological Association from 1949 until 1960.[51] Naturally enough, a woman held this position at the AHEA, where the whole staff was female. Executive officer Mildred Horton, who had come from the Texas state extension service in 1947, served until 1960. Among her many accomplishments were the purchase of the new headquarters, establishment of the new tax-exempt educational foundation, and the successful staging of an international congress on home economics at the nearby University of Maryland campus in the summer of 1958. Horton's successor in 1960, A. June Bricker, was later remembered as the director who brought employee benefits, especially a pension plan, to the members of the staff.[52]

In one case in the mid-sixties a woman even served as temporary executive director of one of the major scientific organizations. This was at the GSA, where the previous executive director, a rather autocratic individual, was ousted by the governing boards; following this, a reorganization was deemed necessary, and a substantial donation made possible a move from New York City, with its ever-rising rents, to a set of new buildings in Boulder, Colorado. Because of the great uncertainty about the society's future as well as the expected resignation of its staff of nearly fifty persons (due to the likely move to Colorado), the governing board decided to entrust the position of executive director to Agnes Creagh, the long-term managing editor who also served as editor, for the interim. As this temporary situation still existed after two years, the overburdened Creagh realized that only her resignation would force the society's leaders to make some final decisions. Therefore, she accepted a job as assistant director for publications at the Educational Testing Service in New York City, and the society, once settled in Boulder, resumed its pattern of male executive directors. Years later the GSA's historian (Creagh's successor as executive director) remained fully aware of her heroic labors on behalf of the society at its gravest hour.[53]

Even more remarkable were the career and success of Eunice Thomas Miner, first the "executive secretary" and later the "executive officer" of the New York Academy of Sciences. When she took over in 1939 the society was moribund and had a local membership of about three hundred; when she retired in 1968 it had twenty-six thousand members worldwide. In a sense this institution needed her at least as much as she needed it; they both grew together. A former research assistant in invertebrate zoology at the AMNH, where her husband was a curator and which in the 1930s housed the academy in two free rooms, she grew out from under such dependent status into a person of consequence in her own right as she built up the academy, its membership, its resources, and its reputation. It benefited from her energy and commitment, while she used it to build an independent identity. She not only recruited members but also raised donations, including in 1949 the substantial gift of a new headquarters, the Woolworth mansion in Manhattan. Each year Miner ran about twenty international meetings, including many in the late forties through the sixties on the medical aspects of the new pharmaceutical drugs, especially the antibiotics. She was able to capitalize upon the considerable metropolitan expertise in biomedical matters, nearby headquarters of several pharmaceutical companies, and extensive press coverage. It is often said about founders and builders that the institution is the shadow of the man. Far less often can this be said about women, who even if they do the work rarely get the public credit for it.[54] Miner was one woman who did get the credit.

Less visible but perhaps as influential were the few women on the staffs of the rich and powerful grant-giving foundations. There the usual pattern was for a board of trustees (nearly all men) to select a male director, often a social scientist formerly in academic administration, to head its nearly all-male staff.[55]

Few women ever penetrated these enclaves. On occasion in the fifties and sixties, however, on the staffs of the largest and most prestigious foundations there was a lone female, camouflaged by the title "associate," "secretary," or even "associate secretary." Though an unsuspecting outsider might initially think, as in the case of Florence Anderson, long an associate secretary and then secretary of the Carnegie Corporation of New York, that she was just a part of the clerical staff who simply "sat in" on important meetings, an astute applicant for a possible grant would quickly discover that she in fact handled most of the proposals and played an important role in the foundation's ambiguous collective decision making, while the then president, ensconced in his suite down the hall, dealt with the trustees and signed the final paperwork.[56] At the Ford Foundation's daughter foundation, the FAE, economist Elizabeth Paschal in the 1950s and early 1960s held the various titles "executive associate," "assistant to the president," "associate program director," "secretary," and "treasurer"; or as the official history of the FAE described it, "Mr. Faust [the president] greatly valued her advice and looked to her for criticism."[57]

But even these mom-and-pop arrangements were rare at the larger foundations in the fifties and sixties. At Ford itself women were rare even at the middle level of program director in the 1950s. Economist Mariam Chamberlain, who had received her doctorate from Radcliffe in 1950, was a mere "program assistant" when she went to the Ford Foundation in 1956, though she returned as a full program officer in 1966.[58] Meanwhile, at the notoriously sexist Rockefeller Foundation, endowed by its founder with hundreds of millions of dollars and charged with the lofty mission of improving the lot of mankind throughout the world, the only female official or staff member ever mentioned in the forties through the sixties was librarian Dorothy Parker, whose fame rested on setting up libraries in Mexico and the Philippines for the Rockefeller-run "green revolutions" there.[59] Evidently the foundations, on whose funds so many visions for the future depended, were not leaders in the employment of women in the fifties and sixties.[60] One exception might be Virginia Apgar, M.D., famous for the "Apgar score" for newborns, who in 1959 at age fifty became an official at the National Foundation for Infantile Paralysis. But it, unlike other "foundations," raised new funds for its purposes, in this case birth defects.[61]

Some other women, finding themselves blocked in various ways or simply seeing a need in research, education, or clinical services that was not being met by existing institutions, started and maintained their own. One such rebel and innovator was paleontologist Katharine Van Winkle Palmer, who held a doctorate from Cornell University but, because of its strict antinepotism rules, was unemployable there after marrying a faculty member. In 1932 she and a professor nearing retirement decided to set up a private, nonprofit research institution near Cornell that would serve both their and others' needs: provide a workplace, affiliation, and titles for paleontologists without such professional necessities and also house and preserve paleontological collections and libraries of persons, such as retirees or the deceased, who had no fireproof storage places for them. Such an institution might also publish a journal and other reports. Thus was born the Paleontological Research Institution (PRI) of Ithaca, New York, chartered by the board of regents of the University of the State of New York and still in existence. It had a second beginning in 1951, when Helen Jeanne Plummer, a micropaleontologist for the state of Texas, bequeathed to the PRI almost $116,000 as well as her library. (A gift of $234,000 in 1969 also helped with construction costs, after the NSF had twice refused such aid.) Long headed, even dominated, by the energetic Palmer, who served as director from 1952 until 1978, the PRI published hundreds of reports and its own journal, the *Bulletin of American Paleontology*, and also provided facilities for local amateurs interested in fossils and gemstones. In many ways the PRI was a creative solution to the needs of many kinds of scientists and amateurs that were not being met by academic institutions with other priorities in mind.[62]

A somewhat different research institution was the National Biomedical Research Foundation, established in 1960 by the biochemist Robert S. Ledley, the reentry biophysicist Margaret O. Dayhoff (who had received her doctorate from Columbia in 1948, possibly under Edith Quimby in radiology), and others in a suburb of Washington, D.C. Interested in studying the role of computers in biomedicine, they needed such an institution in order to receive grant funds from NIH, NASA, NSF, and others. Once it was set up, they became employees, Ledley became the director (though a year younger than Dayhoff), and Dayhoff was the head of her department, associate director, and editor in chief of an innovative reference work, the *Atlas of Protein Sequence and Structure*. After a very productive decade, the foundation moved to the campus of the Georgetown University Medical School, and Ledley and Dayhoff became professor and associate professor, respectively, in the department of physiology and biophysics. By then Dayhoff's work on protein sequences had led her to evolutionary studies of how the basic proteins in similar species differed, cosmological studies of the origin of life, and oncological studies of how cancerous cells differed from normal ones. She was still busy on these topics when she was stricken by a fatal heart attack at age fifty-seven in 1983.[63]

There were at least four female-founded and female-run nonprofit institutions in the realm of innovative science education in these years. The oldest was that on Nantucket Island, off the coast of Massachusetts, run by the Nantucket Maria Mitchell Association, founded in 1902. Its permanent part-year director from 1916 until 1957 was Margaret Harwood, who taught summer classes at the observatory and then spent her winters elsewhere, usually at the Harvard College Observatory. After World War II Harwood was able to supplement her Nantucket income with grants for an extensive astronomical bibliography, which she worked on in the off-season. When she finally retired in 1957 (with only a modest pension), she was succeeded by fifty-year-old E. Dorrit Hoffleit, a Radcliffe Ph.D. and a former astronomer at the Harvard College Observatory. By arranging to spend her winters as a research associate at the Yale Observatory, she was able to carry on the Nantucket tradition of women astronomers until her retirement in 1978.[64]

In the 1950s three members of the Vanderbilt family who owned some waterfront property south of Tampa, Florida, decided to devote part of it to a small foundation or research institute for scientific purposes. Impressed with the recent autobiography *The Lady with a Spear* (1953), by young Japanese-American ichthyologist Eugenie Clark, a research fellow at the AMNH, they offered her the directorship of the Cape Haze Marine Laboratory in 1955. Excited by the prospect of developing a marine station so near an ecologically rich grazing area for sharks, one of her specialties, she accepted, thus becoming at age thirty-three one of the youngest directors, and the only female director, of a marine biological station. Accompanied by her physician husband and several small children, Clark proceeded to set up research and "harvest" programs (for medical researchers who needed fresh fish and shark organs), train volunteers, run classes, and obtain funds from the ONR, the NSF, and the Mary Selbey Foundation, activities she later described in some detail in a second autobiography, *The Lady and the Sharks* (1969). In 1966, however, after a divorce and an offer of an associate professorship at City College of New York, she gave up her directorship and returned to New York City. Then after remarrying in 1969, she moved to a position at the University of Maryland, whence in the 1990s she still takes every opportunity to dive after sharks. In the meantime, her original station has flourished, with financial contributions, new construction, and a new name (Mote Marine Laboratory of Sarasota, in honor of one of the local enthusiasts and large donors). Able to afford a male director, it has become one of the better-situated and better-equipped small marine laboratories in the South.[65]

The third educational innovator was Dixie Lee Ray, a tenured associate professor of marine biology at the University of Washington, who discovered in the late 1950s that despite her Guggenheim Fellowship and other qualifications, her colleagues were not about to promote her any further. Thus, she deliberately pursued outside opportunities, including a three-year stint at the NSF in 1960–63, first as a consultant on biological oceanography and then as a special assistant to the director. There, as she later described it, she became aware of the large communications gap between scientists and the public, especially Congress. Upon her return to Seattle she joined a group that was trying to transform the facilities left over from the 1962 World's Fair there into a permanent science complex of some sort. Ray became director of this new Pacific Science Center in 1963 and served until 1972, devoting much attention to its then innovative hands-on exhibits, its perpetual fund-raising, and her week-

ly television show, *Doorways to Science*. Her fame in Seattle, her skill in telecommunications, and her Washington experience later propelled her into a third
career, in federal and state politics. When Richard Nixon was looking around
in 1972 for a female concerned about the environment to appoint to the AEC,
he found Ray, and then he promoted her to chairman in 1973. In 1975 President
Ford appointed her to the new position of assistant secretary of state for oceans
and international environment and scientific affairs. She resigned after six
impossible months under Henry Kissinger and returned to Washington State,
where she ran successfully for governor in 1977.[66] Thus, the innovative Pacific
Science Center was the right place for Ray at the time: it enabled her to move
beyond her blocked position at the university (which as governor she did not
favor) as well as to serve the people of her region in a new and innovative way.

    Finally, Mamie Phipps Clark and her husband Kenneth were black psychol-
ogists in New York City who held doctorates from Columbia. (As she later put
it wryly, "He was the first in 1940 and I was the second and last in 1943.") They
soon observed that not only was it nearly impossible for her to find a respectable job that would utilize her professional skills but there was no organization
in Harlem or all of New York City providing psychological services for black
children. Nor was there any demand for any, since those parents who were
aware of their possible value feared the stigma then attached to being treated.
Thus, in 1946 Clark and Clark, aided financially by her family, renovated a
basement and opened the Northside Testing and Consultation Center, later
renamed the Northside Center for Child Development. Its clientele grew,
despite the initial suspicions, when parents learned that public school counselors were routinely labeling their children as retarded and sending them off to
special classes. After a few visits to Mamie Clark's clinic, however, such children's test scores might rise enough to justify a return to the regular classroom.
But these successes only fueled the ambitions of Clark and her staff, as increasingly in the 1950s they saw themselves as part of the larger civil-rights movement. They therefore moved to provide such unremunerative extra services as
remedial classes for children and political advocacy of better housing in Harlem. Over the years there were some triumphs (e.g., the building of Schom-
burg Plaza) and even honors, but by 1970 no one, least of all Mamie Clark,
would have called the task complete.[67]

    In addition to these four entrepreneurs who followed a nonprofit route,
there were several similarly energetic and imaginative women scientists in these
years who became self-employed, either as consultants or by starting their own
for-profit businesses in fields and markets where this was possible. Among
these were bestselling writers, such as biologist Rachel Carson; textbook and
reference book compilers (chemist Mary Fieser); reference workers or "literature chemists," such as Cornelia Snell of Foster D. Snell, Inc., of New York
City;[68] consultants of various sorts, most often psychologists but also an occa-

sional meteorologist; and, of course, Cynthia Westcott of New Jersey, the famous "plant doctor" or phytopathologist who from the 1930s into the 1960s ran a business spraying rich women's gardens in the warm months and then lecturing to garden clubs and writing reference works for home gardeners, a growing market at the time, during the winter months.[69] Westcott was such an unusual person that she was one of the very women scientists ever "profiled" in a *New Yorker* article. Some other women scientists set up small laboratories, for example, to analyze seeds or perform routine testing. One of these, a pathologist in Albany, New York, who directed her own laboratory, contributed to local public health by detecting such a high incidence of Hodgkin's disease in the Albany High School graduates of 1954 that further investigation was undertaken.[70] Perhaps the most exotic and colorful, possibly even dangerous, jobs were those of the filmmakers, especially of wildlife and spectacular scenery, sometimes for television and occasionally for motion picture studios like the Walt Disney studio.[71]

Thus, many women scientists found a welcome at a variety of nonprofit institutions, some flourished there, and a few even started their own institutions. Rarely promoted and often underpaid there as elsewhere, they were at least allowed to stay and do their own best work. For the brilliant and tenacious, it did not take more than a toehold—such as Libbie Hyman's seat at the AMNH or Barbara McClintock's cornfields on Long Island—to build a career of distinction. They did not necessarily need an academic or other "real" job, with all its demoralizing constraints, to do their best science, and do it they did! Often rejected elsewhere, they came all the more determined to take the crumbs that were offered and, given some autonomy, to shape this opportunity to prove their superiority and thereby to get some revenge. Then decades later, the sheer massiveness of their accomplishments might, just might, convince some within the system to acknowledge finally the magnitude of their powers, for certainly a large proportion of the top honors going to any women in science went to those, especially biologists, from the nonprofit sector.

12

Corporate Employment

Research and Customer Service

Although the number and proportion of women scientists and engineers employed in business and industry rose in several fields during the 1950s and 1960s, most still worked, despite their limited incursions during World War II, in the feminized areas of routine testing, chemical librarianship, and home economics, and a growing number were employed in the new field of computer programming. Whole other industries, such as aerospace and electronics, as well as management, were essentially closed to them. Compared with other sectors (e.g., academia and government), these areas employing large numbers of women placed less emphasis on doctoral degrees and more on youth and attractiveness. In fact, the term "planned obsolescence," which applied to some industrial products of the time (e.g., cars with big fins), also fit the prevailing personnel practices. Management hired women for specific, usually service, tasks but then, to judge from such statistical indicators of advancement as median age, salary, or years of experience, rarely promoted them. This created a certain emotional strain in those women trying to stay and rise. About the only way for a woman in industry to protest was to write a letter to the editor, as the two young chemists at Hercules Powder Company had done in 1946 (see ch. 1), or to quit, or both—the one could lead to the other. Others went quietly, and "systematic underutilization," to use the manpower analysts' term, persisted largely unchallenged. Even after the civil-rights legislation of the mid-sixties, management's thinking remained full of stereotypes, platitudes, and rationalizations; managers expected women to leave, planned for it, and then blamed them for the turnover!

According to the few publications on women's prospects in business in the 1950s, the best jobs open to them would be as chemists in research laboratories, statisticians in insurance companies, and home economists in the area of customer service. A 1952 article, for example, euphemistically entitled "Opportunities for Women at the Administrative Level," published in the *Harvard Business Review* by two women associated with the Harvard-Radcliffe Program in Business Administration ("program" because Harvard's business school did not accept women at the time), pointed to research positions and customer service as likely growth areas for women. Based on extensive interviews with businessmen in several eastern cities (as far south as Greensboro, North Carolina), the authors found that though few admitted to being prejudiced or discriminatory themselves, most felt that they had to behave in traditional ways in order to defer to the wishes of others—those customers who might be upset or insulted at being served by a woman bank officer, or other businessmen who might not accept such a woman. A few persons interviewed felt, however, that some women customers might like to be served by a woman, who would understand their problems better than a man, and that some businessmen might be particularly eager to seek out businesswomen, especially attractive ones. Most, however, feared that the potential difficulties and loss of business were too great to justify any pioneering experiments, even though the few businesswomen they did know, had met, or had dealt with personally had been more than satisfactory. In general, they felt that women in business above the clerical level would find their best opportunities conforming to prevailing sexual stereotypes: either hidden from sight in laboratory work or huddled together in feminized areas of customer service, both of which kinds of employment were bound to grow.[1]

The women's invisibility in industrial science and technology is also evident in their near total absence from the large volume of writings on personnel problems there. (Only one entry in a lengthy bibliography of the subject had the word "women" in the title.)[2] This literature was large because scientific and engineering personnel presented many continuing problems for management, for whom running an industrial laboratory was a relatively new venture, full of risks and unknowns. One basic tenet, however, was that engineering and industrial science were progressing so rapidly that technical obsolescence in the staff was a serious drawback in competition. It dictated that a company should hire large numbers of young engineering graduates, promote very few of them, and then within a few years replace most of the rest with even more recent graduates. If this was brutal for the men, who seemed not to have expected such an early end to their well-paying careers, the cost of training all the new men must also have been considerable. Nevertheless, those who associated high turnover exclusively with women employees never seemed to notice its centrality to the men's pattern as well. Meanwhile, a series of ever larger surveys of scientists and engineers in industry by social scientists (none of whom reported any women in their samples) repeatedly documented the stress, tension, "role conflict," and lack of communication between management and the "bench" as disillusioned scientists struggled to remain "professionals."[3] As long as there were enough fresh young male engineers available, however, industry had no need for reforms or innovations. Despite all the official talk in the 1950s and 1960s of shortages and the need to train and hire more women, very few companies attempted to hire women scientists or engineers at even the lowest levels.

Although none of the BLS periodic counts of scientists and engineers employed in industry (which showed that their number increased from 519,900 in January 1954 to 1,062,500 in January 1969) was broken down by gender, they did reveal that most of the scientific and technical jobs of the time were concentrated in a few industries, such as electrical equipment, transportation (including aerospace), and chemicals and allied products, and, what was quite unusual, that most engineers and scientists worked for very large companies, those with more than a thousand employees. Unfortunately for women seeking their first job or desiring a promotion in the private sector, these particular industries were, except for chemicals, reluctant to hire them, and large corporations tended to have rigid personnel practices that rarely innovated regardless of the so-called shortages. Two solid-state physicists were exceptions: Esther Conwell, who spent twenty years at GTE on Long Island (1953–72), including eleven as manager of its physics department, and Betsy Ancker-Johnson at the Radio Corporation of America (1958–61) and then Boeing (1961–73), where she reached the level of manager.[4]

Providing more useful information, though of a limited and even misleading sort, on women scientists and engineers in industry were the biennial surveys by the National Register of Scientific and Technical Personnel. Its coverage was about half that of the BLS surveys, since it omitted most of the nation's several hundred thousand engineers. Between 1956–58, the first time it included data on industrially employed women, and 1968, the end of our time period here, it reported that overall the number of industrially employed and self-employed women scientists and engineers rose substantially, from 2,098 in 1956–58 to 3,381 in 1968, or about a 61.2 percent increase, from 3.07 percent to 3.53 percent (see table 12.1). Yet if, for the sake of consistency, we delete both engineers and social scientists from both counts, the percentage actually dropped in these years, from 3.76 percent to 3.51 percent. As usual, however, when broken down by field, the picture is much more varied, with very large growth in some fields (computer sciences, social sciences, statistics) and actual declines in several others.[5] The majority of women by far in both years were in the one field of chemistry.

Women had been a part of industrial chemistry in America at least since World War I. Unfortunately, a traditional pattern of hierarchical segregation and circular reasoning had emerged and persisted, despite some criticism from young chemists during and just after World War II (see ch. 1). Management continued to hire women for particular jobs but rarely promoted them and

Table 12.1.

Scientists and Engineers Employed by Industry or Business and Self-Employed, by Field and Sex, 1956–58 and 1968

Field	1956–58			1968		
	Total	Women	Women as % of Total	Total	Women	Women as % of Total
Other specialties	703	144	20.48			
Psychology	1,658	188	11.34	1,747	179	10.25
Education	100	9	9.00			
Biological sciences	2,748	236	8.59	4,521	296	6.55
Mathematics	4,249	297	6.99	7,289	245	3.36
Statistics	77	5	6.49	692	49	7.08
Astronomy	36	2	5.56			
Medical sciences	481	22	4.57			
Social sciences	94	4	4.26	1,925	91	4.72
Linguistics				41	9	21.95
Sociology				121	20	16.53
Political science				116	7	6.03
Economics				1,636	55	3.36
Anthropology				11	0	0.00
Chemistry	25,856	920	3.56	53,291	1,910	3.58
Geography	121	2	1.65			
Physics	5,708	86	1.51	9,436	133	1.41
Earth and marine sciences	9,194	99	1.08	9,809	74	0.75
Engineering	14,458	72	0.50			
Meteorology	417	2	0.48			
Agricultural sciences	2,466	10	0.41	2,016	6	0.30
Atmospheric and space sciences (astronomy and meteorology)				537	4	0.74
Computer sciences				4,513	394	8.73
Total	68,366	2,098	3.07	95,776	3,381	3.53

Source: NSF, *American Science Manpower, 1956–58,* 81, 83; *1968,* 32, 252.

then, when they left, blamed them for their "instability." Because the field had by the 1950s, as earlier, a backlog of middle-aged or older women, the issue of advancement (or lack of it) was acute for industrial women chemists. One such chemist interviewed by surveyors of the federal WB for a 1950 report on women in "higher-level jobs," that is, jobs above the clerical level, complained that despite her higher degrees, she was still treated like "an errand girl."[6] Another, mentioned in Margaret Cussler's 1958 book *The Woman Executive,* was resentful that after forty years on the job she had so little to show for it.[7]

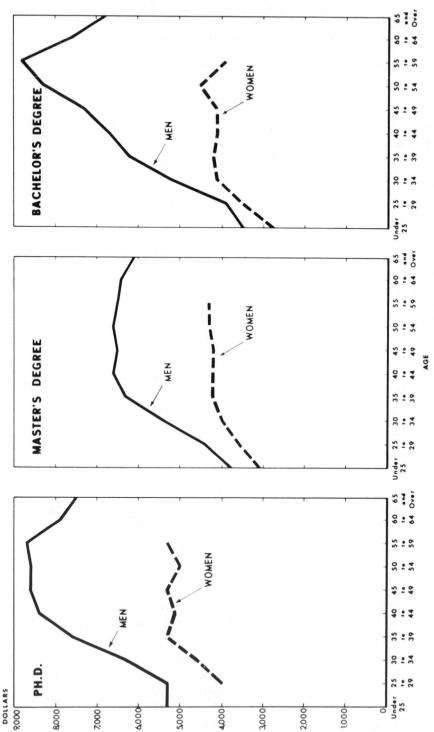

Fig. 6. Median annual incomes of chemists, by age, sex, and level of education, 1951. Reprinted from *Manpower Resources in Chemistry and Chemical Engineering*, BLS Bulletin 1132 (Washington, D.C., 1953), 37 and table A-29.

Income data, some subdivided by sex, were more abundant for chemistry than for most other fields, because the ACS did several comprehensive surveys in the 1940s and 1950s and then published interpretive summaries of the National Register's biennial surveys through 1970.[8] These surveys make it easy to demonstrate vividly the demoralizing situation facing a woman in the presumably well-paying chemical industry, as shown in fig. 6, for 1951. Women were paid less than men at all degree levels and at all stages of their careers. Women with doctoral degrees were paid only slightly more than women at the bachelor's level, and as time went on, they were paid *less* than men with bachelor's degrees. Nor was there much prospect of a raise after years of service, regardless of a woman's degree level, largely because she would not be promoted into the well-paying management positions. (A few made it into the lower management positions of supervising the younger women chemists in the ubiquitous "testing and analysis" work.) Management continued to rationalize this policy of nonpromotion into the 1970s under the catchall common to graduate deans that women were "bad risks": none were worth training for promotion, because most left within a few years for other professions, such as schoolteaching.

In 1964 chemist Claudine Carlton, angered by a recent editorial in *Science*, wrote a letter to the editor explaining why she had left industry for high school teaching. The work was so boring, she claimed, that it was more suitable for a technician than for a professional chemist. Women, she insisted, did not major in chemistry because it was easy; they were often superior to men at the master's and bachelor's degree levels and were often employed in analytical laboratories and technical libraries, "where [their] intelligence is not respected and [their] enthusiasm killed." She concluded that "the short tenure of women [in industry] may not be the fault of the women only," as the piece had claimed, but was the result of the management's systematic underutilization of women chemists.[9] Unfortunately, this policy, which penalized all for the expected behavior of some, had the consequence of further expediting the departure even of those who might have stayed on if they had been given some incentives or encouragement. Of those who stayed a few became "exceptions," as owners, executives, and, occasionally, research directors and managers, but untold others suffered depression and five times the normal rate of suicide.[10] The remainder were urged, often by older women chemists, to head for the traditional female ghettos of "chemical service" work, such as librarianship, literature chemistry (bibliography), technical writing, and secretarial work.

Of the many articles published between 1945 and 1955 on "opportunities for women in chemistry" (a euphemism for "overcoming the limitations of women in chemistry"), most urged them to head for three sex-typed and often conflated areas: chemical librarianship, literature chemistry (including abstracting and often patent searching), and technical writing or journalism. Some included secretarial work, part of which was tending a company's secret technical files.[11] A chemical background was essential for these service jobs, but most men trained in chemistry scorned such "desk work" as suited only to "misfits" and held onto their laboratory jobs, which usually paid better anyhow. Women chemists, however, who would not be promoted or be paid as well for their laboratory work, might more readily see such jobs as offering a desirable opportunity, for the demand was always great, the salaries were at least competitive, and advancement was sometimes possible. As Else Schulze of the Procter and Gamble Company of Ivorydale, Ohio (producers of Ivory soap), put it in 1945,

> The work offers wide opportunity especially to the woman chemist, because women, by nature, are usually good at detail work, interested in writing, and equipped with the curiosity, intuition, and patience needed to hunt for and find facts. Their chances for rising to administrative positions with correspondingly higher salary ranges are greater this way than via the laboratory. True, the work is entirely of the desk variety, but many of the administrative positions held by men in chemical fields are desk jobs. As pointed out by [Lois] Woodford . . . , young men chemists often frankly admit that they consider the laboratory a mere stepping stone to an administrative desk job. By offering vocational guidance and encouragement to women students who show interests in or qualifications for the tasks of the library chemist, the educators will not only be doing a service to industrial libraries; they will also be assuring the future career of the student.[12]

The "literature chemist" was a reference worker trained in chemistry who also knew bibliography and languages and could write well. In fact, to judge by the qualities and skills described in the articles written about these jobs, it would help if she were a nearly omniscient mind reader, for only then could she expect to fulfill the wide range of requests she would encounter. Moreover, she should be very modest, helpful, and tactful with chemists and even foresee their future needs. In short, she should personify the service ideal.[13] Such a person might be an employee on the library staff, attached to a particular division of the company, or, like Cornelia Snell, who wrote several articles on the woman chemist, be part of her husband's consulting firm, which answered questions from chemical companies around the world.[14]

Often confused with the literature chemist was the "chemical librarian," who was, as Irene Strieby of the Eli Lilly Company in Indianapolis often pointed out, a manager with heavy administrative burdens. More and more companies were forming libraries in the 1950s, often at their new corporate settings in the suburbs, and desperately seeking someone to run them. Oftentimes all they could do was hire a librarian, preferably a female between the ages of twenty-six and forty-two (age discrimination was rampant in industry), away from another company or pick someone from the secretarial ranks who could be trained for the position. The ACS job survey of 1955 provides evidence that women were indeed preferred for these positions: 71 percent of those listed under the occupation "library and information service" whose sex was given were women both in 1941 (47 of 66) and in 1955 (223 of 314). The median salary in 1955 for women in such positions was $614 per month, nearly the $615 that was the median for the top 10 percent of all women in chemistry.[15] Chemists who were given these positions quickly turned to the Special Libraries Association's science and technology division, whose many publications and workshops helped them learn their new responsibilities. They in turn became prominent members of the organization and leaders in the movement for training programs, clear-cut job descriptions, and higher salaries.[16] Among the better known of these women chemical librarians—who, regardless of whether or not they were married or mothers, spent long careers with their companies— were Betty Joy Cole, at the American Cyanamid Company of Bound Brook, New Jersey,[17] Irene Strieby of the Eli Lilly Company in Indianapolis,[18] Catherine Deneen Mack of the Corning Glass Works,[19] Gertrude Munafro at Texaco for at least four decades,[20] and Erna Gramse, a Mount Holyoke graduate and with a Ph.D. from Cornell, who worked first for Texaco and then, for most of her career, for the FMC Corporation of Chicago.[21]

But in the mid-fifties the suitability, even the preferability, of women for positions as chemical librarians and literature chemists ceased to be a major theme in articles on women in chemistry or on the future needs for chemical librarians.[22] A 1955 ACS meeting on the training of literature chemists may have marked a transitional point in this possibly deliberate mainstreaming of the field, for the symposium's chairman, M. G. Mellon of Purdue University, said that he had made it a point to have an equal number of men and women speakers, and most of them said that the need for such personnel was so great and yet the jobs appealed to so few individuals of either sex that chemists of both sexes were needed.[23] Thereafter, most articles on the continuing need for still more chemical librarians were written by men, usually from the larger chemical companies (e.g., Herbert Skolnik of the Hercules Powder Company in Wilmington, Delaware), and after 1961 published in the new *Journal of Chemical Documentation,* put out by the ACS's Division of Chemical Literature, rather than in the undergraduate-oriented *JCE.*[24] What real change took place is not clear, for job advertisements for chemical librarians, such as those in *Special Libraries,* continued into at least the late 1950s to contain a statement that one sex or the other was preferred, and the ACS did not collect any data on specialities and sex again until the 1970s.[25]

A vocal advocate for a third area of women's jobs, that of technical writing, editing, and journalism, was feisty Ethaline Cortelyou of Chicago and later Washington, D.C., who started her career as a chemist and then became an editor on the famed Manhattan Project. Since then she had worked for the Armour Research Foundation in Chicago, Argonne National Laboratory, and then, as a "science information specialist," at NASA.[26] In 1955 Cortelyou felt that women had a special skill (or at least willingness) for such work, or as she put it, "Technical editing, writing, literature and patent searching, and library and administrative work have in common certain qualifications that are traditionally attributed to women: desire to serve or help others, patience with routine, capacity for handling meticulous details, willingness to serve without glory and facility with words."[27] But the ACS's statistics indicate that although technical writing was substantially feminized—32.14 percent women (9 of 28) in 1941 and 38.85 percent (54 of 139) in 1955—it was less so than chemical literature or library work.[28] Although the women in the field earned lower salaries than the men, Cortelyou was sure that these women were better paid than they would have been if they had remained in the laboratory.[29] By 1958, however, Cortelyou, whose consciousness was rising, was deploring those college chemistry departments that channeled all their women students into literature searching, patent work, and technical writing and editing; she felt that many of these women should be encouraged to become full-fledged chemists.[30]

The most prominent woman chemist in journalism in the decades after World War II was not Cortelyou but Jane Stafford, a chemistry major at Smith College, class of 1920, who after a few years as a laboratory technician in a hospital moved into medical journalism. In 1928 she joined the Science Service, whose staff wrote stories that were distributed to hundreds of newspapers nationwide, and became one of the country's most prominent medical reporters. In 1944–45 she was president of the National Association of Science Writers. In 1956, after President Eisenhower's first heart attack had increased public interest in heart disease, and by which time the Science Service had begun to falter financially, Stafford moved to the National Heart Institute, where she was promoted to associate director of information in 1964.[31]

Among the exceptional women in industrial chemistry was Hazel Bishop, a 1929 graduate of Barnard College, who after working as a dermatologist's assistant in the thirties and for petroleum companies during the forties, invented at home the first nonsmear "Lasting Lipstick," as she called it, in 1949. By 1954, when it had captured one-quarter of the lipstick market, she sold out to her chief stockholder to get capital for other (as it turned out, less successful) inventions.[32] Several other unusual women chemists rose to the level of director of research or even vice president for research and development in small chemical companies in these years. Among these women, not all of whom held doctoral degrees, were biochemist Betty Sullivan, a Ph.D. from the University of Minnesota, who rose though the ranks to become director of research at the Russell Miller Milling Company of Minneapolis, while also winning a research prize from and serving as president of the American Association of Cereal Chemists.[33] Perhaps even more unusual was chic Dorothy Martin Simon, a physical chemist with a doctorate from the University of Illinois, who after jobs at Du Pont, Oak Ridge National Laboratory, and the National Advisory Committee on Aeronautics (predecessor of NASA), where she became an expert on rocket fuels, joined (along with her husband, also a scientist) AVCO Corporation as principal scientist and technical adviser on research and development to the president. By 1968, after two further promotions, Simon was vice president and director of research of this major aerospace corporation.[34] Yet a third "exception" outdid them all. Hunter College graduate Gertrude Elion determined after the death from cancer of her beloved grandfather that she would cure the disease by working for a pharmaceutical company. After earning a master's degree from New York University in 1941 and job-hunting for three years, she finally got her start at Burroughs Welcome in 1944. Staying on after the war, she and a colleague did such pioneering and successful work on a series of innovative and effective drugs for leukemia, gout, and other diseases that by 1968 she was head of experimental therapeutics for the company (and thus one of the highest-ranking women, if not *the* highest-ranking, in the whole pharmaceutical industry). But, to skip ahead twenty years, the best was yet to come, for in 1988 Elion, her colleague, and an Englishman all shared the Nobel prize in medicine, a rarity for anyone in industry, let alone a woman.[35]

Several foreign-born women chemists also found mid-level management positions in American chemical companies. Rosemarie von Rümker, with three degrees from the University of Bonn, in plant pathology and entomology, came to the United States in 1952 on a seven-month mission for her employer, Farbenfabriken Bayer A.G., and returned two years later as director of research for Chemagro, a large manufacturer of farm chemicals that was partially owned by Bayer. In 1959 she became vice president in charge of research and development for five divisions and the president's "right hand man," as he put it in an interview.[36] Hertha Skala, from Austria, came to the United States in 1946 as a war bride with a master's degree in pharmacy. After several jobs, in 1954 she convinced the Universal Oil Products Company of Des Plaines, Illinois, to hire her as a research chemist, and by 1967 she was assistant director of research, specializing in catalysis.[37] Maria Telkes, born in Hungary and earning a doctorate in Budapest in 1924, was a specialist on solar power and held a variety of positions at universities (MIT, NYU, Pennsylvania, and Delaware) as well as industrial laboratories (Curtis-Wright, Cryo-Therm, and Melpar of Westinghouse Air Brake). She held such positions as project director, director of research, manager, and senior research specialist, indicating that she moved laterally rather than vertically into top management.[38] Finally, Giuliana Tesoro was born in Italy but fled to America with her mother in the thirties. By 1943, at the age of twenty-one, she had earned a doctorate in organic chemistry from Yale University. After a brief stint at the American Cyanamid

Company (part of the time as a literature chemist, because of her knowledge of languages, but which she found boring), she moved on to the smaller Onyx Oil and Chemical Company of Newark, New Jersey, which was run by a fellow Italian and was located nearer to her husband's job in New York City. There she raised two children and got involved in polymer and textile chemistry, specializing in the static electricity in fabrics. In 1958 she joined the research department of the J. P. Stevens Company, a large textile manufacturer, and soon became its assistant director of research, in charge of twenty-five to fifty persons. But in the late 1960s she shared in the decline of the American textile industry, for when in 1968 as a cost-cutting measure Stevens reduced the size of its research department, it fired Tesoro. Within a few months she became director of chemical research at the immense Burlington Industries, but in 1972 it too terminated its research department, causing her to become a consultant and by 1973 an adjunct professor at MIT.[39] Industrial jobs have their risks as well as their rewards.

For most women chemists with doctoral degrees the best that could be expected was to be promoted to the position of research associate or group leader about ten years after arrival and to remain at that level for the rest of one's career. Thus Marion Pierce, with a Ph.D. in physical chemistry from Bryn Mawr in 1934, eight years at U.S. Steel's research laboratory, and three years on the faculty at Barnard, joined the E. I. Du Pont de Nemours and Company in 1946 as a senior research associate and remained at that level into the 1980s.[40] Mildred Rebstock received a doctorate from the University of Illinois in 1945 and went to work as a research chemist at Parke, Davis and Company, where in 1947 she isolated the synthetic form of Chloromycetin, the first antibiotic to be produced this way and thus the first to be made available in large quantities. But despite the vast profits that this must have brought to Parke, Davis, Rebstock was not promoted to research leader until 1959, and she remained at that level into the 1980s.[41] Similarly, Mary Root, who received a Ph.D. in chemistry from Radcliffe in 1950, joined Eli Lilly and Company in Indianapolis. When she was promoted to research associate thirteen years later, she was the first woman at the company to receive this title, an achievement that was acclaimed in the *AAUWJ*. Twenty years later she was still at that level.[42] These were all mid-level positions that had not been open to women before World War II, but they were also far from the burgeoning layers of management stretching from laboratories to divisions to corporate headquarters. A few women left such companies despite a possibly substantial pay cut. Gladys Hobby, for example, the microbiologist who had worked on penicillin for Charles Pfizer and Company during World War II, was in 1950 a co-discoverer of its Terramycin, the "most profitable antibiotic" of the 1950s,[43] yet she was not promoted. Finally in 1959 she left for the VA, where she headed its research institute on infectious diseases.

Among the large chemical companies in these years, the most receptive to women scientists seems to have been the immense E. I. Du Pont de Nemours and Company of Wilmington, Delaware, and elsewhere. At least it claimed to employ some two hundred women scientists and engineers among its forty-three hundred "technically-trained people engaged in research and development," or about 4.65 percent, in 1960, as described in a four-page spread in its in-house newsletter, which pictured twelve scientists (a geologist, a physicist, a biologist, an engineer, and a patent specialist as well as a variety of chemists) with doctoral, master's, and bachelor's degrees. Four were even said to have husbands working for Du Pont, a rather liberal labor practice in 1960, when many colleges and universities still had strict antinepotism rules.[44] The highest-ranking of the women pictured was toxicologist and section head (at the Haskell Laboratory in Newark, Delaware) Dorothy Hood, a Bryn Mawr graduate in the 1930s, who had come to Du Pont in 1946 after a year at Merck and Company and four years in the naval reserve. She supervised a staff of twenty.[45] Not pictured was polymer chemist Stephanie Kwolek, who had also come to Du Pont in 1946, immediately upon graduation from the Carnegie Institute of Technology, and who pioneered in the field of low-temperature polymers, patenting Kevlar, an aramid fiber as strong as steel. In 1959 she won the first of her many prizes, a publications award from the ACS, becoming thereby one of the few women to win anything but the Garvan Medal from the ACS.[46]

Also of note is some decidedly strange behavior in these years (1940–70) at the General Electric Company of Schenectady, New York, toward the few women scientists on its staff and its attempt as late as 1964 to prevent employing any more! By far GE's best-known woman scientist—in fact she was featured on occasion in GE's wartime advertising—was physicist Katharine Burr Blodgett, the daughter of GE's one-time patent lawyer. Hired in 1919 upon graduation from Bryn Mawr and rehired in 1926 after becoming the first woman to earn a doctorate at the Cavendish Laboratory in England, Blodgett worked for many years with GE's Nobelist Irving Langmuir. In 1938 she announced her discovery of "invisible glass," or glass covered with several layers of nonreflecting thin films. The publicity that followed this brought her to the attention during World War II and after (1939–51 chiefly) of the several groups interested in giving awards to women in science, especially such an unusual one as an industrial or GE employee.[47] In time, if one is to believe an oral history interview with a fellow GE employee, Blodgett came to believe her own publicity and thought that she was a combination of Madame Curie and Greta Garbo, and each spring she had to be sent to a nearby psychiatric hospital for six weeks to recover from her delusion and depression.[48] Finally in 1963 she retired, and in 1979 she passed away.

Also spending her whole career (1945–80) at GE was polymer chemist Edith

Boldebuck, who had received her Ph.D. from the University of Chicago. She joined GE after writing the highly publicized Blodgett about her ideas about silicone. Despite her eventual thirty-one patents on silicones and related substances (the second highest number held by any American woman, according to her 1981 obituary), Boldebuck was never promoted at GE, and unlike Blodgett, she went unrecognized in her lifetime by any women's or scientific group.[49] It may be that the company tried to keep her work out of the limelight, or perhaps she, seeing what had happened to Blodgett, shied away from similar publicity.

In the early 1960s, as rhetoric about encouraging more women to go into science began to increase, eager reporters came to Schenectady to talk to Dr. Blodgett and to seek information on other women scientists at GE. Yet they got very little assistance; in fact one reporter was told in 1964 that the GE public relations staff "reacted negatively to any story idea that might encourage an increase in the number of women scientists."[50] Whether GE was being protective of Boldebuck and her industrial secrets, was exasperated by Blodgett, who had finally retired, was just chauvinistic, or all three, not only was it no longer featuring its best-known woman scientist in its ads but it was not even giving modest help to writers of career books for young girls.

By comparison, the young and iconoclastic Polaroid Corporation was proving to be a refreshingly innovative and relatively enlightened employer of women scientists. A recent history-biography of the company and its founder Edwin Land says several times that he hired many, very attractive alumnae of Smith College, which his wife had attended, and put them into responsible positions. These included Meroë Morse, who earned fourteen patents and headed a research division before her early death, Eudoxia Miller (later Mrs. Robert Woodward), and several others referred to collectively as "the Princesses." Some reportedly stayed on the job part-time after marriage and motherhood. Another unusual feature at Polaroid was that in a field still nearly lily-white it employed at least two women scientists who were black, including Esther Hopkins, a Yale Ph.D., and Caroline Hunter, who in the late sixties dared to criticize Polaroid's contract with South Africa for photo identification passes for blacks.[51]

Despite the numerous and even relatively well-paid women mathematicians and statisticians in industry in the 1950s and 1960s,[52] far more attention was paid both at the time and in retrospect to those women hired by the early computer companies, because of one colorful, long-lived, and important contributor, and the many others in the industry's new and highly publicized kind of "women's work." By far the best-known woman in computer science was Grace Hopper, a Yale doctorate and former Vassar faculty member, who had joined the U.S. Naval Reserve in World War II (*not* the WAVES, she insisted in numerous interviews) and been assigned to work with Howard Aiken at Harvard University on his IBM-supported Mark I computer. She remained after the war as a civilian and research fellow in engineering and applied physics but left in 1949 to join the Eckert-Mauchly Computer Corporation, whose UNIVAC offered challenging problems in computer languages and what later came to be called "software." She stayed with UNIVAC until 1967—through its acquisition by Remington Rand and then a merger into Sperry Rand in the mid-1950s—reaching the level of chief engineer for automatic programming in 1959 and director of research in systems and programming for the UNIVAC division in 1961. Before her recall by the Navy to active duty in its Information Services Division in 1967, she had written about fifty articles on computer programming, including several that became classics in the field.[53]

Meanwhile, at giant IBM (International Business Machines), where Jean Sammet designed and developed the COBOL compiler in the 1960s,[54] most college women were being channeled into a lower-level kind of semiprofessional "women's work" enthusiastically developed there in the forties and fifties. Because in these years IBM was trying to catch up with Remington Rand, the front-runner in the emerging computer industry, it chose a somewhat unusual sales and employment strategy, to which it clung long after it had achieved dominance in the industry. At some point in the forties IBM leaders realized that many potential customers were afraid of their machines and unsure how to use them to the utmost. Such persons needed instruction and reassurance, especially when, as happened rather often with the early models, the machines broke down or otherwise simply baffled their users. Thus IBM officials decided that rather than selling the machines and then leaving the buyers to cope with them on their own, they would lease them and provide long-term service contracts guaranteeing that a knowledgeable, calm, tactful employee would be available on short notice when problems arose.[55] For such jobs, bright, "personable" young women, recent graduates in mathematics from good colleges, from which they were specially recruited, seemed to fit the bill. As early as 1943, IBM hired hundreds of women for these jobs, first as "customer service" and later as "systems service" representatives. (Turnover was high, since almost all married in a few years and then left to have children.) The innovator of this new form of relatively well-paying "women's work" was Ruth Leach, an attractive 1937 graduate of the University of California, who had been hired by IBM to demonstrate one of its early machines at the 1939 World's Fair in San Francisco. There she caught the eye of Thomas Watson, Sr., company founder, who put her in charge of the company's women's programs during the war and in 1944 even promoted her to a vice presidency, such a high level that she was often disparaged by others. In her ten years on the job (before "retiring" in 1954 after her marriage at the age of thirty-eight to a Philadelphia banker), Leach gave many speeches and wrote many articles, for women's magazines especially, on these glamorous new jobs at IBM.[56] Psychologist Rita McCabe, who succeeded her, though at the level of a manager rather than a vice president, was still recommending such jobs for bright women seeking employment in the area of computers at a historic meeting on women in science and engineering at MIT in 1964.[57] But by then the limits of such jobs were becoming apparent.

Actual data on the position of programmer, the main entry job to computer science in the fifties and sixties, are hard to find.[58] A 1958 report by the BLS on the employment outlook for office workers, stated, however, that "men are preferred as programmer trainees in most areas of work. Although many employers recognize the ability of women to do programming, they are reluctant to pay for their training in view of the large proportion of women who stop working when they marry or when they have children. Opportunities for women to be selected as trainees are likely to be better in government agencies than in private industry."[59]

Additional reasons for hiring men preferentially included the likelihood of nighttime work (from which women were prohibited in some states by protective legislation), familiarity with the field for which the programs were being written (engineering and accounting especially), and certain personality traits, including patience and "the ability to work with extreme accuracy, paying close attention to detail," traits that under most circumstances employers usually sex-typed as feminine.[60] In 1957, for example, the Systems Development Corporation of California, already one of the nation's largest employers of programmers, suddenly needed five hundred more to fulfill the needs of a contract with the Air Force for an early warning system against missile attack. Articles describing the corporation's heroic efforts to test and recruit suitable personnel (up to seventy-five per month) made no mention of hiring women.[61]

By the early sixties, because the cost of training programmers had been declining and some job applicants had even studied computer science in college, it became worthwhile for a company to hire women for these same positions. In fact some found it a better investment to hire "bright girls" for such positions even temporarily than to hire the available men. As M. Ostrofsky of Westinghouse summarized it in an article entitled "Woman Mathematicians in Industry," in, of all places, the *AAUWJ*, "From the point of view of an employer, it is perhaps better to hire bright girls for two to three years rather than dull boys for an indefinite period of time. An additional point in favor of hiring girls as programmers and coders seems to be that girls are more capable of patiently doing detailed work, a highly desirable characteristic for this type of job."[62]

The sexism surrounding this new interest in hiring women was all too evident, however, in a 1962 article entitled "How to Hire a Programmer," which in trying to be humorous about the frantic pace and dubious credentials of many new recruits described one female in less than flattering terms:

> The second candidate to arrive is female—Miss Sallyann Bunch from East Passerk, New Jersey. Sallyann has had a lot of computer-related experience: two years in the key punch pool of the Unforgivable Assurance Association of North America, Newark, and seven months in charge of tab board wire storage with Automobile Catastrophic Statistical Society, Orange. Also she is a graduate of Princeton (South Princeton Philosophic Junior College) with a major in Oriental Basketry.
>
> Sallyann wears flat shoes, and she is a little cross-eyed. Her figure resembles a full potato sack. Her dress and make-up indicate that she is a solid-plain-thinking person with no frills at all. Miss Bunch is the spitting (she chews Copenhagen) image of a lady programmer.
>
> A munificent offer is made to Sallyann, and she goes home to ask her mother about it.[63]

This description suggests that the prevailing view in the leading journal *Datamation* around 1962 was that "lady programmers" were easy targets for humor, not entirely welcome newcomers, and possibly not as attractive or feminine as the carefully selected IBM systems service personnel had been. By then, however, management's prevailing frustration with women programmers had moved onto the realm of job tenure or turnover. What was particularly maddening to Westinghouse's Ostrofsky was the "irresponsibility of these same women to leave after a few years." He suggested that perhaps the company should invite a woman who had left back to the job after a few months, but he said that this was an "admittedly questionable solution," since many would think that she should stay home with her children. Another possibility was to wait until she had "substantially completed the task of raising her family," but "with rare exceptions such a woman almost certainly will not be creative, simply because most of us have by the age of forty passed our creative age" (a common belief at the time). One consequence that women who chose this last solution "will have to recognize [is that] professional advancement is likely to be limited when any person, man or woman, begins a new career in middle age." Ostrofsky could thus insist that there was little or no discrimination in hiring in the field of mathematics but willingly admit that women's advancement in management (and thus in salary) would necessarily be very limited. Despite this evidence of sexism and ageism, Ostrofsky concluded on as bright a note as possible: sooner or later a few aggressive women might be able to manage their hormones and learn how to pick up restaurant checks and thus move into management. On the whole he felt that he could neither be blamed for the current impasse nor be optimistic about its solution.[64]

As early as 1956 the WB, unlike the BLS, had been telling women that there *were* opportunities in the new computer field, as evidenced by the fact that even married women were being hired on occasion.[65] But not even this 1956 report by the WB was optimistic about advancement for women. It was still rare in 1961, for when Mary Hawes of RCA's electronic data processing division, wrote to John Parrish about one of his recent articles on professional woman-power, she reported that she had been at RCA ten years, was "the lone female manager" in the whole division, and, not coincidentally, had been a widow for eleven years. Thus, unlike many others, she probably had been promotable because as a single parent of two she was in clear financial need.[66]

There were at least two other attempts in the late 1950s and early 1960s to remedy the lack of opportunities for employment, if not advancement, for married women in the computer field. In the late fifties Douglass College, in northern New Jersey, started one of the first "continuing education" programs in the country to retrain suburban housewives (some of whom had had scientific employment during World War II) for jobs as computer programmers. By 1965 Bell Laboratories had hired so many graduates of this Mathematics Retraining Program that its in-house newsmagazine had an article on four of them. All praised the program and their jobs, but one dared to say that she thought full equality lay in the future.[67] Thus the present and future role of women, including married women and mothers, was being thrashed out publicly and privately between 1955 and 1965 in the new and explosively growing field of computers. After a few emotional articles by male executives caught in various mental impasses, for example, exploiting young women and then blaming them for leaving, a few innovative and flexible programs quietly accommodated to the realities of some women's lives seemed to represent a positive step toward a partial solution of the "womanpower" problem in one field at least.

A second innovation was a new company, Computations, Inc., of Harvard, Massachusetts (outside Route 128), formed in 1958 by Elsie Shutt and several other programmer-mothers who worked part-time and largely at home on problems contracted out to them by their former employers, such as Minneapolis-Honeywell and Raytheon. When it was described in *Business Week* in 1963, the innovative but possibly short-lived company, whose founders were nicknamed the "pregnant programmers" by their clients, was looking forward to a profitable future, since by not paying benefits to its workers nor even permitting coffee breaks, it could charge less than the going rate for the product of experienced workers.[68]

Another field whose many women graduates were considered well suited for positions in industry was home economics. Many companies specializing in food and household products were expanding greatly after 1945 as part of the suburban "back to the home" movement. Among these products were utilities;

processed foods of all types, including especially frozen foods, "instant" mixes, and baby foods; soaps and detergents for the new synthetic fabrics; and household appliances such as gas stoves and water heaters as well as electrical irons, mixers, refrigerators, freezers, washing machines, and the first dishwashers. Because most users of such items were women, who might not listen to a man's demonstration of the product, many companies hired female home economists for the area of their sales or advertising staffs known as "customer service." One particularly large and well-known group was the staff of forty home economists at General Mills in Minneapolis, publicized under the trade name "Betty Crocker" and long headed by Marjorie Husted (1923–48) and then Janette Kelly (until her death in 1958).[69] Other well-known women worked for corporations such as Corning Glass Works and General Foods (producers of Birdseye Quick-Frozen Foods); on particular products, such as Sealtest ice cream or "Frigidaire" refrigerators; or for distributors like First National Stores of Boston.[70] Though one of the aims of this type of job was no doubt to increase sales, such home economists might also think that they were improving American (or, when they demonstrated products at World's Fairs, foreign)[71] homes and lives by helping mothers feed their families efficient, attractive, and nutritious, if not cheap, foods. (The social value of "customer service" work at such "junk food" companies as, perhaps, Coca-Cola of Atlanta might be less evident.)

One step removed from this sort of direct product promotion or advertising were positions as food (or other) editors with women's magazines, city newspapers, and local radio and television stations. Elizabeth Sweeney Herbert, for example, was for many years the household-equipment editor for *McCall's* magazine,[72] and in Los Angeles Corris Guy, in addition to being director of consumer services for Helms Bakeries, was food editor of *Western Family* magazine, and the producer and star of a series of cooking shows on KTIA-TV from 1947 into the seventies.[73] But the best-known nationally and probably the most influential was the group at the Good Housekeeping Institute in New York City, run by *Good Housekeeping Magazine,* which tested home products for the coveted Good Housekeeping Seal of Approval, a well-publicized kind of information and "limited warranty" for consumers. The Institute's longtime directors were Katherine Fisher (1924–53), who had previously taught home economics at McGill University and Teachers College, whom the *New York Times* referred to upon her death in 1958 as the "dean of home economists" in America,[74] and her successor and protégée Willie Mae Rogers (1953–75), who was one of the few women to be so well paid that in 1969 she turned down a high-ranking position in the Nixon administration partly because the salary was too low![75]

There were, in fact, so many women home economists in business that in 1923 sixty-two formed a subgroup within the AHEA to represent their interests. Known first as Home Economics Women in Business (HEWIB), later changed, when the first man joined, to Home Economists Working in Business and then just Home Economists in Business (HEIB), the group flourished; by the late sixties it had more than 2,000 members.[76] When in the spring of 1966 Earl McGrath was preparing his report on the status of the field of home economics, he surveyed the 1,945 members of the HEIB section to determine their current positions. (One had to be currently employed to be a member of the section; it was not enough to be a housewife who had formerly worked in business.) The 1,218 responses revealed a progression of jobs held by women in this category: young women tended to work for utility companies, but older women, those who had entered the field in the twenties and, often, had taught in the public schools, tended to end up in journalism or consulting as well. Whether this shift came because of opportunities for older women with established reputations in journalism and self-employment or because they were pushed out of more glamorous, youth-oriented positions in customer service is not clear. Although only 10.3 percent held master's degrees and 0.3 percent held doctorates, salaries were rather high both overall (almost two-thirds earned more than seven thousand dollars a year) and at the top. (Nine reported salaries of more than twenty-five thousand dollars, which impressed McGrath, since this was more than the salary of any dean of home economics in the land-grant system, five of whom earned about twenty thousand).[77] No wonder the HEIBs were enthusiastic about their field and frequently wrote articles in the *JHE* and elsewhere urging young women to prepare themselves for positions in business. The demand was so great, in fact, that there was a danger that unqualified persons might get the jobs that trained home economists ought to hold.[78]

Yet if the HEIBs felt that they had an image problem with young women thinking about their future, they knew they had one (which they less often discussed publicly) with their employers. Although the usual pattern in business was for promotions and advancement to go to persons in sales (rather than research or personnel work, for example), this did not hold for women in home economics. If a well-known few did attain positions such as director of customer service, supervising a staff of other women, they held these positions until early retirement or departure to form their own consulting companies. Very rarely did they move up to positions in the corporate hierarchy, such as to a vice presidency. (When bacteriologist Mary Horton, long the director of customer service at Sealtest, Inc., was promoted to vice president of its holding company, Sheffield Farms, Inc., this was newsworthy, and it was mentioned in her obituary in the *New York Times*.)[79] Marjorie Husted, the former "Betty Crocker," who on the eve of her retirement was made a vice president of General Mills, addressed part of this issue in a hard-hitting talk (for its time) to the Midwest Regional Gas Sales Conference in Chicago in 1952. Under the title "A Critical

Evaluation of Modern Home Service," she told them what they already knew, namely, that they expected very little from their staff of home economists, that they gave them very little training in salesmanship, public speaking, or business skills, that they considered their work mere advertising, and that they gave them very little credit for increasing sales. With better treatment, the home economists in customer service could do even more for the company, she felt, because of their insights into the psyche of the female consumer. Husted herself, for example, had started with General Mills by distributing recipes that required a lot of flour (the company's main product then), but in time, after learning that women liked to please their husbands and that men loved chocolate fudge cake, she had helped move the company into the more profitable dessert market. (And, she reminded them, think of how much gas that had sold!)[80] If the name of the game really was selling, then the women in the customer service department could contribute more to product development and long-range planning. But it seems that little had changed even twenty years after Husted's talk, for in 1971 the HEIBs did a survey comparing management's views of the role of the home economist in business with their own. Although the women wanted to be seen as part of the management and administration of the company and involved in corporate planning, most employers still saw them as suitable only for public relations and product promotion.[81]

This persistence of "traditional" attitudes about women's "place" (or "non-place") in management was well documented in a classic survey tauntingly entitled "Are Women Executives People?" published in the *Harvard Business Review* in 1965, one year *after* the passage of the Civil Rights Act of 1964, which banned discrimination in employment on the basis of either race or sex. The authors, Gerda Bowman of the National Conference of Christians and Jews and N. Beatrice Worthy of Bell Labs, surveyed the attitudes of nearly one thousand men and more than nine hundred women in business and found that little had changed since previous surveys. Although the women were more optimistic than the men about women's opportunities in traditional areas such as personnel and production, they were less cheerful than the men about their future in engineering, research, and development. Both agreed, however, that women's best chances would not be in business per se, but in education, social services, health, and the creative and performing arts, followed by governmental agencies. In all of these, unlike in business (including their own), women would be at least equal and might even have an advantage in management. One reason for such complacency a year after passage of the landmark legislation was that the women in business (themselves "exceptions") were not very angry or even organized. As one California businessman put it,

> I believe that the main push for a change in women's status will have to originate with women themselves; no outsider can really carry off a social revolution for

another group. At present, most women do not seem to be actively concerned about a change in their status: the conversation is still mainly from outsiders. Though I am convinced that Negroes will eventually take over and legitimize the so-called Negro revolution and win in it, I am not at all convinced that women will ever want a woman's revolution. Tremendous changes in traditional imagery have to be made—first by women themselves—in order to create a new place for women in the activities of society.[82]

The women respondents tended to think of themselves as managers rather than as women and saw their role as one of providing greater opportunities for younger women rather than of fighting for themselves, even though several reported discriminatory incidents that had crippled their own careers.[83] Few expected the new law to change anything, even though it might embolden a few aggrieved women to speak out. Certainly in 1965 none of these knowledgeable business people saw a major revolution on the horizon.

13

Governmental "Showcase"?

After 1950 the federal government expanded greatly and employed many women scientists, including married and minority women and even couples. The women were employed more in some fields and at some agencies than others but rarely advanced to the higher levels. When they did it was at just a few agencies, those in the traditionally female-dominated areas of home economics, women's labor, rehabilitation, and public and child health. But the most the CSC was willing to do to spur female recruitment, even as late as 1960, was to publicize outstanding contributions by current employees with a special women's prize. Then, after certain paper reforms in 1962–63, President Lyndon Johnson, wishing to use the civil service as a kind of showcase for enlightened equal opportunity for women and blacks, recruited and appointed several women, including some scientists, to high-level positions in his administration. When in 1967, after race riots in several cities and the formation of NOW, some winners of the women's award recommended further changes, he signed a stronger executive order. Thus by 1967 the federal sector was the only one to respond in any way either to the manpower planners or to the women's movement. But the women scientists' proportion of the total did not increase, for a number of reasons: the female pioneers of World War II retired after twenty years of service, certain longstanding areas of "women's work" in the government were reorganized out of existence, and other expanding areas, such as grants administration, scientific information, and drug regulation, were overshadowed by immense areas like atomic energy and space flight, which employed relatively few women. Less is known about the women scientists and engineers in the state and local governments.

When the CSC and the then new NSF published their first numbers of scientists and engineers employed by executive agencies in the early 1950s (summarized in table 13.1), women were a mere 4.28 percent of the total (4,303 of 100,458). Their proportion ranged from highs of (84.81 percent) in the combined field of nutrition and food technology, to 22.60 percent in the biological sciences and 20.45 percent in astronomy, on down to less than 3 percent in several fields, including the particularly large ones of engineering (with 43,272 total) and agricultural sciences (12,946). Unlike the women's position in

Table 13.1.
Women Scientists and Engineers Employed by the Federal Government, by Field and Agency, 1954

Field	Total Scientists and Engineers	Women	Women as % of Total	DOD	HEW	Agriculture	Commerce	Interior	NACA	Other
Nutrition/food technology	158	134	84.81	15	8	88	—	15	—	8
Mathematics and statistics	3,366	857	25.46	411	62	46	79	15	111	133
Mathematics	(1,318)	(431)	(32.70)	265	—	1	37	10	111	7
Statistics	(1,994)	(419)	(21.01)	146	59	45	42	5	—	122
Actuary	(54)	(7)	(12.96)	—	3	—	—	—	—	4
Biological sciences	1,925	435	22.60	143	162	49	—	1	—	80
Microbiology	(990)	(245)	(24.75)	86	75	17	—	—	—	67
General biology	(401)	(102)	(25.44)	26	65	2	—	—	—	9
Zoological sciences	(212)	(43)	(20.28)	20	14	5	—	1	—	3
Animal physiology	174	36	20.69	20	13	2	—	—	—	1
Zoology	38	7	18.42	—	1	3	—	1	—	2
Botanical sciences	(203)	(33)	(16.26)	6	3	23	—	—	—	1
Botany	52	19	36.54	1	—	18	—	—	—	—
Mycology	37	11	29.73	5	3	3	—	—	—	—
Plant taxonomy	19	2	10.53	—	—	1	—	—	—	1
Plant physiology	95	1	1.05	—	—	1	—	—	—	—
Pharmaceutical sciences	(73)	(11)	(15.07)	5	5	1	—	—	—	—
Genetics	(46)	(1)	(1.05)	—	—	1	—	—	—	—
Astronomy	44	9	20.45	9	—	—	—	—	—	—
General social sciences	3,606	646	17.91	135	158	10	3	4	—	336
Economics	2,437	334	13.71	9	2	41	52	4	—	226
Psychology	1,978	244	12.34	69	76	2	—	—	—	97
Chemistry	4,671	561	12.01	162	132	97	32	40	3	95
Anthropology, archaeology, and linguistics	66	3	4.55	—	—	—	—	2	—	1
Earth sciences	6,828	278	4.07	77	—	1	61	133	—	6

Field	Total									
Geography	(112)	(25)	(22.32)	8	—	1	2	13	—	1
Geology	(1,494)	(97)	(6.49)	—	—	—	—	92	—	5
Cartography	(3,023)	(113)	(3.74)	64	—	—	23	26	—	—
Geophysical exploration	(100)	(3)	(3.00)	1	—	—	—	2	—	—
Meteorology	(2,099)	(40)	(1.91)	4	—	—	36	—	—	—
Medical officers	8,916	358	4.02	33	66	—	—	5	—	254
General physical sciences	3,132	94	3.00	40	3	20	14	1	16	3
Physics	5,351	123	2.30	72	3	4	40	1	1	2
Physics	(2,859)	(112)	(3.92)	65	3	4	36	1	1	2
Electronics	(2,492)	(11)	(0.44)	7	—	4	—	—	—	—
Metallurgy	526	6	1.14	4	—	2	—	—	—	—
Engineering	43,272	175	0.40	131	—	—	2	21	2	—
Agricultural sciences	12,946	44	0.34	3	4	34	—	1	1	19
Plant pathology	(233)	(15)	(5.44)	1	—	14	—	—	—	2
Nematology	16	2	12.50	—	—	2	—	—	—	—
Plant pathology	217	13	5.99	1	—	12	—	—	—	—
Entomology	(471)	(19)	(4.03)	1	3	13	—	—	—	—
Husbandry	(102)	(1)	(0.98)	—	—	1	—	—	—	2
Fisheries/wildlife	(262)	(2)	(0.76)	1	—	—	—	1	—	—
General agricultural sciences	(657)	(3)	(0.46)	—	—	—	—	1	—	—
Veterinary medicine and surgery	(1,604)	(3)	(0.19)	—	—	3	—	—	—	—
Forestry/range management	(3,288)	(1)	(0.03)	1	—	2	—	—	—	—
Soils/irrigation	(5,893)	(0)	(0.00)	—	—	1	—	—	—	—
Agronomy	(354)	(0)	(0.00)	—	—	—	—	—	—	—
Horticulture	(82)	(0)	(0.00)	—	—	—	—	—	—	—
Dental sciences	1,236	2	0.16	—	—	—	—	—	—	2
Total	100,458	4,303	4.28	1,313	673	392	285	243	133	1,264
			99.99%	30.51%	15.64%	9.11%	6.62%	5.65%	3.09%	29.37%

Source: NSF, Scientific Manpower in the Federal Government, 1954 (Washington, D.C., 1957), app. G table 4.

academia, the largest fields for women in the federal government were mathematics and statistics (857);[1] general social sciences, which included "foreign affairs," international relations, and military intelligence research (646); chemistry (561);[2] and the biological sciences (435).

The new and immense Defense Department, formed in the late 1940s by the merger of the Army, Navy, Air Force, and former War departments, employed the most women (1,313, or 30.51 percent of the total). Among these were specialists in almost every field, including astronomy (all 9 women in the federal government), biology (more than at the Department of Agriculture), cartography (more than half the government's total), physics, electronics, metallurgy, engineering, and mathematics and statistics (nearly half the federal total). But because the Defense Department also employed 45,021 male scientists and engineers, the women constituted a mere 2.87 percent of its total technical workforce. Ranking second numerically, with just more than half as many women scientists and engineers (673), was the newly renamed Department of Health, Education, and Welfare (HEW), the most feminized of the large governmental science agencies, with 17.76 percent of its scientists women. Most of these were in biology, the social sciences, and chemistry (probably including biochemistry), and most of the scientists (men and women) were at the USPHS, and especially its famed NIH. There were so many women bioscientists at the NIH in the 1950s and 1960s that a chart is needed just to list their names (see table 13.2). By far the most women (and this was true of NIH scientists in general) were at the large and rapidly growing National Cancer Institute, which since its founding in 1937 has had a long series of important women scientists, including Thelma Dunn, who in 1961 became the first woman president of the American Association for Cancer Research. In fact the more one reads about the distinguished women at NIH and their pioneering work in a number of specialties, the more surprising and even shocking it becomes that not until 1977 was one of them (Elizabeth F. Neufeld, long of the National Institute of Arthritis and Metabolism) elected to the NAS.[3] In third place in 1954 was the USDA, with 392 women scientists (just 2.43 percent of its total), including 34 of the 44 women agricultural scientists in the federal service and its only woman geneticist. Although widely known before World War II as the agency employing the most women in science (half of all those in the federal government), the USDA had not shared the explosive wartime and postwar growth of the Defense Department and HEW.[4]

In addition to the many governmental jobs in greater Washington, D.C., women scientists and engineers found federal jobs at the numerous installations run by various agencies all over the United States. These included the many laboratories and other units of the sprawling and decentralized Department of Defense;[5] the four "utilization" laboratories of the USDA;[6] the so-called national laboratories of the AEC (Los Alamos, Argonne, and

Table 13.2.
Selected Women Scientists Employed at the National Institutes of Health,
1940–1970

Allergy and Infectious Diseases
  A. Elizabeth Verder
  Janet Hartley

Arthritis and Metabolic Diseases
  Evelyn Anderson
  Louise Marshall
  Marjorie Stetten
  Maxine Singer
  Elizabeth Neufeld

Cancer
  Thelma Dunn
  Sarah Stewart
  Margaret G. Kelly
  Mary Maver
  Virginia Evans
  Katherine Sanford
  Willie Smith
  Elizabeth Weisburger
  Mary Fink
  Helen Dyer
  Katherine Snell
  Lucia Dunham
  Delta Uphoff
  Margaret Deringer
  Betty Sander
  Elizabeth Anderson

Dental Research
  Rachel Larson
  Marie Nylen

Division of Biologics Standards
  Sara Branham
  Bernice Eddy
  Margaret Pittman
  Ruth Kirschstein

Division of Research Grants
  Marie A. Jakus
  Ernestine Thurman
  Katherine Wilson

Heart and Lung
  Martha Vaughan
  Marion Webster
  Thressa Stadtman

Library of Medicine
  Ruth Davis

Mental Health
  Nancy Bayley
  Marion Kies
  Margaret Rioch
  Charlotte Silverman
  Audrey Stone

Neurological Disease and Blindness
  Marjorie Whiting
  Elsa Orent Keiles

Office of the Director
  Jane Stafford
  Marjorie Wilson

Sources: NIH, *Scientific Directory, 1968* and *Annual Bibliography, 1967* (Bethesda, Md., 1968); *AMWS;* Thelma Dunn, "Intramural Research Pioneers, Personalities, and Programs: The Early Years," *Journal of the National Cancer Institute* 59, no. 2 (suppl., Aug. 1977): 605–16.

Brookhaven, among others), run by contractors who paid industry-level wages; and the hospitals and mental hospitals run by the VA in almost every state.[7] Usually conveniently located in or near big cities or college towns, where many women's husbands worked, these federal agencies, unlike many universities, often hired married women, divorced women, reentry ones, and even couples. Although some persons might complain that the federal government paid less than industry did, for many women who were unwelcome in industry and had restricted mobility, the federal government offered reasonable jobs at competitive wages. To former research associates it offered secure employment, regular pay raises, and benefits, including a guaranteed pension.

Among the numerous couples at the federal and especially "national" laboratories were some who might have preferred university posts but were barred from them by the antinepotism rules of the time. Others had unconventional marriages and needed to move. When the crystallographers Isabelle Lugosi and Jerome Karle went to the Naval Research Laboratory (NRL), at Anacostia, Maryland, outside Washington, D.C., in 1946 because no university was willing to give them both faculty positions, they felt that they were making what she later called "a big compromise." But the NRL, which had few women at the time, made an exception for them, and over the years it provided them with such good facilities, equipment, and support that they spent their entire careers there. Their work culminated in Jerome Karle's sharing the Nobel prize with a German man in 1985.[8] Similarly, in 1950 physicist Gertrude Scharff Goldhaber and her husband Maurice moved from the University of Illinois, where he was a professor of physics but she was only a "special research assistant professor of physics," to the AEC's new Brookhaven National Laboratory on Long Island, New York, where he became a "senior scientist" and she became an "associate physicist."[9] Among the several scientist couples at Oak Ridge National Laboratory, in Oak Ridge, Tennessee, were Liane Brauch Russell, the second wife of William L. Russell, formerly of the Roscoe Jackson Memorial Laboratory in Bar Harbor, Maine, where his first wife, the distinguished geneticist Elizabeth Shull Russell, and their four children long remained.[10] Another scientist-wife in search of an interesting career near her husband's was physicist Rosalyn Yalow, who in 1950, after four years as a temporary assistant professor at Hunter College, joined the staff of the VA hospital in the Bronx, somewhat near her husband, who taught physics at Cooper Union. At the VA hospital she and Saul Berson developed and perfected techniques for detecting the minute amounts of the essential hormones that can make the difference between normal and abnormal development. For this important and pioneering work in radioimmunoassay, they started winning prizes in 1960, culminating in her receiving the Nobel prize in medicine in 1977 (Berson had died in 1972).[11]

Differing from the rest of the federal government was the partly private Smithsonian Institution, where, as in academia, wives worked without pay, with at best an honorary title and space grudgingly provided in an out-of-the-way corner or attic, where they worked at their own pace for decades. Thus one of the best-known scientific couples in Washington, D.C., was Betty Meggers, unpaid research associate in anthropology at the Smithsonian, and Clifford Evans, head of anthropology there and master grantsman for a variety of international projects. Together they did much fieldwork in South America, published many joint reports, and trained many Latin American anthropologists and archaeologists.[12] Likewise, coleopterist Doris Holmes Blake, wife and long-lived widow of Smithsonian botanist Sidney Fay Blake, worked for decades in attic space generously provided by Doris Mable Cochran, for a time the only woman curator at the Smithsonian.[13]

Because of the low governmental salaries (especially compared with those in the pharmaceutical industry) in the 1950s, the VA and the Food and Drug Administration (FDA) found it difficult to attract and retain competent and dedicated laboratory workers. On occasion a branch responded by hiring a minority woman scientist for a laboratory position, where she would not come into contact with actual patients or clients. For example, biochemist Chi Che Wang, born in China in 1894 but a 1914 graduate of Wellesley with a doctorate from the University of Chicago in 1918, moved after World War II to a position as biochemist-in-charge of the research laboratories at the Hines, Illinois, VA hospital and later, in 1954, to a similar position at the Winter, Kansas, VA hospital.[14] While at Hines she may have known or even hired Leila Green Cochrane, a black graduate of Howard University and a Rosenwald Fellow with a master's degree from Radcliffe College, who had been the acting head of the chemistry department at a black college. A biochemist at Hines, Cochrane was later proud to have some of her research on tranquilizers published.[15] Among the early minority women scientists working in Washington, D.C., was Alma Levant Hayden, who, with a master's degree in chemistry from Howard University and a few years at the NIH, joined the FDA in the mid-1950s and in 1963 became chief of the new Spectrophotometer Research Branch in the Division of Pharmaceutical Chemistry.[16] She may have been one of the first black scientists at the FDA, for when a representative of the WB had interviewed an official of the FDA information division on this very subject in December 1946, she had received a rather pessimistic reply: "Negroes—There are no colored scientists employed by the agency. Those working in the field [i.e., outside the Washington, D.C., area] occasionally have to appear in court to support a case or offer testimony on findings. A colored person might prejudice the case in court, in certain sections of the country. The Administrator of the agency is opposed to any racial discrimination, and their [sic] present policy is only one of expediency."[17]

At some federal agencies the trend in the 1950s and 1960s was less toward expansion and areas of growth and more toward erosion and retrenchment through retirements, deaths, cutbacks, and reorganizations that undermined both the women's former footholds and their few bastions. At the Bureau of Indian Affairs, for example, where women scientists had never been numerous, their number dropped to zero in 1952 when anthropologist Ruth Underhill, one of the last students of Columbia's Franz Boas and the first woman anthropologist at the agency in the 1930s, tired of the rigors of traveling from reservation to reservation and took a position at the University of Denver.[18] Similarly, in the National Park Service there were no women scientists after Jean McWhirt Pinkley, chief of the interpretive division at Mesa Verde National Park from 1943 to 1966, one of only three women anthropologists or archaeologists in the entire federal government (see table 13.1), passed away at her desk in 1969. As the widowed daughter-in-law of a park official, she had been close to "family" within the National Park Service and thus merited an exception that did not extend to others.[19]

At the Forest Service even the modest gains of earlier years were wiped away in the fifties and sixties, as several of the early pioneers retired or passed away and few new women were hired. Eloise Gerry and C. Audrey Richards, for example, both long at the Forest Products Laboratory in Madison, Wisconsin, retired in the mid-fifties, and two women plant pathologists hired by the Forest Service were relatively short-lived: Lucille Bomhard, one of the very few hired by the USDA in the thirties, died in 1952 after eighteen years in range and dendrological research in Washington, D.C.; and Catherine Gross Duncan, a specialist in the wood-destroying bacteria and fungi at the Forest Products Laboratory since 1942, died in 1968,[20] whereupon even this modest foothold the early pioneers had carved out in forestry science was eradicated. Because women were rarely even admitted to schools of forestry until the late 1960s, the Forest Service did not add many new women professionals until the 1970s, and those it did add had earned their degrees in other fields.

Meanwhile, other realms where women had once had a substantial role and considerable advancement were reduced in size and reorganized out of existence. Because the USDA's research budget was actually decreasing in the early and mid-1950s, the Bureau of Home Economics, established in 1923 and long headed by women biochemists—Louise Stanley from 1923 to 1943 and Hazel Stiebeling from 1945 to 1953—was under frequent attack. Its separate existence ended when all the USDA research bureaus were consolidated into one large Agricultural Research Service (ARS) in 1953.[21] In 1957 the three remaining home economics divisions were combined to form an "institute" of home economics under Hazel Stiebeling and her associate Ruth Leverton. But this lasted only until 1962, after which Stiebeling and some of her longtime associates began to retire, and in 1965 Agriculture Secretary Orville Freeman again proposed that research on clothing and textiles be eliminated.[22]

At about the same time, the cluster of important women in the Home Economics Branch of the Division of Vocational Education at the USOE began to retire or leave. Under Edna Amidon, branch chief since 1938, the long-serving staff members (including research assistant Virginia Thomas, its first black woman, 1956–65) not only compiled and published numerous reports and biennial statistics but also kept a sharp eye on all matters affecting home economics education in these difficult years, when it was under attack on many campuses. Unfortunately Amidon's retirement in 1965 and the departure of most of her coworkers about then coincided with such vast other changes in the USOE as a whole that officials proposed that they not be replaced at all and that their work, now seen as special services for one particular group, be "dissipated among the other units." Upon hearing this, the AHEA, with which the Home Economics Branch had worked closely over the years, protested in a sharply worded resolution passed at its annual meeting, reminding the USOE that its twenty-four thousand members now needed more rather than fewer services.[23]

The WB, established in 1920, never really recovered from the cuts imposed by the "Meat Axe" Congress of 1947–49, although its reduced staff continued to write reports, often for other agencies, which helped pay their salaries. Marguerite Zapoleon, for example, who had herself earned an engineering degree in the 1920s, wrote several of the WB's bulletins and reports on women in science and engineering in the 1950s, before her early retirement in 1960. But the successive secretaries of labor of the time were slowly merging women's activities with those of the other bureaus. When, for example, Esther Peterson was appointed director of the WB in 1961, she was also given the position of assistant secretary for labor standards; she was thus one of the highest-ranking women in the Kennedy administration. One consequence of this mainstreaming was that in the late sixties, as the women's movement grew more vocal, the WB was increasingly bypassed and would be given none of the new enforcement powers emerging from "equal employment" legislation.[24]

Other parts of the federal government were growing in the postwar decades. One new kind of scientific (or parascientific) work in the federal government that attracted several women was grants administration, performed first by the ONR and the NIH and then by the NSF. The first woman in the area was mathematician Mina Rees, among the best-known and most important of the women scientists in the Navy during and after World War II. On leave from Hunter College, she spent a decade (1943–53) in Washington, D.C., first as a "technical aide" and executive assistant to Warren Weaver, the head of the applied mathematics panel of the OSRD, and then, after the war, as the head of the ONR's Mathematics Branch and later its Division of Mathematical Sciences. In these capacities she made grants of millions of dollars to promising mathematicians, statisticians, and computer scientists around the country, thus providing them (as well as herself) a prominent place in the postwar world of Big Science.[25]

The NIH hired many women bioscientists for its Division of Research Grants, which administered a very large number of "extramural" grants and contracts for research to be done at universities, especially medical schools. Over time the evaluating and processing of such applications came to require a great deal of technical paperwork. While hardly any women served on the NIH review committees ("study sections") of outside experts who evaluated the proposals, there were several women among the NIH employees (called "executive secretaries") who handled the correspondence, forms, and budgets. Mid-career scientists whose jobs elsewhere had come to an end, such as research associates leaving academia, were preferred for these positions, since they generally called for someone with a doctorate, some specialized research experience, an ability to distinguish between good and mediocre science, the patience to deal with often tedious paperwork, and satisfaction in vicariously sharing in others' research. When the division published a booklet describing its operations in 1963, it listed three women among the eleven officials (27.27 percent) who ran its thirteen fellowship review sections (one woman and one man each ran two) and thirteen women among its forty-three study-section executive secretaries (30.23 percent).[26]

Women were, however, rarely hired as program directors at the NSF, established in 1950, which had a much smaller budget than that of the NIH and was heavily dominated by former ONR officials and physical and other nonmedical scientists. Biochemist Marian Wood Kies said about her year (1952–53) as one of the NSF's first women assistant program directors: "I left the Department of Agriculture for the National Science Foundation. I was not enthusiastic about the position but it was the only one available at the time. For the first time in my life I had an uneasy feeling that I might have had a better offer had I been male. At the Science Foundation, all the Program Directors were men, and the only two professional women were *Assistant* Program Directors" (Kies's italics).[27] Finding the constant talk of research budgets "really boring," Kies moved on to a successful research career in biochemistry at the NIMH. NSF's first woman program directors were Margaret C. Green in developmental biology (1953–55) and astronomer Helen Sawyer Hogg (1955–56).[28] Aside from these rare cases, it was not until the late sixties that there began to be even an intermittent trickle of women program officers in the research directorates at the NSF. But titles can be deceptive, and in fact, as in many private foundations, some NSF women with the title "secretary" or "special assistant," such as Vernice Anderson of the National Science Board or Virginia Sides (later of Wellesley College) and Lee Anna Embrey, assistants and speechwriters for NSF director Alan T. Waterman, may have been fairly powerful.[29]

Because one increasingly important though largely unsung role of the federal government was to prepare and disseminate major standardized reference works on a variety of technical topics, many women scientists found challenging work and led careers of distinction in the general area of "science information," or the myriad ways in which science is communicated both formally and informally. For some this meant long careers in editorial or bibliographic work, ranging from indexes to massive reference works. For example, Lois Olson was the chief geographic editor at the Central Intelligence Agency from World War II until her retirement in 1962;[30] nutritionist Bernice Watt directed the preparation and revision of the USDA's famous Handbook No. 8, which contained tables of food composition used by dietitians around the country, for about thirty years; Margaret C. Schindler oversaw the preparation of the annual *Bibliography of Agriculture* (since computerized as AGRICOLA); and Luella Walkley (Muesebeck) coauthored with her boss (whom she later married) the *Catalog of Hymenoptera of North America: North of Mexico.*[31] At Oak Ridge National Laboratory in the 1960s was Katharine Way, a physicist who had turned to data collection and, under the auspices of the NRC, prepared a series of volumes of newly discovered data about the energy levels and decay rates of nuclei. Similarly, Charlotte Moore Sitterly, an astrophysicist and longtime research associate at Princeton University, joined the NBS in 1945 and spent twenty-six years editing a four-volume work on the collection and analysis of atomic spectra for the various elements.[32]

Among other innovative women moving into this new world of "science information" was Stella Deignan, with a doctorate in anatomy and several years at NIH's Division of Research Grants. In the early fifties she devised a "biosciences information exchange," whereby each funding agency could find out what the others were already supporting. Because the Smithsonian had a long tradition of exchanging scientific books and periodicals with foreign nations, its officials agreed to house this new project, which was funded by NSF and NIH grants until 1961, when Deignan and her husband (the Smithsonian's curator of birds) retired to Europe.[33]

In the field of geology, technical writing and bibliography were a kind of indoor "woman's work" for persons who were not permitted to do fieldwork. This concentration on indoor, especially paper and library tasks was the result of the widespread feeling, held even by some women, such as Jewel Glass of the USGS, that women should not and probably could not do fieldwork successfully, because it required a rugged physique and the ability to endure many discomforts. Among those women geologists who made long careers in bibliographic work were Emma Mertins Thom (*Bibliography of North American Geology*), Leona Boardman (*Geologic Index* maps), M. Grace Wilmarth and Alice S. Allen (*Lexicon of Geologic Names of the United States*), Marjorie Hooker (several works), Catherine Campbell (earthquake reports especially), and Anna Jespersen (several works, including obituaries of other women).[34] The consequence for these women editors and bibliographers of accepting such sex differences, of course, was that though they were praised for their patience with detail and their work was acclaimed for its exactitude, they were rarely promoted into management, since successful fieldwork long remained a chief requirement, written or unwritten, for distinction within geology and for advancement within the Survey.

A few signs of dissatisfaction and protest concerning the prevailing division of work occurred among the women at the USGS in the fifties and sixties.

Geologist Alice Weeks (for whom the mineral weeksite is named) had left her faculty position at Wellesley College to join the Survey in 1949; there she was one of the very few women to do extensive fieldwork, on occasion dressing as a man to get into mines.[35] Another innovative woman geologist devised a more systematic way around the Survey's traditional ban on women's doing field-work. It was quite a breakthrough when Medora Hooper Krieger, who was hired by the USGS in the forties and stationed in California, simply started her own field camp specifically for women geologists, who could then learn and do fieldwork together. Many women geologists across the nation took advantage of her breakthrough and were graduates of her "school."[36]

One of the federal women scientists' few chances to manage something substantial was in the area of libraries, for in these years many of the govern-ment's extensive science libraries were stretching to become "data centers" and "information centers." At the NRL, for example, Ruth Hooker, a physicist at Hood College in the 1920s, was given the chance to head the library in 1930. She accepted and served until her retirement in 1965, except for three years (1951–54) when she served as coordinator of all the Navy's libraries. In 1964 she was awarded a Navy Meritorious Civilian Service Award, but what was even more remarkable, in 1975 the NRL library was renamed in her honor and her portrait was hung at the entrance, a very rare honor for any woman in the govern-ment.[37]

The highest-ranking woman physicist at the NBS was J. Virginia Lincoln, a Wellesley physics major who had earned a master's degree in home economics from Iowa State in the 1930s. After three years on the Iowa State faculty in the field of household equipment, Lincoln joined the NBS staff in 1942 as a physi-cist and never looked back. With the coming of the International Geophysical Year in 1957–58, the NBS expanded, in 1959 adding a new unit on radio warning services, which Lincoln headed, the first woman to become a section chief. When the agency opened its new headquarters in Boulder, Colorado, Lincoln moved there, and in the 1960s she became director of its new World Data Center–A for Solar Terrestrial Physics, which made her the highest-ranking woman in the new National Oceanic and Atmospheric Administration.[38]

One other growing area of the federal government in these years was regula-tion, especially in the realm of food and drugs. Because of certain divided loyalties and insufficient vigilance, however, it was all too often an area of scandal, mismanagement, and some heroic whistle-blowing. The FDA ex-panded greatly in the 1940s and after as penicillin and other antibiotics that required extensive testing were developed. Among its dedicated women em-ployees was microbiologist Frances Bowman, who in 1954 was promoted to branch chief in the Division of Antibiotics. There she was responsible for, among other products, eye ointments, which when contaminated could cause eye damage, including blindness. She struggled in the face of unresponsive superiors to prevent such lapses and to amend federal regulations to require ever safer procedures. It was an exhausting, highly embattled task, closely fought by teams of lawyers from pharmaceutical companies and severely frustrated by delays and inattention from layers of officialdom within the FDA. Final victory, in the form of revised and strengthened statutes, was for dedicated civil servants like Bowman a high point of their careers and lives. However, the stress and tension led Bowman to choose early retirement in 1972.[39]

Out at the new Bethesda "campus" of NIH, whose Division of Biologics Standards (DBS) tested the nation's vaccines to assure their safety, another female microbiologist, Bernice Eddy, could be classified as a whistleblower. After uncovering contamination in the new Cutter polio vaccine in 1955, she was assigned to other duties, where in 1957 she showed that, contrary to prevailing beliefs, some viruses do cause cancerous polyomas. When she was removed from this again controversial project and put back on vaccines, Eddy quickly discovered that an adenovirus that the Army was routinely administering to all its personnel as a way to prevent the common cold actually caused cancer in hamsters. When she mentioned this at a professional meeting, her horrified superiors insisted that in the future she clear all remarks with them in advance, and in 1961 they took away her animal colony to prevent further work in this area. (The Army continued to use the vaccine until 1964.) Yet years later Eddy was proven correct and publicly honored: by 1971 she was on the cover of *Cancer Research,* and her work was hailed in a General Accounting Office report reprinted in Senate hearings in 1972. Rarely does a courageous and persistent woman get such sweet revenge.[40]

But the most famous woman scientist in regulatory history, and, one imagines, an inspiration for all the others, was, of course, pharmacologist Frances Kelsey, M.D., the determined medical officer who in the early sixties resisted much pressure from the industry and her own bosses to speed approval for widespread use the sleeping pill thalidomide, which had been linked to the birth of limbless babies in Germany. Because the publicity about her courage in denying the drug certification appeared in July 1962, the month after Rachel Carson's bestseller *Silent Spring* started to appear in the *New Yorker,* the two reinforced each other in making the public aware of the dangers of uncritical approval and indiscriminate use of the various postwar "wonder drugs." Together they helped to launch the modern environmental era. But rather than being demoted or reassigned, Kelsey, who had come to the FDA just two years earlier, was immediately honored by President John F. Kennedy with the nation's highest award for a federal employee, the Distinguished Service Medal, usually given to old men (or their widows) after long careers as agency heads. Yet Kelsey, who was the second woman ever to receive this medal, after Hazel Stiebeling of the Bureau of Human Nutrition and Home Economics at the USDA in 1959, was fully aware that she had received this recognition not from her own bosses, who had all along been displeased with her independence and refusal to keep the drug manufacturer happy, but as the result of a recommendation by Estes Kefauver, whose Senate committee on governmental operations was at the time running highly publicized hearings on the laxity of FDA controls. After months of disturbing revelations, the media, the politicians, and the nation were hungry for a heroine, and Dr. Kelsey fit the role.[41]

Because salaries were so low in the federal government in the 1950s, the CSC sought nonmonetary ways to reward high-achieving federal employees. Under the Government Employees' Incentives Award Act of 1954 the CSC created a series of prizes for distinguished, superior, and meritorious achievement at governmental agencies. (Those in the military and naval agencies also specified the word "civilian.") So many were there that most of the women mentioned in this chapter won at least one of these in their long careers. One, Catherine C. Campbell, a doctoral geologist turned technical editor for first the Navy and then the USGS, may have set a record by winning twelve of these awards, some after her retirement in 1969.[42] In addition, women occasionally won other awards from their own agencies: when Donna Price, a physical chemist at the Naval Ordnance Laboratory, received its Fliedner Trophy Award in 1962 for her "superlative contributions," she was the first woman to do so.[43]

Women remained such a small part of the visible governmental workforce, however, both numerically and proportionally, that it was widely assumed that most of them were just secretaries at fairly low GS levels. To combat this misimpression, Barbara Bates Gunderson, a former president of the Rapid City, South Dakota, AAUW chapter, a Republican National Committeewoman, and a supporter of Dwight D. Eisenhower, who in 1958 appointed her to a three-year term on the CSC, in 1960 suggested a new honor, the Federal Woman's Award. When she announced the first winners in 1961, Gunderson expressed the hope that the new prize (which would be subsidized by the Woodward and Lothrop department store in Washington, D.C., rather than any federal funds), by focusing attention annually on these top women, would both inspire those others already in the government to greater exertions and attract fresh women to federal positions, which could now be seen to offer challenging work. Gunderson wanted all federal women to be eligible, but especially those in GS-9 positions or above that dealt with technical or professional concerns, since these were the jobs the public felt women were least likely to hold.[44] Thus it should be no surprise that from the start in 1961 until the last awards in 1976, when they were quietly discontinued as no longer necessary and even a bit demeaning, more than half of the winners were scientists, social scientists, or medical women. Among its first prizewinners was Rosalyn Yalow of the VA. Altogether these three to five women scientist winners (of the six annually) constituted a very impressive group (see table 13.3). Such was the growing political clout of federal employees as well as

Table 13.3.

Scientist Winners of the Federal Woman's Award, by Agency and Year, 1961–1970

Central Intelligence Agency	Federal Trade Commission
Penelope Hartland Thunberg, 1965	Barbara Moulton, 1967
Civil Service Commission	Health, Education and Welfare
Evelyn Harrison, 1962	Ida Craven Merriam, 1966
Council of Economic Advisors	NIH: Thelma Dunn, 1962
Frances M. James, 1968	Sarah Stewart, 1965
Defense Department	Margaret Pittman, 1970
Air Force: Gertrude Blanch, 1964	Housing and Urban Development
Army: Katherine Mather, 1963	Ann Mason Roberts, 1967
Navy: Dorothy Gifford, 1965	NASA
Kathryn Shipp, 1967	Nancy Roman, 1962
Joanne Smith Kinney, 1969	Eleanor Pressly, 1963
Department of Agriculture	Evelyn Anderson, 1964
Allene Jeanes, 1962	Jocelyn Gill, 1966
Ruth Benerito, 1968	National Bureau of Standards
B. Jean Apgar, 1970	Charlotte Moore Sitterly, 1961
Department of Interior	National Security Agency
Lucille F. Stickel, 1968	Ann Z. Caracristi, 1965
Department of Labor	Post Office
Aryness J. Wickens, 1961	Beatrice Aitchison, 1961
Department of Treasury	Veterans Administration
Margaret Wolman Schwartz, 1964	Rosalyn Yalow, 1961
Federal Aviation Administration	Marjorie Williams, 1967
Rogene L. Thompson, 1968	Mabel K. Gibby, 1968

Source: Press releases, 1961–70 (I thank Marie Robey Wood, of the Civil Service Commission, for providing copies).

women in the 1960s that on occasion the president would come to the award dinner, and Lyndon Johnson would even give some of these indispensable women a whirl around the dance floor.

The main reason for the invisibility of women in the federal government was, of course, their absence at the top. President Truman had appointed no women to his cabinet, and Eisenhower had appointed just one, Oveta Culp Hobby, who served only two rather embattled years, 1953–55, as secretary of the newly consolidated HEW, her term cut short by the polio vaccine scandals.[45] President Kennedy appointed no women to his cabinet. Nor were more than a handful of women coming up from the career levels within the civil service; most women scientists remained at GS-12 or below. This was due principally to the CSC's own rule allowing "appointing officers," that is, agency heads or other supervisors, to specify in advance just which sex was preferred for a position. (For this the CSC maintained two separate lists, one of "female eligibles" and one of males.) Thus a woman might rank atop the female list for years but be passed over repeatedly for an executive position.

Thus, when W. Lloyd Warner and his associates studied federal executives in 1959–63, they were quite surprised to discover any women at the GS-14 level or above. Those they found were concentrated in just a few departments: HEW, Labor, USDA, and State.[46] Among HEW's women executives were several legendary (because so very dedicated, tireless, and effective) bureau chiefs: Mary Switzer, director of the Office of Vocational Rehabilitation from 1950 to 1970, who first convinced Congress to pass the historic Vocational Rehabilitation Act of 1954 and then spent the rest of her career administering its ever-rising federal appropriations;[47] Martha Eliot, M.D., pediatrician, rickets expert, and public health educator, who capped a long career at Yale and the Children's Bureau by serving as its chief from 1951 to 1956;[48] and her successor from 1957 to 1968, social work dean Katherine Brownell Ottinger.[49] The faltering Department of Labor had a series of women chiefs in the WB and two acting commissioners at the BLS. The latter included statistician Aryness Joy Wickens (1954–55), who had come to the BLS during World War II, and Clara Beyer (1957–58), associate director from 1934 to 1957, whom Secretary Frances Perkins had felt unable to promote any higher, since two women bureau chiefs at Labor would have been politically impossible.[50]

At the next level down, that of division director or chief, once again many of the women were in HEW, Labor, or related areas. These included economist Ida C. Merriam, director (later assistant commissioner) of the Division of Research and Statistics at the Social Security Board (later Administration) from its start in 1936 until her retirement in the 1960s,[51] economists Helen Wood, chief of the BLS's Occupational Outlook Division, 1946–61,[52] and Beatrice Aitchison, a mathematician who specialized in transportation economics during World War II and headed the Post Office's Division on Transportation Rates from 1953 to 1971.[53] One step below was aeronautical engineer Katherine Stinson, the first woman graduate of North Carolina State University, who, after meeting aviatrix Amelia Earhart in the thirties, was in 1941 the first woman engineer hired by the then Civil Aviation Authority (later the Federal Aviation Administration [FAA]). By the time of her retirement in 1973 she had been promoted to the position of technical assistant chief of the FAA's Division of Engineering and Safety.[54]

Meanwhile, Kennedy's momentous decision in May 1961 to best the Soviets by sending a man to the moon by 1970 unleashed great expansion at NASA, established in 1958. Among its tens of thousands of scientists and engineers were several women, most of whom had worked for other governmental agencies previously. Perhaps the best-known was astronomer Nancy Roman, initially a junior faculty member at the University of Chicago but after 1955 on the staff of the NRL, which was often the first step toward a career in the government for formerly academically oriented physical scientists. In early 1959 she heard that NASA was looking for someone to be chief of its Astronomy and

Solar Physics Branch in its Office of Space Sciences. She jumped at the chance and has been there ever since.[55] Similarly when in 1959 Marjorie Townsend, an electronics engineer at the NRL, heard of a more interesting and more responsible position at NASA's Goddard Space Flight Laboratory in nearby Maryland, she rushed to be part of the new agency. Later she worked on the manned flight program.[56] Jocelyn Gill, who had taught astronomy at Smith and Mount Holyoke in the fifties, before receiving her Ph.D. from Yale in 1959, joined Nancy Roman's astronomy unit at NASA, where, despite a long battle with multiple sclerosis (she was the MS Woman of the Year in 1966), she spent two decades working on the problems of manned flight.[57] Even NIH biochemist Evelyn Anderson, whose husband Webb Haymaker was transferred to NASA's Ames Laboratory south of San Francisco, found a job there as well, where her research on stress led to the discovery of a new hormone.[58]

But NASA officials, who were mostly former military and naval personnel, showed themselves to be among the most sexist and discriminatory in the federal government when in 1962 they refused to let women become astronauts. Although Jerrie Cobb, an experienced Pilot of the Year in 1959, and others protested a NASA ruling that astronauts had to be males, and a subcommittee of the House Committee on Science and Astronautics held two days of hearings on the subject in July 1962, NASA officials, including astronaut John Glenn, and administrator George Low held firm. When the Russians launched Valentina Tereshkova into space a year later, in June 1963, and she orbited the earth forty-eight times before returning safely, a NASA spokesman told the press that the thought of an American woman's doing the same made him sick to his stomach. With attitudes like this at NASA, it would take twenty years and a major legal revolution in the area of equal opportunity before astronaut Sally Ride would fly in space in June 1983.[59]

Yet few women scientists, and certainly not those included in the space program, protested, publicly at least, the fact that NASA adamantly refused to test or train women to be astronauts. Nancy Roman, for example, repeated in a 1963 speech at Marymount College (cleared by her supervisors) George Low's reassurances that there were plenty of good jobs on the ground, and not even the rather feisty and generally outspoken psychologist Mildred Mitchell took on NASA officials. But members of the NFBPWC did not remain silent: at their conventions in 1963 and 1964 they passed resolutions protesting women's continued exclusion from the astronaut training program and sent them on to space enthusiast, vice president, and then president Lyndon Johnson. The issue did not fade away; parts of NOW pursued it later.[60] Yet, strangely, almost eerily, the President's Commission on the Status of Women (PCSW), which was meeting at the time, did not take up this case of deliberate and blatant discrimination against women in the federal service. Like the Equal Rights Amendment, which the commission also carefully sidestepped, the "astronauttes" were beyond the limits of organized, measured compromise in the early 1960s.

In December 1961 President Kennedy, whose appointees included few women, established the PCSW, which presented its final report in October 1963. On the Commission were five cabinet members, the head of the CSC, and representatives of many women's groups, carefully chosen to represent Catholic, Jewish, and Negro women.[61] Although one can debate the impact and importance of this commission, which made certain bland recommendations but sidestepped certain other minefields of controversy, its knowledgeable committee on federal employment accomplished several reforms even before submitting its final report. Chaired by the still influential Margaret Hickey, public affairs editor of the *Ladies Home Journal,* former president of the NFBP-WC, and during World War II chair of the Women's Advisory Committee to the WMC, the committee was familiar with, and heard reports by officials on, the longstanding lack of advancement facing women in the armed services, the foreign service, and the civil service. In September 1962 it also met with a panel of six recent winners of the Federal Women's Award (a "high point," according to the final report), including scientists Beatrice Aitchison, Thelma Dunn, and Frances Kelsey.

It quickly became clear that there were two distinct procedural barriers within the Defense Department and the CSC to the promotion of women to the top jobs. One was a deliberate quota on the number of women who could be in the highest ranks of military and naval officers; the second was the practice mentioned above of maintaining separate lists for men and women "eligibles" and allowing the appointing officer to specify which sex he (rarely she) preferred for any position. Knowing the PCSW's interest in this established practice, CSC Commissioner John W. Macy and Evelyn Harrison, deputy director for programs and standards since 1953 (and back in 1932 the first woman to graduate from the University of Maryland in engineering),[62] had taken a spot survey that showed that for 94 percent of the jobs at the GS levels 13–15 the requests were for men only (see fig. 7). At only a few agencies, primarily those concerned with the needs of women and children, was it the case that women were preferred or no preference was given. When, however, Macy had asked such appointing officers to explain in writing why they had to have men, most dropped the specification. Once these two longstanding practices were brought to the attention of the members of the PCSW, they were quickly eliminated. Defense Secretary Robert MacNamara ruled that each branch of the armed forces could decide for itself how many women officers to promote, and the CSC's John Macy, who, not coincidentally, was a member of both the full PCSW and its Committee on Federal Employment, asked Attorney General Robert Kennedy, also a member of the commission, to eliminate the separate lists. (Unfortunately, the CSC's procedures still included the potentially dis-

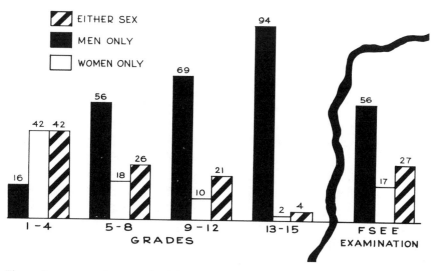

Fig. 7. Percentage of requests for federal appointments and promotions, by sex preferred and general schedule grade, 1962. Reprinted from "Requests for Certificates Specifying Sex in Selected Groups of General Schedule Grades, 1962," doc. IV-29 in "Employment Policies and Practices of the Federal Government (Report No. 2)," by John Macy, Apr. 9, 1962, in Papers of the President's Commission on the Status of Women, box 5 (materials of the Committee on Federal Employment), John F. Kennedy Presidential Library, Boston.

criminatory "rule of three," under which appointing officers were given the top three names on the unified list, not just the top name, and chose among them.)

These reforms, so readily enacted, were possible because, as Hickey and her committee sensed, Macy felt pressure to make the federal service a "showcase" for enlightened employment practices. When Lucy Howorth of the AAUW complained that the committee had "acted on the premise that more was to be gained by being suave than by being aggressive," Hickey agreed and promised that the tone of her final report would be harder-hitting.[63] Therefore the committee recommended other goals in its final report: that more part-time jobs be created and opened to women, that more women be appointed to "policymaking" (the old NFBPWC term) positions, and that the president establish a federal interdepartmental committee on women's employment.[64] Evidently, Hickey and the PCSW's other leaders had reasoned that there was little to gain in criticizing NASA, which was seen as a superpatriotic, all-American agency stretching to beat the Russians, and they might even lose or jeopardize other more important potential reforms (e.g., the abolition of quotas for women officers in the military). Thus, rather than capitalizing on all the publicity about the "astronauttes" to make their point about inequalities facing women in the federal service even stronger, they kept very quiet on the issue but did get other reforms they wanted more.

Since one of the ongoing activities of the staff of the PCSW was to meet with the representatives of other governmental agencies, one such encounter is of interest here. In December 1962 two staff members from the PCSW met with the panel on scientific and technical manpower of the President's Scientific Advisory Committee (PSAC) to discuss the decade-old concept of "woman-power," or the need to train more women scientists and engineers and to utilize them more fully. The PSAC men, chosen for their expertise in science and educational administration, revealed themselves to be not only about a decade behind in their thinking on the issue but also so defensive as to be a big part of the problem themselves. Present were nine members of the panel (all men), one observer from NSF (also a man), eleven staff members (all men), plus Antonia Chayes and Sam Morgenstein from the PCSW. Their message, so obvious at PCSW meetings, proved here rather controversial: young girls needed to be encouraged to pursue science; vocational guidance played a critical role; the graduate schools should be more flexible; some employers discriminated against women, and few gave them an equal chance for high-ranking positions. James Shannon, the director of the NIH, which employed many women scientists, demurred and "maintained," according to the minutes of the meeting, "that it cannot be expected that women will be treated equally in employment because of their potential loss during child-bearing years." Then Eric Walker, president of Pennsylvania State University, "pointed out that there are greater opportunities for women as technicians than as engineers." After this inhospitable reception, "it was suggested" that the panel just table the issue or, more diplomatically, consider it in more detail at some later point, perhaps after the PCSW had published its final reports. But not even this was voted on: according to the minutes, "Panel action, however, was deferred."[65] Thus the staff of the PCSW, carefully chosen for its political moderation on women's issues, was far ahead of the generally misogynist scientific and technical male advisers on manpower issues. Their predecessors James R. Killian, Lee DuBridge, and especially Arthur Flemming had been far more forward-thinking back in the 1950s. Evidently the two presidentially appointed bodies were moving at unequal rates of speed on the issue of women's utilization in science.

In early 1964, a few months after becoming president, Lyndon Johnson announced that he intended to support the recommendations of the PCSW by appointing at least fifty women to top governmental positions, and Mrs. "Lady Bird" Johnson held a series of well-publicized luncheons, possibly as auditions, for those she called "women doers." Among those appointed were economist Elizabeth S. May of Wheaton College, to the board of the Import-Export Bank, and biologist Mary Bunting, to the AEC, which had once sponsored her research.[66] Toward the end of her year at the AEC, Bunting convinced its personnel office to hire professional women, including lawyers, scientists, and engineers, for part-time positions. This was sufficiently newsworthy to merit a paragraph on the front page of the *Wall Street Journal*.[67]

Then in February 1966 President Lyndon Johnson took the occasion of the presentation ceremony of the Federal Women's Awards to announce the formation of a new study group on careers for women in the federal service, to be composed of all the award's winners to date. Although there is evidence that he expected the group's initial project to be the preparation of a booklet on vocational guidance for young women, its preliminary report, in March 1967, was harder-hitting than that. The women winners recommended strong executive action: that the CSC keep better data on women's position in the federal government, that each federal agency be reviewed annually with regard to its recruitment and promotion of qualified women to professional, technical, and administrative positions, and most important, that Johnson issue a new executive order banning sex discrimination in federal employment.[68] Also pressured a bit by then by activist Betty Friedan and NOW (established in October 1966), Johnson accepted the federal women's recommendations and seven months later, in October 1967, signed Executive Order 11375, which created within the CSC a new Federal Women's Program to collect data and monitor women's progress in the federal agencies and, more broadly, to outlaw, effective October 1968, sex discrimination in all federal employment as well as by federal contractors.[69] In time this executive order would have such wide repercussions in American life that even future astronauts would benefit.

But the National Register's data for the late sixties revealed that despite the reforms and the rhetoric, little changed for women scientists in the federal government after all. Because in the late sixties neither the NSF nor the CSC broke down its data on federally employed scientists as fully as in 1954, table 13.4 can provide only government-wide totals (rather than several agency subtotals) by field and by sex. These reveal that the total number of women scientists and engineers had increased by October 1968 to 6,945 (up 61.40 percent since 1954), but since the number of men had increased almost as much (up from 96,155 in 1954 to 153,874, or 60.03 percent, for a total of 160,819 for both sexes), the women's percentage of the total rose only slightly, from 4.24 percent in 1954 to 4.32 percent in 1967![70] Later studies would show in addition that despite Kennedy's and Johnson's reforms, there was hardly any gain at the top for women, for men were still generally at the higher ranks. Congress had raised federal salaries in the 1960s and increased the number of top-level jobs in the civil service, but few of them went to women before the 1970s.[71]

As for the numbers of scientists and engineers employed by state governments in the 1950s and 1960s, all reports and statistics agreed emphatically that just two states, California and New York, employed by far the most scientists and engineers—California the most overall, with many engineers, and New York the most "engaged in research." Within these two states there were also different patterns of concentration that reflected regional interests and needs: New York, for reasons that have never been explored but may relate to its large

Table 13.4.

Scientists and Engineers Employed by the Federal Government,
by Field and Sex, 1968

Field	Total	Women	Women as % of Total
Mathematics and statistics	8,858	1,705	19.25
Social sciences	6,894	1,126	16.33
Psychology	2,033	292	14.36
Geography and cartography	3,403	306	8.99
Physical sciences	29,517	1,821	6.17
Urban planning	222	11	4.95
Biological sciences	27,127	1,256	4.63
Engineering	82,765	428	0.52
Total	160,819	6,945	4.32

Sources: NSF, *Scientific and Technical Personnel in the Federal Government, 1968,*
Surveys of Science Resources Series (Washington, D.C.: GPO, 1970), 4, 11; CSC,
Occupations of Federal White-Collar Workers, October 31, 1968 (Washington, D.C.:
GPO, [1969]), 54–68.

and ethnically diverse population or strong physicians' lobby, employed more
scientists in its health departments and state hospitals than any other state (and
more in this one area than forty-eight other states employed in all areas), while
California, second in health scientists, had far more than New York in "natural
resources" (geology, the fish and game department, and state forests).[72]

Once again, the only data collected at the time that are broken down by
gender are those reported by the National Register. One problem with these
data, which are presented in table 13.5 by field and by sex for 1956–58, the first
year such categories were published, and 1968, the end of this period, is that the
totals are for a residual category—"state and local" governments in 1956–58 and
the vague but broader "other government" in 1968, which included international as well as state and local units. The most important finding for women
scientists employed by state (and local and possibly international) governments was that they were highly likely to be psychologists (61.38 percent of all
the women scientists and engineers in table 13.5 in 1956–58 and 44.89 percent, still the highest for any field, in 1968) or biologists (21.82 percent and
20.11 percent, respectively). The number of biologists, in fact, increased
tremendously—from 134 to 228 in the two years between 1966 and 1968—
adding fuel to the idea suggested in chapter 5 that the sudden jump in the
number of employed or underemployed women biologists in these years may
have been one of the stimuli to later efforts toward "liberation." It may also
have been a part of the emerging environmental movement.

The New York State Department of Health continued to be the premier
state science agency. It also employed so many women scientists of importance

Table 13.5.

Scientists and Engineers Employed by State and Local Governments, by Field and Sex, 1956–58 and 1968

Field	1956–58			1968[a]		
	Total	Women	Women as % of Total	Total	Women	Women as % of Total
Other specialties	12	4	33.33			
Education	16	5	31.25			
Psychology	1,533	391	25.51	1,977	509	25.75
Computer sciences				102	19	18.63
Social sciences	5	1	20.00	928	154	16.59
Linguistics				(15)	(6)	(40.00)
Sociology				(232)	(78)	(33.62)
Political science				(150)	(14)	(9.33)
Economics				(517)	(49)	(9.48)
Anthropology				(14)	(7)	(50.00)
Statistics	6	1	16.67	149	27	18.12
Biological sciences	925	139	15.03	2,020	228	11.29
Mathematics	73	9	12.33	308	27	8.77
Medical sciences	142	17	11.97			
Chemistry	398	38	9.55	1,221	128	10.48
Meteorology	22	2	9.09			
Geography	38	2	5.26			
Earth and marine sciences	420	15	3.57	1,064	36	3.38
Atmospheric and space sciences				72	2	2.78
Physics	29	1	3.45	86	2	2.33
Engineering	2,294	9	0.39			
Agricultural sciences	1,300	3	0.23	2,104	2	0.10
Astronomy	4	0	0.00			
Total	7,217	637	8.83	10,031	1,134	11.30

Sources: NSF, *American Science Manpower, 1956–58*, 81, 83; *1968*, 69–72, 252.

[a]Data for 1968 are from category "Other [i.e., nonfederal] Government" and so may include scientists employed by international government agencies, such as the United Nations.

in the fifties and sixties that only some of them can be listed here (see table 13.6). Of these perhaps the most renowned in the work of discovering the causes of diseases, developing tests for them, and, if possible, developing remedies as well, were Mary Pangborn, who in the 1940s discovered the cardiolipins, which could detect the presence of syphilis; Grace Sickles, who with the head of the state laboratory, Gilbert Dalldorf, in 1947 discovered the Coxsackie virus (named for the New York town where it was found), which *Time* magazine called "polio's little brother" because it produced a polio-like disease in mice;

Table 13.6.
Notable Women Scientists at the New York State Department
of Health, by Field, 1945–1968

Microbiology
  Julia Coffey
  Sophia Cohen
  Rebecca Gifford
  Elizabeth Hazen
  Myrtle Shaw
  Grace Sickles
  Gretchen Sickles
	Statistics
	  Hilda Freeman Silverman
	  Margaret Hoff
	  Sandra Kinch
	
	Pathology
	  Frances Bouchard
	  Doris Collins

Chemistry and biochemistry
  Rachel Brown
  Sally Hipp
  Sally Kelly
  Mary Pangborn
	Economics
	  Mildred Shapiro
	
	Epidemiology
	  Julia Freitag

	Medical librarianship Anna Sexton

Sources: Anna Sexton, *A Chronicle of the Division of Laboratories and Research, the New York State Department of Health: The First Fifty Years, 1914–1964* (Lunenburg, Vt.: Stinehour, 1967); Biographical Files, Library, New York State Department of Health, Albany, NY.

and chemist Rachel Brown and microbiologist Elizabeth Hazen, who in 1950 developed an effective antifungal agent for yeast infections that they called Nystatin in honor of their state. The royalties on this product, which they turned over to the nonprofit Research Corporation, amounted to $13 million by the time the patent expired in 1974. The fund's income was used to help science students and instructors, including in the later years many women entering science.[73]

Despite the presence of many women scientists in the New York State Department of Health, one has the feeling that there were fewer at the higher levels in the 1950s and 1960s than there had been earlier, a pattern that extended to other New York agencies as well. For example, there was a series of retirements in the 1940s and 1950s of women who had either headed or been assistant directors of their agencies and were not replaced by other women. Among these were Emmeline Moore, head of the New York State Biological Survey, who retired in 1944; Mary Kirkbride, associate director of the Department of Health's Division of Laboratories, and Ruth Gilbert, assistant director of its diagnostic laboratories, who retired in 1946 and 1949, respectively; Ruth Andrus, chief of the Bureau of Child Development and Parent Education at the State Education Department, who retired in 1950, and Ethel Cornell, associate head of educational research there, who also retired in 1954; and last but not least, Winifred Goldring, chief of the State Geological Survey, who retired in 1954. Perhaps the state civil-service rules, which emphasized "veterans preference," meant that the top jobs went to men, or else the virulent antifeminism of the time, directed at single or married career women, meant that few women were as eager as earlier to devote themselves to a visible position of great responsibility and stress.[74]

As for other states, some women worked for public health departments, including Pearl Kendrick, an expert on whooping cough vaccines, who retired from the Michigan department in 1951; Helen H. Gillette, chief of the Massachusetts Department of Public Health's Diagnostic Laboratory, where she spent forty-four years, until her retirement in 1961; Nell Hirschberg, for many years the North Carolina inspector of local laboratories; and Myrtle Greenfield, director of the state laboratory in New Mexico from 1920 until her retirement in 1956.[75] Many psychologists worked for state agencies; for example, Dorothy Park Griffin for twenty years (1951–71) was chief of the psychological services section of the North Carolina Department of Public Welfare, and Alberta Turner, a black doctorate in psychology from Ohio State in 1935, taught home economics at a series of black colleges until she was hired in 1944 as a psychologist by the Ohio Bureau of Juvenile Research. By the time of her retirement in 1971 she had become director of research for the Ohio Youth Commission.[76]

A few other women worked for agricultural agencies,[77] state museums,[78] state geological surveys,[79] the California state highway division,[80] and even for state fish and wildlife agencies. Among these were Frances Hamerstrom, who (along with her husband Frederick), after studying wildlife ecology with Aldo Leopold at the University of Wisconsin, worked for the Wisconsin Department of Conservation (she at 60 percent time). In her amusing autobiography, *Strictly for the Chickens,* she described their busy lives observing the mating behaviors of prairie chickens. (Once suspicious townspeople thought they were Communist spies.)[81] The only woman scientist in what Margaret Hickey of the PCSW might have called a policy-making position was Frances N. Clark, director of the California State Fisheries Laboratory from 1941 until 1957. Clark tried valiantly to prevent the decline of the California fishing industry; despite repeated warnings that the fish population was finite, fishermen refused to limit the size of their catches, with the result that there were not enough salmon and sardines to repopulate the waters, and these industries collapsed in California.[82]

Among county governments that hired scientists and engineers, that of Los Angeles County in California was the most ambitious and impressive. The Natural History Museum of Los Angeles County alone employed several important women, including one of the few paleoornithologists in the world,

Hildegarde Howard, an expert on the animals in its famous La Brea tarpit.[83] Elsewhere medical anthropologist Sophie Aberle, also on the National Science Board for a time, was from 1953 until 1966 chief of nutrition at Bernalillo County (New Mexico) Indian hospital;[84] and Mildred C. Kidd was a statistician for the Jefferson County (Colorado) department of public welfare in 1954.[85]

At the city level, New York City employed many scientists in its famous department of health, including two important women researchers: Anna Goldfeder, who had come to the United States from Prague in 1931, was by 1934 in charge of the cancer division for the New York City Department of Hospitals and stayed there for the rest of her career. She won many awards for her pioneering work on adjusting the radiation dosages for various cancer therapies.[86] A second eminent and slightly younger medical researcher in the employ of New York City in these years was Sarah Ratner, who had earned her Ph.D. from Columbia University in biochemistry in the late 1930s and spent her whole career at the Public Health Research Institute of the City of New York on the Columbia campus of its College of Physicians and Surgeons. Reminiscing about her years at the institute, she said gently that some things might have been easier had she been a man.[87] The best-known and highest-ranking woman scientist at the city health department was, of course, its commissioner from 1954 to 1962, Leona Baumgartner, an M.D. as well as a Ph.D., who had formerly been associate director (under the famed Martha Eliot) of the Children's Bureau.[88]

Other large cities also employed women scientists and engineers; for example, H. Marie Wormington was long (1937–68) curator of archaeology at the Denver Museum of Natural History, engineer Elizabeth MacLean worked for Chicago's Bureau of Street Traffic, and in Detroit psychologist Helen Flinn was employed in the city court system. Smaller cities could offer unusual opportunities as well: paleontologist Billie R. Unterman built (with her husband) and later directed a natural history museum in Vernal, Utah, and laid out an instructive scenic drive through the surrounding dinosaur country on Utah highway 44.[89]

In conclusion, substantial numbers of women scientists and engineers worked for federal, state, and local governments in the fifties and sixties. Many performed work and research that won them prizes from their own agencies, the Federal Woman's Award, Inc., and on occasion outside professional groups. But they were rarely promoted into the higher ranks of better-paying managerial positions, and then only at a few agencies. The pattern was more often one of nonpromotion, underpayment, and early retirement, especially for those women who had started during World War II and completed twenty years in the federal service by the early 1960s. One serious bottleneck to their promotion was removed by the reforms instigated by Margaret Hickey's subcommittee on federal employment of the PCSW in 1962, the high point of organized womanpower in these years, but the full effect of this "reform" was undercut by the "rule of three," which permitted subjective factors in promotion decisions, and a subsequent expansion in the higher levels without many women. Possibly more effective in the short run were the president's widely publicized willingness and actions to make the federal workplace a "showcase" for enlightened practices. Certainly no one in Washington since Arthur S. Flemming had been such a forthright and well-placed advocate for women's advancement as Lyndon Johnson. Though it would take more than presidential pressure to change the federal bureaucracy (and beyond it the private sector as well), Johnson's commitment positioned the federal workforce to continue to lead the way in the more substantive reforms that started in 1968.

14

Invisibility and Underrecognition

Less and Less
of More and More

*E*xcept for the ubiquitous Margaret Mead and the outspoken Rachel Carson, women scientists were practically invisible to the public, to other scientists, and even to each other in the 1950s and 1960s. This was particularly odd since by the late sixties there were, by various calculations, tens of thousands of them, and scientists and engineers in general were increasingly known to the public and even took political stands. Yet the women among them remained out of sight. This near invisibility was the result of a lesser proportion of women at each level of formal recognition or scientific prominence—from a high of 33 percent of membership in some scientific societies to about 5 percent of postdoctoral fellows, 1–2 percent of presidencies of scientific societies or memberships in national academies, and their even rarer appearances on the proliferating number of advisory committees and boards where scientists set policy and met each other. Thus very few of the ever-growing numbers of honors and awards scientists were creating and allocating to each other were going to women, young or old, in the fifties and sixties. Although they presumably were awarded strictly on the basis of merit or accomplishment, as the burgeoning field of the sociology of science seemed to depict in numerous studies showing the "cumulative advantage" of successful careers, additional, less "objective" criteria were also involved. It is striking how few (perhaps under 100) of the 30,000 women appeared on the honor rolls, even in highly feminized or rapidly growing fields; how little linkage or overlap there was between awards, as few winners of fellowships percolated upward to other honors or appointments; how often the privately stated reasons for exclusion had more to do with the social awkwardness of a female presence rather than the scientific quality of her work; and how even those women who were honored within this system rarely "belonged" in a meaningful way to the inner circles or "invisible colleges" at the forefront of their specialties. There was also a measurable "chill" or dropoff in certain realms in the mid-1950s, as older women, completing their careers or terms in office, were succeeded by young men rather than by young women, creating a generational gap of about fifteen years before a few women, even then undercut as "tokens," were allowed a seat at the table. All too often even the top women of the time remained outsiders, wielding little actual power and unable or unwilling to do much for other women if they wanted to, which is not clear.

Most scientific societies grew by leaps and bounds in the 1950s and 1960s. Old societies reached record sizes—the ACS and the AAAS both had more than 100,000 members—and new ones formed whenever emerging specialties reached sufficient size for a formal society.[1] Although systematic data on the sex as well as the number of each society's members would be useful, few bothered to collect or publish them, which is still another sign of how little interest there was in women scientists before 1968.[2] One therefore has to rely on other indicators. A sample survey, for example, of sixteen pages of the last published directory of the AAAS (1947) reveals that 11.7 percent were identifiable as females,[3] and the 1955 membership survey of the ACS reported that 7.78 percent of its chemist-members were women.[4] Three other scientific societies for which data have come to light include the American Astronomical Society (with 17 percent women in 1945),[5] the APA (33 percent in 1949),[6] and the American Association of Anatomists (12.11 percent in 1960).[7]

Although these percentages were well over 5 percent, this did not necessarily mean that the women members were thriving, as the situation at the Association of American Geographers reveals. Between 1949 and 1964 the total membership more than doubled, from 1,291 to 2,625; yet the number of these who were women remained nearly constant, 341 in 1949 and 353 in 1964, indicating that the society was increasingly one of (young) men. The proportion of women in the group dropped by half in just fifteen years, from a vigorous 26.4 percent in 1949 to just 13.2 percent in 1964 (see fig. 8).[8] How widespread such a trend was remains to be seen.

Leaders and board members of scientific societies do not seem to have been alarmed about or even aware of this possibly declining proportion of women members. In fact as they sought to discover just where they and their burgeoning organizations were headed so precipitately, some commissioned reports by social scientists, who often read their own preoccupations into the resulting studies. In 1952, for example, the APA, which Mildred Mitchell had publicly criticized in 1951 for the low proportion of women at its higher levels, authorized a survey (funded by the NSF) of its structure and future needs. In the final report, however, the authors admitted that such a task in its full complexity was so overwhelming that they had deliberately limited themselves to those factors

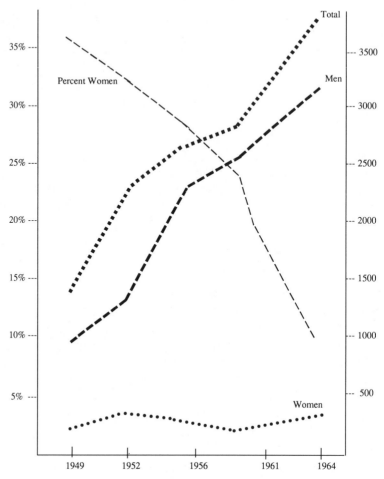

Fig. 8. Number and percentage of members of the Association of American Geographers, by sex, 1949–1964. Adapted from Kay E. Leffler, "The Role of Women in Geography in the United States" (master's thesis, University of Oklahoma, 1965), 3, which was based on data in directories of the Association of American Geographers.

affecting one manageable portion of the whole: the ideal type, or the most productive researchers. Only 8 of these 150, or just 5.33 percent, were women.[9] Similarly, in the late 1950s the ACS commissioned a sociological study of the goals and aspirations of its members. Although the surveyors sent question-naires to nearly 10,000 chemists, 6.55 percent of whom were women, they excluded the women's responses from the final report so that, as they explained, they could concentrate on variations among the men. These factors included the standard, somewhat sexist sociological preoccupations of the time—the relative prestige of the chemical profession as measured by father's occupation (somewhat problematical, since many chemists had working-class origins) and the constraints facing chemists in industry, where the norms and values conflicted with the lofty ones of independent professionals. Thus, at a time when expanded educational opportunity was diversifying both of these large occupational groups, the official vision, spread by studies that claimed to cover the whole but in fact deliberately focused on the most successful and most male-dominated subportions of it, was almost a version of wishful thinking or public relations. Such studies sharply and artificially reduced women's presence or obliterated it altogether, perhaps as a way of inflating the field's (or the authors') prestige still further. In any case, whatever special views or needs the women members might have expressed were ignored and dismissed.[10] Thus, even at this lowest level of society membership they were rendered invisible and silenced.

Some societies continued or instituted in these years a form of stratification with two or even three levels of membership: fellows, members, and associate or junior members. In some societies the upper designations were merely honorary, but in others only the higher-ranking members could hold an office. Usually the proportion of women was considerably lower at the upper levels than at the lower ones, and thus this organizational device, presumably a sex-neutral one that honored and rewarded the top achievers, also served to weaken the women's visibility and power: though they might help pay for the meetings and all the journals, when it came to holding a top office or seeming in any way to represent the society publicly, they were overlooked and ignored. Within the APA, as Mildred Mitchell pointed out in her famous article (1951) on the status of its women, as of 1949 there was a very sharp discrepancy in the sex ratios at the two levels: women were 33 percent of the overall membership but just 21 percent of the fellows. This nearly 3:2 ratio was still the pattern in 1957, when a partial report of membership in the APA calculated that 55.6 percent of the men but just 31.6 percent of the women were fellows. The requirement of a doctorate for fellowship status did not explain the difference, for most of the women associate members held doctoral degrees.[11]

The situation was worse at the Marine Biological Laboratory at Woods Hole, Massachusetts, whose many summer investigators constituted a kind of scientific society. There almost any user of the laboratory could, upon paying a fee, be elected a "member of the Corporation." A spot check of the laboratory's 1959 annual report reveals that 13.08 percent, or 64, of the 474 "regular" members were women, many of whom were faculty members at women's colleges.[12] But women were not elected to the more powerful board of trustees in proportionate numbers. In fact of the 248 trustees elected in the thirty-one years from 1940 to 1970, just 4 were women, 2 of whom served a second term, making their proportion (6 of 248) just 2.42 percent. One of these four women was Sister Florence Marie Scott of Seton Hill College in Pennsylvania, 1963–66, one of the few nuns ever elected to office anywhere.[13]

Sometimes such stratification was a way for a society that had formerly been highly selective to accept, perhaps for financial reasons, an impending flood of members less accomplished than in the past. Rather than capitulate totally to this newly necessary democratization, however, two societies added a second level of membership. In this way the GSA expanded sevenfold after 1947, when its membership reforms went into effect, with certain beneficial consequences for women in the field. Whereas formerly very few women had been elected at all (just 27 out of several thousand persons) in its first fifty-nine years, in the next twenty-two years women were about 4 percent of the members elected and under 2 percent of the fellows. Although this was still a 2:1 ratio, and not equality, it was a big improvement over the near total exclusion earlier.[14] By contrast, the American Institute of Nutrition had always had a substantial number of women in its membership: when it was founded in 1933, 41 of the 178 charter members, or 23.03 percent, were women. In 1958, when its leaders decided that the institute must expand, they decided to add a new category of "fellow" that would honor distinguished members who were over age sixty-five. After an initial group of ten distinguished nutritionists (all men), three were added each year. As of 1967, 7 of the 37 fellows, or a reasonable 19.44 percent, were women; but as no women were added in the next three years, in 1970 they numbered 7 of 46, or 15.22 percent.[15]

Meanwhile, some of the nation's engineering societies, which usually had strict criteria for each level of membership, were still at the stage of electing their first woman: Edith Clarke, Vivian Kellems, and Mabel Rockwell were in 1948 the first women fellows of the American Institute of Electrical Engineering (since renamed the Institute of Electrical and Electronics Engineers, or IEEE), and Gertrude Rand, an expert in physiological optics, in 1952 became the first woman fellow of the Illuminating Engineering Society.[16] The American Institute of Chemical Engineers lagged even further behind, electing its first woman member, Margaret Hutchinson Rousseau, in 1945, and Vera Burford became the first woman member of the American Society of Safety Engineers in 1946.[17]

One other group that elected a lone woman in these years was the Society of Experimental Psychologists (SEP), formed in 1904 and long proud of its small, select membership. (Its membership actually dropped slightly between 1948 and 1968.) In 1929 the SEP had elected its first two women members, but it had not replaced them after their deaths in the 1930s. Then in 1958, after two decades without any women members, its men finally elected Eleanor Jack Gibson, research associate at Cornell University and a wife of member James Gibson, its third woman member. This selection may have played a quiet but particularly important role in her subsequent rise within the psychological profession, since it was the first in an eventually long series of other honors, including election to the NAS in 1971. But the reintegrated SEP did not elect its fourth woman until 1970. Evidently the room at the top for women psychologists, who were so numerous in most counts of women in science, was very small indeed.[18]

A few organizations were not even this grudgingly open and still refused to admit women to membership. The opposition revolved at times around their intrusion and feared misuse of club facilities. At many clubs women were not even allowed to enter the premises, even, as Lillian Gilbreth, Icie Macy Hoobler, and others learned, when they were members of committees meeting there or the invited speaker. Biochemist Hoobler encountered this in 1942 at the Chicago Club, where because of her unusual first name no one had expected her to be a woman. When she arrived with her letter of invitation and asked to be admitted, she was not allowed in. Having expected some such rebuff, her husband, who was with her, had the club manager poll the club's trustees by telephone, and eventually she did get in. And some women reporters and photographers got headlines and pictures for a front-page story in the Chicago newspapers, which changed nothing but provided some satisfying revenge.[19]

Elsewhere the members of the Chemists' Club of New York City, limited to men since its founding in 1898, in 1958 discussed admitting women. Because there was considerable opposition to women's presence at the ten-story clubhouse in Manhattan, especially at the bar, their admission to membership was voted down.[20] Among those other private clubs that still did not elect women members was the notorious Cosmos Club of Washington, D.C., founded in the 1870s, which offered meals and lodging to distinguished members and guests at its ornate mansion clubhouse. It finally elected its first black member, distinguished historian John Hope Franklin, in 1964, but members remained adamant about not admitting women until well into the 1980s.[21]

As for holding an office, one sign of acceptance, women occasionally headed a local or regional subunit of one of the larger scientific associations. This lowest level of responsibility might be a reward for many years of faithful service or loyal attendance or merely a necessity when no one else was available. In some societies this was the only honor or recognition women received, for their participation was greater at the local or regional level than at the national level, where very few women held visible positions. Thus, within the very large ACS, which had about 143 local sections by the early fifties, some women were chosen chairmen. In fact by 1952, when the number had reached at least twenty (Icie Macy Hoobler of the Detroit section had been the first in 1930), the society's executive director published a list in the *C&EN*. One woman had headed two different sections (Esther Engle, Lehigh Valley and Wabash Valley in Indiana), and four sections (Florida, Virginia Blue Ridge, Mid-Hudson, and Wabash Valley) had each had two different women chairmen.[22] By comparison, in these years only nine women headed one of the three major regional affiliates of the APA, which had so many women members: six of the relatively receptive Western Psychological Association, three of the Eastern Psychological Association, but oddly not even one of the Midwestern.[23] Within the national AAAS zoologist Lora M. Shields of New Mexico State University headed the Southwestern and Rocky Mountain division (SWARM) in 1959–61 and thus was a vice-president of the whole association.[24] At the local or even the regional level of some national organizations, then, some women held leadership positions and a few even rose to the very top.

These same large scientific associations also tended to have nationwide subdivisions for particular specialties that on occasion were headed by women. Often but not always these were subdivisions in which women preponderated. Eight women served as ACS division heads between 1940 and 1970, including three in its feminized Division of Chemical Literature (founded in 1948) and two in its Division of the History of Chemistry.[25] At the APA, division 7 (childhood and adolescence) was headed by Florence Goodenough of the University of Minnesota in 1947, Nancy Bayley in 1953–54, and Dorothea McCarthy in the late 1950s; division 12 (clinical psychology) was headed by Anne Roe in 1957–58, and division 1 (general psychology) was headed by Elizabeth ("Polly") Duffy in 1961–62.[26] At the AAAS, biochemist Gladys Emerson was in 1952 the first woman to chair one of its selective Gordon Research Conferences,[27] and eleven women headed sections in these three decades (see table 14.1), which meant, before the reforms of 1973, that they were also vice-presidents of the whole association. There were three in anthropology, two in psychology (both in the late 1940s), and one each in seven other fields, including scientific information, where one might have expected more than one. Of some interest was Essie White Cohn in chemistry, for she, a professor at the University of Denver, owed her visibility and perhaps her election to her recent chairmanship of the Women's Service Committee of the ACS and her past presidency of Iota Sigma Pi, the fraternity for women chemists. At least nine other sections, including education, statistics, and zoological sciences, had no women presidents between 1940 and 1970, however.[28]

The AAAS's board of directors in the fifties and sixties served, moreover, as a platform for the abundant abilities of two of its later women presidents, anthropologist Margaret Mead (1955–62) and mathematician Mina Rees (1958–60 and 1962–72), the only women to serve on the board before the 1970s.[29] There both flourished, and Mead, the first social scientist on the board, made, it could be argued, a substantial difference in the evolution of what has been called the social role of the scientist. She was, for example, such an important advocate of racial integration that in 1955, when the AAAS encountered much criticism for persisting in holding its national meeting (to which it had agreed years before) in still segregated Atlanta, Mead, advised by her many friends at

Table 14.1.
Women Section Chairs, American Association for the Advancement
of Science, 1940–1970

Section	Field	Year	President	Institution
A	Mathematics	1954	Mina Rees	Hunter
B	Physics			
C	Chemistry	1961	Essie White Cohn	University of Denver
D	Astronomy	1952	Charlotte Moore Sitterly	National Bureau of Standards
E	Geology and geography			
F	Zoological sciences			
G	Botanical sciences	1964	Harriet Creighton	Wellesley
H	Anthropology	1950	Margaret Mead	American Museum of Natural History
		1966	Cora DuBois	Harvard
		1970	Margaret Mead	American Museum of Natural History
J	Psychology	1945	Florence Goodenough	Minnesota
		1947	Edna Heidebreder	Wellesley
K	Social and economic sciences	1969	Eleanor Glueck	Harvard Law School
L	History and philosophy of science	1946	Dorothy Stimson	Goucher
M	Engineering			
N	Medicine, dentistry, and pharmacy			
O	Agriculture			
P	Industrial science			
Q	Education			
R	Dentistry			
S	Pharmaceutical sciences			
T	Information and communication	1967	Phyllis Parkins	Biological Abstracts
U	Statistics			

Source: [Catherine Borass], "Women Officers of AAAS, 1885–1976" (1977).

black colleges there and elsewhere, was instrumental in devising a realistic and positive solution. She did not recommend moving or boycotting the meeting (as her own anthropological section did), because the leaders of the several black colleges in Atlanta wanted the considerable publicity of a national convention, but instead she reveled in the chance to hammer out an association policy, liberal for its time, stating that in the future the AAAS would never meet at a place that provided segregated facilities. If this was a first step in the eventual politicization of many scientific associations, especially in the sixties, Mead helped override the objections of the more cautious board members, because she felt such statements and actions were an important part of the scientists' social responsibility.[30]

Even if some women headed regional subgroups and others rose to the middle ranks of some scientific societies, proportionally very few made it to the top level, the presidency, of national associations in the period 1940–70. Not one woman, for example, headed the major chemical, physical, economical, psychological, mathematical, geological, geographical, biochemical, astronomical, physiological, anatomical, or entomological societies, despite the presence of qualified women within most of these groups. With this overall dearth of women presidents in mind, one can examine with some interest those sixty elite women who headed forty-five different scientific societies between 1940 and 1969, half before 1955–56 and half afterward (see table 14.2). Although this was a substantially larger number than had headed such groups in the previous thirty years, there were so many more societies in the postwar decades, all needing officers, probably annually, that it is hard to know whether proportionally the women were gaining much ground. Nevertheless, seventeen of the forty-five different groups listed in table 14.2 had more than one woman president in the years 1940–69; it was as if once the door had been opened, other women in the field, long overdue for their share of recognition, could now be nominated and brought forward. Among the organizations most receptive to women leaders were the American Anthropological Association, with five women presidents between 1941 and 1969; the American Statistical Association, the American Ethnological Society, and the Population Association of America with three each; and nine others, including the large Botanical Society of America and the American Public Health Association, with two apiece. Four remarkable women, at major institutions in their respective fields, each headed two organizations in this period: anthropologist Margaret Mead of the AMNH in New York City, microbiologist Rebecca Lancefield of the Rockefeller Institute also in New York City, sociologist and demographer Dorothy Swaine Thomas of the University of Pennsylvania, and statistician Gertrude Cox, long at North Carolina State University in Raleigh.

The presidency of a national scientific society was open to women in a variety of employment situations: of these sixty women, one was employed in industry (Sullivan), another was a wealthy dowager (Parsons), ten worked full-time and one half-time for nonprofit organizations (Hoffleit was at both Yale and the Nantucket Maria Mitchell Association) in either paid or unpaid positions, and ten more worked for governments, including seven who worked for federal agencies. But the overwhelming number (36.5) worked for colleges and universities, with four at women's colleges, three at other colleges (including two anthropologists at Queens College in New York City), six at medical schools, and one in a school of public health; the remaining twenty-one worked for universities in a variety of ranks and roles ranging from instructor

Table 14.2.
Women Presidents of National Scientific Societies, 1940–1969

Year	Society	President
1940		
1941	American Anthropological Association	* Elsie Clews Parsons
1942	American Gastroenterology Association	* Sara Murray Jordan, M.D.
1943	Society of American Bacteriologists	Rebecca Lancefield
1944	American Association of Cereal Chemists	Betty Sullivan
	American Ethnological Society	Marian W. Smith
	American Institute of Nutrition	Icie Macy Hoobler
	American Statistical Association	* Helen Walker
1945	Genetics Society of America	Barbara McClintock
1946	American Anthropological Association	Ruth Benedict (resigned)
	American Ethnological Society	Hortense Powdermaker
1947	American Public Health Association	* Martha M. Eliot, M.D.
	Society for Research in Child Development	Florence Goodenough
1948		
1949	Paleontological Society	* Winifred Goldring
	Psychometric Society	Dorothy Adkins (Wood)
	Society for Applied Anthropology	* Margaret Mead
1950	Ecological Society of America	* E. Lucy Braun
1951	American Association of Variable Star Observers	Martha Stahr (Carpenter)
	American Bryological Society	Geneva Sayre
	American Industrial Hygiene Association	Anna Baetjer
	American Society for Experimental Pathology	* Frieda Robscheit-Robbins
	Botanical Society of America	Katherine Esau
1952	American Sociological Association	Dorothy Swaine Thomas
	American Statistical Association	Aryness Joy Wickens
	Paleontological Society	Julia Gardner
1953	History of Science Society	* Dorothy Stimson
	Population Association of America	* Irene Taeuber
1954	American Radium Society	Edith Quimby
	Population Association of America	Margaret J. Hagood
1955	American Association of Physical Anthropologists	Mildred Trotter
	American Phytopathological Society	* Helen Hart
	Society for the Psychological Study of Social Issues	* Marie Jahoda
1956	American Association of Immunologists	Martha Chase
	American Bryological and Lichenological Society	Winona Welch
	American Society of Parasitology	Helen Cram
	American Statistical Association	Gertrude Cox
	Botanical Society of America	Harriet Creighton
1957	American Crystallographic Association	Elizabeth Armstrong Wood
1958	Population Association of America	Dorothy Swaine Thomas
1959	American Public Health Association	Leona Baumgartner, M.D.

(*continued*)

Table 14.2.   (*Continued*)

Year	Society	President
1959	American Society of Human Genetics	Madge Macklin
	Health Physics Society	Elda E. Anderson
	Society for Systematic Zoology	Libbie Hyman
1960	American Anthropological Association	Margaret Mead
	American Malacological Union	Katherine V. W. Palmer
1961	American Association for Cancer Research	*Thelma Dunn
	American Association of Immunologists	Rebecca Lancefield
	American Association of Variable Star Observers	E. Dorrit Hoffleit
	American Thyroid Association	Virginia K. Frantz, M.D.
	Society for Research in Child Development	Nancy Bayley
1962	American Association of Cereal Chemists	Majel McMasters
	Teratology Society	Marjorie M. Nelson
1963	American Institute of Nutrition	Grace Goldsmith, M.D.
	Society for the Study of Social Problems	Jessie Bernard
1964	American Pediatric Society	Hattie Alexander, M.D.
	American Society of Plant Taxonomists	*Mildred Mathias
	Society of Vertebrate Paleontology	Tilly Edinger
1965	American Heart Association	Helen Taussig, M.D.
	American Speech and Hearing Association	Mildred Templin
	Phycological Society of America	Janet Stein
1966		
1967	American Anthropological Association	Frederica de Laguna
	American Ethnological Society	Ernestine Friedl
1968	Society of American Archaeology	*H. Marie Wormington
1969	American Anthropological Association	Cora DuBois
	Biometric Society	Gertrude Cox

Sources: Obituaries; *AMWS.*

*Known to be first woman president.

during World War II (Marian Smith) to research associate (Martha Chase and Frieda Robscheit-Robbins at the University of Rochester, Tilly Edinger at Harvard, Hoffleit at Yale, and Nancy Bayley at Berkeley) to full professor (Goodenough, Templin, and Hart at Minnesota, Braun at Cincinnati, and Bernard at Penn State, among others). This honor was also more open to married women than might have been expected, for twenty-one of the sixty presidents were married at some point in their lifetime. In only one case, however, were the wife and husband in the same specialties: both Conrad and Irene Taeuber were sociologists and demographers in Princeton, New Jersey. In fact their careers were so closely related that his could easily have eclipsed hers. In 1948 he was elected president of the Population Association of America, which she then also headed in 1953, one of the very few times when the female half of a distinguished married scientific couple has received equal recognition.[31]

Besides these mainstream scientific societies, which were headed on occasion by one of the exceptional women of the time, several science teachers' associations and science education groups also had one or more women presidents between 1940 and 1970. These societies tended to be of two types. The American Association of Physics Teachers, established in 1932, represents one type: it was dominated by males for so long that it did not elect its first woman president, Melba Phillips, then at the University of Chicago, until 1965. Somewhat similar was the National Science Teacher Association, established in 1944, with just three women presidents by 1970. The second type had a long tradition of female leadership, often going back to its founding. Among these were the Society for Public Health Education, established in 1950, which had nine women presidents by 1970, and the National Council of Geography Teachers (renamed in 1957 the National Council for Geographic Education), which had eight women presidents between 1940 and 1962. There women, often faculty members at state teachers colleges, chaired committees and trained each other (especially at the Universities of Chicago and Pittsburgh), and at least one wrote her dissertation on the career of a female predecessor or role model.[32] Yet with the systematic upgrading and masculinization of their colleges in the fifties, these women were pushed aside, and after 1962 men cornered all the society presidencies, creating a definite "chill" in the field of geography. Somewhere between these two "types" was the National Association of Biology Teachers, established in 1938, which in 1944 had its first woman president and in 1962 had its fourth and last until 1974. Like the first type, it took a while to elect its first woman president, and like the second, it stopped in the early 1960s.[33]

Although a host of professional, social, psychological, and political factors would seem to be involved in determining who is elected president of a scientific society, a key factor was being an "insider" in some sense, especially having been a student at a dominant graduate program or being otherwise connected to a strong subgroup. This was certainly a determinant in the selection of Helen Hart of the University of Minnesota as president of the American Phytopathological Society in the 1950s, for that association had long been controlled by one man, Elvin C. Stakman of Minnesota, who arranged everything a few days before the annual meeting.[34] Dorothy Stimson of the History of Science Society said both modestly and accurately when she was honored at the fiftieth anniversary of that organization that she had been chosen because, as she put it, she was by the 1950s the only one of the early members left who had not yet been president.[35] At least one other centrally placed woman played the role of "kingmaker" herself. The beloved head of the nation's largest training program in radiation protection at the Oak Ridge National Laboratory,

Elda Anderson, knew everyone in the fledgling field and nurtured in them the vision of their own Health Physics Society. When there were enough to establish the society in 1955, she served as its charter secretary and later became its fourth president.[36] Yet such insider status was not enough for some women to become presidents, for surely Cecilia Payne-Gaposchkin of the Harvard College Observatory and anatomist Gertrude van Wagenen, close associate of John F. Fulton of Yale Medical School, were centrally located, but they never headed their professional associations.

Serving as president was not necessarily a pleasant experience, as Ruth Benedict learned in 1947 while at the helm of the presumably quite receptive American Anthropological Association. Like many of the social science associations after the war, the AAA was trying to become more active. Although Benedict was duly elected under the AAA's old rules in the fall of 1946, her view of the society's future was rejected by the voting members at the Christmas meeting. Then when the constitution that was adopted reduced the term of the current president to six months, she declined to run again to fill out her full year. Margaret Mead, who was privy to some correspondence and other goings-on, saw it all as a thinly veiled maneuver to discredit Benedict, who was at the time negotiating with the ONR about her very large "contemporary cultures" project, since most other professional societies were able to reorganize themselves without pressing such indignities upon their leaders. Benedict's refusal to run a second time was as strong a statement of dissatisfaction as any woman took publicly in this period.[37] Very aware of this barely hidden minefield of resentments that Benedict had faced in 1946–47, Mead herself did not run for the presidency of the AAA until twelve years later.

In the large multidisciplinary scientific and professional organizations there was a similar dearth of women at the top. Although Sigma Xi expanded greatly between 1940 and 1970, even forming chapters at several women's colleges, it did not have a woman president until the mid-1980s.[38] The AAAS might have had a woman president, in the 1950s, since its council members put Margaret Mead on its list of nominees several times between 1955 and 1960, but each time she withdrew, preferring to stay on its board of directors (and head the anthropological association first).[39] The ACE, made up of college and university presidents, and the AAUP each had a woman president in the mid-1950s. Psychologist-turned-president Katherine McBride of Bryn Mawr was the second woman to head the ACE, in 1955,[40] and Helen C. White, professor of English at Wisconsin and past president (1941–47) of the AAUW, headed the AAUP during some of the tumultuous later years of McCarthyism (1956–58).[41] But rather than being the first of a fresh tradition of women leaders within the AAUP, White's term in office marked the end of a tradition that went back several decades.

Women had been active in the AAUP since its founding in 1914, and by the fifties they had headed local chapters, even at coeducational colleges and universities such as Indiana, DePauw, and Duke. They had also held national office and even long had had something like a lockhold on the treasurership and the second vice-presidency. But starting in the mid-fifties their positions there were taken over by men, and thereafter only two women (Dorothy Bethurum, professor of English at Connecticut College, and Frances C. Brown of Duke in chemistry) were on the executive committee (both as second vice-president) before the 1970s. This definite "chill" also swept over the AAUP's council: of the hundred persons elected in the decade 1940–49 a total of eleven were women (11 percent); of those hundred between 1950 and 1959 ten were women (with nine before 1956); and of the hundred between 1960 and 1969 just seven were women (7 percent). Seventeen of these twenty-eight women (60.71 percent) were in science or social science.[42] Thus the chill in higher education that was affecting women's recognition in the science education groups, in geography especially, was also reaching the governing board of the AAUP by the mid-fifties.

In these years the postdoctoral fellowship became an essential step in the training and careers of scientists in most fields.[43] Although many of the stipends were provided by professors' research grants and would thus count as a kind of employment, the many fellowships awarded through national competitions can be considered as a form of recognition. Yet although these awards were open to all who applied, in fact strikingly little of even this lowest form of scientific recognition—less than 5 percent, as far as can be calculated—went to women in the 1950s and 1960s. Most of the postdoctoral fellowships for women were in the biological sciences, but the highest percentages were in the tiny, generally underfunded, social sciences. As noted earlier with regard to predoctoral fellowships, all too often women were in the fields that had the least money.

Several of the largest postdoctoral fellowship programs were run by agencies of the federal government.[44] In 1952 the young NSF started its program of (regular) postdoctoral fellowships, which continued at various levels until it was terminated in 1971. The program covered most fields of science, though the number of awards in each varied greatly (see table 14.3). Of the more than 3,000 fellowships awarded in the twenty-year period 1952–71 almost half were in the life sciences and chemistry. The third largest field was physics, followed distantly by mathematics. Overall women got just 131, or 4.33 percent, of these fellowships, with more than half of these in just two fields, life sciences (60) and psychology (17). At the other extreme women got only 1 fellowship in all these years in astronomy and none at all in engineering, economics, and "other." Thus the proportion of a field's fellowships that went to women varied widely, from the very high 28.30 percent in anthropology (15 of 53) and 25 percent in

Table 14.3.
NSF Regular Postdoctoral Fellows, by Field and Sex, 1952–1971

Field	Total	Women	Woman as % of the Total
Anthropology	53	15	28.30
Sociology	12	3	25.00
Political science	15	2	13.33
Psychology	131	17	12.98
Linguistics	16	2	12.50
History and philosophy of science	31	3	9.68
Life sciences	905	60	6.63
Astronomy	34	1	2.94
Chemistry	600	14	2.33
Mathematics	351	7	1.99
Earth sciences	141	2	1.42
Physics	524	5	0.95
Engineering	142	0	0.00
Economics	31	0	0.00
Other	37	0	0.00
Total/average	3,023	131	4.33

Sources: NSF, *Annual Reports for Fiscal Years Ending June 30, 1952–63,* (Washington, D.C., 1953–64); idem, *Grants and Awards for Fiscal Years Ending June 30, 1964–71* (Washington, D.C., 1965–72) as corrected and identified by Terence Porter and the author.

sociology (3 of 12, of which 2 went to Dorothy Zinberg) down to less than 1 percent in physics (just 5 of 524) and, of course, zero in engineering, economics, and "other." Among the talented survivors of this rigorous selection process were paleoecologist Margaret Bryan Davis, astronomer Nannielou Dieter, physicist Elizabeth Baranger, and psychologist Susan Ervin (later Ervin-Tripp).[45]

But even this showing was far better than that for women in the NSF's smaller program for established "senior postdoctoral fellows," where women received just 1.87 percent of the fellowships, or less than half their proportion of the NSF regular postdoctoral fellows. Possibly designed to provide Guggenheim-like fellowships, the program was started in fiscal year 1955–56 and continued, except for 1968–69, until 1971. Of its 1,125 eventual winners three-fourths were in the four largest fields (though in a slightly different order than for the regular postdoctoral fellowships), life sciences, physics, chemistry, and mathematics; and eleven other fields shared the rest (see table 14.4). Of these 1,125 winners only 21 were women. The largest number (8) were in the social sciences, including psychology, 6 were in the biological sciences, and 4 were in mathematics. There were 2 women in chemistry (Sara J. Rhoads of the

Table 14.4.
NSF Senior Postdoctoral Fellows, by Field and Sex, FY 1956–1971

Field	Total	Women	Women as % of Total	Women Fellows
Linguistics	4	1	25.00	Ethel Albert
Anthropology and archaeology	22	3	13.64	Marija Gimbutas (2), Sally Binford
History and philosophy of science	16	2	12.50	Elsa Allen, C. Doris Hellman
Mathematics	97	4	4.12	Joanne Elliott, Josephine Mitchell, Dorothy Stone, Mary Weiss
Economics	26	1	3.85	Irma Adelman
Psychology	41	1	2.44	Nancy Anderson
Biological sciences	370	6	1.62	Virginia Mayo Fiske, Evelyn Cox, Jane Oppenheimer, Betty Twarog, Helen Stafford, Catherine Duncan
Chemistry	163	2	1.23	Sara J. Rhoads, Darleane Hoffman
Physics	192	1	0.52	C. S. Wu
Engineering	94	0	0.00	
Earth and marine sciences	60	0	0.00	
Astronomy	14	0	0.00	
Sociology	9	0	0.00	
Other social sciences	9	0	0.00	
Political science	8	0	0.00	
Total/average	1,125	21	1.87	

Sources: See table 14.3.

University of Wyoming and Darleane Hoffman of the Los Alamos National Laboratory), 1 in physics (C. S. Wu of Columbia University), and none in engineering and earth sciences. Percentagewise these women did better than might have been expected in mathematics (4.12 percent) but much worse in psychology (just 1 woman, or 2.44 percent) and abysmally so in the biological sciences (just 1.62 percent), two fields that were by all other measures quite highly feminized. Another oddity worth noting is the great variability in the number of awards made to women: in just three years the women received 11 awards (or half of their total), five in 1964 and three each in 1959 and 1965, but in twelve other years they received only one or none of the approximately 75 awards made annually. This makes one suspect that the members of the NSF selection committees were a bit more broad-minded in some years than in others.[46]

Meanwhile, several private postdoctoral fellowship programs expanded the number of their awards. Although the women scientists among their winners were, where quantifiable, still generally under 5 percent, they were still higher than the 1.87 percent at the NSF senior postdoctoral program. By far the largest nongovernmental fellowship program in these years was that of the John Simon Guggenheim Memorial Foundation, established in 1925 by a family bequest after the death of a beloved son. Because the foundation's endowment increased greatly in these years, as more family members added to its already substantial holdings, the number and size of its fellowships also increased. The number given to U.S. and Canadian citizens jumped almost fourfold from 68 in 1941 to 253 in 1970. Of the 5,532 first-time fellowships given to Americans and Canadians in these years in all areas of knowledge (renewals have been omitted for the sake of simplicity), about half (2,754, or 49.78 percent) were in the sciences. When the cumulative data on the Guggenheim scientists is broken down by field and gender, as in table 14.5, one can see that once again there was a wide variation, both numerically and in percentages, by field. The largest numbers went, as with the NSF postdoctoral fellowships, to persons in the three big fields of life sciences (596), physics (398), and chemistry (334), which together accounted for 1,328, or nearly half (48.22 percent), of all the 2,754 fellowships given in the sciences. The rest were divided, as usual, among a host of smaller fields.

But only 113 of the scientists, or 4.10 percent, were women. They were, moreover, most numerous in the social and biological sciences (57, or just more than half of the women, and 40, respectively), with 20 in anthropology and archaeology. Yet of the just 13 in mathematics and the physical sciences the 4 in earth sciences represented a far better showing than women had had in this field at the NSF, and the same was true for the 9 in political science and the 7 in psychology.[47] It is somewhat surprising that only 2 women Guggenheim fellows had also held an NSF senior postdoctoral fellowship: plant physiologist Helen Stafford of Reed College and zoologist Jane Oppenheimer of Bryn Mawr. Among the more timely of these 113 first-time awards was that in 1946 to linguist Alice Kober of Brooklyn College, who was assembling texts containing the Minoan script known as "Linear B" as a first step in an eventual decipherment of the ancient Greek dialect. She made important progress but died in 1950, two years before the Englishman Michael Ventris finally succeeded, using her materials.[48] Another important award was that in 1951 to biologist and nature writer Rachel Carson for "the ecological relations of seashore animals on the Atlantic coast." It was during the Guggenheim year that she decided to quit her governmental job permanently and turn to writing full-time. If her *Silent Spring* (1962) can be seen as having ignited the modern environmental movement, then this award, which set her on that path, could

Table 14.5.
U.S. and Canadian Guggenheim Fellows in the Sciences,
by Field and Sex, 1941–1970

Field	Total	Women	Women as % of Total
Anthropology	68	13	19.12
Archaeology	41	7	17.07
Ecology	10	1	10.00
Other	10	1	10.00
Psychology	77	7	9.09
History of science	77	7	9.09
Political science	118	9	8.33
Linguistics	72	6	8.33
Life sciences	596	35	5.87
Sociology	53	3	5.66
Medicine	41	2	4.88
Astronomy	24	1	4.17
Philosophy of science	24	1	4.17
Earth sciences	101	4	3.96
Economics	129	4	3.10
Biochemistry	201	4	1.99
Physics	398	5	1.26
Mathematics	195	2	1.03
Chemistry	334	1	0.30
Engineering	149	0	0.00
Geography	29	0	0.00
Demography	7	0	0.00
Total/average	2,754	113	4.10

Source: John Simon Guggenheim Memorial Foundation, *Reports of the President and Treasurer* (New York, 1941–70).

arguably have been one of the most significant awards the foundation ever made.[49]

If the Guggenheim program looks relatively enlightened in comparison with the NSF senior postdoctoral program, part of the reason may be its timing. If the period 1941–70 is divided at its midpoint, 1955, then 65 (or 57.52 percent) of the Guggenheim awards to women were made in the first half, with just 48 (42.48 percent) in the second half. This imbalance is even more striking when one recalls that each year the total number of awards given in science was generally (but not uniformly) higher than in the previous year. Only one-third (883 of 2,754, or 32.06 percent) of all the fellowships given in the whole period were given in the first fifteen years, and two-thirds (1,871, or 67.94 percent) in the second half. Thus the women's proportion in the first half (65 of 883, or 7.36

percent) dropped remarkably thereafter to just 2.67 percent (48 of 1,871). Presumably as the Guggenheim program grew, in science as well as overall, the number and proportion of its awards going to women decreased! In fact, after the high points of 1952 (when 9 awards went to women scientists), 1953 (11), and 1955 (8), only in 1964 did as many as 6 go to women; more often the number was 4 or just 2.

Since no information is available about the applicants for Guggenheim fellowships, about all we can do is speculate on the basis of the composition of the selection committees. During the years 1941 to 1946, when anatomist Florence Sabin of the Rockefeller Institute was the only woman on the five-person selection committee, women scientists won 12 of 188 awards, or 6.38 percent. After her replacement by Carl O. Sauer, an ecologist who had been on the committee before, the number and proportion of women winners went up markedly and reached its high point of the whole three decades; in the next nine years, 1947 to 1955, there were 53 women among the 695 awardees, or 7.61 percent. This proportion dropped sharply in 1956, when there were just 2 women winners out of 116, or a mere 1.72 percent. But what had changed is not at all clear, for the selection committee was the same in both 1955 and 1956 and barely changed thereafter. Thus, as at the science education associations and the AAUP, there was a noticeable diminution of recognition for mid-level women in science, starting not in the late 1940s, when the fervor over the veterans was at its height, but in the mid-1950s.

In addition to the Guggenheim's broad, multidisciplinary postdoctoral program, a few institutes and centers offered their own fellowships to scientists and also welcomed those who came with NSF or Guggenheim grants already in hand. Among these institutions the oldest and most eminent was the Institute for Advanced Study in Princeton, New Jersey, founded in 1930 and, though it was most famous for its two physicist inhabitants, Albert Einstein and J. Robert Oppenheimer, it also had one woman full professor, Hetty Goldman in classical archaeology, who officially retired in 1947. The institute also had a steady flow of visiting women mathematicians through the 1950s (which reduced to a trickle in the 1960s). Among these were Dorothy Stone, Olga Taussky Todd, Dorothy Bernstein, Josephine Mitchell (Schoenfeld), Joanne Elliott, Alice T. Schafer, Marian Pour-El, and Carol Lee Walker. There were also two prominent women physicists, Leona Woods Marshall and Nina Byers, as well as psychologist Eleanor Jack Gibson, the only woman fellow in that large field. All in all, the invitations extended and the welcome received at the institute must have depended on personal and working relationships with the professors and emeriti already there. Although in two fields, mathematics and classical archaeology, many of the more prominent women of their time came and collaborated with institute faculty, this was far less often the case in physics and the social sciences.[50] Nor was any woman appointed to replace Hetty Goldman or to serve in any other professorship.

Table 14.6.

Fellows at the Center for Advanced Study in the Behavioral Sciences,
by Field and Sex, 1955–1969

Field	Total	Women	Women as % of Total	Women Fellows
Other	7	1	14.29	Marija Gimbutas
Psychiatry	29	2	6.90	Frieda Fromm-Reichmann, Charlotte Babcock
Anthropology	90	6	6.67	Ethel Albert, Cora DuBois, Phyllis Dolhinow, Laura Nader, Peggy Golde, Elizabeth Colson
Psychology	135	6	4.44	Else Frenkel-Brunswik, Helen Peak, Alberta Siegel, Eleanor Gibson, Hilde Himmelweit, Mary Ainsworth
Law	24	1	4.17	Herma Hill Kay
English literature	25	1	4.00	Claire Rosenfield
Education	25	1	4.00	Pauline Sears
Linguistics	26	1	3.85	Mary Haas
History	60	2	3.33	Annelise Thimm, Hanna Gray
Sociology	86	0	0.0	
Political science	78	0	0.0	
Economics	53	0	0.0	
Philosophy	33	0	0.0	
Biology	19	0	0.0	
Statistics	13	0	0.0	
Total	703	21	2.99	

Source: List of past fellows of CASBS, Jan. 23, 1989 (courtesy of CASBS librarian, Margaret Amara).

In 1954, perhaps in imitation of the institute at Princeton, the Ford Foundation formed the Center for Advanced Study in the Behavioral Sciences at Palo Alto, California, on the edge of the Stanford University campus. The center had space for almost fifty fellows per year. In the first fifteen years more than half of the fellows were in four fields: psychology, anthropology, sociology, and political science (see table 14.6). The rest of the 703 fellows were in ten other fields plus "other." Because women constituted 15 percent or more of some of these fields, one might have expected that a substantial number and proportion of these fellows would have been women. Yet in practice there was only about 1 woman "behavioral scientist" there per year, 21 in all from 1955 until 1969 (2.99 percent). (The first female fellow was psychologist Else Frenkel-Brunswik, a research associate at the University of California at Berkeley.) Revealingly, there were no women at all from the large fields of sociology, political science, economics, philosophy, biology, and statistics. Nor were any of the center's male or female fellows identified as home economists. Although some taught at colleges of home economics, the field included specialists in child or human development, family life, and consumer economics, and it was at the time seeking new research funding, trying to upgrade its status, and evolving on many campuses into the more prestigious behavioral science of "human ecology" (see ch. 8). Thus, once again major funding, here presented in the form of a prestigious fellowship, was deliberately masculinizing a field: bright young men were being selected, favored with a valuable year of research together in the sun, and thus propelled upward in the expanding social science departments of the time. There are, however, indications of what might have been in the presence in 1961–63 of Alberta Engvall Siegel, who continued her earlier studies at Stanford of, appropriately enough, working mothers and their children, and then in 1965–66 of anthropologists Peggy Golde, a fellow, and Ann Fischer, the wife of a fellow, who spent part of their year at the center working on one of the early "status of women" reports.[51]

The largest of the three major honorary academies was the American Academy of Arts and Sciences, established in Boston in 1780s. Several decades later, in 1848, it pioneered by electing to membership astronomer Maria Mitchell, who remained its sole woman until her death in 1889, whereupon the academy banned women until 1942, when its membership had reached 796 men. Wartime sentiment favored the selection of some women, and a protest move led by economist Sidney Fay, astronomers Bart Bok and Harlow Shapley, and statistician E. B. Wilson finally won over the council members. As part of their protest Harlow Shapley and E. B. Wilson corresponded about the academy's dual standard of membership (one level for people in Greater Boston and a much more rigorous criterion, such as a star in the *AMS,* for all others) and about several possible women candidates, including Florence Sabin, Margaret Mead, Ruth Benedict, Cecilia Payne-Gaposchkin, Ada Comstock, Marion Hines, Dorothy Wrinch, Ann Haven Morgan, Maud Slye, Ethel Browne Harvey, Barbara Burks, Helen Dean King, and Gertrude Cox. Of these only Mead, Gaposchkin, and Comstock were elected. Thus even these rather well-informed and hospitable men missed a great many others, including, for starters, Barbara McClintock and Gerty Cori. But the opposition was evidently less concerned about the qualifications of any particular woman than about the disruption her presence might cause at meetings, or as Wilson wrote Shapley in 1942,

> Of course the main trouble with the people who don't want to elect women [led by retired MIT chemistry professor Arthur Blanchard] is, as I understand it, that they don't want the women at the meetings. It isn't that they don't want them in

Table 14.7.
Women Scientists Elected to Two or More Academies, 1940–1968

Name	American Philosophical Society	American Academy of Arts and Science	National Academy of Sciences
Cecilia Payne-Gaposchkin	1936	1943	—
Marjorie Hope Nicolson	1941	1955	—
Barbara McClintock	1946	1959	1944
Gerty Cori	1948	—	1948
Katherine Esau	1964	1949	1957
Maria Goeppart Mayer	1964	1965	1956
Libbie Hyman	—	1960	1961
Berta Scharrer	—	1967	1967
Rita Levi-Montalcini	—	1966	1968
	+3 other women	+29 other women	+3 other women

Sources: APS, *Yearbook*, 1940–69 (Philadelphia, 1941–70); Office of the Home Secretary, NAS, list, "Women Elected to the Academy," July 26, 1974 (I thank Paul McClure for a copy); American Academy of Arts and Sciences, *Records* (Boston, 1948–70).

the Academy. It isn't really that they object to them while a paper is read on the main floor. They object to them in the room above sitting around the ginger ale, the beer and the pretzels and cheese. This was the specific point so much emphasized by some of the speakers last time I was present when the discussion came up on the general floor of the Academy.[52]

By 1970, 84 women in all fields of learning and the arts had been elected, including 37 scientists or social scientists, some of whom are listed in table 14.7, which names those honored by two of the three major academies. In 1970, 54 of these women were still members, but because the size of the American Academy had by then nearly tripled, to 2,234, their proportion was just 2.42 percent of the total and a mere 1.77 percent of the scientists. The proportion of women varied by section, with highs of 6.94 percent (or 5 of 72) in the section on political science, a field not elsewhere noted for its openness to women, 6.19 percent (or 6 of 97) in the section on cellular and developmental biology, and 5.05 percent (5 of 99) in the section for "educational and scientific administration"; at the other extreme, there was just 1 woman member in each of five large sections in 1970: mathematics, physics, chemistry, molecular biology, and the combined section of physiology and experimental psychology. Two sections, engineering sciences and technologies and economics, still had none. But on two other fronts the American Academy was making some headway. It had both a higher number and a higher proportion of foreign women members than the other academies: 11 of its 369 foreign members, or 2.98 percent, were women; 6 of the 235 scientists among them, or 2.55 percent, were women; 9 of the 11 foreign women, including all 6 scientists, were from the United Kingdom. The American Academy was also the first of the three academies to elect a woman to office—Mary Bunting to second vice-president in 1963. One other curiosity about academy membership was Flemmie Kittrell, the Cornell Ph.D. in nutrition and faculty member at Hampton Institute and later Howard University. Her entry in the *AMWS* listed her as a member of the American Academy of Arts and Sciences, but the academy's records do not indicate that she was ever even nominated. As a black woman home economist, she would have been an unusual member of the academy. Probably someone on the staff of the *AMWS* misinterpreted the abbreviation AAAS as standing for the academy rather than for the American Association for the Advancement of Science.[53]

In the NAS, founded as late as 1863 and restricted entirely to scientists, with only a few social scientists, by 1970 the women's proportion had crept up to nearly 1 percent of the membership. In 1940 the NAS had only one living woman (its first, anatomist Florence Sabin, elected in 1925) among its 313 members, or a minuscule 0.32 percent. (A second, psychologist Margaret Washburn of Vassar, had been elected in 1931 but had died in 1939.) By 1970, 10 more women—8 biologists and 2 physicists—had been elected (most of them are listed in table 14.7). One (Gerty Cori) was not elected to the NAS until after she won the Nobel prize, as has happened to a few men.[54] Because Cori and Libbie Hyman, as well as Sabin had all passed away by 1970, there were then just 8 living women members among the academy's greatly enlarged total of 866, representing a nearly threefold increase, to the still minuscule 0.92 percent. (Margaret Mead, who was elected to the American Academy in 1948, was not chosen for the NAS Academy until 1975, indicating some longstanding reluctance among her colleagues to give her this nearly supreme accolade.) In 1964 the National Academy of Engineering was founded as a near relation of the NAS. By 1970, on what might be termed the eve of "women's liberation," it had 329 members, including 1 woman, Lillian Gilbreth, a psychologist by training but often referred to as the "first lady of engineering," making the percentage of women there a bare 0.30 percent.[55]

By the 1960s, however, election to the NAS was no longer the nation's top honor for a scientist. It had been eclipsed by the new National Medal of Science, established by Congress in 1959 in the aftermath of the Sputnik scare. At first the medal's purpose was to honor scientific contributions that had aided the United States in the Cold War, as indicated by the first award, in 1963, to Théodor von Karman, one of the pioneers of modern aeronautical science and the American space program. Over the years, however, the work of its up to fifteen annual awardees has had less relation to American life or the Cold War. When in 1970, by which time the medal had gone to seventy-seven men, most of whom were already in the NAS, and even Libbie Hyman had been passed over, President Nixon chose as its first woman winner the much-honored Barbara McClintock, the corn geneticist from the Carnegie Institution of Washington.[56]

Of the three major honorary societies for scientists and other notable scholars, the American Philosophical Society (APS), founded in Philadelphia in 1769, was by far the smallest; and unlike the other academies, it showed considerable restraint in the period 1940–70 in retaining its small size. One result was that one had to be nearly a Nobel prize winner to be elected to membership! Of its 437 members in 1940, 4 (0.92 percent) were women, including 2 novelists and, reflecting the influence there of Harvard's Harlow Shapley, 2 Harvard astronomers, Annie Jump Cannon and Cecilia Payne-Gaposchkin. In the next thirty years just 12 women were elected, seven of them scientists or near scientists: 3 biologists (Barbara McClintock, Gerty Cori, and Katherine Esau), physicist Maria Goeppart Mayer, sociologist Dorothy Swaine Thomas, historian of science Marjorie Hope Nicolson, and classical archaeologist Mary Swindler. Several of these women were also elected to the other major academies in these years (see table 14.7). The APS was so selective that 2 of the 4 women in the natural sciences had actually won the Nobel prize *before* being elected to the APS, and another became a Nobelist years later. Obviously this honor was quite remote from the careers and expectations of most working scientists. Yet despite all this restraint, by 1970, when the size of the society had inched up to 489 members, 10 of whom were women, their proportion was 2.04 percent, double what it had been in 1940, a bit higher than the 1.77 percent of the scientists in the much larger American Academy and twice their proportion in the NAS.[57]

The actual meetings of male-dominated scientific societies in these decades included certain sex-segregated practices that reveal how awkward an invited woman might be made to feel on the basis of her sex. A few months after Margaret Mead was elected to fellowship in the American Academy of Arts and Sciences in 1948, she was invited to address the members on "ladies' night," when wives would be present.[58] Lest Rebecca Lancefield wonder whether she should accept an invitation in 1955 to a "black tie" dinner at the Harvey Society in New York City, the host added a jocular postscript: "but we don't expect you to wear one."[59] Applied mathematician Hilda Geiringer, a fellow of the American Academy, received a particularly insensitive invitation in 1959, addressed to "Sir," to a special dinner in honor of past president E. B. Wilson. Doubt whether she was really welcome may have been raised by the additional statement that "Reservations will be limited to Fellows and their gentlemen guests." Several days later Geiringer got a second letter explaining that no exclusion of female fellows had been intended and that she was especially invited because another woman (Jane Worcester, associate professor of biostatistics at the Harvard School of Public Health) would also be present.[60] Evidently everyone would be more comfortable, relieved in fact, if she came, since then the two women could eat with and talk to each other!

The postwar decades also saw the flowering of the whole area of "scientific advice," given chiefly to governmental agencies in Washington, D.C. In fact there were so many scientific advisory posts that there was a hierarchy among them, and the tens of thousands of scientists and engineers flying to Washington in these years to give advice formed, in effect, a kind of ladder from the myriad lesser, narrower committee assignments on up to the top ones, such as to the PSAC.[61] But there were very few women among them, and none at the very top. When sociologists in the late 1960s analyzed just what criteria had been operating, they settled on five, the first four of which would have excluded most women:

(1) experience, contribution and distinction during World War II;
(2) distinguished service and prominence on departmental or agency advisory committees in the Department of Defense or the Atomic Energy Commission;
(3) the presidency of or high office in the National Academy of Sciences;
(4) distinguished service with one of the outstanding private research foundations; and
(5) major theoretical or technical contributions to basic science.[62]

Even if the men making the selections had known the few women with the suitable background and expertise for their committees and panels, they would not necessarily have invited them, for, as Detlev Bronk, long head of the NAS-NRC and consummate Washington insider, wrote in 1971, long after the whole system had begun to be criticized on many grounds, since not everybody was temperamentally suited to high-level advisory work, it was in fact best left to proven insiders.[63]

Thus the few women who did serve on some of these advisory groups are of interest, especially since they were generally found in just a few areas. One continuing place was the Food and Nutrition Board at the NRC, established just before World War II and still in existence today. It always had women among its members. Of the eighty-three people who had served on the board by the time of its twenty-fifth-anniversary report in 1965, eleven (or 13.3 percent) had been women. Their presence had, however, been tapering off since World War II, when six of the eleven served; in 1965 there were just two, one of whom was the remarkable Grace Goldsmith of Tulane Medical School. She served by far the longest of any of the women, from 1948 until 1969, and was even its chairman from 1958 until 1968.[64] A second spot for women advisers, especially social scientists, AAUW leaders, and women's college presidents, was on the Board of Foreign Scholarships at the State Department, which starting in 1946 selected the new Fulbright fellows. Serving on this committee between 1946 and 1966 were Sarah Gibson Blanding, Helen C. White, Margaret Clapp, Katherine Blyley (of Keuka College), Bernice Brown Cronkhite, and, from the world of state politics, Mount Holyoke alumna and Connecticut secretary of state Ella Grasso.[65] Beginning in 1947 a third niche in the Washington advisory apparatus was a seat on the Committee of Nine, which advised the Secretary of Agriculture on research priorities. The Research and Marketing Act of 1946 had stipulated that one appointee to this committee must be a home economist; accordingly, a series of senior women from major land-grant institutions, starting with Agnes Fay Morgan of the University of California at Berkeley, followed by Pearl Swanson of Iowa State, Catherine Personius of Cornell, and several others, were appointed.[66]

When the NSF was established in 1950, President Harry Truman appointed to its first National Science Board representatives of the several sexual, racial, religious, and geographic constituencies of the time: two women, a Negro (as one was called then), a Catholic priest, and a southerner. The two women were physician and anthropologist Sophie Aberle of New Mexico, whose husband was on Truman's White House staff, and biochemist Gerty Cori from Washington University in Truman's home state of Missouri. It also helped that Cori was a Nobelist (the only one in the entire group).[67] Over the years these first two women were replaced by a trickle of others, including biochemist Jane Russell (Wilhelmi) of Emory University, representing both womankind and the South, and the more traditional trio, better known as administrators than as scientists, all of whom had served several years on lesser committees and panels at the NSF and elsewhere by the time of their appointment, Katherine McBride of Bryn Mawr, Mina Rees, back at Hunter College and CUNY, and Mary Bunting, by then of Radcliffe.[68] The women appointed to the National Science Board generally were not the same women who were being elected to the top science academies, such as Barbara McClintock, Libbie Hyman, and Katherine Esau, although Gerty Cori and Mary Bunting, two very different women, did win recognition in both realms. But not even they had held some other assignments that one might have expected, such as, for example, a place on the President's Committee on the Development of Scientists and Engineers (1956–58) or the education and manpower panels of the PSAC.

As for the supreme accolade, the Nobel prize itself, which was awarded in only three fields of science (plus economics starting in 1969) to persons in any nation, "American" women did rather well, despite all the odds against them. Two of the three women winners in the years 1940–70 were American immigrants: biochemist Cori in medicine in 1947[69] and Maria Goeppart Mayer in physics in 1963 (the first woman since Marie Curie in 1903).[70] (The third female winner was Dorothy Crowfoot Hodgkin of Britain in chemistry in 1964.) The two American women shared certain personal characteristics. Both were foreign-born and trained (Czechoslovakia and Germany, respectively) and had come to the United States with their husbands. Both had encountered great difficulties in continuing their work, but Cori shared the facilities of her husband/collaborator, and Mayer made do without, as a theoretical physicist can do to a certain extent. They did not allow themselves to complain (though Cori admitted to tears on occasion), and they were grateful for whatever opportunities came their way. Both seem to have been immersed in work and family, and they were not active in scientific or professional groups. (Neither, for example, served as president of her main professional society.) Yet the excellence of their work was apparent to those who wrote letters to Stockholm.

Although little publicity accompanied Cori's award in 1947, there was greater coverage of Maria Mayer's accomplishments in 1963–64, as then the local and national media not only compared her to Madame Curie but also, as Dorothy Nelkin has noted, delved much more intrusively into her home life than is usual in stories of (male) Nobelists. "At Home with Maria Mayer" in *Science Digest*, for example, stressed how beautiful she had been back in Göttingen in the 1920s and what a good wife her husband thought she was. *McCall's*, a woman's magazine, said several times that she was an elegant hostess and described in detail the moss-green brocade gown she wore for the Nobel presentation ceremony. But perhaps most revealing of the status of even accomplished women in the 1960s was the headline in the *San Diego Evening Tribune*, "S.D. Mother Wins Nobel Physics Prize." The fact that she was a mother of two seemed more important to the local newspaper editor than her stature as an internationally known physicist or even her professional position as a professor of physics at the University of California at San Diego.

Besides these two actual American winners in the 1940s and 1960s, a more diverse group of four other women were busy in these decades with the work that would later win them Nobel prizes. Medical physicist Rosalyn Yalow was at work on radioimmunoassay at the VA hospital in the Bronx, geneticist Barbara McClintock was studying her corn plants at the Carnegie Institution on Long Island, Italian-born biochemist Rita Levi-Montalcini was in and out of Washington University in St. Louis puzzling over the "nerve-growth factor,"[71] and chemist and pharmacologist Gertrude Elion was designing new drugs at Burroughs-Wellcome in North Carolina. They must have known their work was important and been glad when someone else said so, but by 1970 only McClintock and Montalcini had received any of the honors that might be expected to foreshadow a Nobel prize (e.g., election to the important academies [see table 14.7]).

Besides these four women who were eventually honored, there were at least two others that should have been. It is conceivable, for example, that when the 1957 physics prize was awarded to T. D. Lee and C. N. Yang for the theory of the nonconservation of parity, C. S. Wu of Columbia University, who had performed the crucial experiment proving their theory, should have been included.[72] Nor did Helen Taussig, designer of the famous "blue-baby" operation in the 1940s, which saved many lives, and author of an important article warning against the use of thalidomide drugs in the 1960s, ever win the prize for medicine.

At least six others helped men, in three cases their husbands, to win the Nobel prize but were not so honored themselves. Among these were biochemist Viola K. Graham, who in the 1920s had helped James Sumner of Cornell University synthesize urease, for which he shared the 1946 Nobel prize in chemistry with John Northrop;[73] geneticist Esther Lederberg, who helped her then husband Joshua with the microbial research that won him and two other men the Nobel prize in 1958;[74] Marguerite Vogt, the "colleague and close collaborator for 20 years" on DNA tumor viruses and cell growth of Renato Dulbecco of Cal Tech, who shared the prize with two other men in 1975;[75] Anna Schwartz, who labored for decades on the detailed economic data that formed the basis for the work that won Milton Friedmann alone the economics prize in 1976;[76] and Isabella Karle, who worked with her husband, a fellow crystallographer, for a lifetime, though he and two other men shared the chemistry prize in 1985.[77] Biochemist Ruth Hubbard felt that although she had done independent work on the chemistry of vision for many years, her marriage in 1958 to George Wald, who in 1967 won the Nobel prize for work in the same area, had caused many observers to assume that they had always collaborated and that he deserved most of the credit for her earlier independent work as well. Once she married a scientist of greater reputation, a woman's own independent work could all too easily be dismissed as merely a small part of his.[78]

Thus, a variety of evidence indicates that though women participate in the various sciences in varying proportions, they are quite rare at the very top in almost all fields. At each level of the hierarchy fewer and fewer women are visible—as presidents of scientific societies, as governmental advisers, as members of honorary academies, or as winners of major prizes. Starting as at best 33 percent of the members of a scientific society, they held fewer than 1 percent of the top honors.

It thus becomes possible to ponder why certain women have been honored with prizes and elections to academies. Certainly the winners' work had to be outstanding, but so was that of many other women who were not chosen for any or many of these awards, such as Dorothy Horstmann, Olga Taussky Todd, and Gertrude van Wagenen, to name just three. Certainly the bioscientists, including Cori, McClintock, Hyman, and Esau, dominated almost all lists of important women, with theoretical physicist Maria Goeppart Mayer outstripping experimentalist C. S. Wu, and Eleanor Jack Gibson, the lone psychologist, winning the most prestigious honors in these years. Of the biologists, these top women were so hard-working, so devoted, and so utterly selfless and their work opened up such important new areas that colleagues and contemporaries felt they had to be honored. Although most of them had no particular power base, they had to have at least a "voluntary" professorship at a university or a position at a nonprofit institution (though Hyman had only a title, even in these years of record prosperity for science). The claims of the research associates (except for Cori) and the research assistants could be all too easily dismissed or overlooked. Yet even a professorship at Columbia did not get C. S. Wu a share of the 1957 Nobel prize. Which women were honored and how many there were of them says a lot about them but perhaps even more about the men in the field, especially those in a position to do the nominating or the electing. In that respect these honors take on the coloring of old-fashioned chivalry: after the knights had jousted each other in hard-fought tournaments, they could present their favorite lady with a flower or a prize. She was worthy, but the top prizes were still in the realm of a gift or personal patronage from the men who fought the battles of science politics rather than based on scientific merit impartially assessed.

Computer software pioneer Grace Murray Hopper, shown here with a UNIVAC machine in 1960, was named the first "Computer Sciences Man-of-the-Year" in 1969.

Photograph courtesy of the Smithsonian Institution. Information from Data Management *7 (June 1969): 75.*

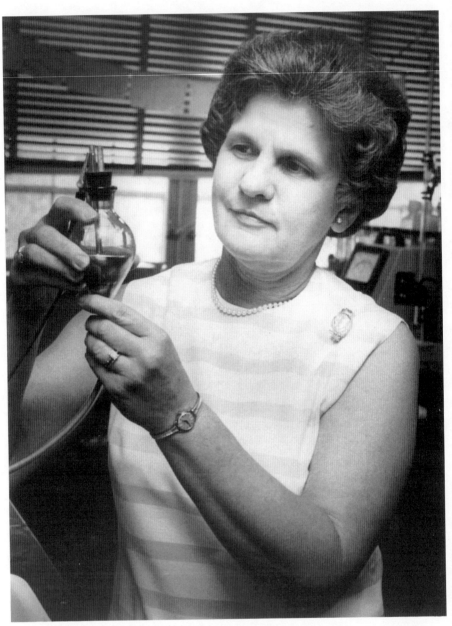

Ruth Benerito of the U.S. Department of Agriculture's utilization laboratory in New Orleans, one of several women chemists trained at Newcomb College, also in New Orleans, won the Federal Women's Award in 1968.

Photograph reprinted with permission from Southern Living *magazine. Roy Trahan, photographer.*
© *1971 Southern Living, Inc.*

Microbiologist and college president Mary Bunting, shown being sworn in as a member of the Atomic Energy Commission, was one of several women "doers" whom President Lyndon Johnson appointed to high positions in 1964.

Photograph reprinted by permission of the Radcliffe College Archives.

After astronomer Mary Shane ran the 1961 meeting of the International Astronomical Union almost single-handedly on a mountaintop in California, the appreciative U.S. National Committee awarded her the special, semi-humorous citation shown here. The inscription reads, in part: "The USA National Committee of the International Astronomical Union resolves to create the post of Permanent Hostess, Honorable Flunky, and Maid of All Work for all future meetings of the Union scheduled to be held in the United States of America. The Committee . . . hereby appoints Mary Lea Heger Shane to this important post, at no increase of salary ($), and with full permission to enjoy all the rights, emoluments, and perquisites of said post." It was signed by Leo Goldberg, Armin J. Deutsch, Lyman Spitzer, Jr., Geoffrey Keller, Frank K. Edmondson, Donald H. Menzel, Olin C. Wilson, Harlan J. Smith, Otto Struve, Donald Shane (her husband), Nicholas U. Mayall, Gerald Clemence, and William Morgan.

Reprinted by permission of the McHenry Library, University of California at Santa Cruz.

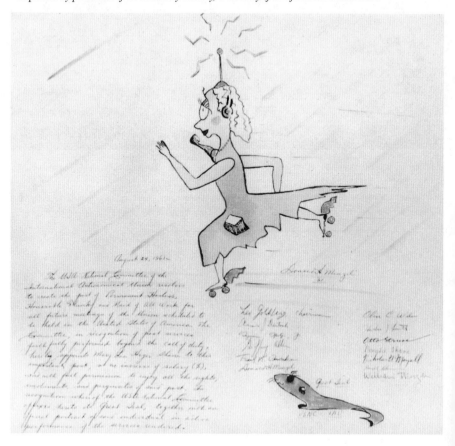

S.D. Mother Wins Nobel Physics Prize

Dr. Mayer 1st Woman in U.S., 2nd in History So Honored

By FRANK HOGAN

Dr. Maria Goeppert Mayer, 57, a research physicist at the University of California here, today was named a 1963 Nobel Prize winner in physics.

The red-haired college professor, mother of two, is the first woman residing in America to win a Nobel physics prize, the second woman in history.

She received the news in a phone call from Sweden at 4 this morning. The caller informed her she had been cited by the Swedish Royal Academy of Science for her work in determining the nature of the shell of the atom's nucleus.

Will Share Award

Mrs. Mayer will share the $51,158 prize with a West German scientist, J. Hans D. Jensen of Heidelberg, who made a similar but independent study, and Eugene Wigner of Princeton University, who helped lay the groundwork for the present advanced study in nuclear physics.

Wigner will get half the prize. Mrs. Mayer and Jensen will divide the rest.

Mrs. Mayer is the second Nobel Prize winner on the university's San Diego campus. The other is Dr. Harold C. Urey of 7890 Torrey Lane, professor of chemistry at large, who won the Nobel chemistry award in 1934.

Caught by Surprise

Mother of a married daughter and a grown son, Mrs. Mayer is the wife of Dr. Joseph E. Mayer, a chemistry professor at the University of California here. They joined the local faculty in 1960, and live at 2345 Via Sienna, La Jolla.

Excited and insisting the award caught her by surprise, Mrs. Mayer said she would go with her husband to Stockholm for the formal award.

While there, she said, she (Continued Next Page, Col. 1)

State Dept. Fires Officer Over 'Leak'

WASHINGTON ⑴ — Otto F. Otepka was dismissed today from his job as a State Department security officer on charges of conduct "unbecoming an officer of the Department of State."

Otepka, 48, has been under suspension since Sept. 23, accused among other things of giving confidential information to the Senate Internal Security subcommittee.

A letter removing him from the ranks of department employes was delivered today.

Signed by John Ordway, chief of the department's personnel division, it replied to Otepka's 12-page-long letter of Oct. 14, requesting that charges against him be dismissed, and upheld the charges.

Otepka had been chief security risk evaluator for the department at a salary of $16,500-a-year.

A qualified source said that Otepka's case was thoroughly studied and the decision to dismiss him was made on a high level.

When Maria Goeppart Mayer, on the faculty of the University of California at San Diego, won the Nobel prize for physics in 1963, the headline in the local newspaper emphasized her maternal rather than her professional status.

Photograph and article reprinted by permission of the San Diego Evening Tribune.

After Austrian nuclear scientist Lise Meitner was denied a share of the 1944 Nobel prize that went to Otto Hahn, her co-discoverer of nuclear fission, she was chosen as Woman of the Year by the Women's National Press Club in Washington, D.C. She is shown here at the award dinner with President Harry S. Truman in 1946.

Photograph reprinted with permission of the Bettman Archive.

Opposite, top: Maria Goeppart Mayer, the first woman to win the Nobel prize for physics since Marie Curie in 1903, at the Nobel ceremony in December 1963.

Photograph reprinted by permission of Reportagebild, Stockholm.

Opposite, bottom: Essie White Cohn (left), of the Women's Service Committee of the American Chemical Society, presented the Garvan Medal to Sarah Ratner, a biochemist with the City of New York, in 1961.

Reprinted with permission from Chemical & Engineering News, *April 3, 1961, 39(14), p. 28.* © *1995 American Chemical Society.*

Two hundred sixty students from 140 colleges, some of whom are pictured here, attended a three-day symposium on women in science and engineering at the Massachusetts Institute of Technology in October 1964, at which sociologist Alice S. Rossi first posed the question "Women in Science: Why So Few?"

Photograph reprinted with permission of the MIT Museum.

Women's Clubs and Prizes

Partial Palliatives

Quite unlike the expanding mainstream forms of scientific recognition, where women were marginal and nearly invisible from the 1940s through the late 1960s, were the largely female realms of women's clubs and prizes. There women members could be as active as they had the time or propensity to be, hold office or not, support certain (especially scholarship and fellowship) projects, and perhaps even run a major activity, such as an international congress for women. In return, participants, who were, except in the fields of home economics and psychology, largely isolated, got some support and companionship, some sense of accomplishment, perhaps some inspiration, as well as the practical advantages of meeting with others like themselves, what later came to be termed "networking." A few of these groups represented the women in their field to outside interests, for example, to the WB of the federal government, which did surveys of women in particular fields, or to the Office of Civil and Defense Mobilization when it ran workshops on encouraging women to become engineers.

Yet there were also certain limits on the effectiveness of these diversely organized and situated groups. Besides the large and comprehensive AAUW, there were only seven, and many women, especially young ones, had never heard of them. (Most were quite weak at the graduate school level, where they were perhaps most needed.) Even those who did belong, especially the psychologists, often had doubts about the political respectability and proper role of a single-sex society in science. Few groups had enough resources of their own or access to foundation grants to provide their members with the services they might have needed. (Sadly, it was in these same years that suicide among women chemists and psychologists was such a disproportionately frequent occurrence—triple that for male chemists, for example—that they, as well as female physicians, were the subject of statistically significant studies.)[1] Moreover, whether or not these groups had the word "women" in their names, few were recognizable by the staff members of those very task forces, such as the

President's Committee on the Development of Scientists and Engineers, that were presumably seeking to recruit women scientists. Yet most members of these associations, especially the alumnae groups, were aware of the larger issues, and their local and national meetings provided some of the first forums for discussions of women's role in science. On occasion this interest could blossom into public sessions at national meetings, which could begin to stir a higher consciousness of the unfairness of much of prevailing scientific practice and even to challenge the sincerity of so much governmental rhetoric. Thus, both the groups' existence and their overall lack of success (except for the AAUW's postdoctoral·fellowships) are useful indicators of the level of interest and the limits and obstacles encountered at the time.

Three of the largest and oldest groups of women scientists were the Greek-letter "fraternities" for women undergraduates and graduate students formed at large, usually land-grant, universities in the period 1900–1925. These often started as local clubs, formed among women undergraduates who, seeing that the men students and faculty had formed single-sex honor societies, wished to have one of their own and found some encouragement from a female faculty sponsor. Home economics was such a fertile field for a strong all-women's organization that once these fledgling groups discovered similar clusters on other campuses, they restructured themselves as local chapters of a national group. This provided for some greater prestige, stability, and coordination, including such central services as standardized pins and rituals and eventually a common newsletter. The largest and best-known such fraternity in the field of home economics was Omicron Nu, established at Michigan State College in 1912 with the purpose of promoting scholarship, leadership, and research in the field. It elected women in the top 20 percent of their class and by 1968 could claim to have initiated 25,350 members in forty-three student and three alumnae chapters across the nation.[2]

Such groups were active at both the national and the local level; they focused on inspiring undergraduates and exhorting them to lead productive lives. Nationally Omicron Nu was lucky throughout these years to have office space and a staff member at Michigan State who put out its quarterly newsletter. Of the several professors who served as presidents in the 1940s through the 1960s, Grace Henderson, the retired dean of the college of home economics at Pennsylvania State University, who presided in the late 1960s, certainly expected the hardest analytical thinking from the members. The field of home economics was, as she well knew, undergoing scrutiny and contemplating a new mission; in several essays in the newsletter she stated that she expected Omicron undergraduates, who presumably were the future leaders of the field, to think about the matter and make some contributions. Also visible in the newsletter were the four or five "honorary members" elected each year, whose lives and careers exemplified important contributions to the field and to Omicron Nu.[3]

For most of the fraternity's thousands of members, its chief activities were those of an honor society—an annual round of fall get-acquainted teas, fund-raising projects, possibly a career day, and then the spring awards banquet, where new members were initiated. The message was educational and inspirational, urging bright undergraduates nationwide to take themselves seriously, make some plans for their future, develop leadership skills, and even think of going to graduate school. (One issue of the newsletter each year was devoted to a listing of opportunities for graduate work in various home economics programs.) These were messages worth voicing and repeating in the 1950s and 1960s. Some chapters were larger and consistently more active than others. Some apparently elected male members, but they were rarely mentioned or visible in the national newsletter, *Omicron Nu Magazine*. What received some attention was the establishment of the first chapter at a black college (Howard University in 1963), whose adviser, Flemmie Kittrell, had served as national secretary of the organization as early as 1952–54. For the times this was quite remarkable.[4]

Although graduate students in home economics were accepted as members and Omicron Nu even had three alumnae chapters by the 1960s, its leaders were unsure how to deal with such persons, who wished to remain in touch with each other and the field. Many felt that they should be sent to other groups, including the AHEA and its subgroups, such as the Home Economists in Business (HEIBs), while Omicron Nu should continue to focus on doing what it did best, namely, initiating ever more undergraduates into its honor society.[5] At this level Omicron Nu was a model organization. Yet for the larger purposes of women in science, it remained locked in the organizationally separate world of home economics, focused almost exclusively on the undergraduates, especially at the land-grant institutions. As that field faced an identity crisis in the 1950s and 1960s, Omicron Nu responded, but its problems were somewhat different from the exclusion and discrimination facing young and mature women in mainstream, male-dominated science.

In chemistry such a separate fraternity for women became a necessity around 1900 when the newly formed honor society for chemists, Phi Lambda Upsilon, established first at the University of Illinois, refused to admit women.[6] Affronted, coeds at several universities set up their own separate honor clubs for chemistry majors at the top of their class. In 1911 several of these clubs formed into one national group, Iota Sigma Pi (ISP), still in existence, which reported a national membership of 7,468 in twenty-five chapters across the nation in 1968. Like other fraternities, it had national officers (but no national headquarters), insignia, rituals, a triennial national convention, and after 1941 a newsletter.[7]

Because chemistry classes contained many men, and chemical departments had hardly any women faculty, the activities of the local chapters (named for the elements) were more social and even decorative than professional or preprofessional. Thus, besides the usual get-acquainted teas and spring initiations, chapter members of ISP might pour tea at the department's weekly colloquium, hold a retirement party for one of the retiring faculty men, or have a joint picnic with the local all-male Phi Lambda Upsilon or Alpha Chi Sigma chapters. In return, they might get a clubroom on campus and, because there would be no female faculty adviser, have a faculty wife as their leader, such as Evelyn McBain of Stanford's ISP, who as national president during World War II managed to visit every ISP chapter, presumably at her own expense. The most feminist activity of any of these chapters was the sponsorship by two of the chapters (Penn State and Illinois) of a lecture by a woman chemist in the spring. Virginia Bartow was proud that her chapter at Illinois had been able to sustain this for more than thirty years (1929–1960), and Mary Willard kept Penn State's series going from 1940 until 1960.[8]

Over the years the organization and its local chapters changed somewhat. Some became inactive when the enthusiastic founder or other driving force retired or passed away. In others, especially those in cities, alumnae and women at other universities wanted a more active role. Agnes Fay Morgan, one of the ISP's founders, who remained active until her death in 1968, sensed this loss of cohesiveness in 1966 and sought to persuade Clara Storvick of Oregon State University to become the group's national president,

> We need a strong steady no nonsense president as we never have before. The society is expanding but I think is not as cohesive as it has been, nor is it doing the important job for all women in the chemical world that it was meant to do. This is no one person's fault. It's just the result of widely scattered varied chapters going in various directions.
>
> . . . the new president should answer the following description: she must be a vigorous middle aged woman with an impeccable reputation and eminent standing in the scientific world. She should be in touch with an active chapter. She should not be tied down by family responsibilities and she should be comfortably off so far as money goes. Thus she might be free to go at her own initiative, on her own time and resources[,] to visit the chapters and to organize Iota Sigma Pi meetings at the ACS and Federation convention.
>
> I think you fill all these specifications. Please think it over and accept.[9]

She did!

Between 1966 and 1969 ISP formalized its drift from an undergraduate honor society into an alumnae professional organization. One of the first signs that ISP might mean more to practicing women chemists than to the undergraduates was the formation in 1960 of a chapter in Chicago (Aurum Iodide) for alumnae only. If too few women at the University of Chicago or Northwestern were interested in joining to keep it going, the many women employed in the area, such as president Hoylande D. Young (Failey) at Argonne National Laboratory, were very eager to have a social or professional group of women chemists. In 1968 this "alumnae" chapter became a "metropolitan" one and was quickly followed by others—Seattle, Berkeley, Cleveland and several more—in the 1970s. Members were very interested in professional issues, visits to each others' laboratories, and a breakfast or luncheon at the national meetings of the ACS (often co-sponsored by its Women's Service Committee) or of the Federation of American Societies of Experimental Biology (FASEB). These women were generally quite ready to discuss the perennial topic of "opportunities for women in chemistry," and they were very interested in choosing women chemists for particular awards, including honorary membership in ISP and the ACS's Garvan Medal (see below). On these topics there were no greater authorities than some of the past ISP leaders, such as Agnes Fay Morgan, Icie Macy Hoobler, Essie White Cohn, Gladys Emerson, Nellie Naylor, and Jane Fraenkel-Conrat. In fact one of Morgan's very last publications, in April 1968, was a perceptive piece entitled "Women in Chemistry or Allied Fields and Professional Opportunities: The Real Purpose of Iota Sigma Pi," which pointed out that women chemists were still underpaid and had achieved only a foothold in the chemical profession. It was up to ISP to encourage women in chemistry to continue to do their best from their undergraduate years throughout their careers. Marriage was already proving to be less of a detriment to careers in chemistry than it had been in the past. In Morgan's view the objectives of ISP, written fifty years before, needed no changing: "a) to promote an interest in chemistry and its related fields among women students, b) to foster mutual advancement in academic, business and social life, and c) to stimulate personal accomplishment in chemical and related fields."[10] Neither Morgan nor her readers were ready to advocate governmental action to change employment conditions facing women in chemistry. They still relied on personal self-discipline, stoicism, and at most mutual support. Yet she had seventy-five hundred women chemists thinking about the issues. In November 1969, in response to recent layoffs at federal installations, the San Francisco Bay Area chapter and the regional office of the Department of Labor's WB cohosted a career day, "Women in Science."[11] Thus, ISP was a longstanding group whose energetic members were reacting to and coping with new currents around them.

Besides these two undergraduate/alumnae Greek-letter fraternities for women in science, there were several organizations that had no student component. The first of these was the small and independent Society of Women Geographers (SWG), established in 1925 by several women explorers and travelers when the all-male Explorers Club in New York City archly refused to admit women. By 1950 SWG had 246 American members and 65 foreign or corresponding members; in 1975 it had a total of 441 in three categories. SWG members struggled to support a national office in Washington, D.C., but managed to do without the men's central clubhouse, full of stuffed bears and other trophies. Instead they formed their own structure and traditions, including the Gold Medal, given on occasion for "original contributions to geographic knowledge," and the SWG flag, which several intrepid members were allowed to take on expeditions to Panama, Africa, exotic islands, and Greenland to mark geographic firsts for womankind. Closer to home, however, SWG members were generally less daring, even traditional, holding a tea for interested women at the annual meeting of the Association of American Geographers and by 1960 amassing a fellowship fund that would support each year a woman graduate student at Columbia University and every second or third year another one somewhere else. The SWG members were not at all interested in criticizing or changing any American institutions, not even speaking out in 1962 on the seemingly related issue of women's rejection as astronauts. (An astronaut and an aquanaut would be the first women finally elected to membership in the Explorers Club in 1981.)

SWG's chief reasons for existing were quite unabashedly social and professional—members simply enjoyed being with each other. Membership was limited to those women who had accomplished something with their travels. A 1950 survey found that the membership included writers, travelers, editors, geographers, and scientists of various sorts and that two-thirds were married. Although members were in fifty-eight countries by 1975, SWG had clusters or chapters in New York City, Washington, D.C., Chicago, Los Angeles, and San Francisco, where they met monthly to hear a speaker, usually a woman, discuss her latest adventures. In addition, all members were expected to send in a few sentences about their latest accomplishments to the society's annual newsletter. One of the SWG's leading members for fifty years was anthropologist and world traveler Margaret Mead (who once said that only by belonging to the SWG had she met her male colleagues from the American Museum, when their member-wives invited them to the occasional "men's night"); others included oceanographer Marie Fish, planktonologist Mary Sears of Woods Hole, nature writer Rachel Carson of Maine, and congresswoman Frances P. Bolton of Cleveland, Ohio. There is no indication that any man ever tried to join this interesting circle; nor is it clear what response he might have received. As a group the SWG members were pleased with their organization's decentralized structure—it was both local and international, personal and apolitical—and quite proud of the fact that its members constituted one of the most diverse and knowledgeable groups of accomplished women one could hope to find anywhere. They had no solutions for the problems facing women in science, but an evening with them would have been good for any woman's mental health and self-esteem.[12]

Another independent group, the SWE, established in May 1950 (see ch. 3), grew rapidly in the 1950s, numbering 350 by 1953 and 510 by 1958 (in eighteen sections, including one "at-large"), though it may have leveled off (at around 750) in the 1960s.[13] From the start the organization's young officers were greatly supported, psychologically and possibly financially, by their patron saint, Lillian Gilbreth, who reassured them at uncertain moments and spoke at innumerable SWE section and national meetings and banquets. Gilbreth not only knew firsthand the problems facing women in engineering (and her own specialty of industrial psychology) but had served over the years on many important governmental and other committees on women's issues. Her inspirational messages and grandmotherly image emphasized the maternal aspects one might not ordinarily associate with women in engineering. She personified the competent but wise and gentle soul they might hope to become, and her frequent presence demonstrated to members as well as others that a woman could be an engineer and still, despite all the public comment and ridicule about how odd it was, be human and even feminine. In 1958, when she turned eighty, SWE established its first scholarship fund and named it for Gilbreth.[14]

Unlike SWG, SWE was not just a social group: it had broad recruitment purposes as well. Its stated goals were to

inform young women, their parents and counselors and the general public of the qualifications and achievements of women engineers and the opportunities open to them;

assist women engineers in readying themselves for a return to active work after temporary retirement;

serve as a center of information on women in engineering;

encourage women in engineering to attain high levels of education and professional achievement.[15]

To these ends one of SWE's first activities was to establish its Achievement Award, which has been given annually since 1952 to pioneers in the field as well as in the related areas of psychology, mathematics, chemistry, and space biology (see table 15.1). To reach its far-flung and often isolated members, in 1951 SWE started a journal that became a quarterly newsletter in 1954 and a bimonthly one in 1966. Another important step was the successful completion of a ten-thousand-dollar fund drive in 1958–61 to pay for the establishment of a new national headquarters with a full-time secretary in the recently completed United Engineering Center, opposite the United Nations Building in Manhattan. Although this drive was quite a burden for the youthful members of SWE, they apparently reached their goal partly by offering five-hundred-dollar lifetime memberships.[16]

From its earliest days SWE's purpose of publicly promoting careers for women in engineering dovetailed more closely with the government's "wom-

Table 15.1.
Winners of the Society of Women Engineers
Achievement Award, 1952–1970

Year	Winner	Specialty
1952	Maria Telkes	Solar energy
1953	Elsie Gregory MacGill	Aeronautical engineering
1954	Edith Clarke	Electric power transmission
1955	Margaret Hutchinson	Chemical engineering
1956	Elise F. Harmon	Printed circuits
1957	Rebecca Hall Sparling	High-temperature metallurgy
1958	Mabel M. Rockwell	Electrical control systems
1959	Desirée Le Beau	Rubber reclamation
1960	Esther M. Conwell	Solid-state physics
1961	Laurel von der Wal	Space biology
1962	Laurence Delisle Pellier	Physical metallurgy
1963	Beatrice Hicks	Sensing devices
1964	Grace Murray Hopper	Automatic programming systems
1965	Martha J. Thomas	Phosphor chemistry
1966	Dorothy Martin Simon	Combustion chemistry
1967	Marguerite M. Rogers	Air-launched weapons
1968	Isabella Karle	Crystal structure analysis
1969	Alice Stoll	Fire-resistant fibers
1970	Irmgard Flügge-Lotz	Aeronautical engineering

Source: SWE, *Achievement Award, 1952–1976* (New York, 1976).

anpower" message than did that of any other women's group in the 1950s. Formed a month before the Korean War started, SWE's early years coincided with heightened federal interest in adequate supplies of technical manpower. The organization produced its own biennial (and then cumulative) surveys of both female engineering enrollments (a gap in the federal statistics) and degrees earned, which members then used to encourage women to enter the field. Their numerous articles appeared in so many diverse magazines, journals, and career guidance booklets that SWE's Patricia L. Brown published a compilation of them in 1955.[17] SWE even cooperated with governmental agencies that were moving into recruitment. In 1953 and 1955 SWE provided the names and addresses of hundreds of women engineers to the WB, which was putting out first a bulletin, *Employment Opportunities for Women in Professional Engineering,* in 1954 and a technical article, "Employment Characteristics of Women Engineers," in 1956. In all of these the public tone was upbeat: you can be an engineer, there are plenty of jobs, and the salaries are high (for women).[18] In the 1960s SWE cooperated with the Office of Civil and Defense Mobilization of the Executive Office of the President on a series of "utilization conferences" that urged guidance counselors, teachers, and others to encourage more women to take up engineering (see ch. 3).

The climax of SWE's first fifteen years was its hosting (with some partial support from NSF, making it one of the very few women's groups to get any NSF funding for a project) the First International Congress of Women Engineers and Scientists in New York City in June 1964, timed to include a day at the World's Fair on nearby Long Island. So well connected were the women engineers that they arranged for President Lyndon and Lady Bird Johnson to send welcoming telegrams. Although most of the five hundred women engineers attending were from the United States, attendees came from thirty-three other nations as well, including twenty-two from England, whose British Women's Engineering Society hosted the Second International Congress at Cambridge, England, in 1967.[19] But if it appeared that women engineers were now seen as the important and valued members of society depicted in the career booklets and the presidential welcome, the short account of the meeting in the *New Yorker* magazine quickly dispelled that notion. The *New Yorker* reporter attended a reception in the basement of the United Engineering Center, where he talked to several women, noting their names, job titles, and the color of their dresses. The resulting account seemed to caricature and ridicule the women engineers, making them sound as if they were animals at the zoo, perhaps birds in full plumage:

> Mrs. Cavanagh turned us over to Miss Barbara Fox, a pretty engineer in a dark-blue dress, who is a sanitary specialist with the City of Chicago Water Department and has done a good deal of tunneling. Miss Fox introduced us to Mrs. Catherine Walshe, also in a dark blue dress, who had helped design a nine-mile dual carriageway for the Dublin County Council, in Ireland, and by being handed on in this fashion for a while we met Miss Anna C. Hanson, in a green and white dress with red trim, who had worked on a sea wall for the City of Detroit.[20]

Only a minor issue for the women engineers was the treatment of men who asked to join or contribute. In 1966 the group created the Rodney C. Chipp Memorial Award, named in honor of the second spouse of SWE's first president, Beatrice Hicks, for men and other nonmembers who had contributed to SWE's goals and presented it to Col. Clarence Davies, the retiring head of the Engineering Manpower Council in New York City. In 1967 SWE organized a small "men's auxiliary," which disbanded in 1975, when SWE began to consider admitting men directly. That it did in 1976, but it did not change its name, which remained the Society of Women Engineers.[21]

Thus SWE was in many ways an effective organization. Like the SWG, its meeting and newsletter brought together many like-thinking but isolated and possibly demoralized individuals for their mutual support and encouragement. Together the members performed many useful functions, such as creating prizes and awarding fellowships, and even rose to the challenges of funding a national office and hosting an international congress. SWE was also the one women's group that was in close working contact with those governmental agencies that were purportedly interested in increasing the numbers of women scientists and engineers. Its membership provided proof that American women could in fact be engineers and thus gave some credibility to the "woman-power" message that was generally so ambivalently and unsuccessfully disseminated. Yet SWE's internal harmony and welcome reception were also signs that it constituted no threat to the established order. Their message that women could do engineering continued to fall on many deaf ears. Nor did women engineers, limping along at less than 1 percent of every countable category (including undergraduate enrollments), face the divisions or raise the fears of the women psychologists (to be discussed shortly), who constituted nearly one-third of their field.

The situation was more complicated for two other women's groups that were, or tried to be, part of mainstream scientific societies in these years. Established in 1927, the Women's Service Committee of the ACS was in the 1950s and 1960s the only official committee on women within a major scientific or professional organization. Evidently the ACS had enough women who felt that they needed such a committee and, unlike the situation in the APA, not much active opposition. Male chemists neither asked to join the committee nor insisted that it be terminated. Possibly its "service" orientation and low cost sufficed to justify its continuation. Its goals, as mentioned in the 1959 annual report to the ACS, were: "1) to improve the status of women in chemistry; 2) to increase the quality of performance of women in chemistry; 3) to attract capable women to the field of chemistry; 4) to increase membership in the ACS of eligible women; and 5) to promote attendance of women at national meetings."[22] In most years these five goals came down to holding luncheons, complete with speakers and presiders, at the two or three national meetings of the ACS and nominating candidates for its "woman's prize," the Garvan Medal. In fact the annual need to find and write up fresh candidates was often the chief item of business for the ACS Women's Service Committee and the one thing that kept it going when consciousness was low.

The prize had been established in 1936 by the wealthy Francis P. Garvan, who, having been overheard to say that there were still no women in chemistry, apologized by funding a prize that highlighted the achievements of those women already in the field. Three awards were made in the first ten years, but starting in 1946 the prize was presented annually. In the years 1948–52 the Garvan Medal winner was featured on the cover of the ACS's weekly news magazine *C&EN*. Then for four years, 1953–56, she was one of a group of prizewinners on a cover. But thereafter, for reasons that are not clear, the Garvan medalist was lucky to have her picture only on the page with the story of the presentation ceremony. Thus, by the late 1950s the initial goal of publicizing women's contributions to chemistry to a wide audience had been lost; the *C&EN* editors had decided that such awards no longer interested their readers and, by denying the winners their former publicity, undercut the effectiveness of even this modest form of recognition.

The very high quality of the work of some Garvan medalists is evident from the astounding fact that two of them have also won the a Nobel prize (in medicine and physiology rather than in chemistry). Biochemist Gerty Cori won the Nobel prize in 1947, a few months *before* winning the Garvan Medal, which she might then have declined but did not. When 1988 Nobelist Gertrude Elion won the Garvan Medal in 1968, it was the first award she had ever received and one that she reported meant a great deal to her, since it came at a moment of great discouragement.[23] All too often, as in the case of biochemist Gertrude Perlmann of the Rockefeller University, the Garvan Medal was the only prize the recipient ever won in a very long career.[24] Without the Garvan Medal, many careers in chemistry would have ended in anticlimax, for few women won any other award from the ACS before 1970. Either the Garvan Medal had become counterproductive, providing an easy excuse for not taking the women seriously for any of the society's more prestigious prizes between 1940 and 1970, or, what is more likely, consciousness of women's contributions was so low among members of other ACS prize committees that without the twenty-plus Garvan medalists there would have been no official recognition at all.[25] In fact when the members of the Women's Service Committee began to protest this continuing omission and to nominate women for the ACS's many other awards in the 1960s, their actions had little impact. Chemistry's occupational segregation and even ghettoization spilled over all too readily into separate men's and women's prizes. It is no wonder that the women chemists, already discouraged and underpaid at work, were committing suicide three times as often as the men in the field and five times as often as other white women.[26]

Occasionally the annual reports of the Women's Service Committee, printed in the *C&EN,* mentioned other projects and ambitions, such as Marjorie Vold's proposed survey in 1948 of women chemists who successfully combined marriage and a family. Discussed again in 1949 as a possible booklet, to be called "One Hundred Women Chemists of 1950 and What We Do," it was abandoned in 1950 for lack of interest, which Vold herself then interpreted too optimistically as "an indication of the welcome fact that women chemists are no longer set apart as a class either by themselves or by the science they serve."[27] In the 1950s the wives of male chemists sometimes joined the women's luncheons as part of the "ladies' programme" arranged for them by the ACS staff.[28] Then, starting in 1963, when the annual reports began to reveal signs of discontent among the women chemists, the Women's Service Committee began to host an open meeting after the luncheon, where such concerns as unequal pay, reentry training, and even discrimination were brought up and discussed. In 1963, for example, "The luncheon was followed by an open meeting of this committee at which were discussed such topics as alleged discrimination against women chemists, improved professional practices, the desirability of surveying women chemists for certain professional and economic information, and ways of making the committee more useful to women chemists."

Usually the committee had no recommendations to make or solutions to propose but felt that it was important to let the issues be raised and discussed; however, it then minimized, denied, and dismissed them or at least deferred them to the ACS's Committee on Professional Relations. Or as the 1963 report concluded, "In discussion of alleged discrimination, there was unanimous agreement that there is not enough discrimination against women to make an issue of it; that women prefer to be considered chemists, not set apart by an arbitrary distinction but judged only by performance, the only sound basis for acceptance. It was unanimously agreed, [however] at least for the present, that the Women's Service Committee should investigate quietly any special cases of discrimination which are brought to its attention."[29] Yet by 1965 and 1966 the committee was backing off from any such investigatory role. As the report for 1966 put it, "In addition, the subject of discrimination against women chemists was again brought up. The committee indicated continuing interest in the professional status of women chemists but reaffirmed its position that the committee should not concern itself with problem cases involving the status of women employees but should refer them to the committee which handles similar problems for men."[30]

These reports of open meetings of the one women's committee of a scientific organization of the time give a glimpse of the issues that were being raised, discussed, and publicized, however guardedly and discreetly, to fellow professionals. They show that despite the daily evidence of sexual segregation and discrimination within the world of chemistry, some of the women chemists were still clinging publicly or at least rhetorically to the ideological notion that women chemists should be treated just the same as men and should not ask for special treatment. To do so would apparently threaten past gains and current footholds. Yet even if the problems had been denied, ignored, and sidestepped for decades, they had not not been solved or gone away. After forty years of "service" to the ACS, its women members still were not getting much recognition or support. What role the scientific or professional society might have in acknowledging or even solving these "political," possibly nonscientific problems was, however, changing rapidly by the late 1960s. Anger was rising, and within two years two irate chemical librarians would denounce the committee in a letter to the editor of the *C&EN* as never having accomplished anything substantial for women chemists.[31]

But even the limited institutional space accorded the women chemists makes them look successful in comparison with women in psychology, where attitudes and circumstances were very different. There women, who made up about a third of the field, had in 1941 formed outside of all existing psychological associations the separate National Council of Women Psychologists, or NCWP (which in 1946 became the International Council of Women Psychologists, or ICWP), partly to provide war work for women in the field. Its substantial membership of 261 in 1943 dropped sharply after the war to just 106 in 1948,[32] whereupon the energetic and committed Mildred Mitchell of Minneapolis and later Dayton, author of the controversial 1951 report on the status of women in the APA, chaired the organization's membership committee. Each year she sent out hundreds of applications to new Ph.D.'s and others recently added to the APA directory. In her eleven years in this position she more than quadrupled the number of members, which reached a high of 488 in 1958. But she must have been disappointed by how uninterested and even afraid of women's issues most of them were. In fact almost annually between 1950 and 1955 she had to explain to the ICWP's governing boards what precautions she had taken to avoid admitting a Communist.[33] This coincided with the refusal in 1954 of six Ohio members to collect data on the status of women in psychology there lest they be labeled subversives. The idea of such a survey was finally dropped, even though one member thought women in the field might need one now more than ever before.[34]

Although the society voted in 1951 to admit men as members, as a possible way of becoming a division of the APA, it retained the word "women" in its name. Evidently this created doubts among some psychologists and others that this was not a serious professional group, as member Katharine Banham of the Duke University psychology department learned in 1952 when she asked for travel funds for an ICWP meeting at the national APA convention. The head of Duke's research council wanted first more information on what sort of organization this was, and even after she explained that despite its name the society had male members, he curtly denied her the funds. When he did the same thing again two years later, Banham, by then the ICWP's president-elect, decided for this and other reasons to withdraw and did not serve as president.[35] When ICWP members voted to drop the word "women" from the their name in 1959, Mitchell quickly resigned as membership chair, and her successors began to recruit men. In 1963 the left-leaning Bernard Riess of Hunter College won a seat on the board of directors, and by 1967, just when the women chemists were becoming more vocal, the ICP elected its first male president, Henry David. (At this point past president Lillian Portenier did what many other displaced women did: she started to write a history of the group's earlier years.) A 1973 study noted that the masculinization of the society had accelerated rapidly after 1966; as more and more men were elected to office and publicized in the newsletter, more and more women resigned. By 1973 this once all-female association, organized to promote the interests of women in psychology nationally and then internationally, was just 56.82 percent female (379 of 667).[36] Although the feminists of the seventies would look back on this and argue that the women of the fifties had made major mistakes, at the time the changes were endorsed overwhelmingly.

Thus, unlike the women geographers, engineers, and chemists, the women psychologists of the forties through the sixties were enmeshed in a kind of thinking that considered an all-women group (in name if not in fact) discriminatory, unprofessional, and probably deviant. This put the group on the defensive even when, as Mitchell's 1951 report had shown, they continued to be badly underrepresented in the upper levels of the APA hierarchy. Yet the male psychologists would not leave the ICWP alone to coexist, and some women took steps to recruit increasing numbers of men, who, as other women withdrew without protest, in time took over the leadership. Thus, although there were thousands of women psychologists in the United States by the 1960s, more than in any other field of science or social science and certainly a higher proportion of the whole than in any other field except home economics, their consciousness was so low that they lost control over the one women's group in the field. Something in their personal or professional socialization held them back from militancy and confrontation, which just a few years later, in 1969, would seem so necessary and so feasible.

The last of the three Greek-letter fraternities for women in science, Sigma Delta Epsilon (SDE), for graduate students in any field of science, was established rather late, in 1921. As its initial members left Cornell for other campuses, sister chapters were formed, so that by 1968 it claimed fourteen hundred members in sixteen chapters, fifteen of them in the east.[37] Despite its small size (perhaps a result of reporting only the currently active members), SDE's members were a potentially forceful group of mostly biologists with a particularly strong interest in the professional issues facing women in science. Its constitution listed its objectives as follows: "(1) to further interest in science, (2) to provide a society for the recognition of women in science, and (3) to bring them together in fraternal relationship."[38]

Among SDE's ways of increasing the recognition of women in science was the creation between 1940 and 1968 of honorary memberships for twenty-seven women (including several at NIH); the awarding of a fellowship, grants-in-aid, and other research awards, depending on the money available; and the election of its own officers. In addition, it did more than any other group, even Iota Sigma Pi, to foster the history of women in science by inviting speakers on the topic to chapter installations, national luncheons, and, at Penn State's Nu chapter, to annual initiations around 1940.[39] Some of these fledgling efforts were among the first such attempts in the United States.

Because SDE members came from a wide variety of scientific fields, the only national meeting that a substantial number might attend was that of the AAAS. (Later the national meeting of FASEB might have served as well.) Accordingly, SDE's leaders sought and finally succeeded, in 1939, on the third try, in becoming a recognized affiliate of the AAAS. This permitted them to run sessions, conduct their annual business meeting, and host an "all-women-in-science" breakfast or luncheon for their own members and any other interested women. SDE also encouraged its own members to apply to become fellows of the AAAS; by 1946 enough of them had been awarded this status for SDE to have two votes on the AAAS Council. Initially content with university-based chapters, over time the leaders of SDE, like the leaders of Omicron Nu and Iota Sigma Pi, added a chapter for alumnae and at-large members who had moved from the region of their initial chapter and several others for particular "metropolitan" regions, whose members were not tied to any one university. The first in this latter category was Kappa of New York City, which was finally given a charter in 1928, on its third request. Another was Omicron in the Washington, D.C., area (1948) followed by still others in Philadelphia, Chicago, and southern California. Yet these urban chapters more than proved their usefulness, for the AAAS frequently met in these very cities (four times in New York City, for example, between 1949 and 1967), and these metropolitan chapters were ready and able to carry the heavy burden of local arrangements.[40]

Its strong alignment with the AAAS made SDE the most prominent of the women's organizations and thus put it in the position to speak out on issues affecting women in science in the fifties and sixties. SDE co-sponsored sessions at three such AAAS meetings in 1958–60, including the first major critical discussions of the "womanpower" message in the post-Sputnik era, as part of the Washington, D.C., meeting in December 1958 and the Chicago meeting in 1959. These led to the creation of the abortive American (or National) Council of Women in Science in 1958–60, whose short history provides an indicator of the level and limits of scientists' incipient feminism in the late fifties. This organization grew out of a day-long (and evening, as it turned out) National Conference on the Participation of Women in Science, sponsored by SDE, the American Association of Scientific Workers, the NFBPWC, its BPW Foundation, and three other organizations and held in conjunction with the AAAS annual meeting in Washington, D.C., in December 1958. (At one point the WB of the Department of Labor was a co-sponsor, but for unknown reasons it withdrew; the staff at the AAUW were informed of the meeting but were too busy with higher-priority tasks to participate.)[41]

The day's main speaker was none other than Arthur S. Flemming, secretary of HEW, which oversaw not only the USPHS, with its NIH and its many grants, but also the USOE, which was just then gearing up to spend the millions made available by the NDEA. Flemming was particularly appropriate, since he had been since the late 1940s a major supporter of the movement to "encourage" women to become scientists (see ch. 3). His speech this time hit just the right themes and electrified the audience of about 150, mostly women. The *New York Times* reported the next day that

> Scientists were told today that more women must get into the science field if the United States is to keep pace with the Soviet Union.
>
> Arthur S. Flemming, Secretary of the Department of Health, Education and Welfare, said that this country's greatest danger was in its personnel shortages, particularly in engineering, science and mathematics.
>
> He denounced the "double standard" under which, he said, women were not welcomed in the graduate schools in these fields. He also said that when they were trained they were not employed at their full potential or with compensation equal to men.
>
> He urged that educators and employers be constantly informed that this "double standard" was without justification, and was a drawback on the national security and the expanding civilian economy.[42]

Another summary of the speech added the crucial mechanism: "Suggesting that a spirit of real urgency was required to solve the problem, Dr. Flemming offered the support of the Department of Health, Education, and Welfare for programs aimed at increasing the participation of women in science."[43]

Flemming's speech was followed by three sets of panels, including presentations by crystallographer Betty Wood of Bell Labs, psychologist Anne Steinman of Hofstra College, and chemist Ethaline Cortelyou, as well as a witty evening address by Betty Lou Raskin of Johns Hopkins University.[44] Delighted at this seemingly strong support for their goals, several women and men met after the conference to form a national organization with three regional sections (in New York City, Philadelphia, and Washington, D.C.) to discuss the issue further. Within a few months they had obtained startup funds of two hundred dollars from the AAAS, and by the spring of 1959 they had prepared a grant proposal for both the NSF and NIH. The group sought about seventy-three hundred dollars for a center whose staff would take all steps possible to urge women to enter science and would work both to liberalize attitudes toward them and to end discriminatory practices against them in employment and in graduate schools. There was also some discussion of holding a White House conference on women in science to dramatize the issue. Before long, however, the group discovered that Flemming's assurances of support were baseless, for their proposal was rejected on unspecified "technical grounds," which is very strange given that one of the active members of the group (of the several from NIH) was Ernestine Thurman of NIH's Division of Research Grants. One would have thought that with her expertise the group would certainly have been able to submit a proposal that met all of NIH's "technical" specifications.[45] Thereupon the group reformulated itself into the American Council of Women in Science and made plans to start a fund drive and resubmit a modified proposal later, which apparently never happened. Active in this short-lived group were several prominent women scientists, including leader microbiologist Mary Louise Robbins from George Washington University Medical School; Helen Dyer, Ernestine Thurman, and Elizabeth Weisburger, biologists from NIH; archaeologist Harriet Boyd from the University of Pennsylvania; and chemist Ethaline Cortelyou, then of the Aerojet Corporation in Frederick, Maryland.

The council survived long enough to hold a "second annual" meeting at the AAAS's next convention, in Chicago in December 1959. Its invited speaker that year was Alan T. Waterman, director of the NSF, who delivered a major address (undoubtedly written by his assistant Lee Anna Embrey) entitled "Scientific Womanpower—A Neglected Resource." With occasional sympathetic flourishes (such as quotations from Virginia Woolf, the English feminist writer), the address presented a sugar-coated version, perhaps even a caricature, of the standard view of manpower thinking about women scientists in 1959. Waterman even went so far as to admit that there were certain problems but then shrank from any remedial action either by the NSF or any part of the federal government in general, since women and society were already correcting them (though no evidence was given). Waterman started off by agreeing that there were prejudices in science and that women deserved our "whole-hearted encouragement."[46] Because, however, today's "opportunities are practically boundless" for them, "it is simply a question of time until women come into their own in science as in other fields."[47] For example, because little girls were not brought up to be scientists, it was necessary to change attitudes. This must begin at home, with teachers, and through vocational guidance. (In this regard, he noted, there had been some women among NSF's visiting lecturers and among its fellows; statistics would be published soon.) In addition private groups, such as this council, should continue their efforts to "educate the public on women in science." Betty Lou Raskin's recent article "Woman's Place Is in the Lab, Too," in the *New York Times Magazine,* was a good example. (It was reprinted in *Reader's Digest* and distributed by Goucher College in a special bulletin.)[48]

On the more controversial subject of employment, Waterman agreed that some employers did not think that women scientists were good risks, since they would leave to marry and raise a family. Some women even believed that there was active discrimination against women in science careers, evidenced in less frequent employment and advancement, separate salary scales and titles, restrictions on employment of couples in the same university, and a ceiling on the advancement of women. But that was "part of the larger social pattern," which was changing, and "marked improvements can be seen."[49] Meanwhile, women were well suited to and being actively recruited for the emerging field of scientific information and for science teaching. They also made good researchers either alone or with their husbands, good assistants, and good laboratory technicians. Of course if a woman was good enough, she would reach the top "regardless of the obstacles that may be put in her way,"[50] as had Lise Meitner and Nobelists Marie Curie and Gerty Cori. Administrators could remove some obstacles, but they could not pave the way to greatness. That was up to the women's own efforts. Soviet women scientists and engineers currently outnumbered American women in many respects. But Waterman was convinced that women did not want preferential treatment in science, just "equal opportunities."[51]

Waterman had thus summed up as gently as possible the federal government's "womanpower" philosophy of the 1950s, namely, that women were to be encouraged to go into science and engineering, but neither institutions nor employers were to be forced to make their reception any easier or to treat them "equally." Male speakers, therefore, placed the burden totally upon the women, especially the young women, by urging them to defy sexual stereotypes and for patriotic reasons to enter traditionally male fields that did not necessarily want them. There was more than a little hypocrisy in such a stance, since the government itself was neither a model employer nor willing to penalize discriminatory employers. By thrusting upon its young people the planners' simplistic and idealistic vision of a world of employment based only upon talent and brains and not upon sex or even race, the government thereby both unleashed some powerful expectations and motivations and postponed any eventual reconciliation. The eventual resolution of the tension between this vision or official "policy" and reality was one of the roots of "women's liberation" a decade later.

The following year, at the 1960 AAAS meeting in New York City, several members of the American Council of Women in Science spoke at a session cosponsored by SDE entitled "Encouraging Women to Select and to Advance in Scientific Careers," but after this the group dropped out of sight. There is no evidence that its grant proposals were ever funded.[52] Although the American Council of Women in Science failed in its stated goals, one wishes that it had survived a while longer or gotten one of its grants funded, in order to see what it might have accomplished in the early 1960s. It had some very influential friends in Washington and could conceivably have emerged as a strong force advocating a greater role for women in science, which was presumably what high governmental officials wanted. But in a sense the group's formation and existence for a brief time helped to expose the policy planners' ambivalence and hypocrisy about wanting women in science. They claimed rhetorically at least that women were needed in science and that an all-out effort should be made to recruit them. One even offered the resources of his funding agency. Yet when a responsible coalition of organizations with strong ties to the affected group came forward with a proposal to do exactly what the government said it wanted, it was cut off with nothing on rather weak procedural grounds. Once again the government's ambivalence and hidden opposition to actual results was all too clear. Even Arthur S. Flemming must have been a bit chastened after this, though the NSF made some grants to the AAUW and in 1961 put out a booklet of statistics on women in science.

After the collapse of the council, the SDE's many chapters moved to work locally toward some of the same goals of increasing women's access to and participation in science. To this end Meta Ellis, then of Aerojet-General Corporation of Sacramento, California, formed and chaired an SDE committee on the encouragement of women in science.[53] She and others brainstormed about how the national group and its local chapters might join in various kinds of outreach to young women interested in scientific careers. Through the national newsletter and correspondence, they urged chapters to take part in local science fairs, to publicize and meet with women winners, and to hold career days, as well as to publicize themselves more in the local press.[54] This revived some chapters, such as the Alpha chapter of Ithaca, New York, that were hovering on the brink of inactivity, because few female graduate students, their traditional clientele, had been joining. But in 1966, under an energetic leader, the Ithaca branch moved into the realm of providing a job roster for trained women in the area. Many were young mothers who sought part-time work or short-term assignments that would utilize their technical skills, but Cornell University departments were often not aware of them and did not know how to reach them. Several SDE members volunteered at the university personnel office, sent out questionnaires to local women's groups, and in time (with some modest support from Cornell) managed to compile a list of more than 100 such women and about 150 such jobs. The chapter's success inspired the national SDE to try to duplicate it, but in its hands (largely in Washington, D.C.) the part-time roster quickly became tied to efforts to advertise federal civil-service positions and by 1968 to the need to find women panelists for governmental advisory committees.[55]

Thus SDE, better than Omicron Nu or even the large Iota Sigma Pi, was well positioned through its links to mature women, the AAAS, and its local chapters to play a prominent role in representing women scientists in general. With the help of other groups, it got the topic of women on the agenda of national meetings, which led to some publicity and rallying together of interested allies. SDE was for a time even part of a national alliance that tried to do even more but whose very failure demonstrated the limits and ambivalence of the government's womanpower campaign. In addition, through its extensive network of local chapters at graduate universities and in metropolitan areas SDE was able to accomplish a lot at the grass-roots level.

Like the Garvan Medal of the ACS, what came to be known as the Annie Jump Cannon Prize of the American Astronomical Society for distinguished contributions by a woman was established in the 1930s and given for several decades thereafter, even though this society did not have a woman's committee. The prize's forty-year history reflected first the usual situation where women of accomplishment were being repeatedly passed over by a society for its major and even minor prizes, then the action of a donor/strategist who understood the political role of the special compensatory women's prizes, but finally (ca. 1970) later women's sense of embarrassment at the patronizing aura of a prize restricted to one sex.

The prize was first given in 1933, when astronomer Annie Jump Cannon, longtime "computer" at the Harvard College Observatory, won the last award given by the Association to Aid Scientific Research by Women. Claiming that women's problems in science had been solved, that association dissolved in 1932; but Cannon, who was not so sure about women's fate in astronomy, donated her award of one thousand dollars to the American Astronomical Society to endow an international prize for "distinguished contributions to astronomy by a woman." Upon her death in 1941, the society renamed it the Annie Jump Cannon Prize in her memory, to be awarded every third year (because the endowment was unable to support an award every year and because, or so it was claimed, there were, alas, too few women in the field to merit an award every year). The first winner was Cecilia Payne-Gaposchkin, also of the Harvard College Observatory, and she was followed by a series of others into the 1960s, when various lapses began to occur. The 1961 prize was not awarded until 1962, whereupon the winner, Margaret Harwood of the Maria Mitchell Observatory on Nantucket Island, complained that the fifty-dollar prize was too small and the accompanying brooch unattractive. The 1964 (or by then 1965) prize was not awarded until 1966, but in that year past winner Helen S. Hogg of the University of Toronto ran a modest fund drive (starting with a canvassing of past awardees) that raised eleven hundred dollars to bring the prize up to the society's standard for other prizes, one hundred dollars. In 1968 Henrietta Swope, once at Harvard but since 1952 at the Carnegie Institution on Mounts Wilson and Palomar, became the twelfth and final recipient. Independently wealthy (she was heiress to a General Electric fortune), she was tactful and unconcerned about the money, writing a friend that the best part of winning the prize was the many letters she got from well-wishers. Yet surely it meant more, since it was the only professional prize (aside from a medal from Barnard College, her alma mater) that she ever got for her four decades of contributions to astronomy.[56]

But then the nominee for the award in 1972, E. Margaret Burbidge, the only woman in the history of the AAS to have won even half of one of the society's other prizes (in 1959 she had won the Helen Warner Prize jointly with her husband Geoffrey for significant achievement in astronomy in the five previous years), surprised everyone by declining the award on the grounds that it was inconsistent to fight for equal rights for women in astronomy (as she was then doing) and at the same time to accept a prize restricted to women. She also complained, perhaps in jest, that there were still so few women in the field that before long they would each have won it. In response the society set up a special committee on the Cannon Prize (as well as the status of women in astronomy in general), which recommended that the prize be turned into a fellowship for a young woman in astronomy, since there the special restrictions of a separate award for women was somehow acceptable. The society then sent the Cannon endowment to the AAUW Educational Foundation to let them administer the new mini-fellowship.[57]

Largely outside science but including it at times were a whole set of other honors for women. Among these were the several "woman of the year" awards (or WOTYs, as journalists called them) given by metropolitan newspapers, chambers of commerce, women's magazines, and the women's national press associations. These often selected several women, each representing a particular area, such as philanthropy, the professions, volunteer work, education, government, and the arts. Usually there was a scientist or physician among them. WOTY awards at the national level grew in part out of public relations efforts by groups of women journalists. In 1946 the Women's National Press Club of Washington, D.C., chose Lise Meitner of Stockholm, famed co-founder of nuclear fission and visiting professor of physics at the Catholic University of America, as its woman of the year. The photograph of her at dinner with President Harry S. Truman, where they may have discussed the atomic bomb, appeared in newspapers around the country.[58] In 1948 Nobelist Gerty Cori and economist Dorothy Brady were among the eight women so honored, and in 1950, when the same group chose biochemist Mildred Rebstock of Parke, Davis and Company, whose synthesis of Chloromycetin, an early antibiotic, had been in the news that year, as one of its six awardees, she was called the "Science Woman of the Year."[59] When in 1951 the U.S. Chamber of Commerce selected twenty-five "women of achievement" "for their accomplishments in what demonstrably is far from being a man's world," a group that included engineer Lillian Gilbreth, General Electric physicist Katharine Blodgett, and penologist Miriam Van Waters, their group photograph appeared in *Life* magazine, which went to millions of homes across America.[60] Three years later the multihonored Gilbreth was among the six "most successful women of 1954" selected by the editors of the *Woman's Home Companion*.[61]

The message beamed to American women and their families—attention to and approval of dignified women working outside the home—was perhaps a bit daring during the "back-to-the-home" movement of the time.

At the local level the *Atlanta Constitution* honored biologist Mary Stuart MacDougall of Agnes Scott College in 1944 and biochemist Jane Russell (Wilhelmi) of Emory in 1960, the *Syracuse Herald Journal* chose economist Marguerite Fisher in 1952, the *Denver Post* featured chemist Essie White Cohn in 1961, the *Los Angeles Times* honored mathematician Olga Taussky (Todd) in 1963 and plant taxonomist Mildred Mathias of UCLA in 1964, and the *Columbus (Ohio) Citizen-Journal* acclaimed chemical abstractor Mary Magill in 1964. The award ceremonies usually involved an elaborate dinner and extensive local publicity, usually on the "society" or "women's" pages, where journalists demonstrated local pride and boosterism and also courted women readers. But the honors benefited the recipients as well, and in ways that more prestigious mainstream prizes did not. Taussky, a research associate at the time, later reported that she particularly enjoyed this award because "I knew that none of my colleagues could be jealous about it (since they were all men), and that it would strengthen my position at Caltech."[62]

Another kind of honor won by some women scientists in these years was an honorary degree from a woman's college. Some such institutions gave them chiefly or only to their own alumnae, retiring faculty, or former administrators; others had no such restrictions. Some colleges awarded them annually, others only on special occasions, such as Radcliffe College upon its seventy-fifth anniversary in 1954.[63] Even the women's colleges that were so intent on hiring male faculty could, as Smith College did, still give from four to seven honorary degrees to accomplished women in a variety of fields.[64] Those women who appreciated such awards the most were alumnae who did not already hold doctorates, such as Mount Holyoke's honorees ornithologist Margaret Morse Nice in 1955 and high school chemistry teacher Dorothy Gifford of Providence in 1957.[65] A few other women felt that certain of these degrees were not worth the cost or the bother. Microbiologist Rebecca Lancefield of the Rockefeller University, for example, turned down Western College for Women in Oxford, Ohio, on two occasions, first in 1966, on the grounds that she would be on sabbatical and possibly outside the country on commencement day, and then in 1968, because her retirement salary was so small that she could not afford the trip to pick up the honor. Yet she was apparently quite pleased later to receive an achievement award from her alma mater, Wellesley College.[66]

In science there is another kind of honor that has not been examined systematically here—that of eponymy, or the naming of things or phenomena for persons. Numerous plants, animals, other organisms, mountains, astronomical objects, and more have been named for women scientists over the years. At least two clubs have been named for women—in Toronto in 1952 the Margaret

Morse Nice bird club for women only (since they were excluded from the Toronto Ornithological Club) and an informal group interested in certain kinds of bacteria (streptococci) that changed its name in 1977 to the Lancefield Club, in honor of microbiologist Rebecca Lancefield. Elsewhere a few all-women science or study clubs persist, such as the Women's Research Club at the University of Michigan (see ch. 6) and the Eistophos Club of Washington, D.C., started in the 1890s, which in 1968 celebrated its seventy-fifth anniversary with a dinner at the Cosmos Club and a volume of essays by members. Among them were women scientists in the government (e.g., astrophysicist Charlotte Moore Sitterly, recent winner of the Federal Woman's Award) and wives or widows of prominent Washington scientists and administrators (e.g., Mary Waterman, Virginia Abbot, and Mary Libbie Dryden).[67]

Although the AAUW's efforts at battling sex discrimination in higher education had largely petered out by 1963 and its board of directors refused to spend AAUW funds on Eleanor Dolan's Roster of Women with Advanced Degrees, it continued to put its money and staff time into pre- and postdoctoral fellowships. This strategy benefited many individual women scholars and scientists, for the AAUW's extensive program of postdoctoral fellowships far outstripped the efforts of all the other organizations and committees mentioned in this chapter. Between 1940 and 1970 the AAUW awarded 318 postdoctoral fellowships, of which 113 (or 35.65 percent) were in natural, physical, or health sciences and another 90 (28.39 percent) went to those in the social sciences, psychology, education, (4), and law (1). These 203 women postdoctoral fellows in the sciences and social sciences would compare very well with the 113 Guggenheim fellowships that went to American women scientists in these same years, the 128 NSF postdoctoral fellowships, and the mere 21 NSF senior postdoctoral fellowships. Among its winners, who ranged from recent Ph.D. recipients on up to older women at women's colleges or research institutes, were biochemist Jytte Muus of Mount Holyoke, mathematician Josephine Mitchell (Schoenfeld), who was leaving the University of Illinois because of antinepotism rules, botanist Geneva Sayre of Russell Sage College, paleontologist Tilly Edinger of Harvard's Museum of Comparative Zoology, political scientist Judith Shklar at Harvard, and economist Carolyn Shaw Bell at Wellesley. Thus through its postdoctoral fellowship program the AAUW was creating and offering professional opportunities to bright, ambitious, and accomplished women at a time when other, larger fellowship programs that were presumably open to all were making 95 percent or more of their awards to men.[68] The AAUW postdoctoral fellowships in a quiet, understated way constituted a kind of female Guggenheim that started or sustained many women scientists in research careers.

The most important women's prize in these years was the annual AAUW Achievement Award, supported by the master fundraisers of the association's

Table 15.2.

Winners of the American Association of University Women Achievement Award, 1943–1969

Year	Name	Field	Previous Awards	Subsequent Awards
1943	Florence Seibert	Biochemistry	Guggenheim, Garvan	
1944	Gisela Richter	Art history	APS	Garvan
1945	Katherine Blodgett	Physical chemistry		Am. Acad.
1946	Ruth Benedict	Anthropology		Am. Acad.
1947	Barbara McClintock	Genetics	Guggenheim, NAS, APS	
1948	Sirarpie Der Nersessian	Byzantine art		Am. Acad.
1949	Helen C. White	English	Guggenheim	
1950	Elizabeth Crosby	Neuroanatomy		
1951	Mary Swindler	Classical archaeology	APS	
1952	Lily Ross Taylor	Latin	Guggenheim, APS, Am. Acad.	
1953	Mabel Newcomer	Economics		
1954	Marjorie Hope Nicolson	History of science	Guggenheim, APS	Am. Acad.
1955	Rosemary Tuve	English		Am. Acad.
1956	Rachel Carson	Nature writing	Guggenheim	

Year	Name	Field		
1957	Cecilia Payne-Gaposchkin	Astronomy	APS, Cannon Prize, Am. Acad.	
1958	Maria Malkiel	Philology	Guggenheim	
1959	C. S. Wu	Physics	NAS	
1960	Lily Campbell	English	Guggenheim	
1961	Cora DuBois	Anthropology	Am. Acad.	Am. Acad.
1962	Gwendolen Carter	Political science		NAS, APS
1963	Helen Taussig	Medicine	Am. Acad.	
1964	Mildred Campbell	History	Guggenheim	
1965	Mina Rees	Mathematics		
1966	Dora Dougherty	Human factors engineering		
1967	Lucy S. Morgan	Public health		
1968	Eveline Burns	Economics	Guggenheim	
1969	Dorothy Weeks	Physics	Guggenheim	

Sources: Idealism at Work: Eighty Years of AAUW Fellowships, A Report by the American Association of University Women (Washington, D.C.: AAUW, 1967), 289–91; *Idealism at Work: AAUW Educational Foundation Programs, 1967–1981, A Report by Doris C. Davies* (Washington, D.C.: AAUW, 1981), 363–64; John Simon Guggenheim Memorial Foundation, *Directory of Fellows, 1925–1974* (New York, 1975); annual report of NAS, American Academy of Arts and Sciences, and APS, as given under table 14.7.

Note: Am. Acad. = American Academy of Arts and Sciences.

northwest central region from 1943 until 1972, when it was taken over by the
AAUW Foundation. The need for this award was explained at its first presentation ceremony by AAUW President Helen C. White (herself later an awardee):
for many years the AAUW fellowship committee had felt frustrated that so
many (nearly all) of their awards were going to young women just embarking
on research projects. There were as yet no general prizes for mature women
who were completing long careers of distinction.[69] Perhaps the AAUW members were influenced by the example of the Annie Jump Cannon and Garvan
prizes as well. The winners of the annual AAUW prize from 1943 to 1969–70
included eighteen scientists and social scientists (see table 15.2); the amount of
the award was $2,500 at first and was raised at some point to $3,000. In the
days before federal research grants this was a princely sum. Ruth Benedict,
the 1946 awardee, immediately put it to use for some ongoing fieldwork,
and neuroanatomist Elizabeth Crosby, the 1950 awardee, intended to use
it, as she explained in her acceptance speech, for two pieces of equipment as
well as a chimpanzee, "which has been a cherished dream of mine for many
years."[70]

The winners of this award constituted perhaps the single best list of top
women scientists and scholars of this period; they would have made an impressive small honorary academy of their own.[71] Several already held other
rare honors: eleven had held Guggenheim fellowships, ten were members of
the American Academy of Arts and Sciences, six were already members of
the APS (which elected one other, Helen Taussig, in 1973), and two were
members of the NAS (McClintock and Wu, and Taussig was added in 1973).
Two had won the Garvan Medal (Seibert and Blodgett) and one the Annie
Jump Cannon Prize (Payne-Gaposchkin). At the other end of the spectrum,
lest one think the AAUW honored only those already well recognized, three
winners—anatomist Crosby, economist Mabel Newcomer, and public health
educator Lucy S. Morgan—had received none of these honors. Perhaps they
had never applied for a Guggenheim fellowship or had been so situated either
institutionally or disciplinarily that they were not noticed by the major
academies. Even the AAUW, however, missed such multihonored notables
as Gerty Cori, Libbie Hyman, Maria Goeppart Mayer, Lillian Gilbreth,
Icie Macy Hoobler, Katherine Esau, and others on previous lists. Nor did
they include any psychologists (not even the research-oriented Eleanor Jack
Gibson) or home economists, despite women's many accomplishments in
these fields. Instead the AAUW concentrated on the accomplished scientists
at research institutions and humanists often at women's colleges. The latter
were generally already fully aware of the importance of AAUW fellowships
either for themselves or for their students. Perhaps one reason for honoring
once again so many of the already multihonored scientists was to remind them

that there was a woman's culture that was supporting many other women aspirants to careers like theirs. Another reason was undoubtedly to remind their own AAUW donors that there were women at the very top of some fields.

Thus between 1940 and 1968 there were several organizations of women scientists, engineers, and home economists and a few prizes and fellowships for them. Some of these groups were reshaping their traditional undergraduate structure to serve alumnae as well. The chief activities of most of them were to hold social events, encourage mutual discussions, support newsletters, elect honorary members, and award prizes and occasionally fellowships. One group, the National/International Council of Women Psychologists felt so uncomfortable with even these traditional activities that it survived between 1945 and 1968 only by allowing itself to be refashioned as an international organization for men and women. Somehow psychologists of both sexes, unlike geographers, engineers, chemists, and others, were so intimidated by a professedly single-sex organization that they let the ICWP be completely integrated by the late 1960s. Two other groups, however, the SWE and Sigma Delta Epsilon, not only had no such paralyzing doubts but instead pressed on to capitalize upon the government's "womanpower" theme. One, small and for the most part young, fit the needs of the manpower analysts by helping governmental bureaus, which already had their own missions and staffs, collect data, write reports, and sponsor sessions, but in return it received some federal funds and a presidential message for its own international congress in New York City in 1964. The second, larger and made up of older, more experienced women and centered in Washington, D.C., even confronted and called the bluff of these very manpower analysts, who were so flush with money for so many other projects in the late 1950s. The women prepared a proposal to carry on the government's message, only to be rebuffed and diverted into local, grass-roots channels.

For the most part, then, these women's groups were, to use a term of the time, "coexisting with" rather than seeking to reform other, more dominant groups within science and engineering. They were struggling to create a little mental space and mutual support (later called "sisterhood") for isolated women doing atypical things. Of all the ventures, however, the most successful was the quiet, continuing, even expanding, philanthropy of the AAUW. Over the years it had considerable impact, as women's need for its postdoctoral fellowships and other awards increased rather than decreased in the age of Big Science and Big Philanthropy.

Yet as attitudes changed in the mid- to late 1960s, these established organizations reacted differently to the sense of protest that erupted so suddenly and angrily in 1969. In some cases these older organizations were criticized and dismissed as trivial and ineffective groups that were themselves part of the problem for women in science; in others they responded sympathetically to the new emotion and pitched in to take part. It is to the roots of this passage from prizes to protest that we turn next.

16

The Path to Liberation

Consciousness Raised, Legislation Enacted

After a few sporadic efforts in the 1950s and some deliberately weak governmental reforms in the mid-1960s, from 1968 to 1972 many women scientists, especially social scientists and research associates, found first their individual and collective voice of protest and then allied to press for political change. Electrified by even the prospect of change, they feverishly signed petitions, formed caucuses and "consciousness-raising" groups, joined marches, wrote checks, collected data and prepared reports, or otherwise worked for institutional reform in the workplace and in their scientific societies. Although often individuals could not say why the old ways were suddenly so intolerable, or the long-accepted rationalizations no longer satisfactory, formerly isolated people rapidly came together to form a women's movement in science and engineering. Energized by their new-found empowerment, the right women suddenly appeared as if out of nowhere in the key places, ready to take the movement the next step. A series of much-publicized gender encounters throughout the years 1970–72 allowed a series of indignant feminists to come forward, express their outrage, and energize still more followers. Among these events were the federal government's first sluggish steps to enforce its much-amended executive orders for federal contractors. In fact the universities' delaying tactics backfired here, for the press coverage of their repeated refusals to comply and endless stalling in providing equal pay convinced many, including some in Congress, that actual legislation was needed. Thus, momentum grew, and many women scientists joined with other professional and working women to push Congress into landmark legislation (including the Equal Rights Amendment) by 1972. Still others, often the best-placed faculty women, could not understand what the excitement was all about and wondered whether their professional society really needed a "committee on the status of women" and why it had put them on it. Yet by the time President Richard Nixon signed the Equal Employment

Opportunity Act of 1972 and the Education Amendments Act that spring, the "Ph.D. glut" had dawned, and for the first time in thirty years physicists and then other scientists could not find jobs. Thus the legislation that promised to transform women's experience in one part of science significantly forever might be threatened by a new depression. One era had clearly ended, and a new, more embattled one was beginning.

The first step was to identify and articulate the structural problem that women scientists were facing. Just how difficult it was for a moderately successful and aware woman scientist around 1960 to have a clear, consistent viewpoint on women's status in her field is evident from Dorothy Weeks's article "Women in Physics Today," which appeared in August 1960. She wanted to be optimistic and loyal to her alma maters, but the limited and outdated data at her disposal documented instead considerable evidence for pessimism. It was hard to know what to applaud, what to accept, what to minimize, and what to criticize. For example, even though the evidence indicated that women had not kept pace with men in the postwar expansion of the field, was it not better, when one was permitted a few pages in a widely read news journal like *Physics Today*, to be as upbeat as possible, unlike the many authors that blamed women for their low participation in science? Should one not try to encourage future women to enter the field, even if this required putting a positive "spin" on the actual data? Weeks tried to show how much better women were doing now than earlier, even though most of her actual data indicated the opposite: The number of women fellows in the American Physical Society was up since 1941 (from 21 to 30) even though their proportion of the greatly increased total (from 901 to 1,624) was down by 20 percent (from 2.32 percent to 1.85 percent). When her educational data revealed that between 1949 and 1953 women earned 4.6 percent of the bachelors' degrees but just 1.7 percent of the doctorates in physics, she tried to find optimism in the relatively high proportion of these undergraduate women coming from women's colleges (including especially her alma mater, Wellesley College) and the recently rising proportion (since the dismal 1.4 percent of 1949–53) at MIT, her doctoral institution. Finally, she was proud to report that MIT had recently received a $1.5 million contribution for an as yet unbuilt women's dormitory.[1]

By contrast, Sylvia Fleis Fava's 1960 article prominently placed in the *American Sociological Review*, "The Status of Women in Professional Sociology," was clear and analytical, though not angry. Unlike other articles of the time (and a prophetic example of what was to come), it contained considerable systematic data and had a serious, consistent, even concerned, viewpoint. Using extensive data for the years 1949–58 (degrees awarded, including comparisons with other fields; participation rates at meetings; books, articles, and reviews published in two major journals; median salary and age by level of education and type of employment), Fava demonstrated women's decreased involvement in sociology at each stage of a career: although women constituted about 55 percent of the undergraduate degrees in sociology, they were only 15 percent of the doctorates and 7 percent of the participants at annual meetings of the American Sociological Association, of which only one woman (Dorothy Swaine Thomas in 1952) had ever served as president. One result was that women overall had lower median salaries than the men despite higher median ages. Fava, who had recently earned her doctorate from Northwestern University (1956) and was on the faculty at Brooklyn College, attributed this progressive disengagement not only to the "pull" of marriage and domestic life but also (and this was done less often in the 1960s) to the "push" of discouragement in graduate school, mentors' lack of interest, and limited job opportunities. But old patterns might be about to change: an expansion in college faculties was expected, and more and more married women were remaining in the labor force. Fava thought that sensible sociologists should be discussing the issue and taking "appropriate action," because the phenomena were of sociological as well as ethical or professional interest. For lack of space, she could not pursue the reasons for the patterns she depicted, which were so familiar to people in the field but "less obvious to students and newcomers."[2] Many women in the field, including Alice Rossi, must have read her article as they tried to make sense of their own experiences. While it may have been reassuring to realize that one was not alone and to see an overall pattern to the "weeding out," it may have been upsetting and radicalizing to learn that it was quite so sex-specific. Within a few years others would build on Fava's pioneering study, for Betty Friedan was among those who cited the article.[3]

Similarly insightful, with a bit of anger as well, was a set of articles in 1962–63 by anthropologist Ethel Alpenfels, a professor at the New York University School of Education. Alpenfels, who had won many awards, including, even though she was white, a woman-of-the-year award from the American Association of Negro Women in 1955, had served as a consultant to an educational subcommittee of the PCSW. Too radical for its reports, her articles, which appeared first under the bland title "Women in the Professional World" in a 1962 anthology and more pointedly as "The 'World of Ideas'—Do Women Count?" in the *Educational Record* of 1963, described and deplored the declining opportunities for women in educational institutions in the previous ten to fifteen years.[4] Theodore Caplow and Reece McGee had accurately reflected their times when in their 1958 book *The Academic Marketplace* they had deliberately left women out, for women faculty had always been rare at prestigious universities. What was even more alarming to Alpenfels was that they were less common in the early sixties in those very positions where they had once been numerous, such as on faculties at women's colleges or schools of education (back when they had been known as "normal schools"), as deans of women, or even as school principals. In fact at a time when women's college enrollments were rising, sharply at New York University, many women could expect only clerical jobs rather than professional ones upon graduation. This was just one more sign of the whole pattern of recent years.

> In a thousand subtle ways, a young college woman learns that women are less capable, less intelligent, less serious, more emotional, and less important than men. It propels her toward "soft" courses, pushes her into "women's" occupations and away from science, and finally, closes the door on many kinds of work for which she may well be suited. Whether recognized or unrecognized, this powerful and pervasive ghost sits at every conference table, in every committee meeting; it hovers in lecture halls and in counseling offices and is part of the college woman's experience, day in and day out.[5]

Everyone's experience, Alpenfels concluded, was diminished by these prejudices, for students needed to work and study with instructors of both sexes. They needed to learn that whether persons were emotionally stable, intellectually curious, and capable of handling ideas had nothing to do with their sex. Yet she herself, one of the very few of her time to see the constricting of opportunity and to write about it, seems to have done little more than identify the problem. Certainly the conclusion to her second article underestimated the enormity of the task she was suggesting, for it merely called on administrators to reform themselves: "College leaders, if they are really serious, can no longer escape their responsibility. A little housecleaning is in order."[6]

Starting about 1963, the level of anger in writings about women in science and women in general began to increase, coalescing before long into a modern women's movement of increasing political clout. One easily identifiable starting point of this new era was, of course, Betty Friedan's best-selling critique of the back-to-the-home movement of the previous fifteen years, *The Feminine Mystique,* published in 1963 but started several years earlier. Friedan, who had graduated summa cum laude from Smith College in psychology in 1942 and had been a graduate student at Berkeley for a year thereafter, had since given up a career in psychology for domesticity in the suburbs. In addition to raising her three children, she had kept up a sideline as a freelance writer and volunteered to write for her college's fifteenth reunion in 1957 a report on what the bright women of 1942 had done since graduation. Horrified by their collective loss of identity ("the problem that has no name"), she published an article entitled "I Say: Women are *People* Too!" in the September 1960 issue of *Good Housekeeping* magazine. Overwhelmed by the response and pushed by a publisher, she commuted biweekly to the New York Public Library to make a full book out of it. Within a few years she had produced a documented and angry critique that externalized and blamed society for making so many of its women "just a housewife," as she found herself responding to the 1960 census taker. Excerpted in several women's magazines, such as *Mademoiselle, Ladies Home Journal,* and *McCall's,* the book had both an immediate and a long-term impact.[7] Women read about the empty lives of other educated women, identified with the phenomenon, talked about it with others, and then one by one began to act on it, as various career days and reentry programs began to spring up. Then once in the labor force, where they faced the traditional, unreformed, sexist, and discriminatory practices of the workplace, some of them began to take the next step and organize for governmental reforms.[8]

A fourth, and here the most important, female social scientist to document and decry in the early 1960s the situation that even well-trained women were not getting the professional opportunities they deserved was sociologist Alice S. (Kitt) Rossi. A 1947 graduate of Brooklyn College, she had escaped a stifling first marriage, remarried in 1951, and while holding a series of temporary jobs managed to complete her doctorate from the prestigious Columbia sociology department in 1957. After giving birth to three children in 1955–59, she took on a series of lecturer and research associate positions around the University of Chicago, where her husband, also a sociologist, was on the faculty and directed its NORC. This sub- or nonfaculty status did not bother her initially, as she recounted later, because it was the norm there and then. But one fateful day in the early 1960s she was quickly radicalized. Because according to the university rules Rossi could not as a research associate submit a grant proposal to the NSF in her own name, she had convinced a male faculty member in anthropology to send it in for her. He had obliged but then betrayed her. When the grant was funded, he decided to keep the money and try to do the work himself, and he fired her! Angry, stunned, and hurt by this deliberate exploitation, Rossi began a wide-ranging and intense rethinking of the current version of men's and women's jobs and roles in society. Already interested in civil rights and something of a reformer/activist in the 1950s, when she conducted a survey on the lack of open housing in Cambridge, Massachusetts, before long she had become a confirmed feminist, convinced that the only way to prevent women's marginalization into childcare and underemployment was to change society drastically.[9]

As a visible sociologist with an interest in sex roles, Rossi was invited to various meetings in the early 1960s where the increasingly fashionable topic of women's "role conflicts" was being discussed. She deliberately made good use of these opportunities to reach an audience outside the field of sociology to embark on what turned into nearly a decade, if not a whole lifetime, of vigorous activism. In particular, when she was invited in 1963 by the editor of *Daedalus,* a prestigious journal of some impact in learned circles, to contribute to a special issue on American women, Rossi spent about a year pulling together and revising six times a lengthy and iconoclastic essay. (She later said that it was a more difficult gestation than any of her pregnancies.) It grew out of her wide reading in historical as well as sociological sources (including Betty Friedan's book), her considerable personal experience as a graduate student and academic wife and mother, and her willingness and ability to reconceptualize standard sociological views. Provocatively entitled "Equality between the Sexes: An Immodest Proposal," her essay criticized the prevailing narrow conservatism of most sociology (including that of Harvard guru Talcott Parsons) and envisioned a future America with androgynous sex roles. There not only would many laws have been reformulated in line with feminist views, but there would be at the very least a network of daycare centers, especially at urban universities, where women would be readily trained in reentry programs; sprawling suburbia would be replaced with apartment buildings closer to both parents' jobs; and reformed primary and secondary schooling would have abolished the widespread sex-typing in personnel and subject matter. Since such a transformation would require many changes in all aspects of daily life, some would have to be brought about by governmental action.[10] When this essay caused a bit of a stir, she revised and reprinted part of it as the deliberately provocative "The Case against Full-time Motherhood" for *Redbook*, a woman's magazine, even though her fellow sociologists criticized her publishing so far outside "the field."[11]

In October 1964 Rossi was one of several speakers at a national symposium at MIT on the specific topic of women in science and technology. The occasion, historic in its own way, was to celebrate the opening a year earlier of MIT's first on-campus dormitory for women students, McCormick Hall. Although MIT had been coeducational since 1870, when it admitted Ellen Richards, a chemist who later founded the field of home economics, it had housed its women in an apartment house across the river since 1945. Because so many of them were unhappy there and in general and subsequently dropped out, MIT had nearly terminated its coeducational status in the mid-fifties. But this was averted by some preventative action by the alumnae, an important decision by top officials at MIT about 1956, and a pledge in 1960 from feminist alumna Katherine Dexter McCormick of first $1.5 million (later raised to $2 million) for a two-hundred-bed dormitory for women.[12] Thus, the opening of McCormick Hall (named for her late husband Stanley), situated on the central part of campus along the Charles River, another case of "creative philanthropy" for women in science and engineering, marked the culmination of one phase of women's history at MIT and the beginning of a new era both there and possibly for women in science and engineering generally.

Perhaps in keeping with the occasion, the symposium was run by the women students of MIT. They chose a variety of speakers and panelists, including Radcliffe College president Mary Bunting, physicist C. S. Wu, Jessie Bernard, Mina Rees, Dorothy Simon, Lillian Gilbreth, Rita McCabe, and various others. But the chief speakers were two male psychoanalysts, Bruno Bettleheim and Erik Erikson, who were sandwiched around Alice Rossi. Among the attendees were 260 women undergraduates from 137 colleges across the country, including 3 black women from Tennessee A&I University. The high point for the audience, to judge by journalists' accounts, came the first morning when, after Bettleheim's opening address on the commitment required of a woman (first to her man and then only secondly to her work or science), he was sarcastically squelched by panelist Vivianne T. Nachmias, M.D., a mother and part-time research associate in anatomy, who thought her commitment to science was equally as strong as the men's in her field. She also questioned the relevance to women in science of recent experiments (which she ridiculed at one point) that showed that some girls liked playing with toys, which demonstrated (to a psychoanalyst at least) their interest in "inner space," while most boys liked to play with toys that seemed to show a predilection for "outer space." Maybe, she asked, Madame Curie should have worked on the interior of the atom and not the emission of radioactivity? And maybe male psychotherapists should not be concerned with love and harmony within the family?[13]

Rossi's lengthy address, buttressed with many data and charts, as well as citations to the literature, seems by contrast to have gone over the heads of the students but later pleased the editors of MIT Press, who considered it the only part of the symposium worth publishing. With the help of an assistant and an NIH career advancement grant, Rossi had dug into the growing literature on the time on scientific careers, interest development, and marriage satisfaction and tried to listen to what the respondents were actually saying as well to notice what interpretation the researchers had been stereotypically imposing on them. In particular she worked from a series of questionnaires sent by the NORC in Chicago to a large number of men and women college graduates of June 1961. From all this she determined that bright girls had a difficult time envisioning or preparing for careers or creative lives, for everyone, from parents to teachers, school counselors, and contemporaries, wanted them to be well adjusted socially rather than the intense, driven, and perhaps lonely (but still happy) people that top scientists tended to be. In 1965 Rossi published an abbreviated version of this paper in *Science*. Under the more provocative title "Women in Science: Why So Few?" she had, possibly influenced by Friedan's slant, reshaped her material into a leaner, more direct, and more powerful political critique. Yes, there were fewer women in science at every successive level, but society as a whole was to blame. Everyone expected them to drop out, felt more comfortable when they did, and even planned for it. This was a doubly wasteful loss: it was bad for the nation, which presumably needed all the bright scientists it could get, and it was bad for the women themselves, who were not using their talents to the fullest. If in the future these women were to develop their potential to the utmost, many parts of society needed to be changed. Continuing education and part-time work, the current panaceas, were not enough.[14]

But even this more pointed version of Rossi's thinking produced little immediate reaction. *Science* editor Philip Abelson, who occasionally discussed women in science in his editorials, printed two letters to the editor three months later, one from a recent graduate of MIT who said she had gone there despite much advice from counselors trying to dissuade her, and the other, from a woman who was apparently a mother, who said that Rossi had omitted women's important role in teaching their children how to use wisely their leisure, of which there would be a great deal more in the future. Yet the MIT symposium of 1964 may have had a certain delayed reaction, for it was still being read about and discussed several years later.[15]

By the mid-1960s a few other women scientists and social scientists were finding their voices and willing to complain and ridicule "the experts" openly. Perhaps it was Rossi's dissatisfaction with the status quo that emboldened Ruth Kundsin, a research associate in bacteriology at the Peter Bent Brigham Hospital in Boston, to write the outspoken "Why Nobody Wants Women in Science," published first in the *Harvard Medical School Alumni Bulletin* and then reprinted in *Science Digest* in October 1965. Kundsin claimed that women in science were not acclaimed for their achievements but rather were singled out for their oddities, were resented by other women, especially subordinates, and were considered socially inadequate if they were unmarried. It would take changes in behavior as well as laws for women to be fully accepted as scientists. As it was, they had to have a "hardy spirit" to stand up to the many obstacles they faced. After decades of articles that usually denied or minimized the problems women faced in science, this was a forthright departure.[16]

A certain defensive awareness was also rising among social scientists, including anthropologists Peggy Golde and Ann Kindrick Fischer, who spent the year 1965–66 at the Center for Advanced Study in the Behavioral Sciences in Stanford, California, Golde as a fellow and Fischer, also a Ph.D. in the field, as the spouse of a fellow. As Radcliffe Ph.D.'s of the 1950s who had experienced a bit of the job market themselves and had read Betty Friedan's book, they disliked both her assumption that because Margaret Mead was so well known the field of anthropology was particularly receptive to women and her blaming the women of the postwar era for withdrawing from graduate school. Taking advantage of the services of a statistician at the center to collect considerable data on the numbers and proportions of doctorates going to men and women in the field over several decades and to calculate various gender ratios, they published a joint report two years later, in April 1968, among the "Brief Communications" in the *American Anthropologist*, the major journal in the field. In it they linked declines in the numbers or percentages of doctorates going to women to discriminatory admissions at the many upwardly mobile graduate departments of the time that did not want to be known for having too many women around. They also collected a bit of information about employment, including a list of those graduate departments with ten or more full-time faculty that had no women. Because they thus spent a lot of time and space on getting the doctoral numbers, on which there was a large literature, precisely right and skimmed over the issues of employment and discrimination, on which there was at the time scanty evidence, their report lacked the focus and bite of the "status of women" reports composed in the angrier late 1960s. Yet their report had some impact among women anthropologists.[17]

Meanwhile, consciousness and even anger were emboldening other social scientists, especially younger women new to the profession and to the job market, as Fava had observed in 1960. In 1967 a pseudonymous article on the role of women at sociology meetings appeared in the *American Sociologist,* a newly created journal with articles about the profession. The short piece must have been written from personal experience, for it spoofed the primarily supportive emotional role women professionals were expected to play at such meetings. It reported several women's ambivalence (first elation but subsequent depression) when men in the field rushed over to greet them and then talked very rapidly about their own recent professional triumphs—articles and books accepted, grant proposals certain to be funded, and, in hushed tones, recent job offers. But somehow there never was time or equal interest for the men to reciprocate by listening to their account of their own productive year. Evidently few women, especially younger ones, were profiting equally from the mutual communication and informal interaction that presumably were the chief purposes of attending the national meetings. Although some of the women attending these sessions had expectations of full equality and were experiencing, noticing, and now criticizing semipublicly those sex-linked convention behaviors that put them at a disadvantage, "Geraldine R. Mintz" had no direct way in 1967 to do more than write a satirical article, send it off to a suitable publication, and hope that its thinly veiled criticisms would have some impact. Perhaps they did, for in retrospect the article can be seen as still another sign of a new outspokenness and simmering discontent among sociologists, whose "consciousness" increased rapidly thereafter.[18]

In October 1966 Rossi became a charter member of NOW and served on the board of governors until 1970. NOW was created by Betty Friedan and others as an "NAACP for women" to pressure the president to broaden Executive Order 11246 to include sexual discrimination and the federal government's new and weak Equal Employment Opportunity Commission (EEOC) to enforce existing provisions about sex discrimination in the Civil Rights Act of 1964. Before long several other women scientists and social scientists had joined NOW—clinical biochemist Inka O'Hanrahan of San Francisco, who organized chapters in both northern and southern California, psychologist Jo-Ann

Gardner in Pittsburgh, and sociologists Cynthia Fuchs Epstein in New York City and Sally Hacker, a Rossi student, in Houston (with husband Bart). Each chapter focused on its own issues; for example, the one in northern California worked on getting women admitted into the astronaut program, while one in New York City sponsored a task force headed by humanist Kate Millett that criticized the women's colleges. A chapter in Ohio, headed by Dr. Elizabeth Boyer, worked on economic and legal issues with educational institutions and later split off to form WEAL, the Women's Equity Action League.[19]

By 1968 there was also some evidence that the number of women in science and engineering, as counted by the NSF's National Register, had risen sharply since 1966, especially at educational institutions, in nonladder ranks, and in certain fields (especially sociology and biology). The National Register detected, for example, a 50 percent increase in the number of women employed full-time at educational institutions between 1966 and 1968, jumping from 9,656 to 14,505. The increases were especially large and disproportionate in the field of sociology (+510, or a dramatic 122.30 percent) and biological sciences (+1,869, or 95.94 percent), and there were lesser spurts in chemistry (+916, or 51.58 percent), mathematics (+627, or 50.98 percent), and psychology (+858, or 41.81 percent). These increases (making the women's proportion more than 12 percent in some fields) may have been a result of improved data collection, or they may been evidence of a "critical mass" behind the formation of women's caucuses in the late 1960s. Although none of the Register's data on academic ranks were broken down by gender, they showed substantial increases between 1966 and 1968 at the levels of research associate (from 219 to 780), research assistant (from 3,902 to 5,909), "other" (from 622 to 1,330), and "no report" (from 4,117 to 7,575), suggesting that many recent additions to the academic workforce were being stockpiled into these soft-money or temporary positions. Many of these may have been women, especially in the fields listed above.[20]

Also by 1968 the student civil-rights movement that had started with sit-ins in the South in the mid-1960s had begun to turn its interest northward, especially to antiwar issues, such as the draft (since deferments for male graduate students were ending), and to universities' involvement in ROTC and military research and recruitment. Women, who did not have draft cards to burn or draft boards to defy, felt marginalized by this new turn, and some began to examine women's role in the movement, at universities, and in their personal lives. For many of them this was an exhilarating experience, a kind of "liberation." Among these young women was psychologist Naomi Weisstein, a participant in several civil-rights protest groups, first as a graduate student at Harvard in 1964 and then as an NSF postdoctoral fellow at the University of Chicago. But it was was discovering that she was nearly unemployable in the

Chicago area, despite all her work and credentials, because of her marriage to a historian (Jesse Lemisch) then at the University of Chicago that caused her to find her articulate and satirical voice. Although she accepted a teaching position at nearby Loyola University, she then did what few women scientists (besides Kundsin) had ever done, publicly at least, before. She began to speak out against scientific attitudes in her field and prevailing employment practices in academia, angrily ridiculing them in such acerbic and humorous pieces as "How Can a Small Girl Like You Teach a Class Full of Big Men, and Other Things the Chairman Said." Her talks and writings began to made the rounds of informal "consciousness-raising" groups, which were springing up everywhere, and were printed and reprinted in the exploding underground literature of the women's movement. Her "Woman as Nigger, or How Psychology Constructs the Female" was published in *Psychology Today*, itself a new kind of popular scientific journal, in October 1969, and both it and her "Kinder, Küche, Kirche, as Scientific Law: Psychology Constructs the Female" were later reprinted in governmental hearings on discrimination against women in academia. In addition, an article in the *Washington Post* in late 1969, "Women Scholars Stymied by System," which described her situation at length, may have been noticed by another dissatisfied psychologist, Bernice Sandler, as well as by governmental officials and congressional staff.[21]

Very quickly women locally and nationally begin to comprehend the enormity of their deprivation and to see the need for change and organization in order to improve their second-class status. Graduate students, research associates, lecturers, and other underemployed and exploited women, already angry about certain aspects of their jobs, began to notice how sex-linked their professional status was and to search for remedies. Consciousness was rising in Chicago and elsewhere, but just what steps to take to effect the change desired were unclear; they needed to be invented and demonstrated. Pioneering in this realm was again the sociologist Alice Rossi. Starting early in 1969, after following her husband to Baltimore, where Johns Hopkins University gave her her nth position as "research associate," Rossi and others planned a multistage campaign to build the women's solidarity and force the issue of women's liberation on the attention of the main professional association in her field, the American Sociological Association, at its next annual meeting, in San Francisco in late August. Perhaps if the issue was on the agenda of the governing boards, certain resolutions could be passed, which might raise the profession's awareness of the issue and encourage members individually and through their home departments to change prevailing behaviors. This had already worked to a certain extent with blacks. (In fact, as she explained, part of her reason for taking on the national association was the growing anxiety among female graduate students and junior faculty in 1968–69 about their future in sociology.

Because most departments were by then seeking blacks as well as the traditional white males for faculty positions, women sociologists, being neither, feared that they would be left out.)

Rossi rather ingeniously planned a three-stage assault. First, foreseeing that an undocumented demand might be met with the usual delaying response of "let's do a study and then decide," she and others prepared a report on the status of women in the field before the meeting; then she would present its results to interested persons at a well-publicized open session ("caucus") early in the ASA meeting; and finally, with this momentum, she would present a list of resolutions to the ASA Council toward the end. Accordingly, with the help of others, she compiled the results of a questionnaire to 188 graduate departments of sociology and wrote one of the first "status of women" report of this later period. Not only did she present, as Fava had done in 1960, the latest impersonal aggregate numbers of the falloff of women—from 43 percent of the college seniors planning graduate work in sociology to 30 percent of the doctoral candidates in graduate school to 14 percent of the full-time assistant professors (in the 78 percent of the departments that had responded) to 4 percent of the full-time full professors—but she went further and even pinpointed the five most prestigious departments that had no women full professors at all.[22]

When, as planned, Rossi presented her data to the recently organized "woman's caucus" in September 1969, the more than five hundred people attending the session voted to endorse her ten resolutions calling for more women on the society's council, editorial boards, and committees, a women's newsletter, and periodic surveys, broken down by gender, of student progress and faculty makeup. A few days later Rossi presented them to the ASA Council, which passed them overwhelmingly, with just two dissenting votes. After an excerpt from her report was published in *Science* in October 1969 (where she deplored the fact that research associates, even those with doctorates and ten or more years' experience, were still not allowed to apply for grants in their own name, whereas any new assistant professor could), other groups formed to discuss the issues (e.g., in "consciousness-raising" groups) and write their own reports.[23]

The value of her advance planning and her strategy of first presenting the report and then calling for a vote becomes clearer if we look at the experience of other groups. At the annual meeting of the AAAS in December 1969, for example, a petition with eight demands for improved status of women was circulated. Despite its "several hundred" signatures, it was not permitted on the council agenda, however, because it had not been submitted thirty days in advance, and it was never published in *Science*.[24] When the newly formed Association for Women in Psychology submitted fifty-two resolutions at an open meeting of the board of directors of the APA in September 1970, including a much-publicized demand for $1 million in reparations to women harmed by the profession, only six were referred to a council meeting a month later. The council then did what Rossi had feared the ASA would do: it voted to set up a task force on women that took almost two years to produce a report with more modest but attainable recommendations.[25] This was just part of the rampaging politicization of scientific societies in the years 1969–71.[26]

On many campuses there was also sudden interest in and pressure to document conditions for women. Current aggregate statistics, such as that there were no women full professors at Harvard (where even the Zemurray professorship was temporarily vacant), were not enough; starting in the academic year 1969–70 university faculties began to set up official committees, composed of the few women on the faculty and some sympathetic men, to examine all phases of women's underrepresentation in campus life and power, to document it in detail, and to suggest intentions of changing it in the foreseeable future. If the original motive behind forming these committees had been to stall until interest had faded or to minimize or whitewash the whole issue with a plethora of statistics and complicating factors, these hopes were dashed, for this cascade of reports revealed all too vividly the results of several decades of certain kinds of employment and promotion policies. Some "reports on the status of women" were more extensive than others, but preliminary and final ones began to appear in a flood in the spring of 1970, just in time to be reprinted in congressional hearings on discrimination on campuses.[27]

In the meantime, however, two landmark events occurred in early 1970. The first was what came to be called the "Ph.D. glut." The signs began to appear as the Nixon administration, which had started cutting the rate of increase in science budgets in 1968, continued to do so into the early 1970s and the Mansfield Amendment stipulated that Defense Department monies could no longer be spent for university research. In 1969–70 the number of scientists and engineers on the federal payroll actually dropped (slightly) for the first time since 1954, when NSF had started to keep count, and for the first time since the 1930s physicists could not find suitable jobs. A "crisis" mentality pervaded their national meetings in April 1970, as special sessions were held to deplore the emergency and hear revised estimates of future manpower needs. Major pre- and postdoctoral fellows programs, designed in an earlier era to meet pressing shortages, were suddenly dismantled, and many hitherto necessary projects, including, ironically, publishing data on scientists, were terminated. (A later critic would blame women for the unexpected oversupply of Ph.D.'s: too many of them—that is, more than the pessimistic experts of the time had predicted—had both gone to graduate school and completed their doctorates in the 1960s, and thus more of them were competing with men for the fewer jobs available. Earlier "experts" had, of course, blamed them for not going on or for dropping out if they did.) Before long the cutbacks and pessimism had affected even fields that were less dependent than physics on governmental appropriations. Colleges and universities, whose faculties had expanded rapidly in the preceding decade or more, were now declared by their academic leaders to be in a "steady state" and would grow no more. By all predictions the "golden age" was over, and by 1971 it was clear that science and higher education faced difficult times, possibly a "new depression," as one author aptly termed it.[28] Just how long it would last remained to be seen, and what impact it would have on those women entering or already in science was not at all clear. If there were to be few jobs for anyone, then even their new-found political awareness might not be worth much.

Yet despite all the resolutions passed and reports written, there was still no legislation that prohibited sex discrimination at the nation's academic institutions. Over the years Congress had exempted colleges and universities from all legislation on civil and women's rights, including the Civil Rights Act of 1964. Congresswoman Edith Green (D.-Oregon), a former schoolteacher, had been so upset with the watered-down Equal Pay Act of 1963, which had excluded all "administrative, executive, and professional" positions, that she had vowed to strengthen it someday. At first the only governmental rulings that might be conceived as covering such powerful and sacrosanct institutions as universities were then were the series of ever broader "executive orders" for federal contractors, issued by presidents since the 1940s, chiefly as a way to seem to be strengthening civil-rights enforcement without having to go through a hostile Congress, controlled by powerful white men from the South. The most recent extension of this series was Executive Order 11375, signed by Lyndon Johnson in October 1967, to ban sexual as well as racial discrimination by federal contractors as well as federal employees. It went into effect a year later, in October 1968, without much fanfare.[29]

Thus, in January 1970, just as the "Ph.D. glut" was making news, another historic event also occurred. At that time another social scientist, clinical psychologist Bernice Sandler of Maryland, took, to adapt a phrase of the time, "a giant step for womankind." Having raised two children and completed a reentry doctorate in clinical psychology, Sandler felt in the fall of 1969 that she was ready to return to full-time work. But when she applied for seven positions at the University of Maryland, she was turned down on the grounds that "she came on too strong for a woman."[30] Upset, she talked with her husband, a lawyer, who pointed out that the University of Maryland was a "federal contractor" and thus subject to the executive orders that prevented racial discrimination and, since October 1968, sexual discrimination in hiring. Thus, though academic officials, including chairmen of departments at the University of Maryland, might think that they were part of a university that had historically been above and outside most laws, this was no longer necessarily the case. After a little research on employment patterns at the University of Maryland, Sandler joined WEAL and filed suit against the University of Maryland in January 1970. Since the pattern was the same "industrywide," a few weeks later she broadened the case to include as well the whole State University of New York system, the City University of New York system, and the nine-campus University of California system. Eventually her "class-action" lawsuit included 250 colleges and universities.[31]

The potential power of such an instrument, hitherto untried, apparently caught the universities unawares. For a short while it even seemed that sex discrimination, so long the norm in higher education, especially at those prestigious universities that were hungry for the millions of federal dollars, was about to end. In April 1970 a team of federal investigators arrived at Harvard University to scrutinize its personnel records but encountered so many delays that it did not produce a letter of finding until eleven months later. Meanwhile, a March 1970 complaint against the University of Pittsburgh filed by a faculty group headed by behavioral scientist Ina Braden led to a visit by HEW investigators, who eventually threatened to withhold $15 million in federal funds until the university presented an acceptable "affirmative-action" plan.[32] HEW's pinpointing of the University of Michigan as its next test case attracted even more attention, including several articles in *Science*. Under such scrutiny, investigators there moved quickly and within a few months had produced a report with evidence of extensive sex discrimination. Faced with a threat to cut off any new federal contracts for the university, Michigan officials quickly came up with an affirmative-action plan, the first at a university. By March 1971 they offered "salary equity" for men and women in the same jobs, as well as back pay to women who had been underpaid. Yet the implementation of this plan, long after the investigators had left, was slow and partial at best.[33]

Thus, Executive Order 11375, the one relevant lever for sexual discrimination in higher education, revealed itself over 1970–72 as quite ineffectual, for the funds never were cut off even at the most notoriously discriminatory employers. Month after month, at campus after campus, even clear-cut findings were not penalized but instead were traded for mere intentions for future partial compliance. HEW, the federal agency responsible for investigating universities, was too understaffed to prolong the task, and its leaders, some of whom came from these very kinds of universities and had spent decades building them up, were not willing to penalize them very heavily.[34] Yet the universities' delaying tactics backfired, for as one by one they stalled in complying with the executive orders and denied equal-pay settlements sought by some employees, the need and pressure for strong, effective legislation became clearer to many.

Meanwhile, increasingly rigorous, scientific evidence of widespread sexual discrimination in hiring and unequal pay and advancement was piling up in *Science* and other scientific and social science journals. Among these were several simple but convincing "vita studies," in which researchers sent sets of paired résumés with similar fictitious credentials under male and female names to department chairmen and other employers, who were then asked to rank or rate them. Statistically significant numbers of employers found the men's credentials more impressive than the matched women's and ranked the male applicants higher than the similar or equal women. These seemed to indicate convincingly that academic departments or at least many of their chairmen, like many nonacademic employers, had definite biases in their employment practices.[35] These studies helped to show that what was generally called "academic freedom," the opportunity to pursue ideas no matter how unpopular or radical they might be, might be so broadly interpreted that under its rubric discrimination could flourish unchecked.

In March 1972 both houses of Congress passed, and President Nixon signed, both the Equal Employment Opportunity Act of 1972, the result of seven years of effort by civil-rights groups and others to broaden and strengthen the Civil Rights Act of 1964, and the Equal Rights Amendment, first proposed to Congress in 1923. Among the former's several important and hard-won provisions was one to drop that portion of Title VII that had formerly exempted all educational institutions from equal-employment-opportunity laws. Led by Republican Senator Jacob Javits of New York and Democratic Senator Harrison Williams, Jr., of New Jersey, substantial numbers of senators overrode several attempts by Southern Democrats Sam Ervin, Jr., of North Carolina and James Allen of Alabama to preserve the old blanket exemption. After four attempts were voted down, the Senate agreed to narrow the exemption to cover religious institutions only, thus expanding the law's coverage of thousands of other colleges and universities nationwide.[36] A few weeks later the Senate voted down nine attempts by Ervin and others to weaken or table the Equal Rights Amendment, which had passed in the House in October. When it passed in late March, the Ninety-second Congress left its mark, one of the most radical of the century.[37]

Over the years, Edith Green had become one of the highest-ranking members of the House's Committee on Education and Labor, which in 1970 was considering an omnibus educational reform bill. Seizing the moment to broaden women's protection in academia beyond the executive orders, she introduced a section "to prohibit discrimination against women in federally-assisted programs and in employment in education; to extend the equal pay act so as to prohibit discrimination in administrative, professional and executive employment; and to extend the jurisdiction of the U.S. Commission on Civil Rights to include sex."[38] She then appointed the knowledgeable Sandler to her staff, which was preparing for hearings on this bill in June and July 1970. Sandler accordingly invited a full panoply of academic women who had been active in the movement to testify and submitted for the printed record many of the recent research and writings of such persons as psychologists Helen Astin and

Naomi Weisstein and political science graduate student Jo Freeman, as well as some of the first status-of-women reports spewing forth from academia (Chicago, Cornell, Berkeley, Harvard, and others) and the professional societies (the American Political Science Association and the APA, as well as Rossi's work for the ASA). These supplemented other testimony and exhibits by the NFBPWC, the AAUW, and groups of women physicians, lawyers, and federal employees. Packed with information and, since they were governmental documents, modestly priced, these volumes quickly became standard textbooks in the proliferating courses on women's studies.

Also in 1970–72 a series of other episodes, each unique in itself but together a kind of morality play of the new gender awareness, unfolded in the media, serving to awaken the public to the full extent of the problem. In some cases governmental agencies took the occasion to stage certain timid "firsts" for womankind, as if to test the waters for other, more daring ventures later. In other cases, as journalists felt emboldened to report behaviors and challenge statements that would have passed unnoticed earlier, a series of cause célèbres, punctuated with occasional high-level resignations, got more publicity than usual. Since this reportage was often sexist, feminists could then criticize it and thus squeeze an extra round of publicity out of it.

Attitudes were changing, for example, in the realm of female fieldwork in formerly forbidden places, such as Antarctica and the ocean floor. Although some women had gone to Antarctica (usually with their husbands) in the first half of the twentieth century and two American wives had even "wintered over" during the privately sponsored Ronne Antarctic Research Expedition of 1947, all women were officially banned in the 1950s, when the U.S. Navy took over American responsibilities there. (Accordingly, the only woman scientist to play even a minor role in events there that were part of the International Geophysical Year of 1957–58 was the Russian marine geologist Marie Klenova, who went as part of a preliminary Soviet expedition in 1955–56.) Officials of the polar program of the NSF, which under the terms of the Antarctic Treaty of 1959 controlled and funded access by American scientists to research facilities in Antarctica, had been routinely turning down qualified women applicants (e.g., Mary Belle Allen, Nancy M. Walls, and Mary Alice McWhinnie) because the U.S. Navy claimed that first its ships and then its land facilities at Antarctica were not suitable for women. However, the constant barrage of stories in Washington newspapers in 1964 and 1965 about President Lyndon Johnson's new priorities and top-level female appointees caused consciousness to rise a bit within the foundation. One such officer even clipped a story about the appointment of Patricia Roberts Harris as ambassador to Luxembourg in February 1965 and, attaching it to a memorandum to his boss, asked, "With this stuff appearing almost daily in the press, how much longer are we going to stall off on Antarctica?" Despite this thawing within one civilian agency, the Navy did not permit American women to go there until the fall of 1969, when, after a decade of resistance and after a team of four Argentine women had successfully spent the winter there, it finally allowed first a couple from Utah State University (Christine and Dietland Muller-Schwarze) to study the penguins near McMurdo Station, the main base there, and then a team of four women, led by geochemist Lois Jones of Ohio State University, to pursue their researches on cold temperature for four weeks together in a remote valley fifty miles away.[39] Needless to say, the newspaper coverage in the *New York Times* and elsewhere portrayed this as women's intrusion into men's final sanctuary on earth.[40]

Then in July 1970, while Green's hearings were under way in Washington, D.C., the Department of the Interior submerged as part of its underwater Tektite II project an all-woman group of five led by experienced aquanaut and marine botanist Sylvia Earle Mead. The five women spent two weeks in a capsule on the sea floor off the Virgin Islands. Although they discovered many new species of plants and fish, most of their widespread publicity above ground included little about their contributions to science and more about mermaids, swimmer Esther Williams, a rumored hair drier on board, and their presumed incompatibility. When pressed to name their greatest breakthrough, one participant mentioned that it might have been "in the minds of the men running the project," which was co-sponsored by General Electric, NASA, the Navy, the Smithsonian, and a few other federal agencies. Others speculated that this event might be a kind of publicity stunt that would help nudge NASA toward letting women become astronauts.[41]

Hardly were the aquanauts above water when in late July 1970 Dr. Edgar Berman, a former physician and currently adviser to former vice-president Hubert Humphrey, leaked to the press his correspondence with Congresswoman Green's colleague in the House, Democrat Patsy Mink of Hawaii, on women's physiological unfitness to hold high political office. At an April meeting of the Democratic Party's committee on priorities Mink had recommended that the party should now put women's issues at the top. To this Dr. Berman had retorted that because of women's "raging hormonal imbalances," they were not fit to hold high political office. Two months later Mink had asked for his resignation from the committee, but he had merely written her a letter restating his beliefs, which in July he passed on to reporters. As a result of the furor, he resigned at the end of the month, whereupon the witty and outspoken physiologist Estelle Ramey of Georgetown University Medical School took up the cudgels, debating him in September at the Women's National Press Club in Washington and subsequently writing an article on the issue for *McCall's*. If leaders of the women's movement, including the few congresswomen of the time, had wanted an example of reactionary and even unscientific thinking at the highest levels of the federal government, they could hardly have found a more suitable one.[42]

Another change started in February 1970, when the AAUP, responding to pressures within its membership, reactivated its Committee W on the status of women in the academic profession, which it had allowed to lapse in the late 1920s. With Alice Rossi as chair, one of the committee's first projects was to prepare an official policy statement (in conjunction with Committee A on academic freedom and tenure) on antinepotism rules. This statement, "Faculty Appointments and Family Relationship," was then approved at the AAUP general annual meeting in April 1971 and by the board of directors of the Association of American Colleges in June 1971. Its language and logic followed that marked out by Josephine Mitchell (Schoenfeld) and Lowell Schoenfeld back in the 1950s: that the criteria for exclusion were based on a factor (marital status) "wholly unrelated to academic qualifications" and thus "limit [faculty members] unfairly" and were in addition "contrary to the best interests of the institution" and even "the community, which is denied a sufficient utilization of its resources." Although some "reasonable restrictions" were proper especially on employment decisions affecting relatives, an overall ban on their opportunities was now seen as constituting "a continuing abuse to a significant number of individual members of the profession." Therefore both associations urged "the rescinding of laws and institutional regulations which perpetrate" such policies and practices.[43] These AAUP recommendations were not legally binding, but an important corner had been turned in professional opinion and preferred administrative procedures. A practice that had blighted untold academic careers might now become a thing of the past.

Likewise advocates of equal pay for professional women had a field day when a discriminatory job advertisement explicitly offering substantially lower salaries for women than for men by the Australian CSIRO (Commonwealth Scientific and Industrial Research Organization) was published in the September 11, 1970, issue of *Science*. The resulting flurry of letters to the editor over the next several months allowed a full exchange of views on this all-too-common practice. Several women were outraged that the journal had accepted and printed such an advertisement; several men then justified the lower salary on the traditional grounds that women were "bad risks" who took time off to raise families; and they in turn were denounced, ridiculed, and presented with counterevidence. Finally Berkeley psychologist Susan Ervin-Tripp ended the debate with a heavily footnoted letter citing Helen Astin's recent book, which had never been reviewed in *Science,* on the substantial employment histories of the American women Ph.D.'s of 1957.[44]

Events such as these stirred even greater consciousness among professional women, who were already busily forming a proliferating number of women's caucuses in 1970 and 1971. Among scientists the early ones included the women's committee of the American Anthropological Association (February 1970) and the American Physical Society (April 1971) and the separate Association for

Women in Mathematics (January 1971) and the Association for Women in Science (AWIS). This last, formed officially in April 1971, grew out of a series of mostly female "champagne mixers" or social hours that had been held at the annual Federation of Scientific and Experimental Biology meetings each year in Atlantic City since 1966. Initially biologist Virginia Upton, then at the West Haven (Conn.) VA Hospital, had sent to women registrants an invitation to a reception sponsored by a pharmaceutical company. The women attending, who had not realized how many other women there were at the meetings and how much they had in common, were pleased to meet each other. Similar meetings were held in 1967, 1968, and 1969 but remained primarily social. In 1970, however, in response to rising interest in the women's movement, Judith Pool of Stanford Medical School suggested that women's issues in science be discussed at the meeting. They were, with important results. Finding that they had a lot to gain by working together and would enjoy doing so, the formed AWIS the next year.[45]

One of AWIS's early projects was to develop and maintain a roster of professionally trained women for use by potential employers and others. Another early area of activity, one that had more immediate results, was the assault on the low proportion (<2 percent) of women on the "study sections" at NIH, which ranked many of their research projects. Mathematical biologist Julia Apter at Rush Medical School in Chicago seized this issue and led AWIS's protest of this discrimination (at first indirectly through her senator) to HEW Secretary Elliott Richardson. When after a meeting Richardson stalled, AWIS instituted a lawsuit (*AWIS v. Richardson* [1972]) charging that NIH had violated the executive order by not appointing women to its advisory committees. Although that action dragged on inconclusively for years, NIH officials quickly put a hold on the appointment of male panelists and appointed so many of the women whose vitas Apter and AWIS had sought and submitted that within six months the number of women scientists on these selection panels increased dramatically, from 1.4 percent to 20 percent. This was still not high enough to suit many, but here was a prime example, hitherto undemonstrated, of the political impact of "scientific womanpower."[46]

Meanwhile, a flurry of other voluntary changes were taking place. In the spring of 1971, when the Chemists' Club of New York may have had a financial crisis, its members voted to admit their first women members. The very first to join was cosmetic chemist turned financial analyst Hazel Bishop, and by 1973 it had eleven women members.[47] In November 1971 the R. R. Bowker Company of New York announced that henceforth its *AMS* would be renamed the *American Men and Women of Science (AMWS)*,[48] and at the December 1971 meeting of the AAAS in Philadelphia, Yale research associate Mary Clutter and postdoctoral fellow Virginia Walbot got Margaret Mead and Hazel Fox, professor of home economics at the University of Nebraska and as president of Sigma Delta

Epsilon that year the only other woman on the AAAS Council (of about five hundred persons), to present a proposal to establish an Office for Women's Equality, later renamed the Office for Opportunities in Science.[49]

But if consciousness was running high and the outpouring of outrage was epidemic in some circles, such feelings were far from universal. Many eminent scientists, women as well as men, did not necessarily agree that there was a problem and wondered what all the fuss was about. Having adjusted to it all years before and believing staunchly in individual virtues such as hard work, they were either oblivious to the problem or, when it was brought to their attention, adamant that it did not exist. They were so much a part of the "system" that had treated them comparatively well that it was difficult for them, as it had been earlier for Jessie Bernard, to see a pattern and think of employers and colleagues, even sexist ones, as villains. Often foreign-born, these faculty women clung to an individualistic view that all that mattered was doing very good work and lots of it; one's sex and marital status were irrelevant. By dint of a lifetime of hard work, considerable self-sacrifice, and perhaps a move to the United States, they had "made it," and they did not wish to criticize American institutions that had made their success possible. Their successful work and high rank on the faculty had blinded them to other views; instead they seemed proof that if, just if, a woman was good enough, she too would be promoted to the highest levels. Their small numbers could be seen as indicators that few women offered this successful combination rather than evidence that stronger credentials might be required for women than for men. For example, German immigrant and Nobel laureate Maria Goeppart Mayer of the University of California at San Diego could not understand why the American Physical Society had created a committee on women in April 1971 or why it had put her on it: she had no interest or expertise in the area.[50] Similarly, Birgit Vennesland, Norwegian-born and long a full professor of physiology and biochemistry at the University of Chicago, ended her autobiographical statement for her fellow physiologists in the early 1970s with some angry remarks about the younger women who now expected to be put on university faculties just because they felt as qualified as men; for women to press too hard in this direction would, she felt sure, lower the quality of the faculty and thus in time endanger the strength of the nation. Academia should hold onto its proven ways and not give in to the merely political pressures of diversifying the faculty.[51] Likewise, when Louise Daniel, long the only woman on the agricultural faculty at Cornell, was interviewed in 1976, she claimed that despite some problems the men there had always treated her equally, even as a "gentleman," and that in the long run it would not help womankind to have women less qualified than men on the faculty.[52]

Some, however, did change their minds, with certain beneficial consequences. Over time, as the issues dragged on and lawsuits multiplied, it became more acceptable to speak out, sign petitions, and sue for equal pay. Sociologist Jessie Bernard, whose 1964 book *Academic Women* had, despite its ambivalence, reportedly inspired some women to think about the issues if not undertake outright political action, continued to waver until she had a well-publicized conversion experience after a small group meeting in 1968. The support of such older, especially tenured women would mean a lot to the younger ones, struggling for jobs and then tenure in the newly politicized atmosphere of the mid-1970s.[53] Biologist Ruth Hubbard, long a research associate at Harvard but oblivious of her underclass status, since she was allowed to sign her own grants, stayed away from the issues for many years before becoming quite radicalized by the early 1970s. Then, after being promoted to full professor, she started one of Harvard's first courses on women.[54] In 1976 Cornell psychologist Eleanor Jack Gibson insisted that she was still not a "woman's libber," but she regretted that, after devoting so much of the 1960s to reestablishing her own career there, she had had so little interest in or energy for its new women's studies program.[55]

Amidst all this turmoil the culmination of Green's action in the House came in March 1972 with the passage of what was finally termed the Educational Amendments Act of 1972. It then went to the Senate, whose version did not include any mention of sex discrimination. Perhaps because of Green's seniority, she was a part of the joint conference committee and was able to preserve her section banning sex discrimination. In June President Nixon signed the compromise bill, whose Title IX finally extended the Equal Pay Act of 1963 to higher education and banned sex discrimination in any program of an institution receiving federal funding, including sports, textbooks, and the curriculum. Just how broadly this would be interpreted (e.g., as extending to the whole institution) remained to be determined in final regulations and guidelines.[56]

Thus, within just a few years, starting in 1968 and essentially complete by 1972, there was a legal revolution in women's education and employment rights. It promised, even seemed to guarantee, broad ramifications for women's careers in science and engineering, but its full implementation would require many battles in the years ahead. One era had ended and a new, more equitable one was beginning.

LIST OF ABBREVIATIONS

AAAS	American Association for the Advancement of Science
AAUP	American Association of University Professors
AAUW	American Association of University Women
AAUWJ	*Journal of the AAUW*
ACE	American Council on Education
ACE-CEW	American Council on Education, Commission on the Education of Women
ACS	American Chemical Society
AEC	Atomic Energy Commission
AHEA	American Home Economics Association
AMS	*American Men of Science* (until 1972)
AMWS	*American Men and Women of Science* (since 1972)
AP	*American Psychologist*
APA	American Psychological Association
APS	American Philosophical Society, Philadelphia
BLS	Bureau of Labor Statistics
CBY	*Current Biography Yearbook*
C&EN	*Chemical and Engineering News*
CSC	Civil Service Commission
CSM	*Christian Science Monitor*
CSP	Committee on Specialized Personnel
ESMWT	Engineering, Science, and Management War Training
FAE	Fund for the Advancement of Education
FDA	Food and Drug Administration
GSA	Geological Society of America
HEW	Department of Health, Education, and Welfare
HUA	Harvard University Archives
ICWP	International Council of Women Psychologists
IW	*Independent Woman*
JCE	*Journal of Chemical Education*
JEE	*Journal of Engineering Education*

JHE	*Journal of Home Economics*
JLP	Jeanne R. Lowe Papers, Special Collections, Vassar College Library
KLCU	Kroch Library, Cornell University
LC	Library of Congress
LCP	Leonard Carmichael Papers, American Philosophical Society Library, Philadelphia
MCRC	Henry A. Murray Center for Research in the Study of Lives, Radcliffe College
MIT	Massachusetts Institute of Technology
MMP	Margaret Mead Papers, Manuscript Division, Library of Congress
NARA	National Archives and Records Administration
NAS	National Academy of Sciences
NASA	National Aeronautics and Space Administration
NAW	*Notable American Women*
NAWDC	National Association of Women Deans and Counselors
NAWMP	*Notable American Women: The Modern Period*
NAWP	*Notable American Women* Papers, Schlesinger Library, Radcliffe College
NBL	Niels Bohr Library, American Institute of Physics, College Park, Md.
NBS	National Bureau of Standards
NCAB	*National Cyclopedia of American Biography*
NCEAB	*National Catholic Education Association Bulletin*
NCWP	National Council of Women Psychologists
NDEA	National Defense Education Act
NEA	National Education Association
NFBPWC	National Federation of Business and Professional Women's Clubs
NIH	National Institutes of Health
NIMH	National Institute of Mental Health
NMC	National Manpower Council
NORC	National Opinion Research Center
NOW	National Organization for Women
NRC	National Research Council
NRL	Naval Research Laboratory

NSF	National Science Foundation
NYT	*New York Times*
ODM	Office of Defense Mobilization
ONR	Office of Naval Research
OSRD	Office of Scientific Research and Development
OSS	Office of Strategic Services
OWI	Office of War Information
P&B	Physical and Biological Sciences series, *AMS* and *AMWS*
PCSE	President's Committee on Scientists and Engineers
PCSW	President's Commission on the Status of Women
PI	Principal Investigator
PSAC	President's Scientific Advisory Committee
RAC	Rockefeller Archive Center
RCA	Radcliffe College Archives
RG	Record Group
SAQ	*Smith Alumnae Quarterly*
S&B	Social and Behavioral Sciences series, *AMS* and *AMWS*
SIO	Scripps Institution of Oceanography
SLRC	Schlesinger Library, Radcliffe College
SWE	Society of Women Engineers
SWG	Society of Women Geographers
USDA	U.S. Department of Agriculture
USGS	U.S. Geological Survey
USOE	U.S. Office of Education
USPHS	U.S. Public Health Service
VA	Veterans Administration
VAM	*Vassar Alumnae Magazine*
WAM	*Wellesley Alumnae Magazine*
WB	Women's Bureau, U.S. Department of Labor
WMC	War Manpower Commission
WSJ	*Wall Street Journal*
WWAW	*Who's Who of American Women*
WWE	*Women's Work and Education*

NOTES

Introduction

1. Joan Kelly, "Did Women Have a Renaissance?" in *Becoming Visible: Women in European History,* ed. Renate Bridenthal and Claudia Koonz (Boston: Houghton Mifflin, 1977), reprinted in Kelly, *Women, History, and Theory: The Essays of Joan Kelly* (Chicago: Univ. of Chicago Press, 1984), ch. 2.

Chapter 1   World War II: Opportunity Lost?

1. See U.S. Bureau of the Budget, *The United States at War: Development and Administration of the War Program by the Federal Government,* Historical Reports on War Administration, no. 1 (Washington, D.C.: GPO, [1946]).

2. Gladys Schwesinger, "The National Council of Women Psychologists," *Journal of Consulting Psychology* 7 (1943): 299.

3. Bureau of the Budget, *United States at War,* ch. 11; Lydia J. Roberts, "Beginnings of the Recommended Dietary Allowances," in *Essays on History of Nutrition and Dietetics,* comp. Adelia M. Beeuwkes et al. (Washington, D.C.: American Dietetic Association, 1967), 107–12 (Roberts mistakenly put Hazel Stiebeling on the Food and Nutrition Board in 1940); H. Mitchell Biographical File, Archives, University of Massachusetts, Amherst (I thank Virginia Vincenti for her help); C. G. King, "The History and Philosophy of the Food and Nutrition Board, Including Its International Activities," *The Food and Nutrition Board, 1940–1965: Twenty-five Years in Retrospect* (Washington, D.C.: NAS-NRC, [1966?]), 3–8; L. A. Maynard, "Contributions of the Food and Nutrition Board to National Nutrition Programs," ibid., 8–12. See also Ethel Austin Martin, "Lydia Jane Roberts, 1879–1965," *Journal of the American Dietetic Association* 47 (1965): 126–28, and idem, "The Life Works of Lydia J. Roberts," ibid. 49 (1966): 299–302.

4. On Stiebeling, see *CBY, 1950,* 548–50, and T. Swann Harding, *Two Blades of Grass* (Norman: Univ. of Oklahoma Press, 1947), 224–25; on Barber, see *CBY, 1941,* 39–40; Mary I. Barber, "You Can Use Army Research," *Omicron Nu* 25, no. 1 (fall 1945): 14–18, and Ida Jean Kain, "Mary Isabel Barber, 1887–1963," *Journal of the American Dietetic Association* 42 (1963): 418; on Lindman, see *NYT,* Sept. 3, 1963, p. 33, col. 1; as well as Frances Maule, *Careers for the Home Economist: Fields Which Offer Openings to the Girl with Modern Training in the Homemaking Arts* (New York: Funk & Wagnalls, 1943); and on Carson, see Paul Brooks, *The House of Life: Rachel Carson at Work* (Boston: Houghton Mifflin, 1972), 69–72. On Hardy's war work on fur fibers for the U.S. Fish and Wildlife Service, see the Thora Plitt Hardy Papers, Smithsonian Institution Archives, Washington, D.C.

5. See "News of University Women," *AAUWJ* 37 (1944): 101; Margaret Mead,

"Ferment in British Education," ibid., 131–33; idem, *Food Habits Research: Problems of the 1960s,* Publication 1225 (Washington, D.C.: NAS–NRC, 1964), esp. ch. 1; and Carl E. Guthe, "History of the Committee on Food Habits," in *The Problem of Changing Food Habits: Reports of the Committee on Food Habits, 1941–1943,* Bulletin of the NRC, no. 108 (Washington, D.C., 1943), 9–19. For more on Mead's wartime work for the OWI in Britain, see boxes E154–E156 in MMP; and for her later assessments of the Committee on Food Habits, see Rhoda Metraux, "The Committee on Food Habits: A Reply" (paper presented at a meeting of the Society for Applied Anthropology, "Across Generations: A Symposium," Tucson, Ariz., Apr. 14, 1973, copy in box E125, MMP). See also Carleton Mabee, "Margaret Mead and Behavioral Scientists in World War II: Problems in Responsibility, Truth, and Effectiveness," *Journal of the History of the Behavioral Sciences* 23 (1987): 3–13.

6. On Benedict, see Eugene Katz to Ruth Benedict, June 2, 1943, and personnel forms, Dec. 26, 1944, OWI, both in Ruth Benedict Papers, Special Collections, Vassar College Library, Poughkeepsie, N.Y.; and Ruth Benedict, *The Chrysanthemum and the Sword: Patterns of Japanese Culture* (Boston: Houghton Mifflin, 1946), esp. preface and ch. 1. For more on Benedict's war work, see Judith Schachter Modell, *Ruth Benedict: Patterns of a Life* (Philadelphia: Univ. of Pennsylvania Press, 1983), 267–71; and Margaret M. Caffrey, *Ruth Benedict: Stranger in This Land* (Austin: Univ. of Texas Press, 1989), 314–18. Anthropologist Katherine Spencer also worked on the OWI's Japanese project (*Survey* 87 [1951]: 84). See also Allan M. Winkler, *The Politics of Propaganda: The Office of War Information, 1942–1945* (New Haven: Yale Univ. Press, 1978); and Leonard W. Doob, "The Utilization of Social Scientists in the Overseas Branch of the Office of War Information," *American Political Science Quarterly* 41 (1947): 649–67.

7. For a general description of the impact of the war on this field, see Gladys Reichard to Dean Gildersleeve, Nov. 30, 1942, Department of Anthropology Papers, Barnard College Archives, New York, and John F. Embree, "Anthropology and the War," *AAUP Bulletin* 32 (1946): 485–95. On Talcott, see *NCAB* 54 (1973): 627–28. On DuBois, see obituary, *NYT,* Apr. 11, 1991, B-14; *AMWS,* 13th ed., P & B, 1976, 2:1093; Judith Walzer, "Interviews with Cora DuBois," Aug. 1981, transcript at MCRC; R. Harris Smith, *OSS: The Secret History of America's First Central Intelligence Agency* (Berkeley: Univ. of California Press, 1972), 13; and "Anthropology—Studied, Taught, and Practiced: A Description of the Department at Barnard and an Outline of the Careers of Two Alumnae," *Barnard College Alumnae Monthly* 40 (Nov. 1950): 5–6. On Japanese-Americans, see Raymond B. Fosdick, *The Story of the Rockefeller Foundation* (New York: Harper & Bros., 1952), 223; *AMWS,* 12th ed., 1973, S&B, 2:2454–55, s.v. "Dorothy Swaine Thomas"; Rosalie H. Wax, *Doing Fieldwork: Warnings and Advice* (Chicago: Univ. of Chicago Press, 1971), 64–66; idem, "In and Out of the Tule Lake Segregation Center: Japanese Internment in the West, 1942–45," *Montana: The Magazine of Western History* 27 (spring 1987): 12–25; and Katharine Luomala, "Fellow Californians . . . Fellow Americans," *AAUWJ* 39 (1946): 208–11. On Colson, see "AAUW News and Notes," ibid. 36 (1943): 113–14.

8. The OSRD eventually bypassed even the Roster and set up its own Office of Scientific Personnel. It also had many battles with the CSC, whose restrictions it had been designed to circumvent. See esp. Merriam H. Trytten, "The Mobilization of

Scientists," and Louis Brownlow, "Successes and Failures," in *Civil Service in Wartime,* ed. Leonard White (Chicago: Univ. of Chicago Press, 1945), 60–63 and 228–46, respectively.

9. Virginia B. Shapley, "Science in Petticoats: In Wartime and in Peace," *AAUWJ* 39 (1946): 148–50; "AAUW Fellows and Defense," ibid. 35 (1942): 92–100 (summarized in Anne Petersen, "Special Roles Vital to Nation Filled by Women Scholars," *NYT,* Jan. 18, 1942, 40); "War Work of AAUW Fellows," *AAUWJ* 36 (1943): 77–81; "Fifty Fellows and the War," ibid. 37 (1944): 86–90.

10. In addition to the generally useful *Scientists against Time,* by James Phinney Baxter (Boston: Little, Brown, 1946), material on the women involved can be found in a variety of places. On Libby, see obituary, *NYT,* Nov. 12, 1986, D-27; her anecdotal book *The Uranium People* (New York: Crane Russak, Charles Scribner's Sons, 1979); "Where Are They Now? The Women at the First Atomic Pile," *Newsweek* 70 (Dec. 4, 1967): 14; Lyn Tornabene, "Fascination of the Unknown," *Cosmopolitan* 148 (Jan. 1960): 29; Walter Sullivan, "Top Scientists Mark 25th Year of the Atomic Age," *NYT,* Dec. 2, 1967, 43; and D. J. R. Bruckner, "The Day the Nuclear Age Was Born," ibid., 30, 1982, pp. C-1, C-3. On Mooney, see *AMWS,* 13th ed., 1976, 4:4139; and John C. Slater to Zach [Jerrold Zacharias], Jan. 14, 1952, John C. Slater Papers, APS (I thank Nathan Reingold for this information). On Grieff, see obituary in *School and Society* 69 (1949): 2169. On Mayer, see Harold C. Urey, "Maria Goeppart Mayer (1905–1972)," *American Philosophical Society Yearbook for 1972* (Philadelphia, 1973), 235; Joan Dash, "Maria Goeppart Mayer," in *A Life of One's Own* (New York: Harper & Row, 1973), 292–94; Robert G. Sachs, "Maria Goeppart Mayer—Two-Fold Pioneer," *Physics Today* 35 (Feb. 1982): 46–51. And on Wu, see *CBY, 1959,* 491. Wu also worked on E. O. Lawrence's cyclotron, from 1936 to 1942, until she was forced to leave by the U.S. Immigration Service (see C. S. Wu folder, E. O. Lawrence Papers, Bancroft Library, University of California at Berkeley), whereupon she taught for a year at Smith and then at Princeton.

11. On Graves, see biographical materials in f. 84, box 25, *NAWP,* and Nuel Pharr Davis, *Lawrence and Oppenheimer* (Greenwich, Conn.: Fawcett, 1968), 250; on Anderson, *NCAB* 50 (1968): 281–82; and on Quimby, *CBY, 1949,* 494–93. There is beginning to be a literature on the women at Los Alamos in World War II. See Lawrence Badash et al., *Reminiscences of Los Alamos, 1943–45* (Dordrecht: Reidel, 1980); Jane S. Wilson and Charlotte Serber, eds., *Standing By and Making Do: Women of Wartime Los Alamos* (Los Alamos, N.M.: Los Alamos Historical Society, 1988), esp. ch. 5, by Serber, who was the Manhattan Project's librarian; Kathleen E. B. Manley, "Women of Los Alamos during World War II: Some of Their Views," *New Mexico Historical Review* 65 (1990): 251–66; and Caroline L. Herzenberg and Ruth H. Howes, "Women of the Manhattan Project," *Technology Review* 96 (Nov.–Dec. 1993): 32–40 (I thank Rosalinda Ratajczak for a copy).

12. On Barlett, see Karl Schwartzwalder, "Memorial of Helen Blair Barlett," *American Mineralogist* 56 (1971): 668–70; and on Foster, Joseph J. Fahey, "Memorial of Margaret D. Foster," ibid., 686–90.

13. See "Woman Scientist Reported a Suicide," *NYT,* Nov. 12, 1963, Western edition, and other clippings in Maria Goeppart Mayer Papers, Mandeville Department of Special Collections, University of California, San Diego. See also Stanislaw Ulam, *Adventures of a Mathematician* (New York: Charles Scribner's Sons, 1976), 110, 213.

14. "Our Fellows Helped with Radar," *AAUWJ* 39 (1945): 51–52; Rosenthal (later Bramley) and Morrow (later Austin) in AAUW, *Idealism at Work* (Washington, D.C., 1967), 57, 73; unidentified clipping, "Dr. Healea Aided Research in Radar," Sept. 27, 1945, Faculty File, Vassar College Archives; obituary, *NYT,* May 15, 1993, 26; "Elizabeth Rebecca Laird," *Optical Society of America Journal* 59 (1969): 1687.

15. On Fennell, see obituary by C. W. Hesseltine, *A[merican] S[ociety of] M[icrobiology] News* 43 (1977): 523; on McCoy, Grace Chatterton, "Wisconsin Women," *Wisconsin Alumnus,* Jan. 1954, 23, and other clippings in her biographical file in University of Wisconsin Archives, Madison; on Hutchinson, "Miss Chemical Engineer of 1955," *C&EN* 33 (1955): 3504. On Hobby, see Gladys L. Hobby, *Penicillin: Meeting the Challenge* (New Haven: Yale Univ. Press, 1985); obituary, *NYT,* July 9, 1993, D-19; and *VAM,* Apr. 1961, 36.

16. W. Mansfield Clark, "History of the Co-Operative Wartime Program," in *A Survey of Antimalarial Drugs, 1941–1945,* ed. Frederick Y. Wiseogle (Ann Arbor: J. W. Edwards, 1946), 1:37, 44; "Dorothy V. Nightingale," *C&EN* 37 (Apr. 20, 1959): 117.

17. On Peak, see Edwin Boring to Helen Peak, Mar. 1, June 11, 23, 1943, in Edwin G. Boring Papers, HUA, and Jane D. Hildreth and Carolyn L. Konold, eds., *American Psychological Association Directory, 1951* (Washington, D.C.: APA, [1951]), 357; on Kelly, see W. A. Noyes, ed., *Chemistry: A History of the Chemistry Components of the National Defense Research Committee, 1940–1946* (Boston: Little, Brown, 1948), 14, and Agatha S. Rider, "In Memoriam: Louise Kelley," *Goucher Alumnae Quarterly* 40 (winter 1962): 16–17. Some of Kelley's wartime letters are in f. 10, box 9, Don Yost Papers, California Institute of Technology Archives, Pasadena. On Young, see Noyes, *Chemistry,* 245.

18. On Anslow, see obituary, *NYT,* Apr. 2, 1969, 47, and Faculty File, Smith College Archives, Northampton, Mass.; see also Lincoln Thiesmeyer and John E. Burchard, *Combat Scientists* (Boston: Little, Brown, 1947), 46–47, 67. On Weeks, see obituary, *NYT,* June 8, 1990, D-16, and Irvin Stewart, *Organizing Scientific Research for War: The Administrative History of the Office of Scientific Research and Development* (Boston: Little, Brown, 1948), 172. On Rees, see *AMWS,* 13th ed., 1976, 5:3632, and *CBY, 1957,* 453–55; see also Mina Rees, "The Mathematical Sciences and World War II," *American Mathematical Monthly* 87 (1980): 607–21.

19. On Telkes, see *CBY, 1950,* 563–64; Betsy Burke, "Woman Engineer Who Tames the Sun," *IW* 34 (July 1955): 10–12; Noyes, *Chemistry,* 381, 386; and John Burchard, *Q.E.D., M.I.T. in World War II* (New York: Technology Press of John Wiley & Sons, 1948), 161. On Morgan, see Agnes Fay Morgan, "The History of Nutrition and Home Economics in the University of California, Berkeley, 1914–1962" (mimeographed), 11–12; "Agnes Fay Morgan: Her Career in Nutrition," *Journal of Nutrition* 91, suppl. 1, pt. 2 (1967): 2; and obituary, *NYT,* July 23, 1968, p. 36, col. 3. On Frantz see *NCAB* 53 (1971): 346.

20. USGSM, *The Military Geology Unit, U.S. Geological Survey and Corps of Engineers, U.S. Army* (n.p., 1945), pamphlet prepared for 1945 GSA meeting, inside front cover, 2–5; On Dorothy Wyckoff, see *AMS,* 11th ed., 1967, 6:6001, and biographical folder, Geology Department Office, Bryn Mawr College, Bryn Mawr, Pa.; on Sears, Susan Schlee, *The Edge of an Unfamiliar World: A History of Oceanography* (New York: E. P. Dutton, 1973), 311; on Stewart, Edmund M. Spieker, "Memorial to Grace Anne

Stewart, 1893–1970," GSA *Memorials* 2 (1973): 112; on Boyd, *CBY, 1960*, 48–49. See also Frank C. Whitmore, Jr., "Memorial to Esther Aberdeen Holm, 1904–1984," GSA *Memorials* 21 (1991): 9–12; "Opportunity for Women in Meteorological Work," *Bulletin of the American Meteorological Society* 23 (1942): 46; "Marge Supplee," *IW* 36 (Mar. 1957): 29; Joanne Simpson, "Meteorologist," in *Successful Women in the Sciences: An Analysis of Determinants,* ed. Ruth B. Kundsin, *Annals of the New York Academy of Sciences* 208 (1973): 41–46; Robert Strausz-Hupe, "Jobs in Geography," *WWE* 15, no. 3 (fall 1944): 4; Joseph A. Russell, "Military Geography," in *American Geography: Inventory and Prospect,* ed. Preston E. James and Clarence F. Jones (Syracuse: Syracuse Univ. Press, 1954), 484–94; Jane H. Wallace, "Women in the Survey," *Geotimes* 24 (Mar. 1979): 34 (I thank Michele Aldrich for this item); and *The Outlook for Women in Geology, Geography, and Meteorology,* WB Bulletin 223-7 (Washington, D.C., 1948), 2–5, 15–19, 29–33.

21. On the women's services in general, see Susan M. Hartmann, *The Home Front and Beyond: American Women in the 1940s* (Boston: Twayne, 1982), ch. 3; and Mattie E. Treadwell, *The Women's Army Corps,* U.S. Army in World War II, Special Studies (Washington, D.C.: Office of the Chief of Military History, Department of the Army, 1954), 28–30. On the WAVES, see Joy Bright Hancock's obituary, *NYT,* Aug. 25, 1986; her autobiography, *Lady in the Navy* (Annapolis, Md.: Naval Institute Press, 1972); and Dean Virginia Gildersleeve's autobiography, *Many a Good Crusade* (New York: Macmillan, 1954), 267–87, quotation from 273. The Navy's Division of Naval History, Washington, D.C., Navy Yard, also has a short unpublished "History" to 1954 (n.d., mimeographed, 7 pp.; I thank Dean Allard for sending me a copy). On de Laguna, see *AMWS,* 9th ed., 1955, 3:161, and Fosdick, *Story of the Rockefeller Foundation,* 274; on Creighton, Biographical File, Wellesley College Archives; on Hopper, obituary, *NYT,* Jan. 3, 1992, A-17, and ch. 12 below.

22. T. G. Andrews and Mitchell Dreese, "Military Utilization of Psychologists during World War II," *AP* 3 (1948): 534, 536. On Stratton, see *CBY, 1943*, 742–44; on Lux, *WWE* 14, no. 3 (Oct. 1943): 12; on Trotter, Glenn Conroy et al., "Obituary: Mildred Trotter, Ph.D. (Feb. 2, 1899–Aug. 23, 1991)," *American Journal of Physical Anthropology* 87 (1992): 373–74, and oral history interview, May 16, 1972, tape at Washington University Medical Center Archives, St. Louis, Missouri. On Mitchell, see *AMWS,* 9th ed., 1955, 3:477; *International Council of Psychologists Newsletter* 3 no. 2 (Jan. 1962): 3, 6; and Mildred B. Mitchell, "Mildred B. Mitchell," in *Models of Achievement: Reflections of Eminent Women in Psychology,* ed. Agnes N. O'Connell and Nancy Felipe Russo, vol. 1 (New York: Columbia Univ. Press, 1983), 121–39. And on van Straten, see Edna Yost, *Women of Modern Science* (New York: Dodd, Mead, 1959), 124–39; *AMS,* P&B, 11th ed., 1967, 6:5558; and CSC, "Appointments under Section 3.2 of Civil Service Rule III," *Sixty-seventh Annual Report of the U.S. Civil Service Commission, Fiscal Year Ended June 30, 1950* (Washington, D.C.: GPO, 1951), 59.

23. Ruth Tolman, "Some Work of Women Psychologists in the War," *Journal of Consulting Psychology* 7 (1943): 129, part of a special issue on women in wartime psychology; Florence Goodenough to Lewis Terman, June 25, 1942, Terman Papers, Stanford University Archives, Stanford, Calif.; Florence Goodenough Papers, University of Minnesota Archives, Minneapolis. On Vanuxem, see "Death of Dr. Mary Vanuxem," *Mental Hygiene* 30 (1946): 148–49; on Hayes, obituary, *NYT,* Apr. 10, 1962, 39. On

Fernald, "Grace Maxwell Fernald, 1879–1950," *Psychological Review* 57 (1950): 319–21, and Andrew Hamilton, "They Think with Their Hands," *Rotarian,* Apr. 1949, 32–34.

24. On Tolman, see *Who Was Who in America* 3 (1960): 857; Hildreth and Konold, *American Psychological Association Directory, 1951,* 468; and OSS Assessment Staff, *Assessment of Men: Selection of Personnel for the Office of Strategic Services* (New York: Rinehart, 1948), vii. On Hanfmann, see Marianne Simmel, "A Tribute to Eugenia Hanfmann, 1905–1983," *Journal of the History of the Behavioral Sciences* 22 (1986): 351; on Peak, Hildreth and Konold, *American Psychological Association Directory, 1951,* 357, and Herbert H. Hyman, *Taking Society's Measure: A Personal History of Survey Research* (New York: Russell Sage Foundation, 1991), ch. 4. See also Lucile Nelson McMahon, "Women and the War of Nerves" (on psychiatric social work), *IW* 21 (1942): 300–302, 315–16.

25. James Capshew and Alejandra Laszlo, "'We would not take no for an answer': Women Psychologists and Gender Politics during World War II," *Journal of Social Issues* 42 (1986): 157–80, and a rebuttal by Alice I. Bryan, "A Participant's View of the National Council of Women Psychologists: Comment on Capshew and Laszlo," ibid., 181–84. There are also relevant materials in the Florence Goodenough Papers, University of Minnesota Archives, as well as in the Margaret Ives Papers, Archives of the History of American Psychology, University of Akron, Ohio.

26. See J. Hillis Miller and Dorothy V. N. Brooks, *The Role of Higher Education in War and After* (New York: Harper & Bros., 1944); and I. L. Kandel, *The Impact of the War upon American Education* (Chapel Hill: Univ. of North Carolina Press, 1948), 160, 164–71. "Retraining and Transfer of College Faculty Members," *Education for Victory* 1 (June 1, 1943): 8–9, provided practical advice to administrators. On Wright's war work, see Frances W. Wright to Dr. Shapley, Feb. 24, 1945, and other items in Harvard College Observatory Director's Correspondence, 1940–50, Wic–Zr file, HUA. On Powdermaker, see *NCAB* 55 (1974): 237–38 and *CBY, 1961,* 372–73. On Phillips, see *AMWS,* 12th ed., 1971, 5:4904; on deMilt, John Mark Scott, "Clara Marie deMilt (1891–1953)," *JCE* 31 (1954): 419–20; on Bilger, *C&EN* 31 (1953): 1094. See also Aaron J. Ihde, *Chemistry: As Viewed from Bascom's Hill* (Madison: University of Wisconsin, Department of Chemistry, 1990), 598–601, one of the few departmental histories to give the wartime women credit for their contributions. See also Julia Wells Bower, "The Effect of the War on College Women and Mathematics," *Mathematics Teacher* 36 (1943): 175–78.

27. [Marguerite Zapoleon], *The Outlook for Women in Science,* WB Bulletin 223-1 (Washington, D.C., 1949), 21.

28. Cynthia Westcott, *Plant Doctoring Is Fun* (Princeton: D. Van Nostrand, 1957), ch. 6.

29. James Bryant Conant, *My Several Lives: Memoirs of a Social Inventor* (New York: Harper & Row, 1970), 347–48; Frances Perkins, *The Roosevelt I Knew* (New York: Viking, 1946), 354 (I thank Sally Kohlstedt for this reference). See also Carroll Pursell, "Alternative American Science Policies during World War II," in *World War II: An Account of Its Documents,* ed. James E. O'Neill and Robert W. Krauskopf (Washington, D.C.: Howard Univ. Press, 1976), 155–57.

30. There are several bibliographies of this recruitment literature: Louise Moore, *Occupations for Girls and Women: Selected References, July 1943–June 1948,* WB Bulletin 229 (Washington, D.C., 1949); Florence S. Hellman, comp., "Women's Part in World War

II: A List of References" (Washington, D.C.: LC, Division of Bibliography, 1942, mimeographed); LC, "Scientific Personnel: A Bibliography" (Washington, D.C., 1950, mimeographed). The best general article on the mobilization of women for war work is Eleanor F. Straub, "United States Government Policy toward Civilian Women during World War II," *Prologue* 4 (winter 1973): 240–54, which uses WMC records in the National Archives; Leila Rupp, *Mobilizing Women for War: German and American Propaganda, 1939–1945* (Princeton: Princeton Univ. Press, 1978), relies mostly on secondary sources. On Hepburn, see Charles Higham, *Kate: The Life of Katherine Hepburn* (New York: W. W. Norton, 1975), 108, and OWI, *A List of U.S. War Information Films* (Washington, D.C.: OWI, Bureau of Motion Pictures, 1943), 5; on Garson, "'Madame Curie' on the Screen," *New York Times Magazine,* Sept. 26, 1943, 18.

31. Harold A. Edgerton and Steuart H. Britt, "The First Annual Science Talent Search," *American Scientist* 31 (Jan. 1943): 55–68; bibliography in Harold Edgerton, *Science Talent: Its Early Identification and Continuing Development* (Washington, D.C.: Science Service, 1961), 86–87. On Bausch and Lomb, see "Grants and Awards," *Science* 108 (1948): 6–7; and [Zapoleon], *The Outlook for Women in Science,* 33.

32. Edna Yost, *American Women of Science,* rev. ed. (Philadelphia: J. B. Lippincott, 1955), xi. On Yost, see *Contemporary Authors* 2 (1967): 225.

33. Quoted in "War Jobs Await 5,000,000 Women," *NYT,* Sept. 25, 1942, p. 16, col. 1. For more on the WMC's weaknesses and difficulties, see George Q. Flynn, *The Mess in Washington: Manpower Mobilization in World War II* (Westport, Conn.: Greenwood, 1979), esp. ch. 8. The papers of Margaret Hickey, the chairman of the Women's Advisory Committee of the WMC, are at the Archives, University of Missouri at St. Louis (I thank Jane Miller for assistance).

34. [CSC], *Civilian War Service Opportunities for College and University Students* (Washington, D.C., 1943), 27, 33, 75, 23–24. All this hiring of women scientists by the federal government was just part of a massive wartime expansion in federal employment. According to data collected by the CSC and the WB, the number of persons employed by the federal government increased from 900,000 in 1939 (of which 18.8 percent were women) to 2.9 million at its peak in June 1944 (37.6 percent female), an increase of almost 1 million men and 1 million women workers. Yet most of these women employees were either clerical or munitions workers. The number of actual women "scientists" (a term used very loosely at the time) involved in the war effort was probably less than 10,000, a large number compared with their previous levels of participation but minute compared with the million women in government (*Women in the Federal Service, 1923–1947,* WB Bulletin 230-1 [Washington, D.C., 1949], 16, 18, 25, 28. See also Lucille Foster McMillin, *The First Year: A Study of Women's Participation in Federal Defense Activities* [Washington, D.C.: GPO, 1941], a list of women holding important defense jobs, and *The Second Year: A Study of Women's Participation in War Activities of the Federal Government* [Washington, D.C.: GPO, 1943], which seeks to recruit women mechanics and college graduates).

35. On Wills, see *NCAB* 53 (1971): 32–33; National Roster of Scientific and Specialized Personnel, *Report to the National Resources Planning Board, June 1942* (Washington, D.C.: GPO, 1943), app. D; and the final data in U.S. Department of Labor, Employment Service, National Roster of Scientific and Specialized Personnel, "Distri-

bution of Roster Registrants by: 1. Professional Field, 2. Sex, 3. Age, 4. Extent of Education, December 31, 1945" (Washington, D.C., Dec. 31, 1945, mimeographed), table 4; Edna May Turner, "Education of Women for Engineering in the United States, 1885–1952" (Ph.D. diss., New York University, 1954), esp. 78 and 148–56 (table 1); "Woman-power Available," *WWE* 13, no. 3 (Oct. 1942): 9; "Demands in the Professions," ibid. no. 4 (Dec. 1942): 11; and "Women Sought in Engineering," *AAUWJ* 36 (1943): 93–94. On Rensselaer, see Lois Graham to author, May 30, 1974; on Columbia, Virginia Gildersleeve, *Many a Good Crusade*, 103–4; and on Curtiss-Wright, "Co-ed Engineers Take Men's Places," *Aviation News*, Dec. 27, 1943, 13.

36. Elsie Eaves, "Wanted: Women Engineers," *IW* 21 (1942): 132–33, 158–59. On the civil service, see "War Jobs," *WWE* 13, no. 1 (Feb. 1942): 9. See also "Women A.I.M.E. Members Contribute Their Share in Engineering War," *Mining and Metallurgy* 23 (1942): 580–81; Lillian M. Gilbreth, "Women in Engineering," *Mechanical Engineering* 64 (1942): 856–57, 859; and Alice C. Goff, "Women CAN Be Engineers," *AAUWJ* 41 (1948): 75–76, based on her book of the same title published in 1946.

37. Henry H. Armsby, *Engineering, Science, and Management War Training: Final Report*, Bulletin 1946, no. 9, Federal Security Agency (Washington, D.C.: GPO, 1946), 15 (quotation), 45–47, 103––11, 132–33. See also Richard W. Lykes, *Higher Education and the U.S. Office of Education, 1867–1953* (Washington, D.C.: USOE, 1975), 124–34; Gladys F. Gove, "Training—The Shortest Distance between Two Points," *IW* 22 (1943): 360–61, 379–80; "Scientific Training for Women for War Jobs," *Monthly Labor Review* 58 (Jan. 1944): 107–8; E. V. Gustavson, "Engineers Made to Order," *Aviation* 42 (Nov. 1943): 167–68, 253–56; "Training for Women in Aeronautical Engineering at the University of Cincinnati," *Science* 97 (1943): 548–49; Frances Tallmadge, "Engineering Training for Women," *Journal of Higher Education* 15 (1944): 379–82; Phyllis Gandee, "Laboratory Ladies," *Flying* 33 (Dec. 1943): 56–57, 130, 136. Several articles discussed the curriculum in these special programs, such as R. H. Baker and Mary L. Reimold, "What Can Be Done to Train WOMEN for JOBS in ENGINEERING," *Mechanical Engineering* 64 (Dec. 1942): 853–55; E. D. Howe, "Training WOMEN for ENGINEERING TASKS," ibid. 65 (Oct. 1943): 742–44; and C. W. Cole, "Training of Women in Engineering," *JEE* 34 (Oct. 1943): 167–84. ESMWT was also well publicized in *Education for Victory* and the *WWE*—it and the National Roster were the most visible governmental agencies concerned with training and recruiting women scientists, mentioned over and over.

38. *The Outlook for Women in Architecture and Engineering*, WB Bulletin 223-5 (Washington, D.C., 1948), 52–62, esp. 61–62.

39. Althea R. Lowe, "Pittsburgh Women Chemists Active in Research," *C&EN* 21 (1943): 1441–44; James Coull, "The Training of Women for the Chemical Industries," ibid., 1707–8; idem, "Training Women Operators for Chemical Industry," *Chemical and Metallurgical Engineering* 50 (June 1943): 116–18; Hannah Garry, "Chemical Woman-power," *Chemical Industries* 452 (Apr. 1943): 445–49; Helen I. Miner, "Women Chemists Play Role in Detroit Production," *C&EN* 21 (1943): 80–83; Marion Cleaveland, "Women Chemists in Cleveland War Industries," ibid. 22 (1944): 438–39; Jane Hastings, "High School Graduates in Industrial Laboratories," *JCE* 22 (1945): 202–3; Lois Woodford, "Opportunities for Women in Chemistry," ibid. 19 (1942): 536–38; idem, "Trends in the Industrial Employment of Women Chemists," ibid. 22 (1945): 236–38;

Cornelia Snell, "Women as Professional Chemists," ibid. 25 (1948): 450–53; idem, "Careers in Chemistry and Chemical Engineering: The Woman Chemist," *C&EN* 28 (1950): 3110–12, reprinted in *Careers in Chemistry and Chemical Engineering* (Easton, Pa.: ACS, 1951), 44–45; "War Demands for Chemists," *WWE* 13, no. 2 (May 1942): 5–6; "Demand for Chemists," ibid. 14, no. 1 (Feb. 1943): 6; *The Outlook for Women in Chemistry,* WB Bulletin 223-2 (Washington, D.C., 1948), 11–13; National Roster, "Distribution of Roster Registrants . . . December 31, 1945," table 4.

40. On Emerson, see *C&EN* 30 (1952): 1514; on Rolland, *NYT,* Apr. 12, 1947, 17; and on Bridgeman, W. Robert Grubb, "Oil Woman of the Year," *National Business Woman* 40 (Feb. 1961): 8, 28 (she was chosen by the Desk and Derrick Clubs, a nationwide group of women employees in the petroleum industry).

41. Woodford, "Opportunities for Women in Chemistry," 536–38; Icie Macy Hoobler to Lois Woodford, Nov. 8 1942, copy in Icie Macy Hoobler Papers, Bentley Historical Library, University of Michigan, Ann Arbor.

42. Alice Kuebel, "Women in Chemistry," *JCE* 29 (1943): 248.

43. Ibid., 249.

44. Walter J. Murphy, "Postwar Opportunities for Women in Chemical Industry," in "War and Post-war Employment and Its Demands for Educational Adjustments" (New London, Conn.: Institute of Women's Professional Relations, 1944, mimeographed), 103–11; idem, "Women in Chemical Industry," *WWE* 15, no. 4 (winter 1944): 4–8; idem, "War Opportunities for Women," *The Chemist* 21 (September 1944): 363–70. See also Margaret W. Rossiter, *Women Scientists in America: Struggles and Strategies to 1940* (Baltimore: Johns Hopkins Univ. Press, 1982), 117–18 and ch. 9.

45. "Women Chemists in New York," *C&EN* 22 (1944): 1673. See also Woodford, "Trends in the Industrial Employment of Women Chemists," 236–38. Woodford's view that management did not want to lay off expensively trained women chemists but that they quit anyway when they got married contradicted the fears of other women at the meeting that they would soon be let go or demoted. This was certainly the case by October 1946 (Walter J. Murphy, "What About the Women Chemists?" [editorial], *C&EN* 24 [1946]: 2597).

46. Florence Seibert to Dr. H. Marjorie Crawford, Oct. 12, 1942, box 7, Florence Seibert Papers, APS.

47. Eleanor F. Horsey and Donna Price, "Science Out of Petticoats," *AAUWJ* 40 (1946): 13–16.

48. Emma Perry Carr, "Letters to Headquarters: Those Petticoats in Science," ibid. 40 (1947): 107. Carr's colleague Mary L. Sherrill expressed the same sentiments in a letter to Mrs. L. B. Zapoleon, Chief, Employment Opportunities Section, July 16, 1947, box 115, WB Papers (RG 86), NARA, Washington, D.C.

49. "From AAUW Headquarters Mail," *AAUWJ* 36 (1943): 97.

50. Florence Codman, "Womanpower 4F," *IW* 22 (1943): 260, 279–80; Priscilla Crane, "Why Won't They Let ME Help? I. Woman Power Unclassified," ibid. 294, 309; Dale A. Stubbs, "Why Won't They Let ME Help? II. To the Aid of the Nation," ibid., 294, 309–10; "Comments on 'Womanpower 4F,'" ibid., 334, 347; "More Comments on 'Womanpower 4F,'" ibid., 367, 380.

51. On Maffett, see *Who's Who in America,* 24th ed. (1946–47), 1482; on Hughes,

CBY, 1950, 267–69. See also Geline M. Bowman, comp., *A History of the National Federation of Business and Professional Women's Clubs, Inc., 1919–1944* (New York: NFBP-WC, 1944). On AAUW, see "Committee on Status of Women Considers Problems," *AAUWJ* 37 (1943): 42–43; "Economic and Legal Status of Women," ibid. 40 (1947): 140–46, summarizes its extensive activities in 1941–47, as does Sarah T. Hughes, "Economic and Legal Status of Women," ibid. 40 (1947): 250–54.

52. "National Group Files Full Data on Membership," *NYT,* Oct. 12, 1941, pt. 2, p. 5, col. 1; "Women Backed for War Posts," ibid., Aug. 23, 1942, pt. 2, p. 4, col. 8, and "21,000 Women on Roster of Specialized Personnel," ibid., Dec. 26, 1942, p. 16, col. 1.

53. Minnie L. Maffett, "We Too Must Fight This War," *IW* 21 (1942): 227–28, 246–47; idem, "Mobilizing Womanpower: An Open Letter to the Manpower Commission," ibid., 356, 380; "In the Mail" (reply), ibid. 22 (1943): 1; [Minnie L. Maffett], "Under-use of Womanpower Slows War Effort," ibid., 230–31, 252; "Few Women Are Making Policies," *NYT,* Sept. 26, 1943, pt. 2, p. 14, col. 8; "Women's Role," *WWE* 14, no. 3 (Oct. 1943): 7; "Conference at the White House," *IW* 23 (1944): 225; [Minnie L. Maffett], "D-Day and H-Hour for Women," ibid., 239–40. Maffett's success in this direction was partial. After she spent months urging chairman Paul McNutt to appoint a woman member of the WMC, he responded by making the NFBPWC's own vice president, Margaret Hickey of St. Louis, chairman of a subordinate Women's Advisory Committee to counsel the commission on women's affairs but not vote on them. See also Straub, "United States Government Policy toward Civilian Women during World War II," 240–54; WMC, *History of the Women's Advisory Committee, War Manpower Commission,* rev. ed (Washington, D.C.: GPO, 1945); "The Federation Is Honored," *IW* 21 (1942): 289, 319; and Virginia Price, "Advisors on Womanpower," ibid. 23 (1944): 104, 122.

54. Women had long been active in peace and antiwar movements (see Gildersleeve, *Many A Good Crusade,* 315–57).

55. "Coordinating the Resources of Women's Colleges with Industries Employing Women," *Education for Victory* 1 (Mar. 15, 1943): 6–7; "Putting College Women in War Jobs Urged as Task for Educators," *NYT,* Feb. 11, 1943, p. 15, col. 2, does not mention Gilbreth's address. On Gilbreth, see *CBY, 1951,* 233–35.

56. "Tentative Legislative Program, 1943–44," *IW* 22 (1943): 80–81; Alice L. Manning, "We Buckle on Our Armour and Return to the Fray," ibid., 338, 342–43; idem, "Legislation—Looking Backward Looking Forward," ibid. 23 (1944): 42–43, 61–62; "Tentative Legislative Program, 1944–45," ibid., 78–79; and the many frequent other articles on pending legislation by Alice Manning. For a more critical assessment, see Susan M. Hartmann, "Women's Organizations during World War II: The Interaction of Class, Race, and Feminism," in *Woman's Being, Woman's Place: Female Identity and Vocation in American History,* ed. Mary Kelley (Boston: G. K. Hall, 1979), 313–28.

57. [Maffett], "D-Day and H-Hour for Women," 239–40.

58. Gilbreth, "Women in Engineering," 857.

59. Straub, "United States Government Policy toward Civilian Women during World War II," 254; see also William Chafe, *The American Woman: Her Changing Social, Economic, and Political Roles, 1920–1970* (New York: Oxford Univ. Press, 1972), ch. 7.

Chapter 2   Postwar "Adjustment": Displacement and Demotion

1. On Zapoleon, see *Who's Who in the South and Southwest*, 16th ed. (Chicago, 1978), 815, and [Marguerite Zapoleon], *The Outlook for Women in Science*, WB Bulletin 223 (Washington, D.C., 1948–49). Notes and correspondence relating to the preparation of this set of bulletins are in the WB Papers (RG 86), boxes 108–20, 597–99, NARA; see also *The Outlook for Women in Dietetics*, WB Bulletin 234 (Washington, D.C., 1950), the notes for which are in box 641. (I thank Rosalind Rosenberg for the reminder about Woolley.)

2. [Marguerite Zapoleon], *Outlook for Women in Science*, WB Bulletin 223-1 (Washington, D.C., 1949), 5; M. H. Trytten, "Coverage of Scientific Personnel in *American Men of Science*, Eighth Edition," *Science* 112 (1950): 265–66; *Employment, Education, and Earnings of American Men of Science*, BLS Bulletin 1027 (Washington, D.C., 1951), 6–7, 11–12, 36, 39; *AMS*, 8th ed., 1949. This was an increase of more than 1,000 from the 1,912 in the sixth edition (1938), but the even greater increase in male scientists in the years 1938–48 meant that the women's proportion dropped from 7.0 percent in 1938 to 6.0 percent in 1948. Almost three-fourths of all entrants (73.1 percent) held doctorates (Ph.D. or M.D.), which were now required of new additions. A higher percentage of women (83.7 percent) than of men (72.4 percent) still held these. Data on unemployment were not presented in the final report, nor were any on the marital status of the scientists, although the eighth edition of the *AMS* was the first to provide them. (A survey of the first 500 women in this edition reveals that 34 percent admitted to being married, compared with 88 percent of the first 500 men.) Trytten's data still exist on microfilm in RG 307, Federal Records Center, Suitland, Md.

3. [Zapoleon], *Outlook for Women in Science*, 52–62.

4. Public Law 346, 78th Cong. (June 22, 1944), *Serviceman's Readjustment Act;* Davis R. B. Ross, *Preparing for Ulysses: Politics and Veterans during World War II* (New York: Columbia Univ. Press, 1969), ch. 4. See also Lt. Col. Mary Agnes Brown, "The Woman Veteran," *WWE* 17, no. 1 (winter 1946): 1–4, and Lt. Col. Clara Raven, "Achievement of Women in Medicine, Past and Present—Women in the Medical Corps of the Army," *Military Medicine* 125 (1960): 105–11.

5. President's Commission on Veterans' Pensions, *A Report on Veterans' Benefits in the United States: Staff Report No. IX, Part A, Readjustment Benefits: General Survey and Appraisal* (Washington, D.C., 1956), 90. Cornell is cited in "Doors Closing for Women Students," *AAUWJ* 39 (1946): 167–68. On the University of Michigan, see *[University of Michigan] President's Report for 1944–45*, 165, *1945–46*, 196, *1946–47*, 235, *1947–48*, 200, *1948–49*, 195 (Ann Arbor, [1945–49]); and Dorothy Gies McGuigan, *A Dangerous Experiment: One Hundred Years of Women at the University of Michigan* (Ann Arbor: Center for Continuing Education of Women, 1970), 112. On the University of Wisconsin, see Keith W. Olson, "World War II Veterans at the University of Wisconsin," *Wisconsin Magazine of History* 53 (1969–70): 86, 88–89. See also idem, *The GI Bill, the Veterans, and the Colleges* (Lexington: Univ. of Kentucky Press, 1974), and Alma Routson, *A Gradual Joy* (Boston: Houghton Mifflin, 1952), a novel about a GI Bill couple at Michigan State University.

6. *New York Times Index for the Published News of 1946* (New York: New York Times,

1947), 2542–49, 2552. See also "Colleges for Women Admit Men," *Higher Education* 3 (Dec. 15, 1946): 8–9. The Bronx campus of Hunter College later became the separate, coeducational Lehman College of the City University of New York. See also "Hunter College Appoints Dean of Men to Aid Veterans," *NYT*, Oct. 3, 1946, p. 56, col. 2.

7. Doak S. Campbell, *A University in Transition*, Florida State University Studies, no. 40 (Tallahassee: Florida State University, 1964).

8. USOE, *Earned Degrees Conferred by Higher Educational Institutions, 1947/48–1952/53* (Washington, D.C., 1949–54).

9. "Doors Closing for Women Students," 167; "Student Pressures in Higher Education," *AAUWJ* 39 (1946): 231; Alice Lloyd, "Women in the Postwar College," ibid., 131–34.

10. Memorandum, Herman Wells to Board of Trustees, Dec. 3, 1945, Housing File, Indiana University Archives, Bloomington, cited in David R. Warriner, "The Veterans of World War II at Indiana University, 1944–1951" (Ph.D. diss., Indiana University, 1978), 101. I thank him for sending relevant portions.

11. "Resolutions Adopted by the Convention," *IW* 25 (Aug. 1946): 254.

12. "Girls' Study Rights Debated on Forum," *NYT*, May 22, 1946, p. 19, col. 5; Alice Lloyd developed a terminal illness in the late 1940s and was not able to continue her project "Alice Crocker Lloyd, 1894–1950," *Journal of the National Association of Deans of Women* 13 (1950): 173–74; obituary, *NYT*, Mar. 4, 1950, 17.

13. Jane Knowles (archivist at Radcliffe College) to author, June 27, 1979, citing enrollment figures for the Radcliffe Graduate School for 1944–62 and annual reports of its dean; Harvard University, *Annual Report of Deans of the Graduate School of Arts and Sciences, 1954–55*, 305 (table 1); Julia Morgan (archivist of Johns Hopkins University) to author, July 10, 1979, with enrollment statistics for its School of Higher Studies of the Faculty of Philosophy for 1946–52. Cornell evidently had no enrollment data by sex for graduate students at this time (Kathleen Jacklin to author, Aug. 10, 1979). Yale did between 1946–47 and 1953–54: from a high of 17.32 percent in 1946–47 (181 women of 1,046), women's enrollment sank to a low of 13.51 percent (152 of 1,125) in 1950–51 and then revived to 16.67 percent (206 of 1,236) in 1953–54 ("Reports to the President by the Dean of the Graduate School," 1944–45 to 1953–54, Yale University Archives, New Haven, Conn.).

14. Kate Huevner Mueller, unpublished autobiography, Apr. 1971, 35–43 (typescript in Indiana University Archives; I thank archivist Dolores Lahrman for providing me with a copy).

15. Alma Lutz, "Women's History—Background for Citizenship," *AAUWJ* 40 (1946): 6–8; M. Eunice Hilton, "On Being a Dean of Women in 1946," *Journal of the National Association of Deans of Women* 10 (1947): 73–74; Louise Walcott Spencer, "Eleven Years of Change in the Role of Deans of Women in Colleges, Universities, and Teachers Colleges," ibid. 14 (1951): 51–83. In 1956 the group changed its name from the National Association of Deans of Women to the National Association of Women Deans [of all levels and titles] and Counselors (NAWDC). See also W. H. Cowley, "The Disappearing Dean of Men," *Occupations* 16 (1937): 147–54; Hurst R. Anderson, "Education for the Dean of Women," *AAUWJ* 48 (1954): 17–18; and Lloyd, "Women in the Postwar College," 133.

16. Lloyd, "Women in the Postwar College," 134.

17. Tom Lewin, "Interview with Fritz and Grace Heider," winter 1985, 50, University of Kansas Archives, Spencer Research Library, Lawrence; Fritz Heider, *The Life of a Psychologist: An Autobiography* (Lawrence: Univ. of Kansas Press, 1983), 167–68. On Grace Heider, see *AMWS,* 12th ed., S&B, 1973, 1:982.

18. Frances K. Graham et al., "In Memoriam: Margaret Kuenne Harlow," *Child Development* 42 (1971): 1314.

19. On G. Ann Magaret Garner, see John A. Iazo, ed., *Biographical Directory* (Washington, D.C.: APA, 1973), 330; on Ross Garner, see *Directory of American Scholars,* 8th ed., vol. 2 (New York: Bowker, 1982), 245. Over the years her interests changed from mental testing of gifted children to the needs of crippled children.

20. Years later, as Mary D. S. Ainsworth, she became a full professor at Johns Hopkins. See *AMS,* 10th ed., S&B, 1956, 6; and her autobiographical account, "Mary D. Salter Ainsworth," in *Models of Achievement: Reflections of Eminent Women in Psychology,* ed. Agnes N. O'Connell and Nancy Felipe Russo, vol. 1 (New York: Columbia Univ. Press, 1983), 200–219.

21. On Einstein and Brandeis University, see Abram L. Sachar, *A Host at Last* (Boston: Little, Brown, 1976), ch. 2; and Hilda Geiringer to Albert Einstein, Nov. 8, 1946, Hilda Geiringer von Mises Papers, HUA.

22. Arnold Dresden to Mrs. Von Mises, Oct. 20, 1947, Hilda Geiringer von Mises Papers.

23. Von Mises to Dresden, Oct. 26, [1947], ibid. Mary C. Graustein, Ph.D., Radcliffe 1917, joined the Tufts faculty as an assistant professor in 1944, was promoted to associate professor in 1950, and retired in 1955 (Alumnae Records, RCA [I thank Jane Knowles for a copy]; "Resolution" and obituaries, Former Faculty File, Tufts University Archives, Medford, Mass. [I thank Robert Johnson-Lally for copies]). Mary Boas was also an instructor there from 1943 to 1948 (*AMWS,* 12th ed., 1971, 1:553. See also Ralph P. Boas, Jr., "Ralph P. Boas, Jr.," in *More Mathematical People: Contemporary Conversations,* by Donald J. Albers et al. [Boston: Harcourt Brace Jovanovich, 1990], 34; and obituary, *Harvard Magazine,* Nov.–Dec. 1992, 97). On Mabel Barnes, see *AMWS,* 12th ed., 1972, 1:295. Edith Bush, the sister of Vannevar Bush (the famous head of the wartime OSRD), was the dean of Jackson College for Women and a professor of mathematics from 1926 to 1952. For more on Tufts in the postwar era, see Richard M. Freeland, *Academia's Golden Age: Universities in Massachusetts, 1945–1970* (New York: Oxford Univ. Press, 1992).

24. J. Donald Kingsley, "A Request for Information Concerning the Effectiveness of Science Teaching in Colleges and Universities," Apr. 1947, 2 (question 3e); summary of responses in box 8, Records of the President's Scientific Research Board (RG 220), Harry S. Truman Presidential Library, Independence, Mo.; President's Scientific Research Board, *Science and Public Policy,* vol. 4 (Washington, D.C.: GPO, 1947), 156. When Ruth Merrill, dean of women at the University of Rochester, summarized the Steelman Report to the AAUW's Committee on the Status of Women in 1948, she was disappointed that it had nothing to say about women or discrimination (Ruth Merrill, "Summary—Report of President's Scientific, Research Board," n.d., *AAUW Archives, 1881–1976* [New York: Microfilming Corporation of America, 1980], reel 122, secs. 101–3,

mentioned in "Minutes of the Committee of the Status of Women of the AAUW," Oct. 30, 1948, 5, ibid., reel 117, sec. 659; sets at Lamont Library, Harvard, and Olin Library, Cornell).

25. Lloyd, "Women in the Postwar College," 133; Geraldine Hammond, "And What of the Young Women?" *AAUP Bulletin* 33 (1947): 298–303.

26. Jane Loevinger, "Professional Ethics for Women Psychologists," *AP* 3 (1948): 551; *AMWS*, 13th ed., S&B, 1978, 740. Loevinger fought most of her life not only for women's issues but also for civil rights in a part of the country where it took considerable courage to do so ([Jane Loevinger], "Jane Loevinger" [autobiography], in *Models of Achievement: Reflections of Eminent Women in Psychology,* ed. Agnes N. O'Connell and Nancy Felipe Russo, vol. 2 [Hillsdale, N.J.: Lawrence Erlbaum Associates, 1988], 155–66). Another victim of the antinepotism rules at Washington University was mathematician Deborah Tepper Haimo (see interview in Aida K. Press, "A Passion for the Beauty of Mathematics," *Radcliffe Quarterly* 65, no. 4 [Dec. 1979]: 10–12). Gerty Cori, there with her husband Carl, got her professorship in 1947, the same year she won the Nobel prize. Helen Tredway Graham had also been a professor for years, despite her husband's professorship there.

27. Marguerite J. Fisher, "Economic Dependents of Women Faculty," *AAUP Bulletin* 35 (1949): 269–73. On Fisher, see *WWAW* 1 (1958): 420.

28. Bernice Brown Cronkhite, *The Times of My Life* ([Cambridge, Mass.]: n.p., 1982), ch. 9; "President's Report to the Trustees, 1946–1947," *Official Register of Radcliffe College* 13 (1947): 7–8; occasional relevant items, including especially BBC to WKJ [President Wilbur K. Jordan], memorandum, Dec. 11, 1946, BBC to Doris Zemurray Stone, Dec. 13, 1946, and minutes of meetings of search committee, fall 1947, in boxes 2 and 22, Papers of the Office of the Graduate Dean, RCA. Sirapie Der Nersessian joined the faculty of the Dumbarton Oaks research facility in Washington, D.C., a part of Harvard's faculty of arts and sciences, in 1946, as an associate professor (J. S. Allen, "Sirapie Der Nersessian [b. 1896]: Educator and Scholar in Byzantine and Armenian Art," in *Women as Interpreters of the Visual Arts, 1820–1979,* ed. Claire Richter Sherman [Westport, Conn.: Greenwood, 1981], 340; I thank Clark Elliott for telling me of this volume). See also ch. 6 below, esp. table 6.4.

29. Belle Krasne, "Harvard's First Lady," *IW* 27 (1948): 357–58, 373; there is a small collection of Helen M. Cam Papers at SLRC. See also Giles Constable to Professor Banham, Mar. 8, 1969, box 13, Katherine Banham Papers, Duke University Archives, Durham, N.C. On DuBois, see *AMWS*, 13th ed., P & B, 1976, 2:1093; obituary, *NYT,* Apr. 11, 1991, B-14; "Anthropology—Studied, Taught, and Practiced: A Description of the Department at Barnard and an Outline of the Careers of Two Alumnae," *Barnard College Alumnae Magazine* 40 (Nov. 1950): 4–6; Judith Walzer, "Interviews with Cora DuBois, August 1981," and idem, "Interview with Emily Vermeule, August 6, 1981," both at MCRC.

30. On Hanfmann, see Eugenia Hanfmann, "Eugenia Hanfmann" (autobiography), in O'Connell and Russo, *Models of Achievement,* 1:147; Marianne L. Simmel, "A Tribute to Eugenia Hanfmann, 1905–1983," *Journal of the History of the Behavioral Sciences* 22 (1986): 352; and *AMWS*, 12th ed., S&B, 1973, 1:931.

31. Thelma G. Alper, "Thelma G. Alper," in O'Connell and Russo, *Models of Achieve-*

ment, 1:189–99, esp. 194–95 (quotation). See also *AMWS,* 13th ed., S&B, 1978, 22; Gordon W. Allport to Dean Bernice B. Cronkhite, Nov. 23, 1945, and reply, Nov. 30, 1945, box 5, Papers of the Office of the Graduate Dean; and Thelma G. Alper, "Psychology in New England: A Retrospective Look, A Woman's Point of View," *Journal of the History of the Behavioral Sciences* 16 (1980): 220–24. Alper needed some kind of visible deliberate act of defiance or protest, and she had no doubt heard from Harvard's gossipy professor of psychology Edwin G. Boring that Margaret Washburn of Vassar had once done such an outrageous thing in the 1930s. See Washburn in *NAW;* Margaret W. Rossiter, *Women Scientists in America: Struggles and Strategies to 1940* (Baltimore: Johns Hopkins University Press, 1982), 215; and Elizabeth Scarborough and Laurel Furumoto, *Untold Lives: The First Generation of American Women Psychologists* (New York: Columbia Univ. Press, 1987), ch. 4. Interestingly, Allport was a social psychologist known for his research discrediting racial and religious discrimination.

32. Wilfred B. Shaw, ed., *The University of Michigan: An Encyclopedic Survey,* vol. 1 (Ann Arbor: Univ. of Michigan Press, 1941), 188, 191; Dorothy McGuigan, *Dangerous Experiment,* 87; *Proceedings of the Board of Regents* (Ann Arbor), Mar. 1899, 355 (I thank Nancy Bartlett for a copy).

33. On Peak, see *AMWS,* 12th ed., S&B, 1973, 2:1904–5; Daniel Katz, "Helen Peak (1900–1985)," *AP* 42 (1987): 510; and Herbert H. Hyman, *Taking Society's Measure: A Personal History of Survey Research* (New York: Russell Sage Foundation, 1991), ch. 4.

34. The report appeared as *Women in Higher-Level Positions,* WB Bulletin 236 (Washington, D.C., 1950), but it was chiefly concerned with women in banking, insurance companies, department stores, and manufacturing, only the last of which included more than a few scientists. See ch. 12 below.

35. "Report of the Committee on the Status of Women to the Biennial Convention of the American Association of University Women, June 19–23, 1949," 27, carton 3, f. 87, Somerville-Howorth Papers, SLRC, published as "[Report of the] Committee on the Status of Women," *AAUWJ* 43 (1949): 57. After its next annual meeting, in December 1949, the Committee on the Status of Women announced another resolution:

> The committee shall work for the wider employment of women faculty members without discrimination as to sex and marital status and without differentials as to rank, salary, departmental chairmanships, and appointments to policy-making committees.
>
> The committee shall continue to concentrate its efforts on securing equality for women in admission policies and practices of graduate, professional, and technical schools and shall continue its endorsement of equal opportunity for training and placement after admission.

See "The AAUW's Resolutions on the Status of Women," *School and Society* 70 (1949) 427; and "AAUW Resolutions on Status of Women in Faculties," *NEA Journal* 39 (1950): 150–51. Strangely, these resolutions seem not to have appeared in the *AAUWJ.*

36. The AAUW long retained a strong Southern influence. See Janice Leone, "Integrating the American Association of University Women, 1946–1949," *Historian* 51 (1989): 423–45. See also "University Women Reject Race Bias," *NYT,* Dec. 11, 1946; and Margaret Mead to Mrs. Harry Crum, President of New York City branch of AAUW, Dec. 11, 1946, and undated reply, all in box 44, MMP; and a flurry of letters between

AAUW President Susan Riley and Lucy Howorth between July 21 and August 4, 1954, in carton 3, f. 105, Somerville-Howorth Papers, about an article in the June 22 *St. Louis Post-Dispatch* that had accused AAUW's national leadership of Communist sympathies. On Kenyon, see *NAWMP;* and "Disloyalty Charges," *AAUWJ* 43 (1950): 224–26. Also cited by the senator was Esther Caukin Brunauer of the State Department, who had worked for the AAUW for seventeen years. See also Leila J. Rupp and Verta Taylor, *Survival in the Doldrums: The American Women's Rights Movement, 1945 to the 1960s* (New York: Oxford Univ. Press, 1987), 136–431, 155–63, for the chilling effects of racism and McCarthyism on women's organizations in the late 1940s and early 1950s.

37. Susan Riley to Mrs. Tryon, July 19, 1954, and Riley to Howorth, July 20, 1954, carton 3, f. 105, Somerville-Howorth Papers; "AAUW Standards Re-Applied," *AAUWJ* 48 (May 1955): 239–42. For more on the militant Committee on Standards and Recognition, see ch. 9 below.

38. Many talented and dynamic women were attracted to the field of psychoanalysis. Almost all specialized in the psychology of women (Karen Horney, Helene Deutsch, Marie Bonaparte) or of children (Melanie Klein, Kate Friedlander, Freud's own daughter Anna Freud), making these two fields important new areas of "women's work" in medicine. Yet no one seemed to criticize these highly trained professionals for telling other women to stay home with the children. See *NAWMP;* Franz Alexander et al., eds., *Psychoanalytic Pioneers* (New York: Basic, 1966); Helene Deutsch, M.D., *Confrontations with Myself* (New York: W. W. Norton, 1973); Suzanne Gordon, "Helene Deutsch and the Legacy of Freud," *New York Times Magazine,* July 30, 1978, 22–25; obituary, *NYT,* Apr. 1, 1982, 22; Brenda S. Webster, "Helene Deutsch: A New Look," *Signs* 10 (1985): 553–71; and Nellie L. Thompson, "Helene Deutsch: A Life in Theory," *Psychoanalytic Quarterly* 56 (1987): 317–53.

39. On Lewis, see *WWAW* 1 (1958–59): 764; and Olive Lewis, "Married Women and Scientific Investigation," *Hygeia* 26 (1948): 338, 366–67. Two other articles at this time also warned working women that employers justifiably preferred men, who would not leave when they got married: Polly Weaver's "Woman, the X Factor," *Mademoiselle,* Feb. 1948, 145–48, 258–63, and John Arrend Timm's "Careers for Women in Science," *Journal of Education* 132 (1949): 8–9.

40. See *Reader's Guide to Periodical Literature,* 1945–51, s.v. "Cori." See also Marcel C. LaFollette, *Making Science Our Own: Public Images of Science, 1910–1955* (Chicago: Univ. of Chicago Press, 1990), ch. 5; and idem, "Eyes on the Stars: Images of Women Scientist in Popular Magazines," *Science, Technology, and Human Values* 13 (1988): 262–75.

41. Frank B. Gilbreth, Jr., and Ernestine Gilbreth Carey, *Cheaper by the Dozen* (New York: Thomas Y. Crowell, 1948) and *Belles on Their Toes* (New York: Thomas Y. Crowell, 1950). A contemporary survey of seventy women scientists in Minneapolis (members of Sigma Delta Epsilon) ranked part-time jobs and more domestic help as the chief needs of mother-scientists (Elizabeth Wagner Reed, "Productivity and Attitudes of Seventy Scientific Women," *American Scientist* 38 [1950]: 132–35).

42. Harry Henderson, "Is Your Wife Starving Alone?" *Nation's Business* 40 (Aug. 1952): 63.

43. Waldo Schmitt, "Kellogg's complaints," undated, in box 7, f. 4, Doris Mable

Cochran Papers, Smithsonian Institution Archives, Washington, D.C. On Cochran, see *AMS,* 11th ed., 1965, 1:910–11. Perhaps as a show of support, the editor of *Herpetologica* devoted an issue (vol. 8, pt. 4 [Jan. 1953]) to her work. There are other Cochran letters, dating from happier times in the 1930s, in the Elisabeth Deichmann Papers, Museum of Comparative Zoology Archives, Harvard. For an unintentionally devastating picture of slipshod Smithsonian personnel practices, see Richard E. Blackwelder, *The Zest for Life, or Waldo Had a Pretty Good Run: The Life of Waldo LaSalle Schmitt* (Lawrence, Kans.: Allen Press, 1979). See also Bertha M. Nienburg, *Women in the Government Service,* WB Bulletin 8 (Washington, D.C., 1920), 7, 14; and Ellis Yochelson, *The National Museum of Natural History: Seventy-five Years in the Natural History Building* (Washington, D.C.: Smithsonian Institution Press, 1985).

44. Coleman J. Goin, "Doris Mable Cochran, 1898–1968," *Copeia,* 1968, 661–62.

45. "AAUW Suggests Appointment of Women to Government Posts," *Penn News,* Oct. 19, 1950, clipping in Althea K. Hottel Alumni File, University of Pennsylvania Archives, Philadelphia.

46. "[Report of] the Subcommittee on Civil Service Study of the Legislative Recommendations Committee," Oct. 7, 1951, Archives, NFBPWC, Washington, D.C. See also Eleanor Lansing Dulles, *Chances of A Lifetime: A Memoir* (Englewood Cliffs, N.J.: Prentice-Hall, 1980), ch. 18.

47. Alice I. Bryan, quoted in *ICWP Newsletter,* Dec. 1949, 3.

48. Evelyn M. Carrington, "History and Purposes of the International Council of Women Psychologists," *AP* 7 (1952): 100; Mary Roth Walsh, "Academic Professional Women Organizing for Change: The Struggle in Psychology," *Journal of Social Issues* 41, no. 4 (winter 1985): 21.

49. Alice I. Bryan, "A Participant's View of the National Council of Women Psychologists: Comment on Capshew and Laszlo," *Journal of Social Issues* 42, no. 1 (spring 1986): 182.

50. Harriett A. Fjeld and Louise Bates Ames, "Women Psychologists: Their Work, Training, and Professional Opportunities," *Journal of Social Psychology* 31 (1950): 89, 93.

51. Gladys Schwesinger, quoted in "The Future of the ICWP," *ICWP Newsletter,* Mar. 1950, 4.

52. Mildred Mitchell, in ibid., June 1950, 3, and Oct. 1950, 3; idem, "Status of Women in the American Psychological Association," *AP* 6 (1951): 193–201.

53. Edwin G. Boring, "The Woman Problem," *AP* 6 (1951): 679–82, reprinted in Boring, *Psychologist at Large: An Autobiography and Selected Essays* (New York: Basic, 1961), 185–93 (see also 72). The quotation is from Edwin G. Boring to Helen Peak, Nov. 14, 1946, in Edwin G. Boring Papers, HUA. Alice I. Bryan and Edwin G. Boring, "Women in American Psychology: Prolegomenon," *Psychological Bulletin* 41 (1944): 447–54; idem, "Women in American Psychology: Statistics from the OPP Questionnaire," *AP* 1 (1946): 71–79; idem, "Women in American Psychology: Factors Affecting Their Professional Careers," ibid. 2 (1947): 3–20, esp. 8 and 13. See also Boring to Maud Merrill James, Sept. 19, 1951, Boring Papers; and James H. Capshew and Alejandra C. Laszlo, "'We would not take no for an answer': Women Psychologists and Gender Politics During World War II," *Journal of Social Issues* 42, no. 1 (spring 1986): 171–74. On Bryan, see obituary, *NYT,* Nov. 4, 1992, D-24; Mary Niles Maack, "Alice I. Bryan:

Psychologist, Library Educator, and Feminist, 11 September 1902–30 October 1992," *Libraries and Culture* 29, no. 1 (winter 1994): 128–32; and "Appreciations and Memorial Tributes from Friends and Colleagues," ibid., 133–41.

54. See esp. Boring to Peak, Nov. 14, 25, 27, 1946, Boring Papers.

55. Boring, "Woman Problem," 682.

56. Fannie Armitt Handrick, "Women in American Psychology: Publications," *AP* 3 (1948): 542 .

57. Boring to Mary W. Calkins, Oct. 5, 1921, and Calkins to Boring, Oct. 9, 1921, Boring Papers.

58. Alberta Gilinsky to Boring, [1954], ibid.

59. Boring, "Woman Problem," 681, 680.

60. Virginia Sanderson, "Progress Report, Committee on the Status of Women Psychologists," Feb. 22, 1955, 2–3, in Doris Twitchell-Allen Papers, Archives of the History of American Psychology, University of Akron, Akron, Ohio.

61. Harriet O'Shea, quoted in *ICWP Newsletter,* Nov. 1954, last page.

62. Margaret Curti, quoted in ibid., June 1950, 4. On the name change, see *International Council of Psychologists Newsletter,* July 1959, 2; the clause in the society's statement of purpose that it was working "particularly with respect to the contributions of women" was dropped in ibid. 3, no. 1 (Oct. 1961): 2.

63. Ravenna Helson, "The Changing Image of the Career Woman," *Journal of Social Issues* 28, no. 2 (summer 1972): 33–46.

64. See, e.g., Gerhart Saenger and Norma S. Gordon, "The Influence of Discrimination on Minority Group Members in Its Relation to Attempts to Combat Discrimination," *Journal of Social Psychology* 31 (1950): 95–120, esp. 118–20. About this time Helen Mayer Hacker applied the concept of minority groups to women in "Women as a Minority Group," *Social Forces* 30 (1951): 60–69.

Chapter 3   "Scientific Womanpower": Ambivalent Encouragement

1. John Edward Wiltz, "The Korean War and American Society," in *The Korean War: A Twenty-five-Year Perspective,* ed. Francis H. Heller (Lawrence: Regents Press of Kansas, 1977), 112–58; "Comments," ibid., 159–80; Dorothy C. Stratton, "Our Great Unused Resource—Womanpower," *New York Times Magazine,* Oct. 1, 1950, 17, 26, 28, 30. See also *New York Times Index,* 1951, s.v. "U.S. Armament—Draft and Mobilization" and "Women—U.S."; and *Education Index,* 1950–53, s.v. "Women and War." General Lewis B. Hershey of the Selective Service, Assistant Secretary of Defense for Manpower Anna Rosenberg, several college presidents, and the NFBPWC leadership favored drafting women.

2. James M. Gehardt, *The Draft and Public Policy: Issues in Military Manpower Procurement, 1945–1970* (Columbus: Ohio State Univ. Press, 1971), ch. 3; John J. Corson, "Making the Most of Manpower," *New York Times Magazine,* May 13, 1951, 12, 37, 39, 42.

3. "What Do You Know about the Beginnings of SWE?" *SWE Newsletter,* May–June 1977, 18; "Women's Council, Western Society of Engineers," *Journal of the Western Society of Engineering* 53, no. 2 (June 1948): 77–79; "Girls Studying Engineering See Future for Women in These Fields," *CSM,* Apr. 16, 1949, 14; Beatrice C. Horneman,

"Engineers Wanted," *IW* 29 (July 1950): 213–14; "Women Engineers See Field Widening," *NYT*, Mar. 11, 1951, p. 47, col. 1; "U.S. Agencies Seek Women Engineers" (report on SWE meeting), ibid. Mar. 12, 1951, l9; "Again Heads Women Engineers," ibid., Aug. 6, 1951, 14; "Women Could Fill Engineering Jobs," ibid., Mar. 16, 1952, p. 79, col. 4.

4. The exact membership of SWE in any given year is hard to determine. In 1953 it was reported as about 350 (*World Directory of Women's Organizations* [London: World Directory, (1953)]). The best sources of information about the SWE are Elsie Eaves, "SWE—25 Years in Perspective," *SWE Newsletter* 22, nos. 3–4 (spring 1976): 14–15, 19; and Marta Navia Kindya with Sudha Dave, *Four Decades of the Society of Women Engineers* (n.p., 1990). On Hicks, see "Career Women," *Coronet* 33 (Apr. 1953): 118–19; *CBY, 1957*, 255–27; Robert C. Toth, "Couple May Beautify 'The Ugly American,'" *New York Herald Tribune*, June 18, 1959, clipping in carton 4, f. 86, ACE-CEW Papers, SLRC; and "Outstanding Engineer, Manager, and Consultant Is a Woman," *Product Engineering* 39 (Apr. 8, 1968): 137–38.

5. [Eugene Rabinowitch], "Scientific Womanpower," *Bulletin of Atomic Scientists*, Feb. 1951, 34, 37; "Help Wanted: Women," *Newsweek* 37 (Feb. 26, 1951): 70–71. On Rabinowitch, see *AMWS*, 12th ed., 1972, 5:5073–74.

6. The president of the ACE, Arthur Adams, seemed to be unembarrassed by this previous exclusion; indeed, he mentioned it, perhaps humorously, as quoted in the *NYT* and in his published introductory remarks: "During the past year we have had a number of conferences on manpower, and always during the course of these conferences some alert woman would rise in the audience and say 'How can you presume to solve manpower questions when you block out the representation of over half the population of the country, namely, womanpower?'" See "Meeting Set Here on Women's Role," *NYT*, Sept. 4, 1951, 24; and Arthur S. Adams, "Introductory Remarks," *Women in the Defense Decade, American Council on Education Studies*, 1st ser., no. 52, ed. Raymond F. Howes (Washington, D.C., 1952), 3, which contains a list of registrants, as well as speeches and reports. See also Margaret Culkin Banning, *A New Design for the Defense Decade* (Washington, D.C.: ACE, 1951); Sarah Van Hoosen Jones, "Report on the National Conference on Women in the Defense Decade," *Proceedings of the Association of Governing Boards of State Universities and Allied Institutions* 29 (1951): 39–46; "Woman's Role in the Defense Program," *IW* 30 (Nov. 1951): 321, 343. The *New York Times* covered the meeting ("Crusade for Home Urged on Women," Sept. 28, 1951, 24, and "'Take Outside Jobs' Homemakers Told," Sept. 29, 1951, 8) and had an editorial, "Women as Partners" (Oct. 1, 1951), 22, which asserted that though many married women were working, their "primary responsibility must continue to be the home." See also Lawrence K. Frank, "The Home . . . Source of the Nation's Strength," *AAUWJ* 45 (1951): 85–88. The Papers of the ACE Conference on Women in the Defense Decade are at SLRC. The success of the occasion also emboldened a group of women deans to convince the ACE to set up what became its Commission on the Education of Women (CEW) two years later.

7. Arthur S. Flemming, "Mobilization," *Scientific American* 185, no. 3 (Sept. 1951): 89–94, 96, 99. On Flemming, see *CBY, 1951*, 207–8, *1960*, 140–42 (he later became secretary of the Department of Health, Education, and Welfare [HEW]); and *Who's Who in America*, 41st ed. (1980–81), 1107. The activities of the ODM can be followed in

the records of the Bureau of the Budget (RG 51) at NARA, and those of the CSP in LCP.

8. "Interview with Chairman, Manpower Policy, Arthur S. Flemming: Planning Nation's Man Power for 10 Years Ahead," *U.S. News & World Report* 32 (Jan. 4, 1952): 29; "Engineering Career for Women Is Urged," *NYT*, Jan. 1, 1952, p. 27, col. 2; Benjamin Fine, "Lack of Engineers for War Stressed," ibid., Nov. 11, 1953, p. 11, col. 5; Arthur S. Flemming, "Memorandum to the Members of the Committee on Specialized Personnel," June 2, 1952, with attachments, and "Summary of the ODM Committee on Specialized Personnel, Meeting No. 10, Tuesday, June 17, 1952," 4, both in LCP; Fred C. Morris, "A Plan for Training Women in Engineering," *JEE* 43 (1952): 174–76.

9. Leonard Carmichael to James H. Taylor, Aug. 19, 1954, and enclosure ("Plan for a Corps of Women for Secondary School Science Teachers"); and Russell M. Stephen, "Utilization of Specialized Personnel Manpower in a Period of Full Mobilization," for draft of ODM CSP Ad Hoc Subcommittee on Labor Management Committee Document, "A Manpower Program for Full Mobilization" (Nov. 1953), 2–3, all in LCP.

10. *Employment Opportunities for Women in Professional Engineering*, WB Bulletin 254 (Washington, D.C., 1954). The SWE cooperated again in 1955, in a second survey, of about 2,800 women engineers, with 874 respondents ("Employment and Characteristics of Women Engineers," *Monthly Labor Review* 79 [May 1956]: 551–54). There are a few relevant items in Alice Leopold Papers, SLRC.

11. ODM, Executive Office of the President, "Defense Manpower Policy No. 8: Training and Utilization of Scientific and Engineering Manpower, Sept. 6, 1952" (Washington, D.C., 1952, mimeographed) also in *Federal Register*. For more on the ODM, see Edward H. Hobbs, *Behind the President: A Study of Executive Office Agencies* (Washington, D.C.: Public Affairs, 1954), ch. 8. The ODM's Health Resources Advisory Committee issued a pamphlet entitled "A Job for Women" (1953), which stressed the idea that healthy women could handle most jobs.

12. Memorandum, W. D. C[arey] to Alger and Milton Turen, June 4, 1956, attached to Turen, memorandum to "the Files," June 1, 1956, unit 175, Bureau of the Budget Papers (RG 51), NARA.

13. NRC, Personnel File, ser. 200/1.2, box 71, files 593–96, RAC. On Wolfle, see *Who's Who in America*, 41st ed. (1980–81), 2:3573; *AMWS*, 17th ed., 1989, 7:731; and Walter S. Hunter, "Dael Wolfle," *Science* 119 (1954): 230–31.

14. Commission on Human Resources and Advanced Training, *America's Resources of Specialized Talent: A Current Appraisal and a Look Ahead, A Report Prepared by Dael Wolfle* (New York: Harper & Bros., 1954).

15. For the reviews, all by men, see L. A. DuBridge in *Science* 121 (1955): 25; Ralph W. Tyler in *Scientific Monthly* 80 (1955): 131–32; Claude E. Hawley in *Annals of the American Academy of Political and Social Science* 298 (1955): 206–7; Leonard Carmichael in *Journal of Higher Education* 26 (1955): 105; Roy N. Anderson in *Personnel and Guidance Journal* 33 (1955): 421–22; L. Wesley Wager in *American Journal of Sociology* 61 (1955–56): 505–6; and Robert Bayless Norris in *Social Education* 20 (1956): 238–39.

16. Phyllis Berger, "Women in Engineering," *NYT*, Oct. 20, 1955, p. 34, cols. 6–7. On the decision in April 1953 to admit "coeds" to Georgia Tech, see Robert C. McMath, Jr., *Engineering the New South: Georgia Tech, 1885–1985* (Athens: Univ. of Georgia Press,

1985), 270–75; *Georgia Tech Alumni Magazine* 58, no. 1 (fall 1982); Leita Thompson, *A History of the Georgia Federation of BPWC, Inc., 1919–1956* (Athens, Ga.: privately printed, 1957), 22–23; and a wealth of material in the archives at Georgia Institute of Technology Library, Atlanta. Despite certain favorable conditions—the Georgia chapter of the NFBPWC had been fighting for coeducation for years, and the wife of Georgia Tech president Col. Blake Van Leer had studied architecture at Berkeley and his daughter had graduated from Vanderbilt with a degree in engineering—the vote by the board of trustees was very close (7 to 5). In 1959 both the Case Institute of Technology and New York University voted to admit women to their engineering schools ("Careers in Engineering for Women Are Urged," *NYT,* Dec. 20, 1959, sec. 4, p. 7, col. 4). For the earlier period, see Edna May Turner, "Education of Women for Engineering in the United States, 1885–1952" (Ph.D. diss., New York University, 1954).

17. *Engineering Degrees and Enrollments, Fall 1963, Final Report,* USOE Circular 785 (Washington, D.C., 1962–63), 61. Extensive data on engineering enrollments and degrees have been published since 1935 in the *JEE* for those schools (ca. 80 percent) approved by the Engineering Council for Professional Development (ECPD). Between the late 1940s and 1963, the USOE supplemented this with data on the non-ECPD schools in its annual *Engineering Degrees and Enrollments.* Since these data were not always broken down by gender, the SWE produced its own biennial (and then cumulative) surveys of female engineering enrollments and degrees earned, for example, *Report on Women Undergraduate Students in Engineering: Biennial Survey, 1959–1969* (New York, 1971). Such data were then used to encourage women to enter the field; see, e.g., Patricia L. Brown, ed., *Women in Engineering* (New York: SWE, 1955). In all of these the public tone was upbeat and propagandistic—you can be an engineer, there are plenty of jobs, and the salaries are high (for women). By contrast, the most honest and realistic article about women engineers of the 1950s described the tremendous pressure two young women engineers at Beckman Instruments, in Berkeley, California, felt every day to prove themselves and their fitness for the job: Harriet Renaud, "Electronics Engineering," *Mademoiselle,* Mar. 1957, 57–58, 60, 121, and note on the non-engineer author, 92. Sally Ann Trieloff, "A Woman in a Man's World," *Journal of College Placement* 21, no. 1 (Oct. 1960): 41, was a plea for understanding by a woman engineering student.

The several works on attrition among engineering students had very little to say about women, and the most comprehensive work on the problem in general, Robert E. Iffert and Betty S. Clarke, *College Applicants: Entrants and Dropouts,* USOE Bulletin 1965, no. 29 (Washington, D.C.: GPO, 1965), did not focus on engineers. Although it had no systematic data on attrition by sex for the twenty institutions (ten private and ten public) that it surveyed, it provided indirect evidence that the attrition rates for men and women students differed: men seemed to quit because of low grades, whereas women left for health or family reasons, especially marriage and motherhood (29–30). Some colleges required married women to withdraw. See, e.g., Helen L. Russell, "The Marriage Policy," *SAQ* 49 (1957–58): 5, 11.

18. "Conference on the Role of Women's Colleges in the Physical Sciences, Held at Bryn Mawr College, Bryn Mawr, Pennsylvania, on June 17–18, 1954" (Bryn Mawr, 1954, mimeographed); William G. Weart, "Women Evalued as Labor Source," *NYT,* June 18, 1954, p. 16, col. 4; "Need in Sciences for Women Cited," ibid., June 21, 1954, p. 18, col. 6;

"Meetings . . . Women in Science," *Physics Today* 7, no. 10 (Oct. 1954): 32. On Michels, see *AMWS*, 12th ed., 1972, 4:4249; and Walter C. Michels, "Women in Physics," *Physics Today* 1 (Dec. 1948): 16–19. C. C. Herd of IBM invited Alan Waterman of the NSF to the conference, but he declined because of a heavy schedule (Alan T. Waterman Diary, Mar. 10, 1954, box 20, A. T. Waterman Papers, LC). Later, upon receipt of a copy of the meeting's proceedings, he was pleased and relieved that the women were being encouraged to become teachers (Waterman to Michels, Mar. 29, 1955, both in NARA, RG 307).

19. Walter J. Murphy, "Women as Scientists," *C&EN* 33 (1955): 1277 (which added that under certain favorable conditions some women might become chemical engineers); "Women Physical Scientists Are in Demand," ibid., 1319; Dorothy W. Weeks, "Woman Power Shortage in the Physical Sciences," *AAUWJ* 48 (1955): 146–49; Katherine Sopka, "Dorothy Weeks: Transcript of an Interview on a Tape Recorder," July 19, 1978, NBL, 200; "Electrical Major," *Mademoiselle* 40 (Mar. 1955): 102–4; Howard Meyerhoff, "Science Careers for Women," *Science Counselor* 18, no. 1 (Mar. 1955): 2, 26. For more on the American perception of Russian women scientists and engineers, see n. 42 below.

20. R.M.B., "Editorial: Women in Physics and Chemistry," *Electrochemical Society Journal* 101 (Aug. 1954, suppl.): 193C.

21. "Courting Distaff Employees," *Chemical Week* 76 (Feb. 11, 1956): 36.

22. See n. 19 above and Howard A. Meyerhoff, "[Annual Report] To the Members of the Commission," Apr. 12, 1956 (mimeographed), 2, LCP; obituary, "Howard A. Meyerhoff, 1899–1982," *Science* 216 (1982): 613; and Dael Wolfle, *Renewing a Scientific Society: The American Association for the Advancement of Science from World War II to 1970* (Washington, D.C.: AAAS, 1989).

23. Otto Kraushaar, "Science and the Education of Women," *Association of American Colleges Bulletin* 43 (1957): 89–94.

24. Lee A. DuBridge, "Scientists and Engineers: Quantity Plus Quality," *Science* 124 (1956): 302. See also his testimony in *Shortage of Scientific and Engineering Manpower: Hearings before the Subcommittee on Research and Development of the Joint Committee on Atomic Energy, Congress of the United States, 84th Congress, 2nd Session, April–May 1956* (Washington, D.C.: GPO, 1956), 258. His own California Institute of Technology did not admit undergraduate women until 1970, but no one seemed to call attention to this.

25. NSF, *The National Committee for the Development of Scientists and Engineers* (Washington, D.C.: NSF, 1956). Eventually the committee had forty-nine members. Part of the Papers of the President's Committee on the Development of Scientists and Engineers are in Washington, D.C., at NARA (RG 307), but the richer portion is in Abilene, Kansas, at the Dwight D. Eisenhower Library.

26. Helen Hill Miller, "Science: Careers for Women," *Atlantic Monthly* 200 (Oct. 1957): 123–38; Allaire U. Karzon, "A Tax Revision Proposal to Encourage Women into Careers," in *Encouraging Scientific Talent,* by Charles C. Cole, Jr. (New York: College Entrance Examination Board, 1956), app. A. On Karzon, see *WWAW* 1 (1958–59): 676. An earlier draft of Cole's volume, with a chapter on women by Jane Blizzard, an MIT Ph.D., is in box 7, f. 100, ACE-CEW Papers.

27. Anne Elston, "Women for Science," *Science Digest* 42 (July 1957): 91.

28. Walker sought figures for a speech to the American Society for Engineering

Education (ASEE) published as "Research and Engineering Education," *JEE* 48 (1957): 70–80. See [Donald S. Bridgman] to Dr. Eric Walker, July 2, 1957 (plus tables), Walker to Bridgman, July 23, 1957, and Bridgman to Walker, Aug. 15, 1957, all three in box 19, and "More Women Technologists Seen Answer to Skilled Brainpower Shortage," press release, June 17, 1958, in box 37, all in President's Committee on Scientists and Engineers Records, Dwight D. Eisenhower Library.

29. NMC, *Womanpower* (New York: Columbia Univ. Press, 1957); Eleanor F. Dolan, review of *Womanpower,* in *Science* 126 (1957): 1249. There is considerable material about the NMC in the Ford Foundation Archives, New York, collection PA 53-66, and in NMC, *Manpower Policies for a Democratic Society: The Final Statement of the Council* (New York: Columbia Univ. Press, 1965), which contains a historical preface and a list of publications. See also David Horrocks and Thomas Soapes, "Interview with Eli Ginzberg, May 15, 1975 and April 11, 1978," esp. 86–90, Dwight D. Eisenhower Library; and, on sex-labeling, "Jobs That Women Don't Get," *New York Times Magazine,* Mar. 17, 1957, 26, 47–49, by Bernard Roshco, a former NMC staff member.

The Council boasted in a memorandum to the Ford Foundation that *Womanpower* had been mentioned in over 250 editorials and reviews, most of them in newspapers, in several radio and even some television interviews, and in numerous news stories, which often included, at the NMC staff's suggestion, local examples of working women. In addition, some women's magazines published excerpts of the report, organized whole issues around the theme womanpower, or, as in the case of the NFBPWC, polled their members to see how much time the mothers had taken off from work for childrearing (over 4,000 responded). Then in October 1957 the NMC held a conference at Arden House, the former Harriman mansion, north of New York City, to hear experts' reaction to *Womanpower*. This led to another NMC volume, *Work in the Lives of Married Women* (1958), and all its attendant publicity. All in all, if the Ford Foundation had wanted maximum fanfare, *Womanpower* got far more than any previous NMC project had (press release, Mar. 13, 1957; NMC, information memorandum no. 114, "Report on Dissemination of *Womanpower,*" June 13, 1957, with excerpts from many reviews, and information memorandum no. 128, "Report on Dissemination," Mar. 7, 1958; and other reviews, all in Ford Foundation Archives, collection PA 53-66. "Womanpower," *AAUWJ* 51 [1958]: 111; Hazel Palmer, "Womanpower Survey Progress Report," *National Business Woman* 37 [May 1958]: 10; NMC, *Work in the Lives of Married Women: Proceedings of a Conference on Womanpower* [New York: Columbia Univ. Press, 1958]. See also Harriet Bixler Naughton, "Work and Married Women," *SAQ* 49 [1958]: 84–85).

30. The best published accounts of Bunting's career are in *CBY, 1967,* 46–48, and in "One Woman, Two Lives" (cover story), *Time* 78 (Nov. 3, 1961): 68–73; there is a discussion of her administration at Douglass (but not her outside activities) in George P. Schmidt, *Douglass College: A History* (New Brunswick, N.J.: Rutgers Univ. Press, 1968), 200–214. See also Jeannette Bailey Cheek, "Mary Ingraham Bunting, Oral Memoir, September–October, 1978," at SLRC; her autobiographical "From *Serratia* to Women's Lib and a Bit Beyond," *A[merican] S[ociety of] M[icrobiology] News* 37, no. 3 (1971): 46–52; and NSF, *Eighth Annual Report,* 1958 (Washington, D.C.: GPO, 1959).

31. Bunting, "From *Serratia* to Women's Lib," 47–48. She used similar wording in a spring 1971 interview with Jeanne Lowe (JLP). A memorandum from Opal David to

Elmer West et al., n.d., encloses Donald Bridgman's report ("Losses of Intellectual Talent from the Educational System Prior to Graduation from College" [June 1959, mimeographed], esp. table B2, p. 66), f. EC 13, carton 1, ACE-CEW Papers, SLRC.

32. Correspondence, proposals, interim reports, and publications about the project are in the Ford Foundation Archives, collection 60-452, including "Interim Report: Ford Foundation Program for the Retraining in Mathematics of College Graduate Women, [Sept. 1961]," and "Report, 1964–65." Publications about the project include Mary Jane Davidson Evers, "Mother's Back in College," *Douglass Alumnae Bulletin,* Jan. 1962, 3; Muriel Katz Glaser, "Testimony to Training, Finding New Satisfactions through the Ford Foundation Retraining Program in Mathematics," ibid., Nov. 1964, 17–18; Janet German Harbison, "Math for Matrons," *SAQ* 54 (1962–63): 17–18, on Helen M. Marston, '37; Helen M. Marston, "The Rutgers Program for Retraining in Mathematics of College Graduate Women," *American Mathematical Monthly* 71 (1964): 1130–31; Leslie Velie, "Come Back to the Work Force, Mother!" *Reader's Digest* 87 (Sept. 1965): 1–2+; and "Mathematics and M'Lady," *Bell Telephone Laboratories Reporter* 14 (Nov.–Dec. 1965): 2–5.

The best sources on the history of the continuing education movement are Helen S. Astin, *Some Action of Her Own: The Adult Woman and Higher Education* (Lexington, Mass.: D. C. Heath, 1976); Carol Ann Fought, "The Historical Development of Continuing Education for Women in the United States: Economic, Social, and Psychological Implications" (Ph.D. diss., Ohio State University, 1966); and JLP. Lowe was writing a report on continuing education for the Alfred P. Sloan Foundation and had amassed considerable material, including interviews with some of the leaders, when she died in 1971.

33. The Woodrow Wilson National Fellowship Program, which started in 1945 and expanded greatly in the 1950s and after (with eventually over $52 million from the Ford Foundation), was notorious in this regard. Until 1962 there was a 25 percent quota on the number of awards that could go to women, regardless of how many applied or were eligible (Woodrow Wilson National Fellowship Foundation, *Report for 1962–1963* [Princeton, (1963?)], 49–50). Such an overall quota must have pit the women applicants in the sciences and social sciences against the many talented women in the humanities, for the proportions of women fellows in the former were well below 25 percent, just 15.43 percent in the sciences and 19.32 percent in the social sciences (Woodrow Wilson National Fellowship Foundation, *Directory of Fellowship Awards for the Academic Years, 1945/46–1960/61* [Princeton, n.d.], appendix, by field, 437–75).

34. Arthur Lack, "Science Talent Hunt Faces Stiff Obstacle: 'Feminine Fallout.'" *WSJ,* Jan. 16, 1958, 1, 11; NMC, *Womanpower,* 183–84. Lack's article may have been inspired by a report or just a remark in late 1957 by Milton Eisenhower, the brother of President Eisenhower and the newly installed president of Johns Hopkins University, that because of the number of women biologists who dropped out of the Johns Hopkins graduate school, it cost about two hundred thousand dollars to educate each fully trained Ph.D. Since this was about eight times what it cost to educate a man who would stay in the field, he presumably favored not training any women, not wasting more federal aid on them, or just being more selective in admissions. James D. Watson at Harvard reportedly agreed and urged his department to admit only those women

applicants with several years of previous graduate study elsewhere, as proof of commitment and seriousness. Statistics later collected by the Radcliffe Graduate School for 1950–62 showed that of its sixteen departments, the biology department had the highest completion rate for women, 47 percent ("Table showing number of students entering Radcliffe Graduate School . . . 1949–50 to 1961–62 . . . ," binder, "Graduate Student Statistics—Degrees Awarded," Records of the Graduate School, RCA [I thank Patricia Palmieri for a copy]).

But Eisenhower's remark was not forgotten; in fact it then took on a life of its own. Apparently unreported in the press or in any of Eisenhower's publications (as far as Johns Hopkins archivists and those at the Dwight D. Eisenhower Library in Kansas can determine), it was nevertheless mentioned, but without citation, years later by Jessie Bernard in both her *Academic Women* (State College: Pennsylvania State Univ. Press, 1964), 283 n. 17, and her 1964 speech at MIT, published as "The Present Situation in the Academic World of Women Trained in Engineering," in *Women and the Scientific Professions: The M.I.T. Symposium on American Women in Science and Engineering*, ed. Jacquelyn A. Mattfeld and Carol G. Van Aken (Cambridge: MIT Press, 1965), 178 n. 19. It was then taken up by Caroline Bird in both a 1965 article ("Women in Business: The Invisible Bar," *Personnel* 45 [May–June 1965]: 29–30) and her 1968 book, *Born Female: The High Cost of Keeping Women Down* (New York: David McKay, 1968), 52–53. Presumably too outrageous to go on the record, the remark was remembered vividly by those who heard it or heard of it. (I thank Judith H. Willis for telling me of this, several librarians for their help, and Jessie Bernard for trying to remember; for more on Milton Eisenhower, see ch. 8).

35. Susan Spaulding, "What Feminine Fallout?" *WSJ*, Jan. 29, 1958, 6; on Spaulding, see *WWAW* 1 (1958–59): 1026. On Morawetz, see *AMWS*, 17th ed., 1989, 5:494; her autobiographical "Cathleen S. Morawetz" in *More Mathematical People: Contemporary Conversations*, by Donald J. Albers et al. (Boston: Harcourt Brace Jovanovich, 1990), 220–38; and Constance Reid, *Courant in Göttingen and New York: The Story of an Improbable Mathematician* (New York: Springer-Verlag, 1976), 255–56. See also "Both Sides of the Fence," *Washington, D.C., Sunday Star*, Feb. 2, 1958, D-10 (clipping in carton 4, ACE-CEW Papers, SLRC).

Also in February 1958, the AAUW's Committee on Education issued a public statement deploring the low proportion of bright high school women graduates who went on to college, but, strangely, it was not published until after the NDEA had been passed and signed ("Higher Education of Women," *School and Society* 86 [Sept. 27, 1958]: 342–43).

36. Public Law 85-864, 85th Cong., 2d sess. (Sept. 2, 1958), *National Defense Education Act of 1958*, Title I, sect. 101. See also Barbara Barksdale Clowse, *Brainpower for the Cold War: The Sputnik Crisis and the National Defense Education Act of 1958* (Westport, Conn.: Greenwood, 1981); LC, Science Policy Research Division, *The National Science Foundation and Pre-College Science Education, 1950–1975*, Report Prepared for the Subcommittee on Science, Research, and Technology of the Committee on Science and Technology, U.S. House of Representatives, 94th Cong. 2d sess. (Washington, D.C.: GPO, 1975), 60, 62, and app. A; and Milton Lomask, *A Minor Miracle: An Informal History of the National Science Foundation* (Washington, D.C.: NSF, [1976]), ch. 8.

37. There are some data on the sex of the student borrowers in *The National Defense*

Student Loan Program: A Two-Year Report (Washington, D.C.: HEW, 1961), 18–19, 24, 27–28 (on medical students); and National Defense Student Loan Program, *Student Borrowers: Their Needs and Resources* (Washington, D.C.: HEW, 1962), passim. See also "Girls Borrow, Too," *NYT,* Sept. 27, 1959, sec. 4, E-11. Quaker Bryn Mawr College led the opposition to the loyalty oath for student loans at the undergraduate level but let its graduate students decide for themselves (Patricia Hochschild Labalmé, ed., *A Century Recalled: Essays in Honor of Bryn Mawr College* [Bryn Mawr: Bryn Mawr College Library, 1987], 49–50; Clowse, *Brainpower for the Cold War,* 154). Someone at Michigan State quipped that if the USOE would not support home economics, the study of the family and the home, under the NDEA, then just "What were we defending?" ("Institutional Application for National Defense Graduate Fellowships for the Academic Year 1963–64," box 385, f. 11, Papers of College of Human Ecology, Michigan State University Archives, East Lansing).

38. Clarence B. Lindquist, *NDEA Fellowships for College Teaching* (Washington, D.C.: HEW, 1971), cited in Lindsey R. Harmon, *Career Achievements of the National Defense Education Act (Title IV) Fellows of 1959–1973: A Report to the U.S. Office of Education* (Washington, D.C.: NAS, 1977), 1.

39. NEA, Research Division, *Teacher Supply and Demand in Universities, Colleges, and Junior Colleges, 1957–58 and 1958–59* (Washington, D.C., June 1959), 25. The president of the NEA in 1959 was Ruth A. Stout, and the associate director of the Research Division was Hazel Davis. The research was supported by a grant from the FAE.

40. Charles Warnath, "Is Discrimination against Talented Women Necessary?" *Vocational Guidance Quarterly* 9, no. 3 (1961): 179–81.

41. Audrey K. Wilder, "Be Good, Sweet Maid," *AAUWJ* 52 (1958): 15–17.

42. Nicholas DeWitt, *Soviet Professional Manpower, Its Education, Training and Supply* (Washington, D.C.: NSF, 1955). On Soviet women scientists and engineers in the American press from 1955 to 1959, see "Education of Women Engineers at Snail's Pace," *C&EN* 33 (1955): 623; and Meyerhoff, "Science Careers for Women." Edward McCrensky, *Scientific Manpower in Europe* (New York: Pegamon, 1958), 155–60, 165, was quite surprised that so many Russian scientists and engineers were women. See also Frances Lewine, "Soviets Are Ahead with Womanpower," *Washington Star,* Feb. 9, 1958, clipping in f. 83, carton 4, ACE-CEW Papers, SLRC; and Gertrude Samuels, "Why Russian Women Work Like Men," *New York Times Magazine,* Nov. 2, 1958, 22–23, 58–59. When a group of nineteen Soviet chemists, eleven of whom were women, toured the United States in 1959, one journalist remarked on this proportion but interviewed only the men and photographed mostly men ("Soviet Chemists Tour U.S." *C&EN* 37 [Aug. 31, 1959]: 20–21).

43. Betty Lou Raskin, "Woman's Place Is in the Lab, Too," *New York Times Magazine,* Apr. 19, 1959, 17, 19–20; J[oseph] T[urner], "Science for the Misses," *Science* 129 (1959): 749; Anna R. Whiting, "Letters—Women Scientists," ibid., 1296; Dorothy Barclay, "For Bright Girls: What Place in Society?" *New York Times Magazine,* Sept. 13, 1959, 126; Ethel Strainchamps, "Plight of the Intellectual Girl," *Saturday Review of Literature* 43 (Nov. 19, 1960): 63–64, 81.

44. Diana Trilling, "Female-ism: New and Insidious," *Mademoiselle* 51 (June 1960): 44, 97–99.

45. *First Jobs of College Women—Report on Women Graduates: Class of 1957,* WB Bulletin 268 (Washington, D.C., 1959); "Women in the Class of 1957," *Science* 130 (1959): 853. The interpretation put on the survey's results may have been due to the cooperation of the National Vocational Guidance Association. Both it and the WB had a vested interest in the public's thinking that more vocational guidance was necessary. See also Bureau of Social Science Research, *Two Years after the College Degree* (Washington, D.C.: GPO, 1963) on the class of 1958.

46. *Careers for Women in the Physical Sciences,* WB Bulletin 270 (Washington, D.C., 1959); *Careers for Women in the Biological Sciences,* WB Bulletin 278 (Washington, D.C., 1961); *Careers for Women as Technicians,* WB Bulletin 282 (Washington, D.C., 1961). See also Judith Sealander, *As Minority Become Majority: Federal Reaction to the Phenomenon of Women in the Work Force, 1920–1963* (Westport, Conn.: Greenwood, 1983). Correspondence and memos regarding Bulletins 270, 278, and 282 in WB Papers (RG 86), boxes 723–26, NARA, esp. memos from Stella P. Manor to Mrs. Alice K. Leopold on women in science, Nov. 19, Dec. 1, 1958, June 19, 1959, all in box 726. In 1960 Leopold just sent Killian's successor copies (Alice K. Leopold to George B. Kistiakowsky, Jan. 20, 1960, and reply, Feb. 5, 1960, box 12, Papers of the Office of the Special Assistant for Science and Technology, Dwight D. Eisenhower Library). Many of the WB *Bulletins* were referred to in Frances Maule's *Executive Careers for Women,* rev. and enl. ed. (New York: Harper, 1961), ch. 11. Finally in 1961 the NSF put out its own booklet of its own and others' statistics: [Bella Schwartz], *Women in Scientific Careers* (Washington, D.C.: GPO, 1961). On Leopold, see *CBY, 1955,* 359–61.

47. Fred M. Hechinger, "Education in Review: Science Is Still a Man's World Despite Efforts Aimed at Attracting Women," *NYT,* Jan. 17, 1960, sec. 4, p. 11, col. 1; Maxine Cheshire, "Distaffers Cutting Capers on Horizons of Science," *Washington Post,* Jan. 17, 1960, clipping in f. EC 83, carton 4, ACE-CEW Papers, SLRC.

48. Bernard Berelson, *Graduate Education in the United States* (New York: McGraw-Hill, 1960), 135; David L. Sills, "Bernard Berelson: Behavioral Scientist," *Journal of the History of the Behavioral Sciences* 17 (1981): 305–11; Carnegie Corporation of New York, *Annual Report for Fiscal Year ended September 30, 1957* (New York, 1957), 45–46, 77. For a discussion of Berelson's evidence, see Margaret W. Rossiter, "Outmaneuvered Again: The Collapse of Academic Women's Strategy of Celibate Overachievement" (paper presented at Ninth Berkshire Conference on the History of Women, Vassar College, Poughkeepsie, N.Y., June 12, 1993).

49. William G. Torpey, "The Role of Women in Professional Engineering," *JEE* 52 (1962): 656–58; *Women in Professional Engineering: A Conference Held under the Auspices of the Executive Office of the President of the United States and Sponsored by the University of Pittsburgh and the Society of Women Engineers, April 23 and 24, 1962, at the University of Pittsburgh* ([New York]: SWE, 1962), included Erwin R. Steinberg, "Women in Technology," 21–26 (also published as "What About WOMANPOWER in the Space Age?" *Space Digest* 45 [Aug. 1962]: 56–58); Donald H. Ford, "Problems in Motivating Young Women toward Engineering from the Viewpoint of University Counselling, Femininity, and Engineering," 55–61 (quotation from 60); and Donald Feight (U.S. Steel), "Remarks," 78, 86. For Walker's speech, see Eric A. Walker, "Women are NOT for Engineering," *Penn State Engineer,* May 1955, 9, 20. On Walker, a major figure in

engineering education in the 1950s and 1960, see *AMWS,* 18th ed., 1992, 7:410; N. J. Palladino, "The Man and the Theme," in *A National Posture on Support of Basic Research and Engineering: The Pennsylvania State University, College of Engineering, Symposium Honoring Dr. Eric A. Walker, University Park, Pennsylvania, June 5, 1970* ([University Park: Pennsylvania State University, 1970]), 1–5; Michael Bezilla, *Engineering Education and Penn State: A Century in the Land-Grant Tradition* (University Park: Pennsylvania State University Press, 1981), 203; and obituary, *NYT,* Feb. 19, 1995, I, 49.

Chapter 4   Graduate School: Record Numbers Despite It All

1. Nancy Lynch, "The Young American Graduate Student," *Mademoiselle* 43 (Sept. 1956): 138, 76, 78, 80 (quotation from 80). For her marriage statistics, Lynch cited *Graduate Education for Women: The Radcliffe Ph.D., A Report by a Faculty-Trustee Committee* (Cambridge: Harvard Univ. Press, 1956). One article that made graduate school sound exciting was "Adventures in Graduate Work," *WAM* 41 (1957): 213–16, 230, with statements from five alumnae in the social sciences.

2. David Boroff, "The University of Michigan: Graduate Limbo for Women," *Campus USA: Portraits of American Colleges in Action* (New York: Harper & Bros., 1961), 177–89 (quotations from 181, 183, 184, 189). Not all women graduate students were sexless and dowdy. Anthropologist Mary Ellen Goodman earned part of her way through Radcliffe Graduate School in the 1940s by working as a fashion model (Margaret Moore, "She Led Dual Life," *Boston Traveler,* Jan. 5, 1956, clipping in Deceased Alumnae Records, RCA).

3. "Bernice Brown Cronkhite: An Interview with Mary Manson," June 1976, 51, RCA. For the Graduate Center, see also Papers of the Office of the Graduate Dean, RCA, and her autobiography, *The Times of My Life* ([Cambridge, Mass.]: n.p., 1982). See also "In Memory, Bernice Brown Cronkhite, 1893–1983," *Radcliffe Quarterly* 69, no. 3 (Sept. 1983): 31.

4. Edgar S. Furniss, *The Graduate School of Yale: A Brief History* (New Haven: Yale Graduate School, 1965), 72–73. See also box 226, Records of President Griswold, and boxes 4 and 30, Records of the Provost, Yale University Archives. By 1967 the Yale women were complaining that they "felt segregated" in the separate dormitory (E. Wight Bakke, "Graduate Education for Women at Yale," *Ventures,* fall 1969, 17). See also "Helen Morris Hadley, Vassar 1883," *VAM* 51, no. 1 (Oct. 1959): 15; and n. 31 below.

5. Joyce Thompson, *Marking a Trail: A History of the Texas Woman's University* (Denton: Texas Woman's Univ. Press, 1982), 182.

6. Wilma Kerby-Miller to Mary Bunting, July 5, 1966, Annual Reports of the Radcliffe Graduate Office, RCA.

7. "Bernice Brown Cronkhite: An Interview with Mary Manson," 43.

8. Audrey K. Wilder, "Be Good, Sweet Maid," *AAUWJ* 52 (1958): 15–17; Arthur Lack, "Science Talent That Faces Stiff Obstacle: 'Feminine Fallout,'" *WSJ,* Jan. 16, 1958, I, II.

9. Isabella Karle, "Crystallographer," in *Successful Women in the Sciences: An Analysis of Determinants,* ed. Ruth B. Kundsin, *Annals of the New York Academy of Sciences* 208 (1973): II.

10. [Ruth Roettinger], *Idealism at Work: Eighty Years of AAUW Fellowships, A Report by the American Association of University Women* (Washington, D.C.: AAUW, 1967), chs. 1, 6, and appendixes, esp. table 14 (312).

11. Ruth Tryon, *Investment in Creative Scholarship: A History of the Fellowship Program of the American Association of University Women, 1890–1956* (Washington, D.C.: AAUW, 1957), 41–44.

12. Mary I. Bunting, "Education in Our Affluent Society," *AAUWJ* 55 (1962): 95–98, esp. 97–98. President Bunting and Dean Cronkhite, who retired in 1960, differed on the necessity for full-time study. Cronkhite justified it on the grounds that the students could only be compared with each other if all were taking a full-time load. The fact that some might have heavy family burdens and so were not able to devote their full energies to graduate study did not enter her calculations. After the first year, students could petition for a course reduction (*Graduate Education for Women*, 82–83). Although she gave the impression that this was almost automatic for mothers, her correspondence indicates that students did not agree, and they still had to pay full Harvard tuition even if they were taking only a part-time course load. Cronkhite and Radcliffe president Wilbur K. Jordan also explained to the dean of the Harvard Graduate School that they permitted very little part-time study (far less than Harvard did), because they had so many applicants that they could limit admission to full-timers; in particular, they intended to discourage the many women around Greater Boston who wanted to pursue advanced courses in a "desultory" manner (Dean J. H. Van Vleck to President Jordan, Oct. 4, 1956; Jordan to Van Vleck, Oct. 11, 1956; and Dean Cronkhite to Van Vleck, Jan. 11, 1957, all in box 4, Papers of the Office of the Graduate Dean).

13. Accounts of the Sister Formation Movement include Sr. Mary Hester Valentine, S.S.N.D., *The Post-Conciliar Nun* (New York: Hawthorn, 1968), ch. 8; Helen Rose Fuchs Ebaugh, *Out of the Cloister* (Austin: Univ. of Texas Press, 1977), ch. 1; and Sr. Regina Clare Salazar, C.S.J., "Changes in the Education of Roman Catholic Religious Sisters in the United States from 1952 to 1967" (Ph.D. diss., University of Southern California, 1971). See also Karen Kennelly, C.S.J., "Mary Evaline Wolff," in *NAWMP*. On nuns who were Ph.D.'s, see esp. Reverend Mother M. Regina, I.H.M., "Can the Mother General Afford to Send Sisters on to Graduate Study?" *NCEAB* 58 (1961): 140–43; Roy J. Deferrari, "The Nun and Research," *Benedectine Review*, summer 1952, 11–15; Rev. John A. Fitterer, S.J., "The Sister PhD: An Essential to the Success of Sister Formation," *NCEAB* 58 (1961): 148–52; and the several papers under the title "The Potential Contribution of the Sisterhoods for Graduate Education" in ibid. 60 (1963): 179–95. On the nuns' fellowship program, see Francis C. Pray, "Fellowship Aid for Sisters," ibid. 58 (1961): 144–47; and Rt. Rev. Msgr. John J. Voight, "Progress and Prospects in Sister Formation," ibid. 60 [1963–64]: 200). See also n. 38 below.

Another female fellowship was (and is) the Amelia Earhart Fellowship, funded by the members of Zonta International and first awarded in 1940. Limited at first to a five-hundred-dollar award to one woman graduate student in the field of aeronautics, it has since been broadened to include any field of science or engineering related to aerospace. In 1969 at least six twenty-five-hundred-dollar fellowships were awarded (Barbara Spector, "Amelia Earhart Fellowships Encourage Women to Study Aerospace Science," *The Scientist* 4, no. 4 [Feb. 19, 1990]: 25–26).

14. Ora Marshino, comp., *Research Fellows of the National Cancer Institute [1938–1958]*, USPHS Pub. No. 658 (Washington, D.C.: GPO, 1959).

15. *Final Report: Atomic Energy Commission Predoctoral and Postdoctoral Fellowships in the Physical and Biological Sciences, 1948 to 1953* (Oak Ridge, Tenn.: Oak Ridge Institute of Nuclear Studies, [1953]); Barbara J. Bachmann, "Statement on Atomic Energy Commission Fellowship," *Science* 112 (1950): 364–65.

16. From 1952 to 1965 NSF awards were published as part of its annual report, e.g., NSF, *2nd Annual Report of the National Science Foundation, Fiscal Year 1952* (Washington, D.C.: GPO, [1952]), app. 2. See Cronkhite to Sides, NSF, Apr. 10, 1953, and replies, Apr. 20, 1953 (with partial data) and Oct. 14, 1953 (with more data), all in box 11, Papers of the Office of the Graduate Dean. In later years more data on the sex of the NSF fellowship applicants became available. They revealed that in fiscal year 1959 women made up 12.37 percent of the NSF applicants (1,522 of 12,304) and 11.89 percent of its awardees (468 of 3,937). But both proportions still varied widely by field (NSF, *Women in Scientific Careers* [Washington, D.C.: GPO, 1961], 6, 7).

17. Lindsey R Harmon, *Career Achievements of NSF Graduate Fellows: The Awardees of 1952–1972* (Washington, D.C: NAS, 1977), 8. It is conceivable that the women's high attrition rate may be at least partly owing to the NRC staff's not knowing their married names and thus not being able to find them in later lists and directories.

At an April 1962 meeting of a committee of the PCSW, Peter Rossi, Alice Rossi's husband and the head of the NORC, which did many studies of federal fellowship winners, proposed that a study be done of governmental fellowship selection practices. These fellowships would be very useful in helping "mature" women reenter the labor market, and discrimination might be occurring (Committee on Employment Practices of Federal Government, "Summary of First Meeting, April 3, 1962," 6, in box 3, Papers of the PCSW, John F. Kennedy Library, Boston, Mass.; on Rossi, see *Contemporary Authors*, vol. 19, rev. [Detroit: Gale Research, 1987], 410–11).

18. Harmon, *Career Achievements of NSF Graduate Fellows*, 10. Another survey of several thousand graduate trainees and fellows supported by institutional grants from the National Institute of General Medical Sciences to 400 departments (including 132 in biochemistry) between 1958 and 1967 revealed that at most 63 percent of the men and 38 percent of the women had attained their doctorate by 1967 (NRC, *Effects of NIGMS Training Programs on Graduate Education in the Biomedical Sciences: An Evaluative Study of the Training Programs of the National Institutes of General Medical Sciences, 1958–67, Conducted by the National Research Council* [Bethesda, Md.: U.S. National Institute of General Medical Sciences, 1969], 64).

19. "Cornell Awards Ph.T.: Degree Given to 30 Wives for 'Putting Husband Through,'" *NYT*, June 24, 1958, clipping in carton 1, ACE-CEW Papers, SLRC; "Student's Wives Get PHT," *Cornell Alumni News* 62, no. 1 (July 1959): 5, Rare and Manuscript Collections, KLCU. One of these winners was Jennie Towle Farley, who later earned her own doctorate with a dissertation on the women's movement in academia (see ch. 15, n. 55). See also Madison Kuhn, *Michigan State: The First Hundred Years* (East Lansing: Michigan State Univ. Press, 1955), 448.

20. See, e.g., the following exchange of letters to the editor: Mrs. Robert T. Keen, "Progress or Progeny?" *C&EN* 35 (Nov. 18, 1957): 9–10; William D. Clark, "Chemists

Need Good Wives," ibid. 36 (Mar. 17, 1958): 10, 12; and Edward H. Vallance, "Speed Up the Bright Ones," ibid. 36 (Apr. 28, 1958): 10.

21. Lindsey R. Harmon, project director, *Career Achievements of the National Defense Education Act (Title IV) Fellows of 1959–1973: A Report to the U.S. Office of Education* (Washington, D.C.: NAS, 1977), 7, 10. See also the caution in n. 17 above.

22. USOE, *Earned Degrees Conferred by Higher Educational Institutions,* published annually 1947/48 through 1962/63 as part of the USOE Circular series; subsequently published separately. Although the usual starting place for data on the science doctorates of the 1950s is the Earned Doctorate File of the NRC in Washington, D.C., its published reports do not specify fields and institutions for women by year (see, e.g., Lindsey R. Harmon and Herbert Soldz, comps., *Doctorate Production in United States Universities, 1920–1962,* NAS Pub. 1142 [Washington, D.C., 1963], 49–53; and Lindsey R. Harmon, *A Century of Doctorates: Data Analyses of Growth and Change* [Washington, D.C.: NAS, 1978], appendixes A, E–H). Despite the differences in the two sets of data, the figures for women's percentage of science doctorates for the years 1947–61 were strikingly similar: 6.92 percent for the NRC and 6.94 percent for the USOE. But see Lindsey R. Harmon, "Production of Psychology Doctorates in the United States," *AP* 16 (1961): 716–17, for an accounting (and minimizing) of the considerable differences between NAS-NRC and USOE data on doctorates in this one field.

Many studies of "career patterns" based on the NRC doctorate data are not only bland and simplistic but also dangerous, since they usually show the women lagging behind the norms set by the men. The women are generally then blamed for such tendencies, while all the factors beyond their control that may have led to such differences (e.g., the admission or hiring practices at the unnamed "top 20 private universities") are omitted and thus minimized. See, e.g., Lindsey R. Harmon, *Profiles of Ph.D's in the Sciences: Summary Report on Follow-Up of Doctorate Cohorts, 1935–1960,* NAS Pub. 1293 (Washington, D.C., 1965) and *Doctorate Recipients from United States Universities, 1958–1966: Sciences, Humanities, Professions, Arts, A Statistical Report,* NAS Pub. 1489 (Washington, D.C., 1967).

23. Edward F. Taylor, "Joanne Simpson: Pathfinder for a Generation," *Weatherwise* 37 (Aug. 1984): 182–83, 206–7; Joanne Simpson, "Meteorologist," *Successful Women in the Sciences,* 44; "The *Bulletin* Interviews: Dr. Joanne Simpson," *W[orld] M[eteorological] O[rganization] Bulletin,* 35, no. 1 (Jan. 1986): 3–14. There is a small collection of her papers with more biographical material at SLRC.

24. Charles S. Yentsch, letter to author, June 22, 1983. See also Ben McKelway, "Women in Oceanography," *Oceanus* 25, no. 4 (1982–83): 75–79.

25. USOE, *Earned Degrees Conferred, 1958–59, Bachelor's and Higher Degrees,* USOE Circular 636 (Washington, D.C.: GPO, 1961), 123; Raymond F. Crossman, "The Students," in *Forestry College: Essays on the Growth and Development of New York State's College of Forestry, 1911–1961,* ed. George R. Armstrong and Marvin W. Kranz (Syracuse: Alumni Association, State College of Forestry at Syracuse University, 1961), 135–36.

In the field of crystallography, the first woman to earn a Ph.D. was Gabrielle Hamburger Donnay at MIT in 1949 (Clifford Frondel, "An Overview of Crystallography in North America," in the immensely useful volume *Crystallography in North America,* ed. Dan McLachlan, Jr., and Jenny P. Glusker [New York: American Crystallographic

Association, 1983], 7. See also Robert Rakes Shrock, *Geology at M.I.T., 1865–1965: A History of the First Hundred Years of Geology at the Massachusetts Institute of Technology,* vol. 2 [Cambridge: MIT Press, 1982], 409–11 [I thank Michele Aldrich for telling me of Shrock's chapter on women]; and Maureen M. Julian, "Women in Crystallography," in *Women of Science: Righting the Record,* ed. G. Kass-Simon and Patricia Farnes [Bloomington: Indiana Univ. Press, 1990], 348).

26. *The Outlook for Women in Science,* WB Bulletin 223-1 (Washington, D.C., 1948), 58–61. See also Ruth W. Howard, "Ruth W. Howard," in *Models of Achievement: Reflections of Eminent Women in Psychology,* ed. Agnes N. O'Connell and Nancy Felipe Russo, vol. 1 (New York: Columbia Univ. Press, 1983), 54–67; and Kenneth R. Manning, "Roger Arliner Young, Scientist," *Sage* 6, no. 2 (fall 1989): 3–7 (special issue on black women in science and technology). On Kittrell, see *WWAW* 5 (1968–69): 662. On Reddick, see ibid. 1 (1958–59): 1057, and obituary, *Atlanta Journal,* Oct. 12, 1966, clipping in Deceased Alumnae Records, RCA; on Woods, *AMWS,* 15th ed., 1982, 7:734–35; on Daly, ibid., 17th ed., 1989, 2:515; on Granville and Young, Patricia Clark Kenschaft, "Black Men and Women in Mathematical Research," *Journal of Black Studies* 18 (1987): 182–84; on Wallace, *AMWS,* 13th ed., 1978, S&B, 1244, and obituary, *NYT,* Jan. 13, 1993, A-19; and on Shockley, S. D. Lewis, "Professional Woman—Her Fields Have Widened," *Ebony* 32 (Aug. 1977): 116.

27. See Sharon Traweek, "High Energy Physics: A Male Preserve," *Technology Review* 87, no. 8 (1984): 42–43, and idem, *Beamtimes and Lifetimes: The World of High Energy Physicists* (Cambridge: Harvard Univ. Press, 1988).

28. Anne Roe, "Women in Science," *Personnel and Guidance Journal* 44 (1966): 786. Roe had omitted this prejudice in her *Making of a Scientist* (New York: Dodd, Mead, 1952). In 1956 she had said that because women lacked mathematical ability, very few could be expected to become scientists (cited in Marguerite Zapoleon, "The Identification of Those with Talent for Science and Engineering and Their Guidance in the Elementary and Secondary Schools" [Washington, D.C.: PCSE, (1957), mimeographed], 39). As late as 1963 she commented at a meeting that she was not interested in women in science ("Discussion Period," in Federal Council for Science and Technology, "Proceedings of the First Symposium, Current Problems in the Management of Scientific Personnel, Oct. 17–18, 1963" [Washington, D.C.: GPO, 1964, mimeographed], 77). See also ch. 7, n. 49.

29. Leona Libby Marshall, *Uranium People* (New York: Charles Scribner's Sons, 1979), 29–30; Judith Walzer, "Interview with Emily Vermeule, August 6, 1981," 11–16, and idem, "Interview with Ruth Turner, July 9, 1981," 32, 37 and 83, both at MCRC; Marian Boykan Pour-El, "Mathematician," *Successful Women in the Sciences,* 15–16; Diane M. Jacobs, "Nothing Succeeds Like Success," *Radcliffe Quarterly* 67, no. 2 (June 1981): 21.

30. Nina Roscher, phone conversation with author, June 23, 1993. See also *Remembering HCB: Memoirs of Colleagues and Students of Herbert C. Brown as Prepared for the Symposium in His Honor in West Lafayette, Indiana on May 5 and 6, 1978* ([West Lafayette], Ind.: Department of Chemistry, School of Science, Purdue University, 1978), esp. p. 23; Gordon W. Allport, to Dean J. P. Elder and Dean Wilma Kerby-Miller, Sept. 11, 1962, with attached chart, in box 5, Papers of the Office of the Graduate Dean; Sandra Scarr, "Twenty Years Growing Up," *Psychology Today* 21 (May 1987): 26.

31. Started as a survey of the feasibility of part-time study and an assessment of opinion about Yale's responsibility toward local women, especially reentry ones, by the time it appeared its findings on other issues were of greater interest (Bakke, "Graduate Education for Women at Yale," 18, 19; the full report is in the Graduate School Papers: Records of the Dean John Perry Miller, Yale University Archives).

32. Semenow (later Garwood) taught for a few years at Pomona College and then became a psychologist (*AMWS*, S&B, 12th ed., 1973, 1:777; Dorothy Semenow Garwood, "Work and Me, Woman," in *Experiences in Being*, ed. Bernice Marshall [Belmont, Calif.: Brooks/Cole, (1971)], 257–64; biographical folder, Mt. Holyoke College Archives [I thank Elaine Trehub for copies]; John D. Roberts, *The Right Place at the Right Time* [Washington, D.C.: ACS, 1990], 116, 145–46).

33. On Cathleen S. Morawetz, see ch. 3, n. 35. New York University also topped a list of thirty-two graduate schools awarding the most doctorates to women in mathematics with thirty-four between 1961 and 1974 (I. N. Holmes, "Graduate Schools of Origin of Female Ph.Ds," *Notices of the American Mathematical Society* 23 [Apr. 1976]: 171). Part of the reason may have been the presence of Lipman Bers from 1957 to 1964. On Bers, see obituary, *NYT*, Nov. 2, 1993, B-10; and Lipman Bers, "Lipman Bers," in *More Mathematical People: Contemporary Conversations*, by Donald J. Albers et al. (Boston: Harcourt Brace Jovanovich, 1990), 12–13.

34. Saunders MacLane, "Mathematics at the University of Chicago: A Brief History," in *A Century of Mathematics in America, Part II*, ed. Peter Duren (Providence, R.I.: American Mathematical Society, 1989), 143–48.

35. Illinois's famed organic chemist Roger Adams trained 184 Ph.D.'s between 1918 and 1958, including 7 women (3.80 percent). All 7, however, earned their degrees before 1938 (of 103 or 6.80 percent), leaving none over the next twenty years (D. Stanley Tarbell and Ann Tracy Tarbell, *Roger Adams: Scientist and Statesman* [Washington, D.C.: ACS, 1981], 97 and app. A).

36. See *Cecilia Payne-Gaposchkin: An Autobiography*, ed. Katherine Haramundanis (Cambridge: Cambridge Univ. Press, 1984), intro. Peggy Kidwell; "Cecilia Payne-Gaposchkin: Transcript of an Interview Taken on a Tape Recorder by Owen Gingerich on March 5, 1968," copy in NBL; and "Nanielou Dieter Conklin: Transcript of an Interview Taken on a Tape Recorder," by David DeVorkin, July 19, 1977, esp. 6–9, where she discusses how much Payne-Gaposchkin meant to her and the other women graduate students in the Harvard department in the 1950s (copy in ibid. and NBL). See also "Interview with Andrea Dupree," in *The Outer Circle: Women in the Scientific Community*, ed. Harriet Zuckerman, Jonathan R. Cole, and John T. Bruer (New York: W. W. Norton, 1991), 94–126.

On Bok, see David H. Levy, *The Man Who Sold the Milky Way: A Biography of Bart Bok* (Tucson: Univ. of Arizona Press, 1993). His wife, Priscilla Fairfield Bok, had taught astronomy at Smith College before their marriage. He was part of the discussion group that produced *Graduate Education for Women* (1956) (see n. 1 above) and the author of a letter to the editor deploring the continuing lack of women graduate students in the field ("Graduate Training in Astronomy," *Science* 154 [Nov. 4, 1966]: 590, 592).

37. On McCarthy, see *AMWS*, 12th ed., 1973, S&B, 2:1548, and some letters in Florence Goodenough Papers, University of Minnesota Archives. On Anastasi, see

AMWS, 13th ed., 1978, S&B, 36. Because Fordham University's undergraduate college was still for men only when she came in 1947, most of her teaching was restricted to the graduate level, which was coeducational; it remained largely that way even after Fordham opened a coordinate college for women in 1963 (Anne Anastasi, "Reminiscences of a Differential Psychologist," in *The Psychologists,* ed. T. S. Krawiec [New York: Oxford Univ. Press, 1972], 3–37; "Anne Anastasi," in *Models of Achievement: Reflections of Eminent Women in Psychology,* ed. Agnes N. O'Connell and Nancy Felipe Russo, vol. 2 [Hillsdale, N.J.: Lawrence Erlbaum Associates, 1988], 58–66; "Anne Anastasi," in *A History of Psychology in Autobiography,* ed. Gardner Lindzey, vol. 7 [San Francisco: W. H. Freeman, 1980], 1–37, esp. 25). *American Psychology since World War II: A Profile of a Discipline* (Westport, Conn.: Greenwood, 1982), app. A, 203–6, ranked Anastasi number 49.5 (tied with Karl Menninger) of "The Most Important People in American Psychology during the Post–World War II Period" in a 1981 survey of 4,000 members of the APA. Anastasi's 1937 textbook *Differential Psychology* went through three editions and was translated into German, Italian, Spanish, Portuguese, and Hindi. Another text, *Psychological Testing,* had four editions, and her *Fields of Applied Psychology* (1964) had two editions and was translated into seven languages, including Chinese.

38. On the nuns at the University of Chicago, see Robert Hassenger, "The Future Shape of Catholic Higher Education," in *The Shape of Catholic Higher Education,* ed. Hassenger (Chicago: Univ. of Chicago Press, 1967), 311–12. On the Catholic Sisters College of the Catholic University of America, see Roy J. Deferrari, *Memoirs of the Catholic University of America, 1918–1960* (Boston: Daughters of St. Paul, 1962), 68–80. Under the heading "SPECIAL UNIVERSITY PROBLEMS," Deferrari, a retired administrator, put as number one the admission of women in the 1920s (led by economist and later anthropologist Sr. M. Inez Hilger, O.S.B.), 229–40. Her scientific notes and papers are at the National Anthropological Archives in Washington, D.C. See also n. 13 above.

39. Mack's arrival was just one part of a resurgence of home economics at Texas Woman's University in 1951–53; see ch. 10 below.

40. For more on the Smith botany department, see ch. 10. Similarly, the one science doctorate in these years awarded by the Woman's Medical College of Pennsylvania in clinical medical sciences in 1960–61 went to a man. In 1969 the institution admitted its first male student, and in 1970 it changed its name to simply the Medical College of Pennsylvania (John F. Ohles and Shirley M. Ohles, comps., *Private Colleges and Universities,* vol. 2 [Westport, Conn.: Greenwood, 1982], 958).

41. Psychologist Mary Henle was on the faculty at the New School for many years (Mary Henle, "Mary Henle," in O'Connell and Russo, *Models of Achievement,* 1:220–32.

42. Allyn Rule, "Women in Science: Why the Shortage?" *Science* 148 (1965): 21. A few weeks later Edwin C. Lewis, a psychology professor at Iowa State University, claimed that lower expectations for all women were justified, and he cited as proof Jessie Bernard's recently published *Academic Women* (State College: Pennsylvania State Univ. Press, 1964) (Edwin C. Lewis, "Women as Graduate Students," *Science* 148 [1965]: 893–94).

43. "Janet Guernsey: Transcript of an Interview Taken on a Tape Recorder, by Katherine R. Sopka on June 29, 1988," 18, 19, NBL.

44. Danforth Foundation, *Annual Report for 1963–64* (St. Louis, [1964]), 14. See also Shelley Lancaster, "Quiet Zone: Mothers Studying!" *WAM* 45 (1960–61): 93, 103.

45. Evidently the women deans of the Radcliffe and Bryn Mawr graduate schools had never belonged to this important group. See Mary Bunting, "Some Views of Problems in Graduate Education," *Journal of Proceedings and Addresses of the 17th Annual Conference, Association of Graduate Schools in the Association of American Universities*, 1965, 11.

46. Richard Magat, *The Ford Foundation at Work: Philanthropic Choices, Methods, and Styles* (New York: Plenum, 1979), 97–99 (quotations from 99).

Chapter 5 Growth, Segregation, and Statistically "Other"

1. J. Merton England, *A Patron for Pure Science: The National Science Foundation's Formative Years, 1945–57* (Washington, D.C.: NSF, 1982), 177–79, 253–54. There is also much material on the early days of the the National Register in the LCP and in boxes 4–6, NSF Papers (RG 307), Accession No. 66A-665, Washington National Records Center, Suitland, Md.

2. NSF, *American Science Manpower: Employment and Other Characteristics, 1954–55, Based on the National Register of Scientific and Technical Personnel, 1954–55* (Washington, D.C.: GPO, 1959), v. The early Register also built upon the Navy's experience with the BLS in collecting data through professional societies (*Personnel Resources in the Social Sciences and Humanities: A Survey of the Characteristics and Economic Status of Professional Workers in 14 Fields of Specialization*, BLS Bulletin 1169 [Washington, D.C., 1954]).

3. NSF, "Technical Notes," *American Science Manpower, 1956–58: A Report of the National Register of Scientific and Technical Personnel* (Washington, D.C.: GPO, 1961), 33–34.

4. See Carolyn Shaw Bell, "Women in Science: Definitions and Data for Economic Analysis," *Successful Women in the Sciences: An Analysis of Determinants*, ed. Ruth B. Kundsin, *Annals of the New York Academy of Sciences* 208 (1973): 134–42. In 1964 the American Economic Association, the American Sociological Association, and the Center for Applied Linguistics of the Modern Language Association were added to the societies surveyed; in 1966 the American Anthropological Association was also included. Yet from time to time the Register's statisticians admitted that they did not know how representative of the whole number of scientists their data really were (see, e.g., introduction to NSF, *American Science Manpower, 1964* [Washington, D.C.: GPO, 1967], 1).

5. Alden Emery, executive secretary of the ACS, reported in 1952 that about twenty women members had resigned after marrying chemists (Emery to J. J. Phillips, Feb. 4, 1952, box 26, "Membership—Joint" folder, ACS Papers, LC). It is not always clear who belonged to a scientific society. In 1953 Dael Wolfle, for example, the executive director of the APA, estimated that only about half of the practicing psychologists in the United States belonged to that organization (Dael Wolfle, "Comparisons between Psychologists and Other Professional Groups," *AP* 10 [1955]: 231). The usual assumption is that the nonmembers are less active professionally, that is, that they do not publish or that they are not in academia, but what they do is not clear (see Daryl Chubin, "Sociological Manpower and Womanpower: Sex Differences in Career Patterns of Two Cohorts of American Doctorate Sociologists," *American Sociologist* 9 [1974]: 88 n. 10).

6. Cf. *Employment, Education, and Earnings of American Men of Science*, BLS Bulletin 1027 (Washington, D.C., 1951), 36 (table A). Of the *AMS*'s fifty-one subfields only two were sufficiently different for one to suspect that perhaps some change had occurred between 1948, the date of the *AMS* questionnaires, and the time of the National Register's 1956–58 data: entomology had lost women, their percentage falling from 2.80 percent in 1948 to 1.82 percent in 1956–58, and aeronautical engineering had gained substantially, from no women in 1948 to the 2.33 percent presented here in 1956–58. The tendency to dichotomize the sciences into the "hard" and the "soft" and then to read vague stereotypes into the fields' demographics and practices, as in Norman W. Storer, "The Hard Sciences and the Soft: Some Sociological Observations," *Bulletin of the Medical Library Association* 55, no. 1 (1967): 75–84, and Saul D. Feldman, *Escape from the Doll's House: Women in Graduate and Professional Education* (New York: McGraw-Hill, 1974), ch. 3, oversimplifies and distorts a complex phenomenon.

7. See "Criteria for Inclusion in the National Register," in NSF, *American Science Manpower, 1966* (Washington, D.C.: GPO, 1968), app. B.

8. See, e.g., Lindsey R. Harmon, *Profiles of Ph.D.'s in the Sciences: Summary Report on Follow-Up of Doctorate Cohorts, 1935–1960*, NAS-NRC Pub. No. 1293 (Washington, D.C., 1965); and NRC, Office of Scientific Personnel, *Careers of Ph.D.'s, Academic versus Nonacademic: A Second Report on Follow-up of Doctorate Cohorts, 1935–1960*, NAS-NRC Publication No. 1577 (Washington, D.C., 1968).

9. The Register has data on the sex, degree level, employment sector, and type of work done by scientists in various fields but not on all these variables at once.

10. On Cynthia Westcott, see her *Plant Doctoring Is Fun* (Princeton: D. Van Nostrand, 1957), and obituary, *NYT,* Mar. 25, 1983, B-5.

11. Bell, "Women in Science," 140 n. 9.

12. NSF, *American Science Manpower, 1970: A Report of the National Register of Scientific and Technical Personnel* (Washington, D.C.: GPO, 1971), 87, 96, 102, 104–5. In 1954 the BLS found that federally employed women economists were the best paid of all women humanists and social scientists, but it did not compare them with scientists (*Personnel Resources in the Social Sciences and Humanities,* 35). Similarly, a 1957 AAUW study of its past fellows showed that four of the twenty-two best paid (over $8,000 annually), or 18.18 percent, were federal employees, even though such women were only 4 percent of all fellows (Ruth Tryon, *Investment in Creative Scholarship: A History of the Fellowship Program of the American Association of University Women, 1890–1956* [Washington, D.C.: AAUW, 1957], 24–25).

13. N. Arnold Tolles and Emanuel Melichar, "Studies of the Structure of Economists' Salaries and Income," *American Economic Review* 58, no. 5 (suppl., 1968): 77.

14. "Women Chemists Best Paid Grads," *C&EN* 37 (June 15, 1959): 72–73, summarizing *First Jobs of College Women—Report on Women Graduates, Class of 1957*, WB Bulletin 268 (Washington, D.C.: GPO, 1959), which surveyed eighty-eight thousand women graduates of the class of 1957.

15. David A. H. Roethel, "Employment and the ACS," *C&EN* 36 (Dec. 29, 1958): 74.

16. Patrick P. McCurdy, "A Man and A Woman" (editorial), ibid. 48 (Oct. 26, 1970): 5; "Women Chemists: Concerned over Rights," ibid., 26–28.

17. NSF, *American Science Manpower, 1970,* 73–75, 238 (tables A-10, A-61).

18. NSF, "Scientists, Engineers, and Physicians from Abroad: Fiscal Year 1965," *NSF Reviews of Data on Science Resources,* no. 13 (Mar. 1968): 8–10; NSF, "Scientists, Engineers, and Physicians from Abroad: Fiscal Year 1968," ibid., no. 18 (Nov. 1969): 13–14, 15; NSF, *Immigrant Scientists and Engineers in the United States: A Study of Characteristics and Attitudes,* Survey of Science Resources Series (Washington, D.C.: GPO, 1973), 3. See also Andrea Tyree and Katharine Donato, "A Demographic Overview of the International Migration of Women," in *International Migration: The Female Experience,* ed. Rita James Simon and Caroline B. Brettell (Totowa, N.J.: Rowman & Alllanheld, 1986), 35, 36, 39. From 1962 until at least 1971 the NSF had a special fellowship program for senior foreign scientists that brought about fifty men and very few women to the United States; some became permanent residents. For lists, see NSF, *Grants and Awards,* 1964–71 (Washington, D.C.: GPO 1965–72).

19. NSF, "Scientists and Engineers from Abroad: Trends of the Past Decade, 1966–1975," *NSF Reviews of Data on Science Resources,* no. 28 (Feb. 1977): 1–3. See also Phillip M. Boffey, "The Brain Drain: New Law Will Stem Talent Flow from Europe," *Science* 159 (1968): 282–284; and Katharine M. Donato, "Think of Immigrants as Mostly Women" (letter to the editor), *NYT,* July 23, 1986.

20. Helen Astin, *The Woman Doctorate in America: Origins, Career, and Family* (New York: Russell Sage Foundation, 1969).

21. Deborah S. David, *Career Patterns and Values: A Study of Men and Women in Science and Engineering* (Washington, D.C.: U.S. Department of Labor, Manpower Administration, 1971), 87.

22. Louise Dolnick Solomon, "Female Doctoral Chemists: Sexual Discrepancies in Career Patterns" (Ph.D. diss., Cornell University, 1972), 23–29.

23. Barbara Reskin, "Sex Differences in the Professional Life Chances of Chemists" (Ph.D. diss., University of Washington, 1973), 68, 70.

Chapter 6 Faculty at Major Universities: The Antinepotism Rules and the Grateful Few

1. An earlier, oral form of Lincoln Constance, *Botany at Berkeley: The First Hundred Years* ([Berkeley: University of California], 1978), contained a statement to the effect that the botany department at Berkeley had never had a woman faculty member and he was damned proud of it (Nov. 7, 1975, Berkeley).

2. College and University Personnel Association, "Employee Personnel Practices in Colleges and Universities: A Survey Completed under the Sponsorship of the College and University Personnel Association, Spring 1949" (Champaign, 1949, multilithed); idem, "Employee Personnel Practices in Colleges and Universities, 1951–52: A Survey Completed under the Sponsorship of the College and University Personnel Association, Champaign, Illinois" (Champaign, [1952?], multilithed). See also reports for 1953–54 and 1958, also multilithed; and Harry G. Shaffer and Juliet P. Shaffer, "Job Discrimination against Faculty Wives: Restrictive Employment Practices in Colleges and Universities," *Journal of Higher Education* 37 (1966): 10–15.

3. See esp. Alice Rossi's brief article "Discrimination and Demography Restrict Opportunities for Academic Women," *College and University Business* 48 (Feb. 1970): 74–78.

4. "Report of the Subcommittee on the Status of Academic Women on the Berkeley Campus," in University of California, Academic Senate, Berkeley Division, "Report of the Committee on Senate Policy," May 1970, 11, copy in JLP.

5. For some of those who became research staff members, see ch. 7, esp. table 7.2.

6. Louise Dolnick Solomon, "Female Doctoral Chemists: Sexual Discrepancies in Career Patterns" (Ph.D. diss., Cornell University, 1972), 70.

7. Cynthia Irwin-Williams, "Women in the Field:, The Role of Women in Archaeology," in *Women in Science: Righting the Record*, ed. G. Kass-Simon and Patricia Farnes (Bloomington: Indiana Univ. Press, 1990), 19.

8. On Josephine Mitchell Schoenfeld, see *AMWS*, 15th ed., 1982, 5:420; on Lowell Schoenfeld, ibid., 12th ed., 1972, 5:5596. Personnel files for both Schoenfelds at the University of Illinois Archives, Urbana, are restricted.

9. Josephine Mitchell and Lowell Schoenfeld, "A Statement on the Nepotism Policy of the University of Illinois," enclosed with Josephine Mitchell to Miss Smith, Mar. 23, 1954, AAUW Archives, Washington, D.C. I thank Mary K. Jordan for sending me a copy.

10. Prestige rankings for 1957 were in Hayward Kenniston, *Graduate Study and Research in the Arts and Sciences at the University of Pennsylvania* (Philadelphia: Univ. of Pennsylvania Press, 1959), appendix, 137; and for 1964, in Allan M. Cartter, *An Assessment of Quality in Graduate Education* (Washington, D.C.: ACE, 1966), 66–67.

11. On Bernard, see nn. 59–69 below. On Harald and Rosemary Schraer, see *AMWS*, 18th ed., 1992, 6:568, plus clippings in Harald Schraer Biographical File, Penn State Room, Pattee Library, Penn State University. On the Schoenfelds, see clippings and press release in Lowell Schoenfeld Vertical File, ibid. On the Sherifs, see clippings in Carolyn Sherif Vertical File, Penn State Room, and obituary for Muzafer Sherif in *NYT*, Oct. 27, 1988, sec. 4, p. 23, col. 4. On the Pour-Els, see *AMWS*, 12th ed., 1972, 5:4999; Marian Boykan Pour-El, "Mathematician," in *Successful Women in the Sciences: An Analysis of Determinants*, ed. Ruth B. Kundsin, *Annals of the New York Academy of Sciences* 208 (1973): 16; and idem, "Spatial Separation in Family Life: A Mathematician's Choice," in *Mathematics Tomorrow*, ed. Lynn Arthur Steen (New York: Springer-Verlag, 1981), 187–94. Geochemist Della Roy and materials scientist Rustum Roy were also at Penn State from the late 1940s on, she as a research associate and senior research associate until 1969 (*AMWS*, 16th ed., 1986, 6:335, 336).

The wife of the chair of the mathematics department, Aline Frink, was also a full-time faculty member, though in 1958, after nearly twenty years, she was still only an associate professor ("Husband, Wife Add Answers; Won't Discuss Mathematics," clipping from *Centre [Pa.] Daily Times*, Mar. 19, 1969, Orrin Frink File, Penn State Room). On Orrin Frink, see obituary, *NYT*, Mar. 17, 1988, sec. 2, p. 10, col. 4. See also Kenniston, *Graduate Study and Research*, 137; and Cartter, *An Assessment of Quality*, 66–67. The Penn State Room has some material on Euwema in his biographical file and in the Papers of the Dean of the College of Liberal Arts. The latter reveal that not all couples accepted offers from Penn State: in 1960 Ralph and Mary Boas rejected offers from the math and physics departments, respectively, which offered lower salaries than they were currently getting at Northwestern and DePaul universities (correspondence in "Mathematics, 1959–61" file, box 24).

12. "Advisory Board on Education, Annual Meeting of the National Research Council, March 31, 1958," 3, in carton 4, f. EC 83, ACE-CEW Papers, SLRC.

13. "Advisory Board on Education," *National Academy of Sciences–National Research Council Annual Report, Fiscal Year 1957–1958* (Washington, D.C.: GPO, 1959), 63–65.

14. Among the women scientists who were employed by men's colleges in the 1950s and 1960s were botanist Hannah Croasdale, who after twenty-four years as a research assistant and associate rose from assistant professor to full professor of biology at Dartmouth College in 1959–71 (*AMWS*, 15th ed., 1982, 1:427); Edith Lucille Smith, associate professor, then full professor of biochemistry at Dartmouth Medical School, 1958–78 (ibid., 18th ed., 1992, 6:888); and Fay Ajzenberg-Selove, associate professor of physics (1957–62), professor (1962–70), and acting chairman (1960–61, 1967–68) at Haverford College (ibid., 16th ed., 1986, 1:52). A lengthy vita and some of her correspondence are in the T. Lauritsen Papers, California Institute of Technology Archives, Pasadena, Calif. See also her recently published autobiography, *A Matter of Choices: Memoirs of a Female Physicist* (New Brunswick, N.J.: Rutgers Univ. Press, 1994), ch. 4.

15. Eleanor F. Dolan and Margaret P. Davis, "Antinepotism Rules in American Colleges and Universities: Their Effect on the Faculty Employment of Women," *Educational Record* 41 (1960): 285–95; apparently the AAUW files contain no other correspondence about the report.

16. On Parrish, see *AMWS*, S&B, 13th ed., 1976, 922; part of his work was supported by a faculty fellowship from the Ford Foundation (see f. 57-326, Ford Foundation Archives, New York City). Before 1975 Parrish published at least fifteen articles on women in science and engineering, including, "Professional Womanpower as a National Resource," *Quarterly Review of Economics and Business* 1 (1961): 54–63; "Top Level Training of Women in the United States, 1900–1960," *Journal of the NAWDC* 25 (1962): 67–73, and "Women in Top Level Teaching and Research," *AAUWJ* 55 (1962): 99–103, 106–7, which published in more aggregate form some of the data discussed here. The unpublished worksheets that identify particular institutions are in the John B. Parrish Papers, SLRC.

17. Theodore Caplow and Reece J. McGee, *The Academic Marketplace* (1958; reprint, New York: Arno, 1977). Supported by the FAE, a branch of the Ford Foundation, to help academic institutions cope with the increasing faculty mobility, in the fall of 1957 the authors questioned and interviewed 418 faculty members, including some women, directly involved in 215 specific vacancies in 1954–56 at ten unidentified elite universities, including several of the ones in Parrish's data. When Dolores Burke redid Caplow and McGee's survey in the 1980s, she revisited six of the original universities, which can be identified indirectly as Cornell, Berkeley, Indiana, Northwestern, Pennsylvania, and Texas (*A New Academic Marketplace* [Westport, Conn.: Greenwood, 1988], table 9, 187). Caplow and McGee had explained the rather quaint, almost clubby, and usually secret employment processes of the time in terms of departments' and individuals' attempts to maximize (men's) prestige. Even women who had degrees from prestigious universities, had published, or had served on the lower echelons of the hardworking subfaculty

were not, according to Caplow and McGee's respondents, plausible candidates for desirable jobs. Although an exceptional woman might be hired once in a while, on the whole, the authors observed,

> women tend to be discriminated against in the academic profession, not because they have low prestige but because they are outside the prestige system entirely and for this reason are of no use to a department in future recruitment. . . . The major universities may seek men from abroad before they will seek them from the minor leagues at home. Failing to discover a candidate to their taste in a foreign land, they may decide not to hire at all; or they may even hire a woman, who, being outside the prestige system, cannot hurt them. Not even as a last resort will they recruit from institutions with prestige levels much below their own. (111–12)

While supply and demand might force some to change their practices in the foreseeable future, as was already happening at Penn State, it was more likely that most would either keep on paying more for the highly desired young men or let quality slip and dig deeper into the pool of those men whose degrees were not yet completed. Caplow and McGee concluded matter-of-factly: "Women scholars are not taken seriously and cannot look forward to a normal professional career. This bias is part of the much larger pattern which determines the utilization of women in our economy. It is not peculiar to the academic world, but it does blight the prospects of female scholars" (226).

18. For a discussion and definition of "hierarchical" and "territorial" segregation among women scientists, see Margaret W. Rossiter, *Women Scientists in America: Struggles and Strategies to 1940* (Baltimore: Johns Hopkins Univ. Press, 1982), ch. 3.

19. Diane Johnson, *Edwin Broun Fred: Scientist, Administrator, Gentleman* (Madison: Univ. of Wisconsin Press, 1974), 117, which cites Ruth De Young Kohler, *The Story of Wisconsin Women* (Kohler, Wisc.: Committee on Wisconsin Women for the 1948 Wisconsin Centennial, 1948), 110. See also E. B. Fred, "Women and Higher Education: With Special Reference to the University of Wisconsin," *Journal of Experimental Education* 31 (1962): 158–72.

20. On Stahr, see Carpenter, *AMWS*, 18th ed., 1992, 2:71. See also "Jesse Thomas Carpenter, 1899–1986," *Memorial Statements, Cornell University Faculty, 1986–87* [n.d.], 19.

21. See Frances Clayton, "A Source for College Faculties," *Pembroke Alumna* 37 (Oct. 1962): 6; and on Clayton, *AMWS*, 12th ed., S&B, 1973, 1:406.

22. Besides Olive Hazlett, their distinguished associate professor emerita, long at the Kankakee State Mental Hospital, these included two junior women, Felice Bateman and Irma Reiner, who were unpaid, according to the trustees' published budget, presumably because their husbands were already on the math department payroll (full professors Paul Bateman and Irving Reiner each earning $11,000). On Hazlett's sad later years, see Rossiter, *Women Scientists in America*, 174, 365 n. 20, 369 n. 55; on the others see the listing for the mathematics department in *University of Illinois Bulletin: Undergraduate Study, 1959–60* 56, no. 81 (July 1959): 454–55; and "University of Illinois Internal Budget for 1959–60," published in *Fiftieth Report of the Board of Trustees of the University of Illinois for the Year Ending June 30, 1960, University of Illinois Bulletin* 58, no. 34 (Dec. 1960): 720–21.

23. University catalogs are more useful for specific institutions than are the large volumes of the *AMS* and, after 1971, the *AMWS,* which are not complete and even contain errors. For example, in Rose Goldsen's entry in the *AMWS,* 13th ed., S&B, 1973, a line is missing, thus omitting her nine years as a senior research associate at Cornell and in effect promoting her to associate professor nine years earlier than does her entry in *WWAW* 5 (1968–69): 456. Similarly, the *AMWS* entry for astronomer E. Margaret Burbridge, whom the 1959–60 University of Chicago catalog listed as a research associate, has her as an associate professor there from 1959 to 1962 (*AMWS,* 18th ed., 1992, 1:863).

But using catalogs also presents certain definitional problems. Not every faculty member at a medical school was necessarily a "scientist": some might be anatomists or biochemists, but what about the professors of pediatrics, medicine, surgery, or other clinical fields? Similarly, some of the faculty at a school of public health (e.g., those in biostatistics or nutrition) might classify themselves as scientists, but those in "maternal and child health" might not. Thus the catalogs give the institutions' leaders' views concerning where a faculty person should be put; for women this sorting was often-times at odds with what they had been trained to do. Quite common, for example, was the location of many psychologists in a university's school of education, which could cause considerable mutual resentment, for the captive spouse probably saw this position as less desirable than what she might have had elsewhere had she been mobile, and the hiring unit knew that taking her on reinforced its own already secondary status within the institution. For a discussion of the centrality of faculty gender to schools of education, see Geraldine Joncich Clifford, *Ed School* (Chicago: Univ. of Chicago Press, 1988), 157–58, 334.

24. One big mistake was classifying New York University as a public institution; this was partly offset by considering Cornell University as an entirely private one (it has a private arts college, engineering school, and medical school, but the home economics, agriculture, labor, and veterinary schools are state-supported). Unfortunately, a close comparison of tables 6.2 and 6.3 reveals certain inexplicable and perhaps inevitable discrepancies (of both omission and commission) between Parrish's numbers and the catalogs' lists. Parrish omitted Harvard's famous full professor of astronomy Cecilia Payne-Gaposchkin but included (under sociology) a woman professor of social relations, although there was none at Harvard then. (Sometimes errors are revealing; perhaps someone thought longtime lecturer Florence Kluckhohn deserved to be and was a full professor.) He also seems to have included Vanna T. Cocconi, a "research assistant professor" of physics at Cornell's Laboratory of Nuclear Studies, as a faculty member, but he omitted the similarly titled Donatella B. Adler at the University of Illinois.

25. On Wu, see *AMWS,* 18th ed., 1992, 7:812; Gloria Lubkin, "Chien-Shiung Wu: The First Lady of Physics Research," *Smithsonian* 1, no. 10 (Jan. 1971): 52–57; Chien-Shiung Wu, "Discovery Story I. One Researcher's Personal Account," *Adventures in Experimental Physics* 3 ("Gamma") (1974), 101–2, 104, 108–11, 114–18; idem, "The Discovery of Nonconservation of Parity in Beta Decay," in *Thirty Years since Parity Nonconservation: A Symposium for T. D. Lee,* ed. Robert Novick (Boston: Birkhäuser, 1988), 19–35; and Edna Yost, *Women of Modern Science* (New York: Dodd, Mead, 1959), 80–93.

Marjorie Hope Nicolson had earned a Litt. D. from Princeton in 1946 (*WWAW* 5 [1968–69]: 893).

26. For 1951–63, see NSF, *Annual Reports;* for later years see, idem, *Grants and Awards*. On Low, see *AMWS*, 18th ed., 1992, 4:927; on Benedict, see Judith Schachter Modell, *Ruth Benedict: Patterns of a Life* (Philadelphia: Univ. of Pennsylvania Press, 1983), 292; on DuBois, see obituary, *NYT,* Apr. 11, 1991, B-14, and Judith Walzer, "Interviews with Cora Du Bois," Aug. 1981, MCRC; on Macfarlane, see obituary, *NYT,* Mar. 18, 1989, sec. 1, p. 10, col. 4, and esp. the portion on Jean Macfarlane in "Four Interviews," June 3, 1969, in Milton J. Senn Oral History Collection on Child Development (#45), Historical Library, National Library of Medicine, Bethesda, Md.

27. Two other efforts at "creative" or "coercive" philanthropy were under way in these years at Brown University and MIT, but neither showed substantial results until after 1970. In 1958, at her class's twenty-fifth reunion, Nancy Duke Lewis, alumna and dean of Pembroke College at Brown University, started a fund to endow a professorship at Brown for a woman. Both she and her mother bequeathed it substantial sums. In 1966 its total reached $125,000, and by the 1980s it was sufficient to support the Nancy Duke Lewis Professor (Grace E. Hawk, *Pembroke College in Brown University: The First Seventy-Five Years, 1891–1966* [Providence, R.I.: Brown Univ. Press, 1967], 222, 246–47, 267, 305 n. 39; Jean E. Howard, "Finding a Voice: Women on the Brown Corporation," in *The Search for Equity: Women at Brown University, 1891–1991,* ed. Polly Welts Kauffman [Hanover, N.H.: Univ. Press of New England for Brown Univ. Press, 1991], 234–37, 240; Barnaby C. Keeney, "Women Professors at Brown," *Pembroke Alumna* 37 [Oct. 1962]: 8–11; Michael Diffily, "Barnaby Conrad Keeney: President of Brown University, 1955–1966" [Ph.D. diss., Boston College, 1988]).

At MIT, Laurence Rockefeller and the Rockefeller Brothers' Fund each contributed $200,000 in 1963 to endow a faculty position for a woman to be named in honor of their sister Abby Rockefeller Mauzé, who had not graduated from college. At first it was used for visiting women professors (1963–72), but since 1973 it has been held by a woman professor interested in the women undergraduates (Mildred Dresselhaus) (see *Endowed Professorships at MIT: A History* [n.d], 94–95, copy at Institute Archives, MIT; and Mauzé obituary, *NYT,* May 29, 1976, p. 26, col. 2).

28. On Payne-Gaposchkin, see Andrea K. Dobson and Katherine Bracher, "A Historical Introduction to Women in Astronomy," *Mercury* 21 (Jan.–Feb. 1992), 13–15; Elske Smith, "Cecilia Payne-Gaposchkin," *Physics Today* 33 (June 1980): 64–65; *AMWS*, 12th ed., 1972, 2:2043; and *CBY, 1957,* 421–23. See also her 1968 oral history interview at NBL; Katherine Haramundanis (her daughter), ed., *Cecilia Payne-Gaposchkin: An Autobiography and Other Publications* (Cambridge: Cambridge Univ. Press, 1984); and Peggy Aldrich Kidwell, "Cecilia Payne-Gaposchkin: Astronomy in the Family," *Uneasy Careers and Intimate Lives: Women in Science, 1789–1979,* ed. Pnina Abir-Am and Dorinda Outram (New Brunswick, N.J.: Rutgers Univ. Press, 1987), 216–38. There are many Payne-Gaposchkin letters in the Henry Norris Russell Papers, Princeton University Archives, Princeton University, and some in MMP.

29. On Stewart, see *AMS,* 10th ed., P&B, 1961, 4:3925; and Edmund M. Spieker, "Memorial to Grace Anne Stewart, 1893–1970, GSA *Memorials* 2 (1973): 112. See 1954–55 letters amongst Perlmann, Detlev Bronk, Walter Bauer, and William H. Stein, all in box

77, f. 1, Gertrude Perlmann Papers, RAC. She was an associate professor for fifteen years, 1958–73.

30. Walzer, "Interviews with Cora DuBois," sess. 4.

31. [Matthew N. Lyons], "Rose Kohn Goldsen, 1917–1985: A Biography," in *Rose Kohn Goldsen Papers: A Descriptive Guide* ([Ithaca]: Department of Manuscripts and University Archives, Cornell University Library, 1992), 10–11.

32. "Report of the Subcommittee on the Status of Academic Women on the Berkeley Campus," 29 (table 4.3), lists the date the last woman was hired for thirty departments. Seven departments (astronomy, biochemistry, chemistry, criminology, engineering, physics, and zoology) had never hired a woman (botany and geography were not listed); six others (including sociology and political science) had hired only one; economics had hired its last woman in 1936, sociology in 1925, and psychology in 1924. From this perspective, anthropology (1966) and even mathematics (1953) and statistics (1950) look almost progressive. (Sociologist Dorothy Swaine Thomas's appointment there in the 1940s was in the School of Agriculture [see Dorris W. Goodrich, "Varieties of Sociological Experience," in *Gender and the Academic Experience: Berkeley Women Sociologists,* ed. Kathryn P. Meadows Orlans and Ruth A. Wallace (Lincoln: Univ. of Nebraska Press, 1994), 12].)

By 1961 two members of the Heller family had also died, terminating part of the funding for Huntington's position (Alice Greene King, "Emily H. Huntington: A Career in Consumer Economics and Social Insurance," 1961, esp. 45, 89–90, Bancroft Library, University of California, Berkeley; Mary E. Cookingham, "Social Economists and Reform: Berkeley, 1906–1961," *History of Political Economy* 19 [1987]: 47–65; obituary, in *Radcliffe Quarterly,* 68, no. 3 [Sept. 1982]: 43). Meanwhile, Eveline Burns, another woman economist who made pioneering contributions to the design of social security and other forms of old-age insurance, was on the faculty of the Columbia University economics department with her husband, economist Arthur Burns, until 1942 and then (after 1946) on that of its New York School of Social Work (see *CBY, 1960,* 64–66; *AMWS,* S&B, 11th ed., 1968, 1:213; *WWAW* 8 [1974–75]: 131; and Linda R. Wolf Jones, "Eveline M. Burns and the American Social Security System, 1935–1960: A Model of Professional Leadership" [D.S.W. diss., Yeshiva University, 1985], recently published as *Eveline M. Burns and the American Social Security System, 1935–1960* [New York: Garland, 1991]).

33. Curiously this is not listed in Marshall's later entry in the *AMWS,* which reports her as an assistant professor at the University of Chicago from 1954 to 1957 and off at Brookhaven as a visiting scientist from 1958 to 1960 (*AMWS,* 14th ed., 1979, 4:2978).

34. Maria Goeppart Mayer to Dr. Rita Arditti and Miss Buonaventura, Dec. 10, 1970, copy in author's possession. At the University of Chicago several women faculty members were evidently married to nonfaculty, for example, psychologist Bernice Neugarten in "human development" (*AMWS,* S&B, 13th ed., 1978, 882; "Distinguished Contributions to Education in Psychology Awards," *AP* 31 [Jan. 1976]: 84–85) and Lillian Eichelberger (Cannon) in biochemistry (*NCAB,* vol. M [1978]: 389–90, which merely states that on August 11, 1923, she married Ralph H. Cannon and gives no details about what he did subsequently). "A large private university wrote that it has strong objections to members of the same family being in the same department, but that if

'such a second appointment is made it is usually on a voluntary basis, i.e. without salary and without a vote in the deliberations of the area'" (Dolan and Davis, "Antinepotism Rules," 293). The respondent must have been the University of Chicago, but the Mayers were in two different departments.

35. On Simmonds, see *AMWS*, 16th ed., 1986, 6:755. On Fruton, see ibid., 2:1216; and Joseph S. Fruton, *Eighty Years* (New Haven: Epikouros, 1994) (I thank him for a copy). See also "Garvan Medal, Sofia Simmonds," *C&EN* 47 (Feb. 24, 1969): 100.

36. On Pauline and Robert Sears, see *AMWS*, 13th ed., S&B, 1978, 1072; and [Robert R. Sears], "Robert R. Sears," in *A History of Psychology in Autobiography*, ed. Gardner Lindzey, vol. 7 (San Francisco: W. H. Freeman, 1980), 395–433. On Helen Farnsworth, see *AMWS*, 12th ed., S&B, 1973, 1:665; and "Memorial Resolution: Helen Cherington Farnsworth, 1903–1974," in box 22, f. 42, *NAWP*. On Paul Farnsworth, see *AMS*, 11th ed., S&B, 1968, 1:459.

37. On Elizabeth Miller, see *AMWS*, 16th ed., 1986, 5:386; on James Miller, see ibid., 390; and see also "Elizabeth and James Miller: An Interview Conducted by Laura L. Small," 1985, University of Wisconsin Oral History Office, Madison.

38. Abraham L. Sachar, *A Host at Last* (Boston: Little, Brown, 1976), chs. 9, 12; Richard M. Freeland, *Academia's Golden Age: Universities in Massachusetts, 1945–1970* (New York: Oxford University Press, 1992), ch. 6. On Levine, see *AMWS*, 16th ed., 1986, 4:723; on Van Vunakis, ibid., 7:297; on the Sussmans, ibid., 6:1141. On Aberle, see *Contemporary Authors* 21-24R (Detroit: Gale Research, 1977):11–12; on Gough, ibid. 13-16R (1975): 10. Aberle and Gough had come from Ann Arbor, where Aberle was a full professor but Gough was, after a year as a lecturer at Wayne State University, apparently unemployed because of local antinepotism rules (see below, ch. 15, n. 49). In 1963 they left Brandeis for the University of Oregon, where Aberle was a professor and Gough was a research associate. In 1967 they moved to Vancouver, Canada, where Aberle became a professor at the University of British Columbia and she was employed for a time at Simon Fraser University (Sachar, *A Host at Last*, 197–201). Gough's vocal Marxist views and vigorous antigovernment activities infuriated her colleagues and administrators and shortened several of those academic appointments that she did receive (Joan P. Mencher, "Deaths," *Anthropology Newsletter* [American Anthropological Association], December 1990, 4; I thank Naomi Quinn for a copy).

In addition, Brandeis University hired among its first hundred or more science faculty anthropologists Elizabeth Colson and Helen Codere, psychologist Eugenia Hanfmann (to run the psychological clinic; later she held an endowed chair), biochemist Mary Ellen Jones, and Marie Boas, one of the first women Ph.D.'s in the history of science. See Sachar, *A Host at Last;* on Colson, *AMWS*, S&B, 13th ed., 1978, 237; on Codere, who replaced Colson when she left for Berkeley in 1964, ibid., P&B, 13th ed., 1976, 1:773; on Hanfmann, Marianne L. Simmel, "A Tribute to Eugenia Hanfmann, 1905–1983," *Journal of the History of the Behavioral Sciences* 22 (1986): 348–56, and Eugenia Hanfmann, "Eugenia Hanfmann" (autobiography), in *Models of Achievement: Reflections of Eminent Women in Psychology*, vol. 1 (New York: Columbia Univ. Press, 1983), 140–52; on Jones, *AMWS*, 16th ed., 1986, 4:13; and on Marie Boas (later Hall), the sister of mathematician Ralph Boas, *Directory of American Scholars*, 8th ed., vol. 1 (New York: Bowker, 1982), 303.

39. Joanne Simpson, "Meteorologist," in Kundsin, *Successful Women in the Sciences,* 42; she had earlier lost first a job and then her husband because of antinepotism rules at another institution. An earlier faculty couple at UCLA were Clara Szego of the biology department in the arts college (*AMWS,* 18th ed., 1992, 6:1249) and Sidney Roberts in biological chemistry at the medical school (ibid., 230).

40. On the Willises, see *AMWS,* 18th ed., 1992, 7:692; and Judith H. Willis, "Changing Status of Women in Academic Entomology," *Bulletin of the Entomological Society of America* 33 (spring 1987): 17. Earlier, physicists Gertrude and Maurice Goldhaber had left for Brookhaven (see ch. 13, n. 9; and *AMWS,* 15th ed., 1982, 3:189).

41. On the Cosers, see *AMWS,* S&B, 13th ed., 1978, 248.

42. On Cori, see *NAWMP;* and Joseph Larner, "Gerty Theresa Cori, 1896–1957," *NAS Biographical Memoirs* 61 (1992): 110–35. On Carl Cori, see Mildred Cohn, "Carl Ferdinand Cori, 1896–1984," ibid., 78–109.

43. On Flügge-Lotz, see *NAWMP.*

44. Tom Lewin, "Interview with Fritz and Grace Heider," winter 1985, 62–63, University of Kansas Archives, Spencer Research Library, Lawrence; Fritz Heider, *The Life of a Psychologist: An Autobiography* (Lawrence: Univ. of Kansas Press, 1983), 183. On Beatrice Wright, see *AMWS,* 13th ed., S&B, 1978, 1310; Beatrice Wright, *Physical Disability: A Psychological Approach* (New York, Harper, 1960); David K. Hollinsworth et al., "Beatrice A. Wright: Broad Lens, Sharp Focus," *Journal of Counseling and Development* 67 (1989): 384–93; Franklin D. Murphy to Mrs. Beatrice Wright, Jan. 22, 1959, University of Kansas Chancellor's Files, Department of Psychology, 1958–59; "Regents Asked to Ease Faculty Restrictions," *Topeka Capital,* June 22, 1963; and University of Kansas, *Faculty Handbook, 1964–65* (Lawrence, n.d.), 22. See also Harry G. Shaffer and Juliet P. Shaffer [University of Kansas associate professors of economics and psychology, respectively], "Job Discrimination against Faculty Wives," 10–15, which also mentioned the Mitchell-Schoenfeld case (11).

45. Finding aid, Papers of the University of North Carolina [Men's] Faculty Club, Southern Historical Collection, Main Library, University of North Carolina, Chapel Hill; Joseph C. Wetherby (president) and Earl W. Porter (secretary) to "Dear Colleague," Nov. 23, 1956, box 13, Katherine Banham Papers, Duke University Archives, Durham; clipping, "Brookes Speaks to Faculty Unit; Brown Is Elected," *Durham Morning Herald,* Apr. 3, 1963, in Frances C. Brown Papers, ibid.

46. "Welcome Back Home, Statler Club," leaflet, 1989, Statler Club Papers, Rare and Manuscript Collections, KLCU. Women were first admitted in 1968.

47. Icie Macy Hoobler, *Boundless Horizons: Portrait of a Pioneer Woman Scientist* (Smithtown, N.Y.: Exposition Press, 1982), 161–62.

48. Elizabeth Scott, conversation with author, May 16, 1975, and interview on KQED-TV (San Francisco), May 16, 1977. On Nader, personal communication, Sept. 30, 1994. Meanwhile, in 1966 law professor Barbara Armstrong defied some attendants and marched through the club's Great Hall (for men only) to get to the meeting room.

49. "Anthropologist Cora DuBois Dies," *Harvard University Gazette,* Apr. 12, 1991; Judith Walzer, "Interview with Emily Vermeule, August 6, 1981," 18, MCRC.

50. Lincoln Constance to author, Nov. 23, 1987. Botanist Annetta Carter was the first

woman member of the Biosystematists, elected in the 1970s ("Anetta Mary Carter, 1907–1991," in "University of California at Berkeley Women's Faculty Club Newsletter," June 1991, 2; I thank Suzanne Hildenbrand for a copy). For the Research Club of the University of Michigan, see William H. Hobbs, "The Research Club of the University of Michigan," in *The University of Michigan: An Encyclopedic Survey,* vol. 1, ed. Wilfred B. Shaw (Ann Arbor: Univ. of Michigan Press, 1942), 399–406; and Malcolm H. Surle, "The Junior Research Club," ibid., 407–10. Albert Hazen Wright, *Pre-Cornell and Early Cornell: The Research Club of Cornell University, 1919–1965,* Studies in History, 35 (privately printed, 1967), lists the 421 male faculty members who belonged to this organization over the years 1918–65. In this project Wright, one of the club's charter members, was assisted, ably as always, by his wife Anna Allen Wright, a nonmember but an important herpetologist in her own right.

51. Lauramay T. Dempster, conversation with author, Oct. 13, 1979. The more traditional Catherine Campbell reported in her oral history that despite her doctorate in paleontology, she joined the other wives in drinking tea when the men-only LeConte Club of geologists of the Bay Area held their meetings. The Branner Club of southern California had operated in the same way. ("Ian and Catherine C. Campbell, Geologists: Teaching, Government Service, Editing, An Oral History Conducted in 1988 by Eleanor Swent," 84, Bancroft Library, Berkeley). For more on the "scientific hostess," see Margaret W. Rossiter, "Women and the History of Scientific Communication," *Journal of Library History* 21 (1986): 52–53.

52. For criticism of such a broad membership policy, see Dr. Margaret Reed Lewis to Dean Althea K. Hottel, Apr. 23, 1948, and reply, Apr. 28, 1948, both in Women's Faculty Club Papers, University of Pennsylvania Archives, Philadelphia.

53. Embryologist Dorothea Rudnick, a professor of biology at nearby Albertus Magnus College and a long-term "research guest" at Yale, was a fellow from 1963 to 1969. She even listed it in her entry in the *AMWS,* 14th ed., 1979, 6:4345. For more on her, see Yost, *Women of Modern Science,* 156–70.

54. The most extensive history of any faculty club, let alone both men's and women's clubs, is Gerard M. McGrath, "Collegiality at Columbia: The Origin, Development, and Utility of Two Faculty Clubs" (Ed.D. diss., Columbia University, Teachers College, 1983), which also cites Ida A. Jewett, *Women's Faculty Club of Columbia University, 1913–1963: An Informal History* (New York: Women's Faculty Club of Columbia University, 1963).

55. The Papers of the Woman's Research Club of the University of Michigan are at the Bentley Historical Library in Ann Arbor. One of the more interesting items for this period is Margaret L. Clay to Helen Tanner, Apr. 12, 1969, describing what the club has meant to her, her morale, and her image of the potential of womankind over the years. See also Orma F. Butler, "The Woman's Research Club," in Shaw, *The University of Michigan,* vol. I, 410–11, for the earlier years.

56. For more on Sigma Delta Epsilon in these years, see ch. 15 below.

57. See National Council of Administrative Women in Education of the National Education Association, *Administrative Women in Higher Education* (Washington, D.C.: NEA, 1952).

58. On Payne-Gaposchkin, see n. 28 above; on Dodson, *AMWS,* 13th ed., 1976,

2:1050, and *WWAW* 5 (1968–69): 319–20, with slightly different dates; on Penniman, *Who's Who in America,* 46th ed. (1990–91), 2:2559, and press release, Jan. 14, 1965, Penniman File, University of Wisconsin Archives, Madison; on Scott, *AMWS,* 15th ed., 1982, 6:563; and on Newell, *Who's Who in America* 44th ed. (1986–87), 2:2057.

59. Widowed in 1951, Bernard retired a bit early (at sixty-one) in 1964, affluent from investments in IBM stock, disenchanted with teaching and the academic life, and possibly defeated in department politics, to move back to Washington, D.C., to write full-time, something she has been doing ever since. She has written an abundance of autobiographical material, including Jessie Bernard, "Minnesota Women: One Alumna's Story," *University of Minnesota Alumni News,* Oct. 1971, 8–13; Catherine Breslin, "A Rap on Marriage with Jessie Bernard," *Chatelaine, the Canadian Home Journal* 45, no. 1 (Jan. 1972): 21, 38–41; Jessie Bernard, "My Four Revolutions: An Autobiographical History of the ASA," *American Journal of Sociology* 78 (1973): 773–91; and Jessie Bernard, "A Woman's Twentieth Century," in *Authors of Their Own Lives: Intellectual Autobiographies by Twenty American Sociologists* (Berkeley: Univ. of California Press, 1990), ed. Bennett M. Berger, ch. 14. She has also been the subject of two biographies: Robert C. Bannister, *Jessie Bernard: The Making of a Feminist* (New Brunswick, N.J.: Rutgers Univ. Press, 1991) and Gwendolyn S. Safier, "Jessie Bernard, Sociologist" (Ph.D. diss., University of Kansas, 1972), both of which are based in part on interviews and correspondence with Bernard. There is also a sizable collection of biographical material in the Jessie Bernard Papers, Historical Collections and Labor Archives, Pattee Library, Pennsylvania State University, and there are several clippings in the Jessie Bernard Faculty File in the Penn State Room at the Pattee Library. There is also some material by and about her in the sociology and anthropology department files in box 31, Papers of the Dean of the College of Liberal Arts.

60. Jessie Bernard, "Breaking the Sex Barrier: Our First Woman Professor Finds It Rough Going in This Fortress of Anti-Feminism," *Princeton Alumni Weekly* 61, no. 1 (Sept. 23, 1960): 3–7. There were sympathetic letters to the editor in subsequent issues: "Mrs. Bernard," in 61, no. 5 (Oct. 21, 1960): 3, and "The Sex Barrier," ibid., no. 9 (Nov. 18, 1960), 5.

61. Clipping, "Sociologists Granted Leave to Continue Career Women Study," Jessie Bernard File, Penn State Room; Margaret Matson Faculty File, ibid.; *AMWS,* S&B, 13th ed., 1978, 784. It is not clear just when Matson left the project or why (Bannister, *Jessie Bernard,* ch. 7). There is also a two-page fragment entitled "Academic Woman," in box 1, Jessie Bernard Papers, that indicates that she found it difficult to decide whether the volume was a systematic survey of the subject or an autobiographical tale.

One part of Bernard's research in the early 1960s comprised some pioneering and rather inventive experiments (with the help of the Division of Instructional Services at Penn State) in the realm that would later be called "sexist" practices and evaluations. For example, she compared test scores in a mathematics course of those students who had had female section leaders with those who had had male leaders. She also compared two sociology classes' reactions to a lecture given by a man and a second one by a woman. In what she called her "matched scientists" study, funded by the National Institute of Mental Health (NIMH), Bernard compared the publication rates of 28 male and 28 female zoologists: she found that though overall the men published more, this was largely because of their positions at universities rather than colleges. The few women zoologists at universities published more than either the men or the women at colleges, who had heavy teaching loads and fewer opportunities and incentives to do research. (Interestingly, one respondent was a Negro woman at a teachers college who described to Bernard her "peculiarly isolated position": because the nearby white university was still segregated, she could obtain journals only by having others check them out for her, and she was reluctant to visit a white scientist in his or her laboratory, lest her presence embarrass someone there.) See Jessie Bernard, *Academic Women* (State College: Pennsylvania State Univ. Press, 1964), 310 n. 27.

Finally, Bernard participated in analyzing data on the scientific information practices of 673 bioscientists (including 68 women) on a project run by Charles Shilling and Joe Tyson of George Washington University. Their findings—that though their travel was not restricted, the women attended fewer meetings and visited fewer laboratories and talked to fewer people when there—helped her formulate her thoughts and observations on what she termed the "stag effect," or men's tendency at professional meetings or other occasions presumably open to all scientists to limit their informal communication to other men, with whom they ate and drank, rarely inviting any women to join them. Because the results of these studies were tentative and suggestive rather than definitive, Bernard did not publish them in journals but put them into a set of appendixes to *Academic Women*, where they were essentially buried and were rarely cited even by those later seeking information on the topic. See ibid., app. D; and Jessie Bernard, Charles W. Shilling, and Joe W. Tyson, "Informal Communication among Bioscientists" (Washington, D.C.: George Washington University Biological Sciences Communication Project, BSCP communiqué no. 16–63, 1963).

62. Bernard, *Academic Women*, viii–x.

63. See, e.g., Dorothy E. Smith, "A Berkeley Educator," in Orlans and Wallace, *Gender and the Academic Experience*, 48.

64. Bernard cited a recent NSF report that claimed that at least in 1959 the success rate of women and men applicants for fellowships had been about the same and a recent report from the Woodrow Wilson National Fellowship Program that it was giving 25 percent of its fellowships to women. In fact the Woodrow Wilson staff was quite proud of the many women they had *dissuaded* from going to graduate school, convincing them during what must have been rather stressful interviews that they were not highly enough motivated for the rigors of graduate school (*Academic Women*, 283 nn. 21 and 23, citing NSF, *Women in Scientific Careers* [Washington, D.C.: GPO, 1961], 7, and Woodrow Wilson National Fellowship Foundation, *Report on Activities, 1957–61* [Princeton, c. 1961], IV-7, IV-43, IV-45).

65. See Clayton, "A Source for College Faculties," 4–7; and *AMWS*, S&B., 12th ed., 1973, 1:406.

66. Bernard, *Academic Women*, 6.

67. Mirra Komarovsky, review of *Academic Women* by Jessie Bernard, *Social Forces* 43 (1965): 604–5 (typescript with note attached, Mirra to Jessie, Jan. 6, 1965, box 5, Jessie Bernard Papers). There were at least fourteen other reviews, copies of many of which are in the Jessie Bernard Papers. Among the reviewers were Logan Wilson of the ACE (and

author of *Academic Man*) in *American Sociological Review* 30 (1965): 611; Margaret Marshall of Brooklyn College in *Journal of Higher Education* 36 (1965): 233–34; Henrietta Fleck in *JHE* 57 (1965): 473–74; Jean G. Henderson in *Teachers College Record* 66 (1965): 766–67; and, quite poignant, Mary Ellman in *Commentary* 39 (Mar. 1965): 67–70.

68. Bernard, *Academic Women,* 44.

69. Ibid., chs. 6, 8.

Chapter 7   Resentful Research Associates: Marriage and Marginality

1. Louise Dolnick Solomon, "Female Doctoral Chemists: Sexual Discrepancies in Career Patterns" (Ph.D. diss., Cornell University, 1972), 58; Judith Walzer, "Interview with Ruth Hubbard," July 1981, typescript in MCRC.

2. NSF, *Scientific Research and Development in Colleges and Universities: Expenditures and Manpower, 1953–54* (Washington, D.C.: GPO, 1959), 17–18, 59, 72; idem, *American Science Manpower 1962: A Report of the National Register of Scientific and Technical Personnel* (Washington, D.C., 1964), 89.

3. "Science and Engineering Professional Manpower Resources in Colleges and Universities, 1961," *NSF Reviews of Data on Research and Development,* no. 37 (Jan. 1963): 9.

4. Carlos E. Kruytbosch and Sheldon L. Messinger, "Unequal Peers: The Situation of Researchers at Berkeley," *American Behavioral Scientist* 11 (May–June 1968): 33–34; see also Charles V. Kidd, *American Universities and Federal Research* (Cambridge: Harvard Univ. Press, 1959), 152–54.

5. John B. Parrish, "Women in Top Level Teaching and Research," *AAUWJ* 55 (1962): 106.

6. Kruytbosch and Messinger, "Unequal Peers," 42 n. 11.

7. "Report of the Subcommittee on the Status of Academic Women on the Berkeley Campus," in University of California, Academic Senate, Berkeley Division, "Report of the Committee on Senate Policy," May 1970, app. 15, 77–78, copy in JLP.

8. For a discussion of which sciences have greater or lesser needs for staffs of assistants, see Lowell L. Hargens, *Patterns of Scientific Research: A Comparative Analysis of Research in Three Fields* (Washington, D.C.: American Sociological Association, 1975).

9. NSF, *Sixth Annual Report: Fiscal Year 1956* (Washington, D.C.: GPO, 1956), 46–47.

10. Kruytbosch and Messinger, "Unequal Peers," 42 n. 11.

11. Helen MacGill Hughes, "Wasp/Woman/Sociologist," *Society* 14 (July–Aug. 1977): 77.

12. On Cecilia and Sergei Gaposchkin, see ch. 6, n. 28; on Gerald Hassler, see *AMS,* 11th ed., 1966, 3:2159; and on Mildred Mathias, see *AMWS,* 16th ed., 1986, 5:259. There is also a collection of Mathias's papers (mostly from before World War II but some later) at the Hunt Institute for Botanical Documentation, Carnegie-Mellon University, Pittsburgh, which also has a biographical file on her. See also Mary Terrall, "Among the Plants of the Earth: Mildred E. Mathias," UCLA Oral History Program, 1982, available at UCLA and at the Bancroft Library at the University of California at Berkeley. For Cornell's Patricia and Olin Smith, who left for faculty positions at Bowling Green State University in 1966, see ch. 9, n. 30.

13. E. C. Pollard, "How To Remain in the Laboratory Though Head of a Department," *Science* 145 (1964): 1020. (I thank Joan Burstyn for telling me of this article.) See also the unsigned "How to Get More from Research," *Nation's Business* 46 (Mar. 1958): 34–35, 73–77, which recommended hiring housewives as part-time clerks and technicians.

14. On Pollard, see *AMWS*, 12th ed., 1972, 5:4965.

15. *University of Chicago Announcements, Graduate Programs in the Divisions, Sessions of 1959–1960,* 59, no. 2 (Oct. 1958): 62.

16. On Juhn, see obituary in *Poultry Science* 49 (1970): 1151. On Shorb, see obituaries in *Washington Post,* Aug. 20, 1990 (I thank Michele Aldrich for a copy), and *NYT,* Aug. 21, 1990, B-11; and *AMWS,* 15th ed., 1982, 6:688. There is a collection of Shorb's papers at the University of Maryland Archives, College Park.

17. Ruth Hubbard, review, "*Rosalind Franklin and DNA* by Anne Sayre," *Signs* 2 (1976): 229–37. On Hubbard, see *AMWS,* 18th ed., 1992, 3:921; on Wald, ibid., 7:404.

18. Hans Selye, *From Dream to Discovery: On Being a Scientist* (1964; reprint, New York: Arno, 1975), 25–26. Selye also commented about "twosomes" in research, including married couples (170–72). Often the wife was in his view a technician, secretary, or just a good listener, but sometimes she was "ill-tempered" and nagged her husband too much. On Selye, see *AMWS,* 15th ed., 1982, 6:598.

19. On the Lederbergs, see *AMWS,* 18th ed., 1992, 4:693–94; and *CBY, 1959,* 251–52.

20. On Sitterly, see obituary, *Washington Post,* Mar. 9, 1990, 6; *AMWS,* 12th ed., 1972, 5:5861; and "Charlotte Moore Sitterly: Transcript of an Interview Taken on a Tape Recorder by David DeVorkin on 15 June 1978," 38–55, NBL.

21. See, e.g., Dan H. Campbell to Elizabeth Robinton, Apr. 9, 1969; Campbell to Matthew C. Dodd, Aug. 25, 1969; Campbell to Clayton G. Loosli, Mar. 2, 1970; and Campbell to John R. Porter, May 14, 1970, all in Dan Campbell Papers, California Institute of Technology Archives, Pasadena.

22. On Irene Taeuber, see *NAWMP; AMWS,* S&B, 12th ed., 1973, 2:2425; *NCAB* 58 (1979): 221–23; and Irene B. Taeuber, "The Social Sciences: Characteristics of the Literature, Problems of Use, and Bibliographic Organization in the Field," in *Bibliographic Organization,* ed. Jesse H. Shera and Margaret E. Egan (Chicago: Univ. of Chicago Press, 1951), 127–39, which describes in part her own work on the *Population Index.* Her husband Conrad Taeuber was a statistician and demographer for the federal government.

23. On Edinger, see *NAWMP.* There is a collection of her papers at the Museum of Comparative Zoology Archives at Harvard.

24. On Proskouriakoff, see obituary, *NYT,* Sept. 11, 1985, D-27; and Erik Eckholm, "Secrets of Maya Decoded at Last, Revealing Darker Human History," ibid., May 13, 1986, C-1, C-3. See also ch. 11, n. 15.

25. On Glueck, see *NAWMP.*

26. On Hubbard and Wald, see n. 17 above.

27. On Turner, see *AMWS,* 16th ed., 1986, 7:232.

28. On Tharp, see ibid., 12th ed., 1973, 6:6343; Marie Tharp and Henry Frankel, "Mappers of the Deep," *Natural History* 95, no. 10 (Oct. 1986): 49–62; Manik Talwani, "Bruce Heezen: An Appreciation," in *The Ocean Floor: Bruce Heezen Commemorative*

Volume, ed. R. A. Scrutton and M. Talwani (New York: John Wiley & Sons, 1982), 1–2; and Marie Tharp, "Mapping the Ocean Floor, 1947 to 1977," ibid., 19–31. See also Henry W. Menard, *The Ocean of Truth: A Personal History of Global Tectonics* (Princeton: Princeton Univ. Press, 1986).

Another prominent East Coast female oceanographer was Marie Poland Fish of the University of Rhode Island's Narragansett Marine Laboratory. Supported by the ONR for more than twenty years, Fish pioneered in the field of bioacoustics, detecting and analyzing the underwater sounds of more than three hundred marine organisms, lest the Navy confuse a school of fish with an approaching submarine. On Fish, see *AMWS,* 13th ed., 1976, 2:1303; *CBY, 1941,* 280–81; and obituary, *NYT,* Feb. 2, 1989, B-8. Since she was involved in many ichthyological expeditions before settling down at Kingston, she may merit a full biography.

29. On Robinson, see *AMWS,* 14th ed., 1979, 6:4251; Elizabeth Noble Shor, *Scripps Institution of Oceanography: Probing the Oceans, 1936 to 1976* (San Diego: Tofua, 1978), 244–50, 422, 431–32; and Margaret Robinson Biographical File, SIO Archives, La Jolla, Calif. (I thank Deborah Cozort Day for much help there).

30. See Dorothy H. Eichorn et al., eds., *Present and Past in Middle Life* (New York: Academic, 1981). On Bayley, see *AMWS,* 13th ed., S&B, 1978, 75. On Jones, see obituary, *NYT,* Aug. 21, 1987, B-7; Deana Dorman Logan, "Mary Cover Jones: Feminine as Asset," *Psychology of Women Quarterly* 5, no. 1 (1980): 104–15; and Milton J. Senn, "Four Interviews," June 3, 1969, typescript, Bancroft Library. On Eichorn, see Eichorn, "Dorothy Hansen Eichorn," in *Models of Achievement: Reflections of Eminent Women in Psychology,* ed. Agnes N. O'Connell and Nancy Felipe Russo, vol. 2 (Hillsdale, N.J.: Lawrence Erlbaum Associates, 1988), 207–24. On Jean Macfarlane, see obituary, *NYT,* Mar. 18, 1989, sec. 1, p. 10, col. 4; and Milton J. Senn, "Four Interviews," 1969, Bancroft Library, Berkeley.

31. See, e.g., Selma Monsky, "In Memoriam: Shirley A. Star, 1918–1976," *Public Opinion Quarterly* 40 (1976): 265–66. See also Mildred Schwartz, in *AMWS,* 13th ed., S&B, 1978, 1068; and John M. Allswang and Patrick Bova, eds., *NORC Social Research, 1941–1964: An Inventory of Studies and Publications in Social Research* (Chicago: NORC, University of Chicago, 1964), which lists several women authors and study directors; Peter H. Rossi, "Researchers, Scholars, and Policy Makers: The Politics of Large Scale Research," *Daedalus* 93 (1964): 1142–61; Jean M. Converse, *Survey Research in the United States: Roots and Emergence, 1890–1960* (Berkeley and Los Angeles: Univ. of California Press, 1987); and Herbert H. Hyman, *Taking Society's Measure: A Personal History of Survey Research* (New York: Russell Sage Foundation, 1991).

32. On Kendall, see obituary, *NYT,* Mar. 7, 1990, B-6; and *AMWS,* 13th ed., 1978, S&B, 644.

33. On Vogt, see *AMWS,* 13th ed., 1976, 6:650; H. M. Weaver to U. S. Employment Service, Jan. 16, 1953, Biology Division Papers, California Institute of Technology Archives; and Walter Eckhart, "The 1975 Nobel Prize for Physiology or Medicine," *Science* 190 (1975): 650, 712, 714.

34. Kruytbosch and Messinger, "Unequal Peers," 33–43. See also Harold Orlans, *The Effects of Federal Programs on Higher Education: A Study of Thirty-six Universities and Colleges* (Washington, D.C.: Brookings Institution, 1962), ch. 5.

35. Later surveys at Berkeley found evidence that 25 percent of the women felt that they could obtain outside funding in their own name and that, in fact, 18 percent had written the grant on which they were currently employed ("Report of the Subcommittee on the Status of Academic Women on the Berkeley Campus," 77). There is also considerable material on the status of University of California research staff in a collection at the SIO Archives: University of California, Special Committee on Non-Senate Academic Personnel, 1966–68.

36. On Hoffleit, see *AMWS*, 16th ed., 1986, 3:754. See also Dorrit Hoffleit, "Some Glimpses from My Career," *Mercury* 21 (Jan.–Feb. 1992): 16–18; David DeVorkin, "Dorrit Hoffleit: Transcript of an Interview, August 4, 1979," NBL; and NSF, *Fourth Annual Report: Fiscal Year 1954* (Washington, D.C.: GPO, 1954), 72.

37. NSF, *Seventh Annual Report: Fiscal Year 1957* (Washington, D.C.: GPO, 1957), 125. Elizabeth Ralph was not in the *AMWS* but see Beth Ralph, "Age Is No Longer a Secret," *WAM* 41 (1957): 142–43.

38. Judith Walzer, "Interview with Ruth Dixon Turner, 1981," MCRC, 38, 50, 68.

39. On von Mises, see *NAWMP*. There is a collection of her papers at HUA, which is rather unusual for a mere research associate; more often they are ignored as totally after death (with no obituary in the university necrology series, for example) as they were in life.

40. On Frenkel-Brunswik, see *NAWMP;* and *AMWS*, 9th ed., S&B, 1956, 224.

41. Rose Goldsen to Jessie Bernard, Oct. 30, 1971, box 5, Jessie Bernard Papers, Historical Collections and Labor Archives, Pattee Library, Pennsylvania State University. Rose Goldsen was appointed associate professor of sociology in 1958; there is a collection of her papers at KLCU.

42. On Jones, see *AMWS*, 16th ed., 1986, 4:13.

43. On Ernst Scharrer, see *AMS*, 9th ed., Biol. Scis., 1955, 2:987. On Berta Scharrer, see *AMWS*, 17th ed., 1989, 6:486; and Berta Scharrer, "Neurosecretion: Beginnings and New Directions in Neuropeptide Research," *Annual Review of Neuroscience* 10 (1987): 1–17. On Waelsch, see "Interview with Salome Waelsch," in *The Outer Circle: Women in the Scientific Community,* ed. Harriet Zuckerman, Jonathan R. Cole, and John T. Bruer (New York: W. W. Norton, 1991), 71–93; and Michael T. Kaufman, "A Jew, a Woman, and Still a Scientist," *NYT,* Feb. 6, 1993 (I thank Sara Tjossem for a copy of this article). In 1957 Einstein also hired anatomist Helen Deane Markham, formerly at the Harvard Medical School, who had refused to cooperate with a congressional committee investigating Communism in universities (Ellen W. Schrecker, *No Ivory Tower: McCarthyism and the Universities* [New York: Oxford Univ. Press, 1986], 200–204, 249, 267, 269–70, 292, 298. There is also biographical material about Markham in box 23, f. 58, *NAWP*).

44. On Maria Mayer, see *AMS*, 11th ed., 1966, 4:3537; on Joseph Mayer, see *AMWS*, 15th ed., 1982, 5:274. See also ch. 6 above.

45. On Brazier, see *AMWS*, 16th ed., 1986, 1:691. See also John D. French, Donald B. Lindsley, and H. W. Magoun, *An American Contribution to Neuroscience: The Brain Research Institute, UCLA, 1959–1984* (Los Angeles: UCLA-BRI, 1984).

46. On Caserio, see *AMWS*, 18th ed., 1992, 2:95; and John D. Roberts, *The Right Place and the Right Time* (Washington, D.C.: ACS, 1990), 225–29, 247. On Loeblich,

see *AMWS,* 18th ed., 1992, 4:887. In 1985 she was president of the Paleontological Society.

47. On Langenheim, see *AMWS,* 17th ed., 1989, 4:593 and Ralph Wetmore to William T. Doyle, Feb. 2, 1968, letter of recommendation, box 3, Ralph Wetmore Papers, HUA. In 1985 she was president of the Ecological Society of America. On Sweeney, see *AMWS,* 17th ed., 1989, 6:1176; and Beatrice M. Sweeney, "Living in the Golden Age of Biology," *Annual Review of Plant Physiology* 18 (1987): 1–9. In 1979 she was president of the American Society for Photobiology.

48. On Kernan, see *AMWS,* 18th ed., 1992, 4:308; and Gary Taubes, *Nobel Dreams: Power, Deceit, and the Ultimate Experiment* (New York: Random House, 1986), 43, 94–95, 102, 136, 213. On the Burbridges, see *AMWS,* 18th ed., 1992, 1:863; and Vera C. Rubin, "E. Margaret Burbridge, President-Elect," *Science* 211 (1981): 915. See also E. Margaret Burbridge to Jesse Greenstein, June 16, 1954, his reply, July 2, 1954, and her responses, Apr. 12, 19, 1955, all in Jesse Greenstein Papers, California Institute of Technology Archives. Similarly, when Harvard-trained botanists Barbara and Grady Webster moved to the University of California at Davis in the 1960s, he became a professor in the botany department, and she became an instructor in agronomy (*AMWS,* 17th ed., 1989, 7:487, 488).

49. On Roe, see obituary, *NYT,* June 4, 1991, D-22; Elizabeth L. Simpson (Roe's daughter), "Occupational Endeavor as Life History: Anne Roe," *Psychology of Women Quarterly* 5, no. 1 (1980): 116–26; Gwendolyn Stevens and Sheldon Stevens, *The Women of Psychology,* vol. 2 (Cambridge, Mass.: Schenckman, 1982), 105–12; and "Richardson Creativity Award," *AP* 23 (1968): 870–71. There are also some letters between Roe and Lewis Terman in the Lewis Terman Papers, Stanford University Archives, c. 1946–50. See also *Simple Curiosity: Letters from George Gaylord Simpson to His Family,* ed. Léo LaPorte (Berkeley and Los Angeles: Univ. of California Press, 1987), esp. the section entitled "Harvard, 1959–1967," 303–21 (quotation from 316).

50. [Eleanor J. Gibson], "Eleanor J. Gibson," in *A History of Psychology in Autobiography,* ed. Gardner Lindzey, vol. 7 (San Francisco: W. H. Freeman, 1980), 238–71. See also Gibson's *Odyssey in Learning and Perception* (Cambridge: MIT Press, 1991); and NSF, *Fifth Annual Report: Fiscal Year 1955* (Washington, D.C.: GPO, 1955), 117. Her grant was with Richard D. Walk.

51. In taped interview with Eleanor Jack Gibson, c. 1976, KLCU, she is more outspoken.

52. [Gibson], "Eleanor J. Gibson," 267–68.

53. On Rita James Simon, see *AMWS,* 13th ed., S&B, 1978, 1103; on Julian Simon, see ibid., 1102. See also Rita James Simon, Shirley Merritt Clark, and Larry L. Tifft, "Of Nepotism, Marriage, and the Pursuit of an Academic Career," *Sociology of Education* 39 (1966): 344–58. A later critic thought Simon could have made an even stronger case had she controlled for field, age, and type of employer, because her women published more research with seemingly fewer resources than did comparable men (Solomon, "Female Doctoral Chemists," 5–6).

54. [Alice S. Rossi], "Point of View: Women and Professional Advancement," *Science* 166 (1969): 356.

Chapter 8   Protecting Home Economics, the Women's Field

1. See John Parrish's data cited in tables 6.1 and 6.2 above.

2. In 1963 Jeanette Lee and Paul Dressel, both of Michigan State University, noted that although "top level administrators were, in general, not very knowledgeable about the home economics curricula prevailing at the university," they often had strong opinions about the field's inadequacies. When asked to be specific, vice presidents and deans admitted that they were unaware of recent curriculum changes on their very own campus! (Jeanette A. Lee and Paul L. Dressel, *Liberal Education and Home Economics* [New York: Bureau of Publications, Teachers College, Columbia University, for the Institute of Higher Education, 1963], 69).

3. USOE, Vocational Division, "Home Economics in Degree-Granting Institutions," *Miscellaney* 2557 (mimeographed), revised biennially from 1939–40 to 1963–64, whereupon the AHEA published the final USOE data as *Home Economics in Institutions Granting Bachelor's or Higher Degrees, 1963–64: Tables Selected from Material Prepared by the U.S. Office of Education* (Washington, D.C.: AHEA, 1967) and then a sequel, J. C. Gorman and Laura J. Harper, *Home Economics in Institutions Granting Bachelors or Higher Degrees, 1968–69* (Washington, D.C.: AHEA, 1970).

4. Ernest L. Wilkinson and W. Cleon Skousen, *Brigham Young University: A School of Destiny* (Provo, Utah: Brigham Young Univ. Press, 1976), 504–7, 662. In 1954 Wilkinson hired Marion Pfund, long a professor of home economics at Cornell University, to be co-dean with Royden Braithwaite, BYU's male professor of psychology. A few months after her arrival, Braithwaite left to head the College of Southern Utah, putting an end to the novel experiment with male and female co-deans of a college of home economics. On Pfund, see *AMS*, 11th ed., 5:4158, and Nell Monday, conversation with author, 1991.

5. Flora Rose et al., *A Growing College: Home Economics at Cornell University* (Ithaca: New York State College of Human Ecology, Cornell University, 1969), 215–17. Cornell's entries in the USOE's "Home Economics in Degree-Granting Institutions" fluctuate widely.

6. Sharon Carroll, "Oral History of Elizabeth Lee Vincent," Sept. 10, 1964, KLCU, 147–48, 149, 165, 195, 197, 213. On Vincent, see *AMS*, 11th ed., S&B, 1968, 1655.

7. USOE, *Earned Degrees Conferred*, 1947–48 through 1967–68. For reasons that are unclear, these data do not agree with comparable ones in the biennial figures given in the USOE's "Home Economics in Degree-Granting Institutions." See, e.g., Marguerite Mallon, "Why Men Elect Home Economics," *Omicron Nu* 26, no. 1 (fall 1946): 22–24; Dr. Russell Smart, "Men Need to Know about Children," ibid. 26, no. 2 (spring 1947): 37–41; "Woman Dean Urges Men Study Home Economics," *Oakland (Calif.) Tribune,* Mar. 31, 1957, and Jim Thomas, "Hey, Men! Here's a Calling You're Missing," *Syracuse Herald-American,* Mar. 31, 1957, Clipping File, M. Eunice Hilton Papers, Special Collections, George Arendts Research Library, Syracuse University.

8. USOE, *Earned Degrees Conferred,* 1947–48 through 1969–70; Ercel Sherman Eppright and Elizabeth Storm Ferguson, *A Century of Home Economics at Iowa State University* (Ames: Iowa State University Home Economics Alumni Association, 1971), 258, 312.

9. USOE, *Earned Degrees Conferred,* 1947–48 to 1962–63; Jeanette C. Gorman and Laura Jane Harper, "A Look at the Status of Home Economics in Higher Education," *JHE* 62, no. 10 (1970): 742.

10. NSF, *Graduate Student Enrollment and Support in American Universities* (Washington, D.C.: GPO, 1957), esp. 48, 71, 86, 100, 124, 136. Another deterrent to graduate work in the field of home economics was its exclusion from eligibility for NDEA fellowship funds. Yet *Omicron Nu,* the publication of the home economics honor society, fought back by devoting one issue per year to graduate opportunities in the field (see ch. 15).

11. On Blunt, see *NAWMP.* On Roberts, see ibid.; Ethel Austin Martin, "Lydia Jane Roberts, 1879–1965," *Journal of the American Dietetic Association* 47 (1965): 127–28; idem, "The Life Works of Lydia J. Roberts," ibid. 49 (1966): 299–302; and Franklin C. Bing, "Lydia Jane Roberts: A Biographical Sketch," *Journal of Nutrition* 93 (1967): 3–13. On Kyrk, see *NAWMP;* on Doyle, *AMWS,* 15th ed., 1982, 2:707; on Oldham, *AMS,* 11th ed., 1966, 4:3963; and on Porter, ibid., 1967, 5:4233.

12. On Dye, see *AMS,* 11th ed., 1965, 2:1334; and Marie Dye, "History of the Department of Home Economics, University of Chicago" (Chicago: University of Chicago Alumni Association, 1972, mimeographed), 177, copy in Special Collections, Regenstein Library, University of Chicago. Abraham Flexner had publicly criticized the Chicago program as early as 1930 for the seeming triviality of the dissertation titles in "domestic science." He thought this particularly objectionable at a private, seemingly prestigious university (Abraham Flexner, *Universities: American, English, German* [New York: Oxford Univ. Press, 1930], 152–60).

13. Orrea F. Pye, "The Nutrition Program at Teachers College, Columbia University," in *Conference on Education in Nutrition—Looking Forward from the Past, Teachers College, Columbia University, New York City, February 26 and 27, 1974* (Sanford, N.C.: Microfilming Corporation of America, 1980), pp. 10–11. In 1953 the Department of Home Economics had become the Department of Home and Family Life, stressing family relations. See also the filiapietistic biography *Mary Swartz Rose, 1874–1941: Pioneer in Nutrition,* by Juanita A. Eagles, Orrea F. Pye, and Clara M. Taylor (New York: Teachers College Press, 1979). On Taylor, see *AMS,* 11th ed., 1967, 6:5327; and obituary, *NYT,* Jan. 15, 1988, A-15.

14. A much fuller version of the Berkeley story is told in Maresi Nerad, "'Home Economics Has to Move': The Disappearance of the Department of Home Economics from the University of California, Berkeley" (Ph.D. diss., University of California, Berkeley, 1986), and in her "Gender Stratification in Higher Education: The Department of Home Economics at the University of California, Berkeley, 1916–1962," *Women's Studies International Forum* 10 (1987): 157–64. Nerad points out that Morgan had such heavy science requirements for her majors that relatively few of those who persisted became schoolteachers. When the state education department protested, the university administration used her high standards as a pretext for discontinuing her department, a reverse of the usual reasoning of Berkeley administrators. See also Agnes Fay Morgan, "The History of Nutrition and Home Economics in the University of California, Berkeley, 1914–62" ([Berkeley?, 1962], mimeographed), and Alexander Callow, Jr., "Interview with Agnes Fay Morgan," 1959, Centennial Oral History Project,

Bancroft Library, University of California, Berkeley. Certain facets of the episode can be followed in the Agnes Fay Morgan Papers, Bancroft Library; in the student newspaper, especially "Home Economics Transfer Proposed," *Daily Californian*, Dec. 16, 1955, 11, and "Plan for Home Ec Department Opposed in Resolution to Kerr," ibid., Feb. 29, 1956, 4; and in the oral histories of some of the administrators of the time done by the University of California Regional Oral History Office, such as Willa K. Baum, "Claude B. Hutchison: The College of Agriculture, University of California, 1922–1952" (1961), 306–23; and Malla Chall, "Harry R. Wellman: Teaching, Research and Administration, University of California, 1925–1968" (1976), 96–97. See also Ruth Okey, "A Woman in Science, 1893–1973," *Journal of Nutrition* 118 (1988): 1429–30; and Joann L. Larkey, "Knowles Ryerson: 'The World Is My Campus'" (1977, typescript), 324–27, 522–23, Oral History Center, Shields Library, University of California at Davis (Ryerson had opposed the change). There is also much relevant correspondence, including some brutal memorandums on the Department of Nutrition and Home Economics (1956–61) in the Chancellor's Files. On Briggs, see *AMWS*, 16th ed., 1986, 1:712. On Hurley, see ibid., 3:914–15; obituary, *NYT*, Aug. 1, 1988, D-9; several letters of recommendation, 1951–57, in the Agnes Fay Morgan Papers; and "1981 AIN Award Winners," *Journal of Nutrition* 111 (1981): 1501. On Everson, see Ruth E. Shrader and Frances J. Zeman, "Gladys June Everson: A Biographical Sketch," ibid. 109 (1979): 732–37. See also [Gladys Branegan Chalkley], *The California Home Economics Association, Yesterday, Today, and Tomorrow, 1921–1961* (n.p.: CHEA, 1962).

15. Rose et al., *A Growing College*, 242–43, 314–16; The Grant Foundation, Inc., *Report for the Years Ended October 31, 1951 and October 31, 1952* (New York, 1952), 16–17, through *Biennial Report, 1961 and 1962* (New York, 1962), 48. Stipends ranged from $2,800 to $4,400.

16. On the history of child development, see Hamilton Cravens, *Before Head Start: The Iowa Station and America's Children* (Chapel Hill: Univ. of North Carolina Press, 1993).

17. Eppright and Ferguson, *A Century of Home Economics*, ch. 14, 187–97, ch 16, and p. 286. On Swanson, see *AMS*, 11th ed., 1967, 6:5277; on Eppright, *AMWS*, 12th ed, 1972, 2:1690; and on Brewer, ibid., 15th ed., 1982, 1:679. See also Margaret W. Rossiter, "Mendel the Mentor: Yale Women Doctorates in Biochemistry, 1898–1937," *JCE* 71, no. 3 (Mar. 1994): 215–19.

18. On Kittrell, see *AMWS*, 14th ed., 1979, 4:2669; and Merze Tate, "Interview with Flemmie P. Kittrell, August 29, 1977," typescript, Black Women Oral History Project, SLRC. "In Memoriam," *JHE* 73, no. 1 (1981): 12, is inadequate. There are biographical materials in box 3 of the Hazel Hauck Papers, KLCU, and items about home economics at Hampton Institute in box E69, MMP. Kittrell deserves a full biography. See also ch. 15, n. 4.

19. On Hauck, see *AMS*, 10th ed., P&B, 1960, 2:1682; Helen H. Gifft et al., "Hazel Marie Hauck, July 15, 1900–April 23, 1964," *Necrology of the Faculty of Cornell University, 1963–64* (Ithaca: Cornell University, [1965]), 12–14; "Hazel Marie Hauck, 1900–1964," *Journal of the American Dietetic Association* 47 (1965): 191; as well as the Hazel Hauck Papers, KLCU.

20. *Ninth International Congress on Home Economics* (Washington, D.C.: AHEA,

1958); Mary Hawkins, ed., *The American Home Economics Association, 1950–1958: A Supplement to the AHEA Saga by Keturah E. Baldwin, 1949* (Washington, D.C.: AHEA, 1959), 5–9.

21. Hadley Read, *Partners with India: Building Agricultural Universities* (Urbana: University of Illinois College of Agriculture, 1974), esp. app. 2. Box 3 of the Janice Smith Papers, at the University of Illinois Archives, Urbana, is devoted to home economics in India, especially Baroda, 1956–67.

22. Eppright and Ferguson, *A Century of Home Economics*, ch. 23. Upon retirement an American home economics administrator often spent a year or more abroad at a fledgling college of home economics, as did Helen Swift Mitchell at Hokkaido University in Japan in 1960–62 and Frances Zuill in Pakistan in 1961–62 (obituaries, Biographical File, Archives, University of Massachusetts-Amherst; "Frances Zuill to Retire from University of Wisconsin," *JHE* 53 [1961]: 475–76).

23. Margaret A. Ohlson, P. Mabel Nelson, and Pearl P. Swanson, "Co-operative Research among Colleges," *JHE* 29 (1937): 108–13; Eppright and Ferguson, *A Century of Home Economics*, 226–28; clipping on Nelson in Faculty Files, Department of Foods and Nutrition, Iowa State University Archives, Ames. On Nelson, see *AMS*, 10th ed., P&B, 1961, 3:2949. Apparently the first home economics research project to use a computer is described in Pauline Keeney et al., "UNIVAC and WE," *JHE* 54 (1962): 43–44.

24. Ruth O'Brien, "The Flanagan-Hope Act in Relation to Home Economics Research," *JHE* 39 (1947): 150–52; Ruth O'Brien and Georgian Adams, "RMA Home Economics Research," ibid. 40 (1948): 120–22; Edward D. Eddy, Jr., *Colleges for Our Land and Time* (New York: Harper & Bros., 1956), 232. There is also material on cooperative research projects and the "committee of nine" in boxes 383–85, Papers of College of Human Ecology, Michigan State University Archives, East Lansing. The home economists on the USDA's "committee of nine" over the years were Agnes Fay Morgan (1947–50), Pearl Swanson (1951–52), Catherine Personius (1953–54), Florence MacLeod (1955–56), Elizabeth Dyer (1957–59), Janice M. Smith (1960–62), Virginia Y. Trotter (1963–65), Clara Storvick (1966–68), and Laura Harper (1969–71) (Maryanna Smith, USDA historian, to author, Oct. 29, 1982).

25. Agnes Fay Morgan, ed., *Nutritional Status, USA*, California Agricultural Experiment Station Bulletin 769 (1959); Gladys A. Emerson, "Nutritional Status, USA," *Journal of Nutrition* 91, no. 2, suppl., pt. 2 (1967): 51–54 (this whole issue was dedicated to Agnes Fay Morgan, a rarity for a woman in any field or journal). Correspondence in boxes 1 and 2, Agnes Fay Morgan Papers, indicates that the project's need to publish so many results almost led to the creation of a new research journal in home economics, which had first been suggested as far back as 1930 in *Survey of Land-Grant Colleges and Universities*, 2 vols., USOE Bulletin 1930, no. 9 (Washington, D.C.: GPO, 1930), 1:983. See also n. 41 below.

26. Eppright and Ferguson, *A Century of Home Economics*, 267; Joyce Thompson, *Marking a Trail: A History of the Texas Woman's University* (Denton: Texas Woman's Univ. Press, 1982), 157–58; Rose et al., *A Growing College*, 248–57, 318–27, 362–64, 394–404. See also Dale Lindsay, "Home Economics Research: Interest of National Institutes of Health," *JHE* 54 (1962): 26–31; and Mary Beth Minden, "Home Economics Research in the Experiment Stations," *JHE* 54 (1962): 32–35.

27. See Helen Pundt, *AHEA: A History of Excellence* (Washington, D.C.: AHEA, 1980), 260–61; Hawkins, *American Home Economics Association, 1950–58,* 13–14; box 385, f. 11, Papers of the College of Human Ecology, Michigan State University Archives; and esp. Papers of Catherine Personius, a participant, at the KLCU.

28. Grace Henderson, "Federal Research Related to Home Economics, Given before Home Economics Division of the American Association of Land-Grant Colleges and State Universities, November 15, 1955," typescript in Grace Henderson Papers, University Archives, Penn State Room, Pennsylvania State University; idem, "Reasons for a Research Foundation for the American Home," July 1, 1957, typescript, ibid., published as "A Research Foundation for the American Home," *JHE* 49 (9157): 511–14; Pearl Swanson, "New Resources for Research," ibid. 53 (1961): 161–72; and occasional items in Catherine Personius Papers, Cornell.

29. Arthur S. Flemming, "Education for a Dynamic Society," *JHE* 49 (1957): 508.

30. Pundt, *AHEA,* 288–89, 306, 328; Ruth Bonde, "Federal Research Related to Home Economics," *JHE* 52 (1960): 556; list of committee members, ibid., 599; Luise Addiss, "Federal Research Related to Home Economics," ibid. 54 (1962): 608–9; list of members, ibid., 656.

31. A list, "Groups with Which an Institute of the American Home has been discussed," is appended to Ruth Bonde, "A National Research Institute for the American Home: A Proposal of the Committee on Federal Research Related to Home Economics of the American Home Economics Association, New Haven, Connecticut, August 18, 1961," in box 1, Catherine Personius Papers.

32. Nick Kotz, *Let Them Eat Promises: The Politics of Hunger in America* (Englewood Cliffs, N.J.: Prentice-Hall, 1969).

33. Pundt, *AHEA,* 308; Eppright and Ferguson, *A Century of Home Economics,* 262, 267; Mary Lee Hurt, "Expanded Research Programs Under Vocational Education," *JHE* 57 (1965): 173–75.

34. See, e.g., [Agnes Fay Morgan, comp.], "Areas of Home Economics Research: Reports to the Committee on Research, Division of Home Economics, Association of Land-Grant Colleges and State Universities, 1947–1956" (1958, mimeographed); "A Report on the Workshop on Administrative Management for Home Economics in the Association of Land-Grant Colleges and Universities, Estes Park, Colorado, August 9–14, 1953" (mimeographed); "Workshop Report: Administrative Management for Home Economics, American Association of Land-Grant Colleges and State Universities, August 25–30, 1957, Estes Park, Colorado" (mimeographed), which contained a "Review of Past Workshops," by Florence McKinnery (pp. 6–8), which dated the first workshop, led by Margaret Justin of Kansas State, as October 1943 (all in box 4, Catherine Personius Papers). See also *Home Economics Seminar, July 24–28, 1961, French Lick, Indiana: A Progress Report* (n.p., n.d.).

35. The association first approached the Kellogg Foundation, but it was not interested. See Helen G. Canoyer, "Report of the Home Economics Study Committee," *Proceedings of the AALGCSU* 75, pt. 1 (1961): 182, and other relevant items in the Catherine Personius Papers, e.g., in box 6, Helen G. Canoyer, "Report of the Origin and History of the National Study of Home Economics in Land-Grant Colleges and State Universities, Entitled 'The Changing Mission of Home Economics,' Given to a Meet-

ing of the Home Economics Commission, of the National Association of State Universities and Land-Grant Colleges, Chicago, Illinois, May 2, 1968" (quotations from 2, 6), apparently read by someone else in her absence; and in box 2, "Notes on History of ASULGC Home Economics Study," a chronology to 1962. In an address to her faculty, Grace Henderson recalled that the demand for the new report came from Penn State's own Eric Walker ("Outlook for Home Economics," Sept. 17, 1965, typescript, 4, Grace Henderson Papers). Two similar reports had appeared in 1959: AHEA, *Home Economics, New Directions: A Statement of Philosophy and Objectives* (Washington, D.C., [1959]); and ASULGC, Home Economics Development Committee, *Home Economics in Land-Grant Colleges and Universities: A Statement of Objectives and Future Directions, Prepared by the Home Economics Development Committee, Division of Home Economics, AALGC&SU, November 1959* ([Washington, D.C.], 1959).

36. Paul A. Miller, "Higher Education in Home Economics: An Appraisal and a Challenge," *Proceedings of the AALGCSU* 74 (1960), 244–46. Reports of Miller's address aroused so much interest that the next year the division passed a resolution that henceforth the complete text of any talk presented to the division be made available immediately to each member institution ("Division of Home Economics, Minutes," ibid. 75, pt. 1 [1961]: 182; Reuben G. Gustavson, "The Evaluation of Home Economics," ibid., 75, pt. 2 [1961]: 216–30, esp. 221). See also the hortatory exchange about this time between Arnold Baragar, "Opportunities for Men in Home Economics," *JHE* 52 (1960): 833, and Loyal Horton, "Men in Home Economics," ibid. 53 (1961): 159.

37. See Frederick. L. Hovde to Catherine Personius, Apr. 30, 1964, and her reply June 22, 1964, both in box 6, Catherine Personius Papers; several other reports in this box indicate that Personius was instrumental in setting up the new organization. The changeover can also be followed in the *Proceedings of the ASULGC* 76 (1962) through 81 (1967), esp. in annual "Minutes" of the Division of Home Economics, and "Report of the Committee on Organization" (to the ASULGC), ibid. 78 (1964):41–50, and the annual reports of the new Commission on Home Economics. The association deserves a full history. Hovde had earlier delivered an address to the AHEA entitled "The Importance of Research" (see *JHE* 46 [1954]: 455–58).

38. Lee and Dressel, *Liberal Education and Home Economics;* Earl J. McGrath and Jack T. Johnson, *The Changing Mission of Home Economics: A Report on Home Economics in the Land-Grant Colleges and State Universities* (New York: Teachers College Press, 1968), i–iii, vii, x. See also C. J. Personius to Helen LeBaron, Mar. 3, 1967, and Canoyer, "Report of the Origin and History of the National Study of Home Economics," box 6, Catherine Personius Papers. On McGrath, see obituary, *NYT*, Feb. 5, 1993, A-18; and Edwin Kiester, Jr., "The Education of Earl McGrath," *Change* vol. 9 (Apr. 1977): 23–29. Professor Ruth Eckert of the School of Education at the University of Minnesota might have been a better choice, for she had land-grant experience and had already done studies of academic women (see ch. 9, n. 10). But these very factors might have made her seem less than "objective" to the plight of home economics.

39. For differing accounts compare McGrath and Johnson, *Changing Mission of Home Economics,* ix, with Earl J. McGrath, "The Imperatives of Change for Home Economics," *JHE* 60 (1968): 505.

40. "Question and Answer Panel," *JHE* 60 (1968): 513–14. The AHEA's Commis-

sion on Graduate Programs was in 1968 already at work formulating standards. See S. J. Ritchey and Mary Lee Hurt, "Guidelines for Graduate Programs in Home Economics—Philosophy and Purpose," ibid. 62 (1970): 747; and the final report, *Guidelines for Graduate Programs in Home Economics* (Washington, D.C.: AHEA, 1971). Not all of McGrath's recommendations were exactly new. Helen Canoyer had published a forceful summary of the field's problems even before McGrath started his project (Helen Canoyer, "The Changing Role of Home Economics," in *The Development of the Land-Grant Colleges and Universities and Their Influence on the Economic and Social Life of the People*, published as *West Virginia University Bulletin*, May 1963, 97–114).

41. Geitel Winakor, "Editor's Notes," *Home Economics Research Journal* 1 (Sept. 1972): 71–74. See also "Home Economics Research Journal," *JHE* 65 (1973): 33–34.

42. McGrath and Johnson, *Changing Mission of Home Economics,* 99. See also "Question and Answer Panel," 512; and Patricia Durey Murphy, "What's in a Name?" *JHE* 59 (1967): 702–7. For reactions to the McGrath Report, see Doris E. Hanson, "The Future Is Now," ibid. 60 (1968): 84; "Letters," ibid., 406, 408, 410; and Marjorie M. Brown, "Implications of the McGrath Report and the Structure of Home Economics" (paper presented at the North Central Home Economics Administrators Conference, Chicago, Mar. 28, 1968, mimeographed) (I thank Professor Brown for a copy).

43. Theodore Vallance, "Home Economics and the Development of New Forms of Human Service Education," in *Land-Grant Universities and Their Continuing Challenge,* ed. G. Lester Anderson (East Lansing: Michigan State Univ. Press, 1976), 88–91; Pundt, *AHEA,* 359.

44. On Henderson see "Dean Henderson Announces Retirement," *JHE* 57 (1965): 763; obituary, *Omicron Nu* 38, no. 2 (spring 1972): 4–5; and Rev. A. Jackson McCormack, ". . . However . . . ," ibid., 6–8. For more on Henderson's views, see Grace M. Henderson, "Issues Confronting Home Economics in Colleges and Universities," *JHE* 57 (1965): 759–64. On Houghton, see obituary, *NYT,* Dec. 3, 1975, 48. Details of the reorganization are given by the victors in Vallance, "Home Economics," 91–103, which cites various internal documents. There are also clippings and several relevant speeches in the Grace Henderson Papers. Donald H. Ford had been an undergraduate at Kansas State when Eisenhower was there, and he followed him to Penn State (Stephen E. Ambrose and Richard Immerman, *Milton S. Eisenhower: Educational Statesman* [Baltimore: Johns Hopkins Univ. Press, 1983], 89).

45. Vallance, "Home Economics," 83–88. See also [Cornell University], "Final Report of the President's Committee to Study the College of Home Economics, December 1966" (mimeographed). For an analysis of a similar episode where male workers and male management conspired to take over an established area of "women's work," see Ruth Milkman, *Gender at Work: The Dynamics of Job Segregation by Sex during World War II* (Urbana: Univ. of Illinois Press, 1987).

46. Michael W. Whittier, "Part III. Epilogue, 1965–68," in Rose et al., *A Growing College,* 527–28.

47. Helen Canoyer, "Changing Role of Home Economics," 109, 113; Urie Bronfenbrenner et al., "Helen Gertrude Canoyer, 1903–1984," *Memorial Statements: Cornell University Faculty, 1983–84* (Ithaca: Cornell University, 1984), 7–8, which claims (erroneously) that she was one of the first women in the United States to earn a doctorate in

economics. Many of these male faculty members had also been fellows at the Center for Advanced Study in the Behavioral Sciences at Stanford, California (see ch. 14).

48. Donna Hartshorne, "Oral History of Glenn S. Pound, Former Dean of the College of Agriculture," 1979, 122–35, University of Wisconsin Oral History Office, Madison (the rest of the tape is restricted until 1999). Marshall was, however, soon let go for lacking integrity (Donna Taylor, "Interview with May S. Reynolds, Department of Home Economics, July 1977," tape index, 4, University of Wisconsin Oral History Office, Madison).

49. Paul L. Dressel, *College to University: The Hannah Years at Michigan State University, 1935–1969* (East Lansing: Michigan State University Publications, 1987), 70; College of Home Economics, Michigan State University, "The Report of the Committee on the Future of Home Economics," Jan. 1968, typescript, reprinted in 1980. (I thank Mary Grace for a copy.) It does not contain the "minority report" mentioned in [Jeanette Lee], "Dean's Remarks, Presentation of Report of Committee of the Future, Home Economics Faculty Meeting, January 26, 1968," typescript, 6, box 364, f. 23, Papers of the College of Human Ecology, Michigan State University Archives.

50. Susan Weis, Marjorie East, and Sarah Manning, "Home Economics Units in Higher Education: A Decade of Change," *JHE* 66 (May 1974): 11–15.

51. Eppright and Ferguson, *A Century of Home Economics*, 259, 277, 279. Several were part of a group that has been called "Aydelotte's boys" because they were influenced in one way or another by Frank Aydelotte, associate professor of English at MIT (where James R. Killian was his devoted student), longtime president (1920–40) of Swarthmore College (alma mater of Kerr, Perkins, and A. C. Valentine, later president of the University of Rochester), president of the Association of American Rhodes Scholars (1930–56), head of the Institute for Advanced Study (1939–47), and chair of the educational advisory board of the John Simon Guggenheim Memorial Foundation (1925–50). For more on him, see entries in *NCAB* 43 (1961): 44–45 and *CBY, 1952*, 27–29; James R. Killian, *The Education of a College President: A Memoir* (Cambridge: MIT Press, 1985), 2–3; and Frances Blanshard, *Frank Aydelotte of Swarthmore* (Middletown, Conn.: Wesleyan Univ. Press, 1970).

There were also stories about just which women were appointed deans. See, e.g., Dolores Greenberg, "Oral History of Mary Henry, October 24, 1963," which discusses the appointment of Sarah Gibson Blanding as Cornell's director of home economics in 1941, despite her lack of any previous training in the field. (Perhaps she was willing to hire men.) See also idem, "Oral History with Sarah Gibson Blanding, June 11 and 13, 1964." Blanding's successor as dean, E. Lee Vincent, felt that since there were so few women available, if some women, like herself, did not hold deanships, they would go to men (Carroll, "Oral History of Elizabeth Lee Vincent, September 10, 1964," 148, 165, 213. All in KLCU.)

52. Ambrose and Immerman, *Milton S. Eisenhower*, 89. For more on Justin, see James C. Carey, *KSU: The Quest for Identity* (Lawrence: Regents Press of Kansas, 1977), 214; "Margaret M. Justin (1889–1967)," *JHE* 59 (1967): 488; and Margaret Justin File, Kansas State University Archives, Manhattan.

53. See, e.g., Robert W. Topping, *The Hovde Years: A Biography of Frederick L. Hovde* (West Lafayette, Ind.: Purdue University, 1980), 381–82.

54. Helen LeBaron Hilton, "Foreword," in Eppright and Ferguson, *A Century of Home Economics*. Helen LeBaron Hilton merits a biography. She died in August 1993. Before going to Iowa State in 1952, she had been an assistant dean at Penn State under Grace Henderson, an example of a protégée chain among home economics administrators.

55. See n. 2 above.

56. See Viola J. Anderson, *The Department of Home Economics at the University of Kansas: The First Fifty Years, 1910–1960* (Lawrence: Univ. of Kansas Press, 1964), as well as college and department histories cited above.

57. The redoubtable Agnes Fay Morgan was an exception: one of her last articles for the Iota Sigma Pi newsletter mentioned the new movement (see ch. 15, n. 10). Similarly, Ethel Vatter had become a "women's libber" by 1970 ("Interview with Ethel Vatter," c. 1976, KLCU).

58. Jennie Towle Farley, "Women on the March Again: The Rebirth of Feminism in an Academic Community" (Ph.D. diss., Cornell University, 1970), 243; Edward C. Devereux et al., "Harold Feldman, 1917–1988," *Memorial Statements, Cornell University Faculty, 1987–88* (Ithaca: Cornell University, 1989), 23–24.

Chapter 9   Surviving in "Siberia"

1. See David Riesman and Christopher Jencks, *The Academic Revolution* (Chicago: Univ. of Chicago Press, 1968), 231–36; Harold L. Hodgkinson, *Institutions in Transition* (New York: McGraw-Hill, 1971), ch. 5; and esp. E. Alden Dunham, *Colleges of the Forgotten Americans: A Profile of State Colleges and Regional Universities* (New York: McGraw-Hill, 1969), which refers to this process as "hardening" the formerly "soft" female faculty (47). See also Earl W. Hayter, *Education in Transition: The History of Northern Illinois University* (Dekalb: Northern Illinois Univ. Press, 1974); M. Janette Bohi, *A History of Wisconsin State University, Whitewater, 1868–1968* (Whitewater: Whitewater State Univ. Foundation, 1967); and Walker D. Wyman, ed., *History of the Wisconsin State Universities* (River Falls, Wis.: River Falls State Univ. Press, 1968), which contains essays on the nine branches of that new "system." On funding for new laboratories, see NSF, *The NSF Science Development Programs: A Documentary Report,* NSF 77-17, 2 vols. (Washington, D.C., 1977); and J. Merton England, "Investing in Universities: Genesis of the National Science Foundation's Institutional Programs, 1958–1963," *Journal of Policy History* 2, no. 2 (1990): 132–56.

2. "Teacher Supply and Demand in Degree-Granting Institutions, 1954–55," *NEA Research Bulletin* 33, no. 4 (1955): 127–63 (whole issue). NEA researchers sent their questionnaire to 992 institutions (omitting junior colleges, technical and "other professional" schools, and all institutions run solely by members of a religious order). The response rate was 67.8 percent, ranging from 79.7 percent of the teachers colleges to just 50.9 percent of the private universities. The NEA survey included professional librarians but excluded full-time administrators.

3. Kay E. Leffler, "The Role of Women in Geography in the United States" (master's thesis, University of Oklahoma, 1965), 71; Anthony de Souza, Ingolf Vogeler, and Brady

Foust, "The Overlooked Departments of Geography," *Journal of Geography* 80 (1981): 170–75. For more on geographic education, see ch. 14, n. 32.

4. NEA, Research Division, *Teacher Supply and Demand in Universities, Colleges, and Junior Colleges, 1957–58 and 1958–59* (Washington, D.C., 1959), 72 (table P). For a similar survey, see Helen D. Berwald, "Attitudes toward Women College Teachers in Institutions of Higher Education Accredited by the North Central Association," 2 vols. (Ph.D. diss., University of Minnesota, 1962).

5. Calculated from USOE, *Biennial Survey of Education, 1944/46* (Washington, D.C.: GPO, 1949), ch. 4, p. 43, and *1946/48* (1950), ch. 4, p. 56; and idem, *Faculty and Other Professional Staff in Institutions of Higher Education, First Term, 1961–62,* USOE Circular No. 747 (Washington, D.C.: GPO, 1965), 14, and *1963–64,* No. 794 (1966), 15.

6. Verne A. Stadtman, comp. and ed., *The Centennial Record of the University of California* (Berkeley and Los Angeles: Univ. of California Press, 1967), has information on the nine campuses. Useful for its brief synopses of the transformation of hundreds of similar colleges is John F. Ohles and Shirley M. Ohles, comps., *Public Colleges and Universities* (Westport, Conn.: Greenwood, 1986).

7. Dunham, *Colleges of the Forgotten Americans,* 14. A critical study of the whole mentality and process of upgrading is needed.

8. An occasional oral history or archival collection reveals the former faculty's point of view, as in Gibbs Smith, "Interview with [chemist] Hazel Severy," transcript, 1972, Special Collections, University Library, University of California at Santa Barbara. Social security legislation, which required retirement at age sixty-five, was extended in the early 1950s to some educational institutions (NEA, "Teacher Supply and Demand in Degree-Granting Institutions, 1954–55," 141).

9. Frances L. Clayton, "A Source for College Faculties," *Pembroke Alumna* 37 (Oct. 1962): 6. For advertisements, see, e.g., "Personnel Placement," *Science* 120 (1954): 8A (before p. 325); and "Psychological News and Notes," *AP* 4 (1949): 123, 514. Thereafter the APA published a separate *APA Employment Bulletin,* which William D. Wells and Sandra J. Richer analyzed for 1953 in "Job Opportunities in Psychology," *AP* 9 (1954): 639–41. They reported that 96 (31 percent) of the jobs were overtly restricted to men and that these jobs averaged higher rates of pay than those open to both sexes (by $50) and those restricted to women (by $102). They concluded that "in terms of earning power, the difference between the X and the Y chromosome appears to be just about as important as the difference between the MA and the PhD." (641).

For descriptions of faculty-recruiting practices in the 1950s, see Kevin P. Bunnell, "Recruiting College Faculty Members: A Short-Range View of the Problem," *Educational Record* 41 (1960): 138–42, which surveyed the practices of twenty-three deans and presidents. See also the notorious Theodore Caplow and Reece J. McGee, *The Academic Marketplace* (1958; reprint, New York: Arno, 1977).

10. Ruth E. Eckert and John E. Stecklein, "Academic Woman," *Liberal Education* 45 (1959): 390–97; idem, *Job Motivations and Satisfactions of College Teachers: A Study of Faculty Members in Minnesota Colleges,* USOE Cooperative Research Monograph, 7 (Washington, D.C.: GPO, 1961). Because so many of Eckert's women faculty were nuns who had started out as schoolteachers, the proportion holding doctorates was partic-

ularly low. On Eckert, see *WWAW* 8 (1974–75), 264; and *Contemporary Authors,* vols. 13–16, rev. (Detroit: Gale Research, 1975), 242.

11. On Holt, see *AMS,* 11th ed., 1966, 3:2373. Other women scientists of note in the California State system were Myrtle Johnson of San Diego State, longtime head of the biology department, 1921–46 (ibid., 10th ed., 1960, 2:2023), and psychologist Virginia Block, chair of the department of guidance and counseling at San Francisco State College from 1950 to 1968 (*NCAB* 56 [1975]: 249–50).

12. Dr. Vesta Holt, "The Biology Department, Then and Now," *Scope* (newsletter for alumni of the biological fraternity Omicron Theta Epsilon), 1958, Vesta Holt Papers, Special Collections, Meriam Library, California State University, Chico. Hired in 1949 was Margery Anthony, soon to complete her doctorate in botany at the University of Michigan. She was still at Chico in 1986 (*AMWS,* 16th ed., 1986, 1:151).

13. On Ruml, see *NCAB* 44 (1962): 64–65. See also Beardsley Ruml and Sidney G. Tickton, *Teaching Salaries Then and Now: A Fifty-Year Comparison with Other Occupations and Industries* (New York: FAE, 1955).

14. Ford Foundation, *The Pay of Professors: A Report on the Ford Foundation Grants for College-Teacher Salaries* (New York, 1962). See also NSF, *The NSF Science Development Programs,* 1:5–6; and Ford Foundation, *Toward Greatness in Higher Education: A First Report on the Ford Foundation Special Program in Education* (New York, 1964).

15. BLS, *Personnel Resources in the Social Sciences and Humanities: A Survey of the Characteristics and Economic Status of Professional Workers in 14 Fields of Specialization,* BLS Bulletin 1169 (Washington, D.C., 1954), 1–2, 17–18, 33–35, 65–66, 118–21, and app. B ("Scope and Method of Survey"), 126–31.

16. For a ten-year retrospective analysis on faculty salaries, but with no attention to gender, see "The Economic Status of the Academic Profession: Taking Stock," *AAUP Bulletin* 51 [1965]: 248–69 plus tables.

17. Hayter, *Education in Transition,* 232. Because regional accreditation often required a suitable proportion of doctorates on the faculty, in 1955 the Danforth Foundation started a program of "teacher grants" to help young faculty return to graduate school and complete their degrees, but it helped only fifty persons per year. See Harriet C. Owsley, "The Danforth Foundation," in *Foundations,* ed. Harold M. Keele and Joseph C. Kiger (Westport, Conn.: Greenwood, 1984), 96; and Danforth Foundation, *Annual Report for 1963–64* (St. Louis, [1964], 19).

18. NEA, Research Division, *Teacher Supply and Demand . . . 1957–58 and 1958–59,* 59 (table C).

19. Katherine G. Clark, "Women as a Potential Resource for Relieving Teaching Shortages in College Science and Mathematics Departments, Prepared for the Science and Engineering Program of the Ford Foundation," typescript, n.d., 26–28, carton 4, folder EC 83, ACE-CEW Papers, SLRC.

In 1957 the NSF supported a survey run by Eleanor Dolan of a sample of AAUW members (twenty-five hundred, or every thirty-fifth member) to find out how many might be interested in taking up science teaching in the schools. Not many were interested in that, but several indicated that at some point they might want further training to prepare them for a career in college-level teaching. See "Science Foundation Project," *AAUWJ* 51 (1957): 58; Eleanor F. Dolan, "All Our Thirty-fifths . . . ," ibid.,

187–89; "NSF Follow-up," ibid. 52 (1959): 184–85; NSF, *Seventh Annual Report for the Fiscal Year Ended June 30, 1957* (Washington, D.C.: GPO, 1958), 169; copy of the final report, in carton 4, f. EC 90, ACE-CEW Papers, SLRC, and on reel 97, secs. 147–54, of AAUW microfilms (see Barbara A. Sokolsky, ed., *AAUW Archives, 1881–1976: A Guide to the Microfilm Edition* [New York: Microfilming Corporation of America, 1980]).

When by the 1960s "continuing education" became feasible, Dolan used this earlier, seemingly unsuccessful NSF survey to justify her AAUW College Faculty Program for training reentry women in eleven Southern states in 1962–66. With the help of two grants from the Rockefeller Brothers Fund ($225,000 initially and a $42,000 extension), seventy-seven women completed master's degrees, and seven their doctorates, but only 9 percent were in science or mathematics. See numerous articles in the *AAUWJ*, such as Eleanor F. Dolan, "That Second Career," 55 (1962): 151–54; "A Midpoint Report," 57 (1963): 31, 36; Eleanor F. Dolan, "Q.E.D.," 57 (1964): 136–38; and "Evaluation of College Faculty Program," 62 (1968): 31–32; and reel 99, secs. 254–61, and 132, sec. 76, AAUW Microfilms. There are also leaflets and correspondence about the College Faculty Program in box 2, Papers of the Dean of the Graduate School, RCA.

20. Records of the AAUW's Committee on Standards and Recognition for 1947–58 are on microfilm reels 61 and 62; those for 1958–April 1963, when it merged with the Committee on Higher Education, are on reel 99; and those for 1965–72, when it was revived, are on reel 101. Psychologist Lillian Portenier of the University of Wyoming was a member of the Committee on Standards and Recognition from 1946 to 1952, as were anthropologist Erna Gunther of the University of Washington, 1950–57, and chemist Essie White Cohn of the University of Denver, who served from 1958 until her death in 1963. Geographer Minnie Lemaire served on the committee from 1961 until 1968, and chemist Jean'ne Shreeve of the University of Idaho in 1966–67 and 1969–71. See also "AAUW Standards Re-applied," *AAUWJ* 48 (1955): 239–42; and Flora Rose et al., *A Growing College: Home Economics at Cornell University* (Ithaca: New York State College of Human Ecology, Cornell University, 1969), 225.

21. John R. Overmann, *The History of Bowling Green State University* (Bowling Green, Ohio: Bowling Green State Univ. Press, 1967), 189–90. Chemist Peggy Hurst started her long career there in 1955 (*AMWS*, 18th ed., 1992, 3:968). On Roberts, see *WWAW* 1 (1958), 1084; "Minutes" of March 16–18, 1956, meeting of Committee on Standards and Recognition, 21 (reel 62).

22. "Women and Higher Education," *AAUWJ* 51 (1957): 14, reprinted from "Women in Institutions of Higher Education," a statement issued by the Middle States Association of Colleges and Secondary Schools; a copy also appeared as appendix B to "Minutes" of meeting of Committee on Standards and Recognition, Aug. 27–28, 1956, reel 62, AAUW Archives. See also *The Middle States Association at Age One Hundred, the Last Twenty-five Years: Issues and Challenges* (n.p.: Middle States Association of Colleges and Schools, 1987), 7.

23. The committee's considerable correspondence with particular institutions is arranged alphabetically on reels 58–86. That for Clark University, including Eunice C. Roberts to President Howard B. Jefferson, Dec. 11, 1959, which rejects his application, is on reel 65. See also Committee on Higher Education, "Minutes," Nov. 25–27, 1960, 5, May 19–20, 1961, 15 (sec. 443), Apr. 19–20, 1963, app. E, all on reel 99; and AAUW Board

of Directors, "Minutes," Nov. 16–17, 1963, 5, 25–30, on reel 42, AAUW microfilms. There is a brief note on the policy change in *AAUWJ* 57 (1964): 97. William A. Koelsch's *Clark University, 1887–1987* (Worcester, Mass.: Clark Univ. Press, 1987), mentions the tenuring of the first woman there in 1963 (183) and the long service of one woman trustee who became chair in 1967, the first such woman at a major private research university (204).

24. "Educational Foundation Awarded Grants," *AAUWJ* 55 (1962): 74–75; Lorraine B. Torres, "AAUW's Roster of Women Holding Earned Doctorates: A Report—and a Plan for the Future," ibid., 175–78; AAUW Committee on the Status of Women, "Minutes," Nov. 11–12, 1953, 18–19, reel 117, AAUW microfilms. The records of the AAUW Roster of Earned Doctorates (Advanced Degrees) are on reel 101, including a memorandum of a luncheon conference on January 24, 1955, with Sophie Aberle of the National Science Board, who was enthusiastic about Dolan's proposed roster, and Eleanor F. Dolan's long letter to Dr. Sidney Wallach, July 13, 1965, explaining the need for the Roster. See also Records of the AAUW Educational Foundation, on reels 128 and 132.

Dolan had been a faculty member in government at Florida State College for Women and New York University before becoming professor and dean at Flora Stone Mather College for Women in 1941. For biographical information, see *WWAW* 9 (1975–76): 234; Dolan's brief biographical statement "In the Field of Educational Administration," *WAM* 45 (1960–61), 213; obituary, *Radcliffe Quarterly* 74, no. 2 (June 1988), 45; and a file of clippings on Dolan, Ph.D. 1935, in the Deceased Alumnae Records, RCA.

25. On Irwin-Williams, see *AMWS*, 13th ed., 1976, 3:2093; clipping, "Harvard Scholar's Discoveries in France Shed Light on Stone Age," *Boston Sunday Herald*, Mar. 1. 1959, sec. 1, p. 3(?), in box 4, Papers of the Office of the Graduate Dean, RCA; and Barbara Williams, *Breakthrough: Women in Archaeology* (New York: Walker, 1981), 2. By 1981 Irwin-Williams had been elected president of the Society of American Archaeology, and shortly thereafter she moved to the University of Nevada at Reno.

26. J. O. Brew et al., "Eleanor C. Irwin," *American Antiquity* 39 (1974): 608.

27. On Shields, see *AMWS*, 17th ed., 1989, 6:707; George Gaylord Simpson referred to her disparagingly in 1962 as the "Frau Professor" in George Gaylord Simpson, *Simple Curiosity: Letters from George Gaylord Simpson to His Family*, ed. Léo LaPorte (Berkeley and Los Angeles: Univ. of California Press, 1987), 315; and "AAAS Officers: Committees, and Representatives for 1960," *Science* 131 (1960): 507.

28. Elizabeth M. O'Hern, "Cora Mitchell Downs, Pioneer Microbiologist," *A[mer-ican] S[ociety of] M[icrobiology] News* 40 (1974): 862–65, reprinted in O'Hern, *Profiles of Pioneer Women Scientists* (Washington, D.C.: Acropolis, 1985), ch. 18; transcript, "Interview with Cora Downs by Mrs. Phyllis Lewin," Sept. 1984, plus numerous clippings, press releases, and reprints in Cora Downs Faculty Folder, University of Kansas Archives, Spencer Research Library, Lawrence, Kans., including John Alexander, "Alice in Reverse: The Rabbit Came to Her," *Saturday Review* 44 (Apr. 1, 1961): 48–49.

29. Thelma Thurstone, "Dorothy C. Adkins, 1912–1975," *Psychometrika* 41 (1976): 434–47; there is also a large collection of Adkins's papers at the Archives of the History of American Psychology, University of Akron.

30. [Patricia Cain Smith], "Patricia Cain Smith," in *Models of Achievement: Reflec-*

tions of Eminent Women in Psychology, ed. Agnes N. O'Connell and Nancy Felipe Russo, vol. 2 (Hillsdale, N.J.: Lawrence Erlbaum Associates, 1988), 148–49; *AMWS,* 13th ed., S&B, 1978, 1119. As soon as Patricia Smith left, Eleanor Jack Gibson became a full professor of psychology (see ch. 7, nn. 50–52).

31. On Bilger, see *AMS,* 11th ed., 1965, 1:412 (which gives the dates of her prize as 1950 and the date of her retirement as 1954). When she won the Garvan Medal of the ACS, she was on the cover of *C&EN* 31 (Mar. 16, 1953) as well as in a short article inside (1094). Just a year later, however, she was ousted as department chair because of her "personal style of administration, her emphasis on older, more formal teaching methods, and the rapid advances in chemical research techniques." She then spent the remaining four years before retirement (at age sixty-five) planning the university's new chemistry building, which the trustees named for her and her husband in 1959 (without faculty approval, which again caused a stir). Controversial in many ways, she had a close friendship with a university regent who lived with her and her husband for many years and bequeathed her $3.5 million. See Madeleine J. Goodman, "Leonora Neuffer Bilger," in *Notable Women of Hawaii,* ed. Barbara Bennett Peterson (Honolulu: Univ. of Hawaii Press, 1984), 33–37, quotation from 36; and Victor N. Kobayashi, ed., *Building a Rainbow: A History of the Buildings and Grounds of the University of Hawaii's Manoa Campus* (Manoa: Hui O Students of the University of Hawaii, 1983), 91, 96–97, 100. (I thank Mary Rossiter for copies.)

32. On deMilt, see *AMS,* 8th ed., 1949, 601; and John Mark Scott, "Clara Marie deMilt (1891–1953)," *JCE* 31 (1954): 419–20. There is a collection of Clara deMilt's papers at the Tulane University Archives in New Orleans, along with items about several other Newcomb chemists.

33. On Simmons, see *AMWS,* 12th ed., 1970, 5:5831. Yet others did not find suitable new jobs so readily and had to make do with what was available locally. Carol Anger Rieke, for example, arguably one of the best-trained women scientists of the 1930s, having earned a Radcliffe Ph.D. in astronomy in 1932 and the AAUW Sarah Berliner Postdoctoral Fellowship after that, followed her husband to Chicago, where he was a physicist first at the University of Chicago and then at the Argonne National Laboratory. For a while she was an assistant to Robert Mulliken, the expert on molecular orbitals and a Nobelist in chemistry, but starting in 1957, after raising two children (both of whom became scientists), she was an instructor in mathematics and astronomy at nearby Thornton Junior (later Community) College for 28 years. See ibid., 17th ed., 1989, 6:181; and Robert S. Mulliken, *Life of a Scientist* (New York: Springer-Verlag, 1989), 103, 112, 124.

34. Rufus Harris to Prof. Rose L. Mooney, Dec. 22, 1948, Rose Mooney Faculty File, Tulane University Archives. In 1954 Mooney, a widow, married solid-state physicist John Slater, resigned from Newcomb, and became a research associate at MIT (*AMWS,* 12th ed., 1972, 5:5877). Her papers are at the American Philosophical Society in Philadelphia. See also John C. Slater to Jerrold Zacharias, Jan. 14, 1952, John Slater Papers, also at APS. (I thank Nathan Reingold for telling me of this letter.) Between 1945 and 1970 several universities, generally private—including the University of Pennsylvania, the University of Rochester, Fordham University, the Carnegie Institute of Technology, and Duke—that had formed a "coordinate college for women" as a grudging hesitant

half-step toward coeducation quietly merged the two into one larger coeducational student body. In cases where there had been a separate faculty for the women, it generally had employed some women, who fared less than equitably when the two faculties also merged. Oftentimes their subjects were eliminated. For a fascinating study of one such case (Margaret Morrison Carnegie College), see Joan Burstyn, "Educational Experiences for Women at Carnegie-Mellon University: A Brief History," *Western Pennsylvania Historical Magazine* 56, no. 2 (1973): 141–53.

35. On Smith, see *AMWS,* 16th ed., 1986, 6:827; on Shields, see n. 27 above; on Barclay, see Jean Gerow, "Down to Earth," *Dialog* 13, no. 2 (summer 1989): 6–9; and on Welch, see *AMWS,* 12th ed., 1973, 6:6803.

36. On Gunther, see *AMS,* 11th ed., S&B, 1968, 1:619. There is a substantial collection of Erna Gunther's papers at the University of Washington Archives in Seattle and a small one in the Alaska and Polar Regions Department, Ramussen Library, University of Alaska at Fairbanks.

37. Adkins had worked with L. L. Thurstone at the University of Chicago before World War II. When he and his wife moved to the University of North Carolina after the war to set up a new psychometrical laboratory, she went along as a full professor and department chair and also helped publish the journal *Psychometrika*. Thelma Thurstone, who also held a research professorship at the University, wrote Adkins's obituary (see n. 29 above).

38. On Hunt, see *AMWS,* S&B, 12th ed., 1973, 1:1091; On Gunther, see n. 36 above; and the special issue of *Landmarks, Magazine of Northwestern History and Preservation,* 4, no. 1 (1985). Cornelia Smith (above and in n. 35) also directed the Strecker Museum at Baylor from 1945 to 1967.

39. Richard L. Anderson, "Gertrude Mary Cox, 1900–1978," *NAS Biographical Memoirs* 59 (1990): 117–32; William G. Cochran, "Some Reflections," *Biometrics* 35 (1979): 1–2; Richard L. Anderson et al., "Gertrude M. Cox—A Modern Pioneer in Statistics," ibid., 3–7; Bernard G. Greenberg et al., "Statistical Training and Research: The University of North Carolina System," *International Statistical Review* 46 (1978): 171–207, with several appendixes; Alice E. Reagan, *North Carolina State University: A Narrative History* (Raleigh: NCSU Foundation, 1987), 103, 162 and 195. Largely thanks to the efforts of archivist Maurice Toler, the North Carolina State University Archives has the Gertrude Cox Papers, the Cox Faculty File, and a file on the North Carolina State Statistics Department. Cox Hall is mentioned in Marion Tinling, *Women Remembered: A Guide to Landmarks of Women's History in the United States* (Westport, Conn.: Greenwood, c. 1986), 215.

40. On Rees, see *AMWS,* 12th ed., 1972, 5:5155–56.

41. On Tyler, see the announcement of her election as APA president in *AP* 25 (1971): 751; Leona E. Tyler, "My Life as a Psychologist," in *The Psychologists: Autobiographies of Distinguished Living Psychologists,* ed. T. S. Krawiec, vol. 3 (Brandon, Vt.: Clinical Psychology, 1978), ch. 11; and Gwendolyn Stevens and Sheldon Stevens, *The Women of Psychology,* vol. 2 (Cambridge, Mass.: Schenckmann, 1982), 112–19.

42. On Goldsmith, see *NAWMP; AMWS,* 12th ed., 1972, 2:2205; John Duffy, *The Tulane University Medical Center: One Hundred and Fifty Years of Medical Education*

(Baton Rouge: Louisiana State Univ. Press, 1984), 175, 199; and Grace Goldsmith Faculty File, Tulane University Archives.

43. See NEA, National Council of Administrative Women in Education, *Administrative Women in Higher Education* (Washington, D.C., 1952).

Chapter 10   Majors, Money, and Men at the Women's Colleges

1. These counts are given annually in the USOE, *Education Directory,* pt. 3, on higher education. Its title is, however, a year ahead of its data; that is, the directory for 1963–64 has statistics on the fall of 1962. Because the USOE's data collectors relied on the colleges' own definition of their status, some inconsistencies developed: some clearly "coordinate" colleges for women (e.g., Radcliffe, with its own trustees but not its own faculty) reported themselves as independent institutions and so were classified as "women's colleges," while others that seemed even more like independent women's colleges (e.g., Barnard and Douglass Colleges, with their own trustees and their own faculty) classified themselves as part of the larger university and so were counted not as women's colleges but as parts of coeducational institutions. Similarly, because the Bronx campus of Hunter College had gone coed in 1951, the USOE counted the whole as a coeducational institution, thus omitting the very large downtown campus, which remained all-female until 1964 (John F. Ohles and Shirley M. Ohles, comps., *Public Colleges and Universities* [Westport, Conn.: Greenwood, 1986], 486–88). This inconsistency seriously undermined the validity of numerous studies, such as Donald L. Thistlethwaite's, "College Environments and the Development of Talent," *Science* 130 (1959): 71–76, which ranked the women's colleges as the weakest producers of doctorates in the natural sciences, partly because he had classified all the coordinate colleges (Barnard, Douglass, Newcomb, Duke's Woman's College, etc.) as coeducational ones.

2. See Mary J. Oates, "The Development of Catholic Colleges for Women, 1895–1960," in Oates, *Higher Education for Catholic Women: An Historical Anthology* (New York: Garland, 1987), reprinted in *U.S. Catholic Historian* 7 (1988): 413–28; Sr. M. St. Mel Kennedy, "The Faculty in Catholic Colleges for Women," *Catholic Educational Review* 59 (1961): 289–98. On Burke, see obituary, *Chicago Tribune,* Oct. 6, 1990, 15. On O'Byrne, see obituary, *NYT,* Oct. 7, 1987, D-22. On Wolff, see *NAWMP;* and Sr. Mary Immaculate [Helen Creek], *A Panorama: Saint Mary's College, Notre Dame, Indiana, 1844–1977* (Notre Dame, Ind.: St. Mary's College, 1977). On Grennan see the autobiographical article "How Can Catholic Institutions Secure More Foundation-Research Support? *NCEAB* 60 (1963–64): 216–18. And see Sr. Mary Emil, "Catholic Colleges for Women—An Ongoing Enterprise," *Religious Education* 63 (1968): 20–32. There was some criticism within the Catholic hierarchy of so much dispersion of effort. Responding to such critics were Sr. M. Evarista, "Too Many Small Catholic Colleges?" *Catholic Educational Review* 60 (1962): 185–91; and Sr. M. Adele Francis Gorman, "In Defense of the Four-Year Catholic Women's College," ibid. 63 (1965): 369–75. On Sr. M. Emily, see obituary, *Boston Herald,* Mar. 26, 1991 (I thank Sr. Mary Oates for a copy). On Rogick, see "Necrology," *Ohio Journal of Science* 65 (1965): 238, as well as other material in box 19, f. 17, *NAWP,* SLRC. On Wallis, see *AMWS,* 13th ed., S&B, 1978, 1245; obituaries,

NYT, Jan. 25, 1978, B-2, and *Radcliffe Quarterly* 63, no. 2 (June 1978); and Patricia Case, "Ruth Sawtell Wallis (1895–1978)," in *Women Anthropologists,* ed. Ute Gacs et al. (Urbana: Univ. of Illinois Press, 1989), 361–66. And on Hornig, see *AMWS,* 13th ed., 1976, 3:1998.

3. USOE, *Earned Degrees Conferred, 1962–63,* USOE Circular 777 (Washington, D.C.: GPO, 1965). Here I defined a woman's college as an institution that awarded 90 percent or more of its bachelor's degrees to women. Insistence on a full 100 percent would have eliminated some still essentially women's colleges that were adding men to special evening, weekend, or other programs.

4. See, e.g., Mary Ellen Goodman, "Sociology Field Work," *WAM* 37 (1953): 84–85, 87, 95–96. Although Wellesley College ranked rather high here, its total of 155 science majors in 1962–63 was not only lower than earlier highs of 193 in 1948 and 190 in 1951 but also much lower in terms of percentage, because the size of the college's graduating class had increased almost a third, from 305 in 1945 to 396 in 1956 (President's Office, Statistics Notebook, 1899–1966, Wellesley College Archives, Wellesley, Mass.).

5. The Department of Bacteriology and Hygiene at the Mississippi State College for Women was the result of the driving enthusiasm of Martha O. Eckford, an alumna and Johns Hopkins Sc.D. in 1925 (see Bridget S. Pieschel and Stephen Pieschel, *Loyal Daughters: One Hundred Years at Mississippi University for Women, 1884–1984* [Jackson: Univ. Press of Mississippi, 1984], 103–4, 132; and *AMS,* 6th ed., 1938, 400). Its most famous graduate was Elizabeth Hazen, 1910, of the New York State Department of Health. For more on her see, ch. 13, n. 73.

6. George P. Schmidt, *Douglass College: A History* (New Brunswick, N.J.: Rutgers Univ. Press, 1968), 230.

7. Edward Alvey, Jr., *History of Mary Washington College, 1908–1972* (Charlottesville: Univ. Press of Virginia, 1974), 423, 475.

8. On Michels, see *AMWS,* 12th ed., 1972, 4:4249; and Walter C. Michels, "Women in Physics," *Physics Today* 1 (Dec. 1948): 16–19, which is somewhat autobiographical. Michels was apparently the organizer of the June 1954 Conference on the Role of the Women's Colleges in the Physical Sciences held at Bryn Mawr (see ch. 3). Rosalie Hoyt was also a member of the department from 1941 until at least 1986 (*AMWS,* 16th ed., 1986, 3:8564).

9. "Smith Has First Women's Section of Physics Institute," *Springfield (Mass.) Union,* Dec. 25, 1959, and "First AIP All-Woman Student Section Formed in U.S.," *Daily Hampshire Gazette,* Dec. 29, 1959, both in Physics Department Clipping File, Duke University Archives, Durham, N.C. Dudley Harmon, "Mr. Josephs Believes Women in Physics Have No Bounds Set on Their Possibilities," *SAQ* 51 (Feb. 1960): 85. See, on Jess Josephs, *AMWS,* 17th ed., 1989, 4:148. See also Sr. Mary Therese, "Scientific Womanpower—Our Country's Need and What Women's Colleges Are Doing to Supply Physicists," *American Journal of Physics* 21 (1953): 569–70, which reported the results of her own recent survey of physics at eighty-five liberal arts colleges for women; only thirty-three offered a major.

10. On the earth sciences, see Caroline E. Heminway (later Kierstead), "Training of Women in Geology," Proceedings of Fifth Conference on Training in Geology, *Interim Proceedings of the Geological Society of America,* 1947, pt. 1, report of Mar. 1947, 66–71, includes a brief report of a recent survey of the careers of Smith's geology majors over

the previous twenty years. See also Susan Allen Toth, *Ivy Days: Making My Way Out East* (Boston: Little, Brown, 1984), 103–7. The department was mentioned occasionally in the alumnae magazine, e.g., in Helen Stobbe, "Smith Seniors Make Geologic History," *SAQ* 44 (1952): 17–18, which describes the faculty and students' summer fieldwork in 1952, when they were the first women undergraduates to participate in a program at Princeton University's geology camp in the Beartooth Mountains in Montana.

On astronomy, see, e.g., on Andrea Kundsin (later Dupree), "Student of the Skies," *WAM* 44 (1960): 44–45; *AMWS*, 15th ed., 1982, 2:756; and "Interview with Andrea Dupree," in *The Outer Circle: Women in the Scientific Community*, ed. Harriet Zuckerman, Jonathan R. Cole, and John T. Bruer (New York: W. W. Norton, 1991), 94–126. See also Richard E. Berendzen, "On the Career Development and Education of Astronomers in the United States" (Ph.D. diss., Harvard University, 1968), esp. 458 and numerous tables based on a 1966 questionnaire. (I thank Peggy Kidwell for this reference.)

On physiology, see Toby Appel, "Physiology in American Women's Colleges: The Rise and Decline of a Female Subculture," *Isis* 85 (1994): 26–56.

11. Alvey, *History of Mary Washington College*, 456–57; Virginia Onderdonk, "The Curriculum," in *Wellesley College, 1875–1975: A Century of Women*, ed. Jean Glasscock et al. (Wellesley, Mass.: Wellesley College, 1975), 153.

12. Mary L. Sherrill, "Group Research in a Small Department," *JCE* 34 (1957): 466, 468; Emma Perry Carr, "Research in a Liberal Arts College," ibid., 467–70; John R. Sampey, "Chemical Research in Liberal Arts Colleges, 1952–59," ibid. 37 (1960): 316. Twenty-five years later Mount Holyoke still topped the list for women by a wide margin in Alfred E. Hall, "Baccalaureate Origins of Doctorate Recipients in Chemistry: 1920–80," ibid. 62 (1985): 407.

13. "Chem Ed, Editorially Speaking," ibid. 36 (1959): 533; [Harry F. Lewis and John D. Reinheimer], *Research and Teaching in the Liberal Arts College* (n.p., 1959), 15. There is also some related correspondence between Lewis and Anna Harrison in the Mount Holyoke College Archives. Starting in the late 1940s there was also at Mount Holyoke a Psychophysical Research Unit, long supported by an ONR grant and directed by John Volkmann, professor of psychology (Gail Hornstein, conversation with author, Dec. 1988; Volkmann obituary, *NYT*, July 19, 1990, D-19).

14. Robert H. Knapp and H. B. Goodrich, *Origins of American Scientists* (Chicago: Univ. of Chicago Press, 1952), 20n.

15. See, e.g., NMC, *A Policy for Scientific and Professional Manpower: A Statement by the Council with Facts and Issues Prepared by the Research Staff* (New York: Columbia Univ. Press, 1953), 186–87.

16. "Completely Modern Memorial Telescope," *SAQ* 55 (July 1964): 241; "Frontiers of Physics," *WAM* 40 (1956): 86–87; Edwina Davis Christian, "Oak Ridge Comes to Agnes Scott," *Agnes Scott Alumnae Quarterly* 38, no. 1 (fall 1959): 11; Janet Brown Guernsey, "Nuclear Scientists of the Future," *WAM* 45 (1961): 156–57; "Girls With Geiger Counters," ibid. 47 (1962–63): 217–19, reprinted from *Monsanto Magazine*, Dec. 1962; Jean V. Crawford, "Curriculum Changes . . . ," *WAM* 46 (1961–62): 12.

17. Beale W. Cockey, "Mathematics at Goucher, 1888–1979" (undergraduate honors project, 1980, typescript), 31–32 (I thank Mary Ellen Bowden for a copy); James W.

Cortada, comp., *Historical Dictionary of Data Processing: Technology* (Westport, Conn.: Greenwood, 1987), 209–10. Other early signs of interest in computers at women's colleges include Virginia Sides, "Despite Data Processing, Alumnae are not . . . and Never Will be Just Numbers at Their College," *WAM* 49 (1964–65): 229–30, 251; and Winifred Asprey, "'The Machine' at Vassar," *VAM* 44 no. 1 (Dec. 1958): 9–12, 25. Aided by advice from alumna Grace Hopper and a proximity to the IBM Research Laboratory, Vassar offered a course on numerical analysis in 1957 and 1958.

18. Helen A. Padykula, "Through the Electron Microscope," *WAM* 49 (1964–65): 152–54. On Padykula, see *AMWS,* 17th ed., 1989, 5:845–46.

19. "The Case for the Sciences [at Bryn Mawr College]" (Bryn Mawr, Pa.: Bryn Mawr College, Jan. 1976, typescript), 14 (I thank Patricia King for a copy). The program had been helped earlier by a grant from the Carnegie Corporation. See also J. Merton England, "Investing in Universities: Genesis of the National Science Foundation's Institutional Programs, 1958–63," *Journal of Policy History* 2, no. 2 (1990): 131–56.

20. "Research in Colleges," NSF, *Fourth Annual Report of the National Science Foundation, Year Ending June 30, 1954* (Washington, D.C.: GPO, [1954?]), 24–26, 90, 120–22; W. Rodman Snelling and Robert F. Boruch, *Science in the Liberal Arts College: A Longitudinal Study of Forty-nine Selective Colleges* (New York: Columbia Univ. Press, 1972).

21. NSF, *Annual Reports,* 1952–58 (Washington, D.C.: GPO, 1952–59).

22. "A Summer in Chemistry Research," *WAM* 47 (1962): 14–16; Pepi Allen, "What's New on Campus?" ibid. 48 (1963–64): 16–17; Edward R. Linner, "The Challenge to the Sciences," *VAM* 48, no. 2 (Dec. 1962): 5–27.

23. On Blakeslee, see *AMS,* 8th ed., 1949, 222; Edmund W. Sinnott, "Albert Francis Blakeslee, 1874–1954," *NAS Biographical Memoirs* 33 (1959): 1–38; obituary, *SAQ* 46 (1954–55): 127 (his wife Margaret Bridges Blakeslee was a Smith alumna); "Grants to Smith for Cancer Research: Two Departments Now at Work on Problem," ibid. 39 (1947–48): 20; and "Dr. Blakeslee Gets Coveted Award," ibid. 44 (1952–53): 93. On Satin, see *AMS,* 8th ed., 1949, 2164; and on Avery, ibid., 9th ed., biological sciences, 1955, 38. Blakeslee's papers are at the APS. On the establishment of the William Allan Neilson Chair of Research, see Margaret Farrand Thorp, *Neilson of Smith* (New York: Oxford Univ. Press, 1956), 214–17.

24. See "Naval Research at Smith College," *SAQ* 40 (1948–49): 5; Gladys Anslow, "Once a Coalbin, Now a Spectroscopic Laboratory," ibid. 43 (1951–52): 6–7; Nora Mohler, "Too Many Retirements! . . . Miss Anslow," ibid. 51 (1959–60): 199; Anslow's faculty file, Smith College Archives, Northampton, Mass. Anslow was one of the few women scientists of the 1940s and 1950s with ties to the men in Washington who were running the ONR and trying to get the NSF organized. For example, Frank Waterman, father of Alan T. Waterman of the ONR and NSF, was for many years her department chairman; she had been the first woman to work on E. O. Lawrence's cyclotron at Berkeley, in 1939; and during World War II she had been head of the information section of the OSRD's field service ("The Presidential Certificate of Merit," *SAQ* 40 [1948–49]: 7–8; "Congratulations from Alumnae," ibid., 8) She also wrote articles supporting federal support of basic research, e.g., "Smith Responds to National Trends in Science

Education and Research," ibid. 38 (1946): 133–34, and "Women in Science" *Educational Focus* 20 (Sept. 1949): 6–12.

Despite her brilliance, or perhaps partially because of it, Dorothy Wrinch never held a regular academic post. In the early 1940s Otto Glaser of Amherst College (whom she later married) had Amherst, Smith, and Mount Holyoke piece together a visiting professorship in physics for her. Later this was seen as one of the first joint projects that led to the "Four [now Five] College" consortium in the area. On Wrinch, see *AMS*, 11th ed., 1967, 6:5995; Pnina Abir-Am, "Synergy or Clash: Disciplinary and Marital Strategies in the Career of Mathematical Biologist Dorothy Wrinch," in *Uneasy Careers and Intimate Lives: Women in Science, 1789–1979*, ed. Pnina Abir-Am and Dorinda Outram (New Brunswick, N.J.: Rutgers Univ. Press, 1987), 239–80; Marjorie Senechal, ed., *Structure of Matter and Patterns in Science: A Symposium Inspired by the Work and Life of Dorothy Wrinch, 1894–1976* (Cambridge, Mass.: Schenckman, 1980); and the Dorothy Wrinch Papers, Smith College Archives. On the "Four Colleges," see Dudley Harmon, "Off the Ground and Apt to Stay," *SAQ* 52 (1960–61): 198–200.

25. On Mack, see *AMWS*, 13th ed., 1976, 4:2736; Joyce Thompson, *Marking a Trail: A History of the Texas Woman's University* (Denton: Texas Woman's Univ. Press, 1982), 166–67, 184 nn. 4, 9; and William G. Pollard, *Atomic Energy and Southern Science* (Oak Ridge, Tenn.: Oak Ridge Associated Universities, 1966), 28–31. On Mack's earlier career at Penn State, see Dorothy Rickard, "Gold Medalist in Chemistry Research," *IW* 29 (1950): 201; "Garvan Medal to Pauline Mack," *C&EN* 28 (1950): 1032, as well as picture on cover, Mar. 27, 1950; Harry Henderson, "Is Your Wife Starving Alone?" *Nation's Business* 40 (Aug. 1952): 30–31, 62–63; and Thompson, *Marking a Trail*, 136–38, 157–58. The Pauline Beery Mack Papers and the Papers of the Ellen Richards Institute are both at the Penn State University Archives, University Park, Pa.

26. Nevitt Sanford, "The Mellon Research Program Today," *VAM* 44 (Oct. 1958): 2–5; "Publications of the Mellon Foundation Staff, 1952–1958," ibid., no. 2 (Dec. 1958): 24–25; assorted articles in Nevitt Sanford, ed., *The American College* (New York: John Wiley, 1961); Ravenna Helson, "The Changing Image of the Career Woman," *Journal of Social Issues* 28, no. 2 (1972): 37–38 (I thank her for a copy); David Riesman, "Two Generations," *Daedalus* 93 (1964): 729n (quotation).

27. William Calder, "Astronomy at Agnes Scott," *Sky and Telescope* 9 (1950): 274–75. Clippings in the Bradley Observatory File, Agnes Scott College Archives: Anna Belle Close, "Seeing Stars at Agnes Scott," *Atlanta Journal Magazine,* Feb. 6, 1949, and Eleanor Hutchens, "Agnes Scott Beats Russia to South's Finest Telescope," *Atlanta Journal and Constitution Magazine,* June 18, 1950, 6–7. See also "Agnes Scott— Astronomy Center of the New South," *Agnes Scott Alumnae Quarterly* 27, no. 3 (spring 1949): 2; and Kathryn Johnson, "A Calder Kaleidoscope," ibid. 36, no. 2 (winter 1958): 3–6. Calder in *AMWS*, 12th ed., 1971, 1:847. (I thank Rosalinda Reynolds Ratajczak for assistance in Atlanta.)

28. Anatomist Florence Sabin, 1893, and physician Dorothy Reed Mendenhall, 1895, mother of then Smith president Thomas Mendenhall, had been good friends at Smith and later at Johns Hopkins Medical School (both in *NAWMP*). See Anna Young Whiting, "Quarters for Scientists," *SAQ* 47 (1956): 193–94; "Science at Smith—

Recommendations," ibid. 53 (1961–62): 26; "News from Northampton," ibid. 54 (1963): 163; "Sabin-Reed Hall," ibid. 56 (1964–65): 160; "McConnell Hall," ibid. 57 (1965–66): 153–55; "The Clark Science Center," ibid. 58, no. 3 (Apr. 1967): 6–9; and " . . . at Smith College," *American Schools and Universities* 40 (Sept. 1967): 29–30.

29. See Emma Perry Carr, "Mount Holyoke College," *JCE* 25 (1948): 11–15; idem, "Research Grants to Department of Chemistry, Symposium, May 31, 1952," typescript, College Archives, Dwight Hall, Mount Holyoke College; idem, "Hydrocarbon Research at Mount Holyoke College," typescript in College Archives, n.d., which appends several lists of Mount Holyoke alumnae and master's degree recipients who have gone on in chemistry; "Mount Holyoke College to Co-operate with Industry," *School and Society* 76 (1952): 364; "Holyoke's $1 Million Investment," *C&EN* 33 (1955): 1982; obituary of Newcomb Cleveland, *NCAB* 38 (1953): 123. On Carr, see *NAW* and Bojan Hamlin Jennings, "The Professional Life of Emma Perry Carr," *JCE* 63 (1986): 923–27; on Sherrill, see Wyndham Miles, ed., *American Chemists and Chemical Engineers* (Washington, D.C.: ACS, 1976), 436; and on Pickett, see *AMWS*, 12th ed., 1972, 5:4913. See also nn. 12 and 13 above.

30. See also The Dunmore Foundation, Racine, Wis., Saint Mary's College, Notre Dame, Ind., and John Price Jones Co., "Corporate Support of Women's Colleges: A Survey of Corporate Attitudes, September 1958" (mimeographed); J. Whitney Bunting, "The Relation of Business to Women's Higher Education," *Educational Record* 42 (1961): 287–95; Benjamin Wright, "Corporate Giving and Liberal Colleges," *SAQ* 46 (1954–55): 204–5; idem, "The Ford Foundation Grant," ibid. 47 (1955–56): 81; "Corporations and Colleges," *WAM* 40 (1955–56): 142–44, a useful survey of nineteen corporate foundations and their potential benefit to Wellesley; Virginia Sides, "Can You Pick a Corporation's Pocket?" ibid. 43 (1958–59): 156–57, 165, and "Can You Pick a Corporation's Pocket?" ibid. 45 (1960–61): 94–95. Sides (Wellesley 1944), a chemistry major, had been an assistant to Alan T. Waterman, first at ONR (1946–51) and then at the NSF (1951–55), before returning to Wellesley to assist Miss Clapp with fund-raising ("Wellesley Today—Our New Vice President," ibid. 47 [1962–63]: 148).

31. Sarah Gibson Blanding, "Report of the President," *Bulletin of Vassar College* 54, no. 4 (Dec. 1964): 18–19. The drive was to have included funds for a new biology building, but the trustees dropped that from the plan in 1962. Olmsted Hall was finally completed in 1974, after a second fund drive.

32. The lives, thoughts, and donations of Lamont and Morrow were often written up in the *SAQ* and *WAM* as well as the *NYT*. See, e.g., "Two Women Expressed Their Belief in Women's Colleges" [about Lamont's (d. 1952) recent bequest of $1.2 million to Smith], *SAQ* 44 (1952–53): 81; "Florence Corliss Lamont, 1873–1952, and Henceforth," *WAM* 38 (1953–54), inside back cover; and Harold Nicolson, *Dwight Morrow* (New York: Harcourt, Brace, 1935). Morrow had long chaired the board of trustees at nearby Amherst College. On Rood, see Katharine Timberman Wright, "Dorothy B. A. Rood Retires from Trustees," *WAM* 43 (1958–59): 291, 302; and " . . . There are Diversities of Gifts, but the Same Spirit . . . ," ibid. 49, no. 5 (1964–65), inside back cover.

33. Janet Brown Guernsey described her daily life (and her housekeeper Veronica) in "The Married School Marm," *WAM* 35 (1951): 137–38. See also ch. 4, n. 43.

34. President's Office, Statistics Notebook, 1899–1966, Wellesley College Archives.

35. Virginia Mayo Fiske to Mrs. Cronkhite, Sept. 14, 1954, box 2, Papers of the Office of the Graduate Dean, RCA. Mayo also reported that of the eighty-four women on the Wellesley faculty, sixteen were married, and seven of those had children.

36. On Israel, see obituary, *SAQ* 53 (1961–62): 62; on Siipola, see *AMWS,* 13th ed., S&B, 1978, 1098 (after his death she held the Harold Israel Professorship until her retirement in 1973); and on both Crawfords, see ibid., 13th ed., 1976, 1:874–75. See also Maria Luisa Crawford, "Choosing a Career for the Fun of It," in *Women in Geology: Proceedings of the First Northeastern Women's Geoscientists Conference, St. Lawrence University, Canton, New York, April 26–27, 1976,* ed. Susan D. Halsey et al. (Department of Geology and Geography, St. Lawrence University, 1976), 11–12.

37. Alice Bryan to Edwin Boring, Jan. 26, 1945 and his reply, Jan. 31, 1945, both in Edwin G. Boring Papers, HUA. On G. Seward, see *WWAW* 1 (1958–59): 1157; on both Sewards, see Jane D. Hildreth and Carolyn L. Konold, eds., *American Psychological Association Directory, 1951* (Washington, D.C.: APA, [1951]), 419.

38. "Minutes of the Seventh Meeting of the Study Committee," Nov. 3, 1953, Graduate School Study Committee, Minutes and Correspondence, President Jordan's Papers, RCA, 3. Despite McBride's statements, the more usual pattern at Bryn Mawr in the fifties and sixties was for only the husband to be on the faculty, while the wife (often the second wife, a former student or graduate student) was a research associate (Clelia Mallory, chemistry), lecturer (Frances Berliner, chemistry), or dean (Patricia Pruett, biology) (see *AMWS,* 13th ed., 1976). President Pendleton of Wellesley had years earlier established a rule against hiring married couples in the same department (Mary L. Coolidge to Mrs. Cronkhite, Oct. 20, 1954, box 2, Papers of the Office of the Graduate Dean).

39. "Miss Blanding's College," *Newsweek,* June 11, 1956, 114.

40. See, e.g., John A. Pollard, "Costly They Habit . . . : The Financial Status of Women's Colleges," *AAUWJ* 49 (1956): 98–100, which summarizes results of a 1954 survey of 111 women's colleges; Sarah Gibson Blanding, "HOW Can Colleges Attract and Keep First-Rate Professors?" ibid. 51 (1958): 146–48, which refers to faculty members as "he"; C. Benton Kline, Jr., "Building the Faculty at Agnes Scott," *Agnes Scott Alumnae Quarterly* 43, no. 3 (spring 1965): 2–4; Lucy Killough, "Facts and Figure on Faculty Salaries," *WAM* 41 (1956–57): 153–54, 160–61; G. Kerry Smith, "Smith Alumnae and Faculty Salaries: An Editorial by a Smith Husband," *SAQ* 48 (1956–57): 39; and Dorothy Coyne Weinberger, "A Better Break for Barnard's Faculty," *Barnard Alumnae Magazine* 47 (Feb. 1958): 15–16, which reports the lament of a male faculty member that at current salary levels it was difficult to interest his son in an academic career. See also Lloyd Woodbourne, *Faculty Personnel Policies in Higher Education* (New York: Harper & Bros., 1950), 56–57; and Mabel Newcomer, *A Century of Higher Education for American Women* (New York: Harper & Bros., 1959), 161–65, and 168. Newcomer feared that if salaries were not raised, then only second-rate men would come. In 1965 salaries at northeastern women's colleges were still among the lowest in the nation ("The Economic Status of the Academic Profession: Taking Stock," *AAUP Bulletin* 51 [1965]: 266).

41. [Cecilia Kenyon], "Memorandum on Radcliffe Assistant Professors, April 12, 1954," 4, Graduate School Study Committee, Minutes and Correspondence, President Jordan's Papers.

42. [Myra Sampson], "A Study of the Teaching Faculty of Smith College, 1956–1957," Myra Sampson Papers, Smith College Archives. (I thank the college archivist for bringing these episodes to my attention.)

43. Frances Clayton, "A Source for College Faculties," *Pembroke Alumna* 37 (Oct. 1962): 4–7.

44. [Myra Sampson], "Report [on the] Status of Women Faculty in Academic Departments in Smith College, 1955–56 vs. 1965–66," Aug. 18, 1966, Myra Sampson Papers. In 1965–66 there were more women faculty than men in only six of Smith's twenty-six departments: bacteriology, French, geology, Spanish, speech and theater, and zoology. On Sampson, see *AMS*, 11th ed., 1967, 5:4638. In 1963 an anonymous donor gave $65,000 to the college to endow a professorship in biological sciences in Sampson's honor and requested that it be first held by Esther Carpenter, who was followed by B. Elizabeth Horner, 1968–86. See B. Elizabeth Horner, obituary for Myra Sampson, *SAQ* 75 (summer 1984): 67–68; and "Gifts for Endowment," in "Smith College 88th Financial Report, July 1, 1962–June 30, 1963" (typescript), 109 (I thank Maida Goodwin for sending me a copy).

45. On Smith's first presidents, see Thorp, *Neilson of Smith*, 166 and ch. 11; John Wieler, "Interviews with Professor Marjorie Nicolson," 1975, Columbia University Oral History Research Office, 333 (quotation; see also 273–74, 275–76, 282). See also obituaries in *SAQ* 72, no. 4 (Aug. 1981): 69 and *American Scholar* 50 (winter 1980–81): 81–90. Nicolson merits a full biography.

The leisure-time activities of Smith's male science faculty differed widely with regard to their relationship to women. Geologist Marshall Schalk, for example, spent his summers and at least one sabbatical in Alaska at the Navy's men-only Arctic Research Laboratory, while zoologist Howard Parshley spent his free time translating Simone de Beauvoir's classic *The Second Sex* into English (Marshall Schalk, "Letter from Alaska," *SAQ* 48 [1956–57]: 88–90; Simone de Beauvoir, *The Second Sex*, trans. and ed. Howard Parshley [New York: Knopf, 1953]; on Parshley, *AMS*, 8th ed., 1949, 1898; "In Memoriam, Howard Madison Parshley," *SAQ* 44 [1952–53]: 220).

Another phenomenon that accompanied the high proportion of men on the faculty of the women's colleges was great interest (at Smith College inordinately so) in having a faculty club, which presumably would be open to all, but whose existence could present overtones of a kind of resegregation. Smith's was built in the late 1940s after considerable pressure by Albert Blakeslee (see n. 23), and Wellesley built its College Club in 1963 (Jean Glasscock, "The Buildings," in Jean Glasscock et al., *Wellesley College, 1875–1975*, 328). At Mary Washington College there was even a Faculty *Men's* Club from 1937 until the late 1950s, led by Dean Edward Alvey, Jr., among others; the college provided it with a basement room where the men could smoke and relax alone (Alvey, *History of Mary Washington College*, 222–23).

46. *John Simon Guggenheim Memorial Foundation: Reports of the Secretary and of the Treasurer, 1947 and 1948* (New York: John Simon Guggenheim Memorial Foundation, n.d.).

47. *Graduate Education for Women, The Radcliffe Ph.D.: A Report by a Faculty-Trustee Committee* (Cambridge: Harvard Univ. Press, 1956), 44–45. There are more extensive data in [Cecilia Kenyon], "Memorandum on Radcliffe PhDs who Are or Have Been

Professors," Apr. 1954, and "Minutes of the Twelfth Meeting of the Study Committee, April 19, 1954," in Graduate School Study Committee, Minutes and Correspondence, President Jordan's Papers. Yet Eleanor Gibson later recalled that Harold Israel in psychology had not published at all ([Eleanor J. Gibson], "Eleanor J. Gibson," in *A History of Psychology in Autobiography,* ed. Gardner Lindzey, vol. 7 [San Francisco: W. H. Freeman, 1980], 241). Sarah Gibson Blanding was praised upon her retirement for having promoted more rapidly those who had published (C. Gordon Post, "Sarah Gibson Blanding and the Faculty," *VAM,* June 1964, 8–9). It was Marjorie Hope Nicolson's recollection that despite all the raises and promotions at Smith, most men left within ten years (John Wieler, "Interviews with Professor Marjorie Nicolson," 273). Eleanor Gibson said the reason her husband James left Smith was that he wanted male graduate students (taped interview with Eleanor J. Gibson, c. 1976, KLCU). See also Eric Lampard, "Thoughts on Leaving Smith College," *SAQ* 51 (1959–60): 83–84.

48. Tom Mendenhall to Dr. Alan Waterman, Aug. 2, 1960, Alan T. Waterman Papers, Manuscript Division, LC (I thank Nathan Reingold for bringing this item and collection to my attention). See also "Obituaries, Frank Allen Waterman," *SAQ* 50 (1958–59): 127; and John Van Vleck, "Mendenhall: A Scientist by Heredity," ibid. 51 (1959–60): 10–11. The retirees described in the alumnae magazines were not necessarily "dead wood." Several continued their research, perhaps with the aid of a new grant or fellowship. In 1966–67 Smith physicist Gladys Anslow held a Sophia Smith Fellowship awarded by the college to emeriti engaged in research (see her obituary from *SAQ,* copy in Faculty File, Smith College Archives). See also Mary Dougherty Lenox, "Unretiring Faculty," *WAM* 48 [1963–64]: 20–21, 31–32).

49. Curtis J. Smith, "Charlotte Haywood," *Biological Bulletin* 143 (Aug. 1972): 16–17, quotation from 16. Like Clapp, Haywood had spent nearly every summer at Woods Hole (in Haywood's case, from 1920 to 1970).

50. On Makemson, see *CBY, 1941,* 552–54; *AMS,* 10th ed., P&B, 1961, 3:2632.; "Person, Place and Thing," *VAM* 43, no. 1 (Oct. 1957): 18; and Makemson Faculty File, Special Collections, Vassar College Library, Poughkeepsie. On Roman, see David DeVorkin, "Dr. Nancy Roman: Transcript of an Interview, August 19, 1980," 15, copy in NBL. On Albers, see *AMWS,* 13th ed., 1976, 1:38; and Albers Faculty File, Special Collections, Vassar College Library. In some cases it is not easy to determine who replaced whom, since some departments were being consolidated and retirements were often the occasion for some long overdue reallocation of duties. Martha Stahr, an assistant professor at Cornell at the time, might also have been considered. See also Vera C. Rubin, "Vera C. Rubin," in *Origins: The Lives and Worlds of Modern Cosmologists,* ed. Alan Lightman and Roberta Brawer (Cambridge: Harvard Univ. Press, 1990), 287–89.

51. Florence M. Read, *The Story of Spelman College* (Atlanta: privately printed, 1961), 217–18, 307–8; idem, "The Place of the Women's College in the Pattern of Negro Education," *Opportunity* 15 (1937): 267–70. On Read, see *Who Was Who in America* 6 (1974–76): 338; and "President Read's Twenty-fifth Anniversary," *Spelman Messenger* 68, no. 4 (Aug. 1952): 17–20. The biology department was evidently one of the best at Spelman (Barnett Smith, "Biology at Spelman College," ibid. 83, no. 2 [Feb. 1967]: 2–6). On Albro, see *AMS,* 10th ed., 1960, 1:16; "Spelman Honors Dr. Helen T. Albro," *Spelman Messenger* 76, no. 3 (May 1960): 25–26; Birdie Scott Rolfe, "A Salute," ibid., 26–

29; and "In Memoriam, Helen Tucker Albro," ibid. 78, no. 4 (Aug. 1962): 20–21. On Barnett Smith, see *AMWS*, 14th ed., 1979, 6:4734. On Patterson, see ibid., 15th ed., 1982, 5:871; her untitled address to the Spelman students, printed in *Spelman Messenger* 84, no. 1 (Nov. 1967): 31–37, was a historical account of some previous Spelman alumnae in medicine and science. Oran Eagleson, dean of instruction from 1954 to 1970, was a psychologist of note (Robert V. Guthrie, *Even the Rat Was White: A Historical View of Psychology* [New York: Harper & Row, 1976], 141–43) and husband of Louise Johnson Eagleson, biology instructor, who died in October 1960 (see her obituary in *Spelman Messenger* 77, no. 1 [Nov. 1960]: 28).

52. See, e.g., Benjamin Wright's comments, "Minutes of the Seventh Meeting of the Study Committee," Nov. 3, 1953, 1–5.

53. David Boroff, "The University of Michigan: Graduate Limbo for Women," *Campus USA: Portraits of American Colleges in Action* (New York: Harper & Bros., 1961), 184; the essays originally appeared in *Mademoiselle*.

54. Margaret Clapp, "Journey into the Future," *WAM* 42 (1958–59): 139. Theodore Caplow and Reece J. McGee, *The Academic Marketplace* (1958; reprint, New York: Arno, 1977), describes the hiring practices of the 1950s as follows: "In theory, academic recruitment is mostly open. In practice, it is mostly closed" (109).

55. On Smith's botany department, see, on Ganong, *AMS*, 6th ed., 1938, 501; on Frances Smith, ibid., 7th ed., 1944, 1649; on Helen Choate, obituary, *SAQ* 49 (1957–58): 127; on Bache-Wiig, Kenneth Wright and Myra Sampson, "Too Many Retirements! . . . Miss Bache-Wiig, Botany," ibid. 51 (1959–60): 201; on Kemp, "Commencement for the Faculty," ibid. 52 (1960–61): 208, and obituary, ibid. 55 (1963–64): 71; on Dorothy Day, *AMS*, 11th ed., 1965, P&B, 2:1151; and on Wayne Manning, 1966, P&B, 4:3447. For the search, see the Margaret Kemp Faculty File, Smith College Archives, esp. Howard S. Reed to Kemp, Mar. 24 (no year), July 24, and Aug. 27, 1945; Ralph Wetmore to Kemp, July 25, 1946; and Victor A. Greulach to Kemp, Jan. 11, 1949. On K. Wright, see *AMS*, 11th ed., 1967, P&B, 6:5992. President Wright's comments about searches are in the "Minutes" cited in nn, 38 and 52 above. Two women and Smith's first male graduate student earned doctorates in botany at Smith between 1947 and 1963 (USOE, *Earned Degrees Conferred*, annually 1947–48 through 1962–63).

56. On Williams, see *AMS*, 11th ed., 1968, P&B, 6:5877. Six years after leaving Smith she became a program assistant at the NSF, and in 1960 she became the assistant program director for optical astronomy there. Her departure was not mentioned in the alumnae magazine. Alice Farnsworth, Mount Holyoke's longtime professor of astronomy, may not have been replaced at all, as astronomy was henceforth to be taught at Amherst as part of the new "Four College" consortium. See Helen Sawyer Hogg, "Alice H. Farnsworth, Professor of Astronomy," *Mount Holyoke Alumnae Quarterly* 43 [summer 1959]: 69–70).

57. In the end Smith hired a woman with a master's degree, Waltraut Seitter, not in *AMS* (Benjamin F. Wright to Bart J. Bok, Mar. 8, 1955, box 5, Bart J. Bok Papers, HUA. This collection, has under various headings, a lot of other material on the situation at Smith in particular and Harvard-related women astronomers in general). Benjamin Wright is mentioned in "Minutes of the Seventh Meeting of the Study Committee," Nov. 3, 1953, 1. On Constance Sawyer, see *AMWS*, 15th ed., 1982, 6:441; and Bart Bok to

Marjorie Williams, Sept. 19, 1945, box Wic–Z, Harvard College Observatory, Director's Correspondence, HUA. On James Warwick, see ibid., 7:434. See also Bart J. Bok, "Graduate Training in Astronomy," *Science* 154 (1966): 590, 592. Bok's wife Priscilla had been an astronomer at Smith (see obituary, *Sky and Telescope* 51 [Jan. 1976]: 25; and box 19, *NAWP*, SLRC). See also David H. Levy, *The Man Who Sold the Milky Way: A Biography of Bart Bok* (Tucson: Univ. of Arizona Press, 1993).

58. FAE, *A Report for 1954–56* (New York, 1957), 39–41 and financial data, 118–19; "Interim Report for the Fund for the Advancement of Education's Program for Utilization of College Teaching Resources, June 30, 1957," and Helen Randall, "Smith College: Final Report on a Study of the Feasibility of Using Competent Liberal Arts Graduates in the Community for Teaching Assistance," Aug. 30, 1957, both in carton 4, f. EC-87 Teaching, ACE-CEW Papers, SLRC; *Better Utilization of College Teaching Resources, A Summary Report: A Report by the Committee on Utilization of College Teaching Resources, May 1959* (New York: FAE, [1959]).

59. See Kate Millett, *Token Learning* (New York: NOW, 1968); Fred M. Hechinger, "In the College Presidency, More and More a Man's World," *NYT,* Aug. 24, 1969, 9; Doris L. Pullen, "The Educational Establishment: Wasted Women," in *Voices of the New Feminism,* ed. Mary Lou Thompson (Boston: Beacon, 1970), 115–35; Caroline Bird, "Women's Colleges and Women's Lib," *Change* 4, no. 3 (Apr. 1972): 60–65; and, e.g., Frederic O. Musser, *The History of Goucher College, 1930–1985* (Baltimore: Johns Hopkins Univ. Press, 1980). Probably the strongest feminist on the faculty of a woman's college after the retirements of Parshley and Sampson of Smith was sociologist Mirra Komarovsky of Barnard (Shulamit Reinharz, "Finding a Sociological Voice: The Work of Mirra Komarovsky," *Sociological Inquiry* 59 [1989]: 374–95).

Chapter 11 Nonprofit Institutions and Self-Employment: A Second Chance

1. The women's proportion would have been even higher if the National Register's statistics had been limited to just this kind of "nonprofit" institution, but it also included another sort, invented by the military-industrial complex after World War II to "contract out" lucrative Air Force and other contracts for research and development (see n. 12).

2. Alexander Bunts and George Crile, *To Act as a Unit: The Story of the Cleveland Clinic* (Cleveland: Cleveland Clinic Foundation, 1971). On Brown, see *AMWS,* 15th ed., 1982, 1:723; on Lewis, ibid., 16th ed., 1986, 4:743; and on Dustan, ibid., 15th ed., 1982, 2:762, as well as the oral history by Regina Morantz Sanchez, c. 1977, in the Special Collections on Women in Medicine, Florence A. Moore Library, Medical College of Pennsylvania, Philadelphia.

3. [Elizabeth K. Patterson], "Early History of the Institute for Cancer Research, 1927–1957," and Timothy R. Talbot, Jr., "Recent History of the Institute for Cancer Research, 1957–1976," in the ICR's *Scientific Report* 23 (1977–78), fiftieth anniversary issue, 14–35 and 36–51, respectively. On Medes, see Ethel Echternach Bishop, "Grace Medes, 1886–1967," in *American Chemists and Chemical Engineers,* ed. Wyndham D. Miles (Washington, D.C.: ACS, 1976), 330–31; "Grace Medes: Garvan Medal," *C&EN* 33 (1955): 1515; and obituary, ibid. 46 (1968): 68. On Mintz, see *AMWS,* 15th ed., 1982,

5:410; and Jean L. Marx, "Tracking Genes in Developing Mice," *Science* 215 (1982): 44–47. On Glusker, see *AMWS*, 16th ed., 1986, 3:169; "Garvan Medal," *C&EN* 56 (1978): 43–44; and Jenny P. Glusker, "Some Aspects of Crystallography in Cancer Research in North America," in *Crystallography in North America,* ed. Dan McLachlan, Jr., and Jenny P. Glusker (New York: American Crystallographic Association, 1983), 404–9. Many ICR and other crystallographers contributed their recollections to *Patterson and Pattersons: Fifty Years of the Patterson Function,* ed. Jenny P. Glusker, Betty K. Patterson, and Miriam Rossi (Oxford: Oxford Univ. Press for the International Union of Crystallography, 1987); see esp. 687 and 688.

4. Icie Macy Hoobler, *Boundless Horizons: Portrait of a Pioneer Woman Scientist* (Smithtown, N.Y.: Exposition Press, 1982); Icie G. Macy et al., *The Composition of Milks,* NRC *Bulletin* 119 (1950) and 254 (1953); Harold H. Williams, "Icie Gertrude Macy Hoobler (1892–1984): A Biographical Sketch," *Journal of Nutrition* 114, no. 8 (1984): 1351–62; Sheldon J. Kopperl, "Icie Macy Hoobler: Pioneer Woman Biochemist," *JCE* 65 (1988): 97–98. Hoobler's extensive papers are at the Bentley Historical Library, University of Michigan, Ann Arbor. The unique Children's Fund of Michigan was set up in 1931 by a former senator who had been an orphan and sought a way to repay the people of Michigan for his early upbringing. His stipulation that the Fund be entirely spen in twenty-five years made it difficult for Hoobler to hire staff in the later years.

5. On the Henles, see *AMS,* 11th ed., 1966, 3:2236; his obituary, *NYT,* July 8, 1987, D-23; and Michael B. Shimkin, *Contrary to Nature* (Washington, D.C.: GPO, 1977), 381–82.

6. On Kuttner, see *AMS,* 11th ed., 1966, 3:2953; Rebecca Lancefield, "Foreword to the Second Edition," in *Rheumatic Fever,* by Milton Markowitz and Leon Gordis (Philadelphia: Saunders, 1972); there is also some Kuttner correspondence in box 3, f. 15, and box 6, f. 2, of the Rebecca Lancefield Papers, RAC, North Tarrytown, N.Y.

7. Clement A. Smith, *The Children's Hospital of Boston: "Built Better Than They Knew"* (Boston: Little, Brown, 1983), chs. 14, 15. On Cohen, see *AMWS,* 15th ed., 1982, 2:278.

8. On Smith, see obituaries in *Radcliffe Quarterly* 69, no. 3 (Sept. 1983): 33, and *Harvard Magazine* 85 (July–Aug. 1983): 95; Olive Watkins Smith, "The Men in My Life," *Radcliffe Quarterly* 67, no. 2 (June 1981): 25–27; and Robert Meyers, *DES: The Bitter Pill* (New York: Seaview/Putnam, 1983), chs. 5, 6.

9. Lawrence J. Friedman, *Menninger: The Family and the Clinic* (New York: Knopf, 1990), has a lot on the strange ways women, staff as well as patients, were treated at the Menninger Sanitarium. See also Margaret Mead, "Institutional Expressions of Ideas: The Menninger Foundation and Its Associated Institutions, 1943–1977," in *The Human Mind Revisited: Essays in Honor of Karl A. Menninger,* ed. Sydney Smith (New York: International Universities Press, 1978), 475–92; and Herbert C. Modlin, M.D., "Contributions of the Menninger School of Psychiatry to Psychiatric Education," *Bulletin of the Menninger Clinic* 47, no. 3 (May 1983): 253–61.

10. On Escalona, see *AMWS,* 13th ed., S&B, 1978, 352; and Sibylle Escalona and Grace Moore Heider, *Prediction and Outcome: A Study in Child Development,* Menninger Clinic Monograph Series, 14 (New York: Basic Books, 1959), 17–24. On Heider, see *AMWS,* 12th ed., S&B, 1973, 1:982; and Fritz Heider, *The Life of a Psychologist: An*

Autobiography (Lawrence: Univ. of Kansas Press, 1983). On the Murphys, see *AMWS*, 12th ed., S&B, 1973, 2:1773, 1774; Lois Barclay Murphy, "Roots of an Approach to Studying Child Development," in *The Psychologists: Autobiographies of Distinguished Living Psychologists,* ed. T. S. Krawiec, vol. 3 (Brandon, Vt.: Clinical Psychology, 1978), 167–80; and idem, "Lois Barclay Murphy," in *Models of Achievement: Reflections of Eminent Women in Psychology,* ed. Agnes N. O'Connell and Nancy Felipe Russo, vol. 1 (New York: Columbia Univ. Press, 1983), 88–107. There is considerable correspondence relating to the Menninger Foundation in the Lillian Gilbreth Papers in Special Collections, Purdue University Library, West Lafayette, Indiana, and in MMP. Both Gilbreth and Mead were frequent visitors to Topeka in the 1950s and 1960s. Less well known but also affiliated with the Menninger Foundation was clinical psychologist Margaret Brenman-Gibson (*AMWS,* 13th ed., S&B, 1978, 144), who listed Negro girls as one of her specialties and published a short autobiography, "The War on Human Suffering: A Psychoanalyst's Research Odyssey," in Smith, *The Human Mind Revisited,* 99–137. Several of the individuals interviewed by Milton Senn in his series on the history of child development mentioned the Menninger Foundation.

11. Sue V. Rosser, *Female-Friendly Science* (New York: Pergamon, 1990), 106–9. The bag lady had worked for the Westinghouse Talent Search in the forties, and the unidentified university town may have been Madison, Wisconsin.

12. See, e.g., George A. W. Boehm, *Science in the Service of Mankind: The Battelle Story* (Lexington, Mass.: Lexington Books, 1972); Claude Baum, *The System Builders: The Story of SDC* (Santa Monica: System Development Corporation, 1981), for Beatrice Rome (59) and Sally Bowman (92, 117); Weldon B. Gibson, *SRI: The Founding Years,* vols. 1, *The New Frontier,* and 2, *The Take-Off Days: The Right Moves at the Right Times* (Los Altos, Calif.: Publishing Services Center, 1980 and 1986). The latter pictured Shirley Radding and Margaret Ray in photographs (54, 193), but since they are not in *AMWS,* it is not clear whether they are scientists or assistants. An earlier survey (1960) of mathematics personnel, which was broader than its title, found that 224 of 9,815 women mathematicians were employed in "non-profit institutions," where their median salary was $9,400 (compared with the overall median of $8,500), partly because of the high proportion of doctorates (20.5 percent) (NSF with BLS and Mathematical Association of America, *Employment in Professional Mathematical Work in Industry and Government: Report on a 1960 Survey* [Washington, D.C.: GPO, 1962], 1, 14, 29).

13. On Ramsey and Rawles, see James D. Ebert, "Report of the President," *Carnegie Institution of Washington Year Book* 82 (1982–83): 4; and *AMWS,* 16th ed., 1986, 6:41, and 12th ed., 1972, 5:5128. On Donnay, see *AMWS,* 15th ed., 1982, 2:682; Gabrielle Donnay, "Crystallography: Fifty Years of X-ray Crystallography at the Geophysical Laboratory, 1919–1969," in McLachlan and Glusker, *Crystallography in North America,* 37–41 (reprinted with additions and corrections from *Carnegie Institution of Washington Year Book* 68 [1970]: 278–83); and reminiscences in Glusker, Patterson, and Rossi, *Patterson and Pattersons,* 663–66. On Rubin, see *AMWS,* 17th ed., 1989, 6:350; Sally Stephens, "Vera Rubin: An Unconventional Career," *Mercury* 21 (Jan.–Feb. 1992): 38–45; and Vera C. Rubin, "Vera C. Rubin," in *Origins: The Lives and Worlds of Modern Cosmologists,* ed. Alan Lightman and Roberta Brawer (Cambridge: Harvard Univ. Press, 1990), 285–305. On Witkin, see *AMWS,* 15th ed., 1982, 7:688; and Sharon Schlegel, "Witkin Stalks

Cancer Like a Sleuth," *Rutgers Today* 4, no. 2 (spring 1981): 4 (copy in Biographical Files, New York State Department of Health Library, Albany).

14. See obituaries in *NYT,* Nov. 29, 1980, p. 28, and *Physics Today* 34 (Mar. 1981): 88; "Interview with Dr. Henrietta Swope, by Dr. David DeVorkin," Aug. 3, 1977, transcript at NBL. The Henrietta Swope Papers are at the SLRC.

15. See Erik Eckholm, "Secrets of Maya Decoded at Last, Revealing Darker Human History," *NYT,* May 13, 1986, C-1, C-3. On Proskouriakoff, see obituary, ibid., Sept. 11, 1985, D-27; and Christopher Jones, "Tatiana Proskouriakoff, 1909–1985," *APS Yearbook 1985* (Philadelphia, 1986), 176–80. Her 1960 breakthrough came after the closing down of the CIW's Department of Archaeology, in 1957–58, after more than fifty years of operation (*Carnegie Institution of Washington Year Book* 57 [1957–58]: 22–25, 434–45).

16. McClintock to Beadle, May 8, 1945, Biology Division Papers, California Institute of Technology Archives, Pasadena.

17. McClintock to Bush, Mar. 16, 1953, second letter of that date, and her reply, Mar. 19, 1953, both in Department of Genetics File, Carnegie Institution of Washington, Washington, D.C. There is also some correspondence from and about McClintock for May–July 1953 in the George W. Beadle Papers, Biology Division Papers, California Institute of Technology, such as Beadle to Bush, May 25, 1953, Beadle's reply, May 27, 1953, Demerec to Beadle, June 2, 1953, and McClintock to Beadle, July 9, 1953 (about McClintock's upcoming trip to Pasadena). On McClintock, see *AMWS,* 15th ed., 1982, 22; obituary, *NYT,* Sept. 4, 1992, A-1, D-16; Roger Lewin, "A Naturalist of the Genome," *Science* 222 (1983): 402–5; Nina Fedoroff and David Botstein, eds., *The Dynamic Genome: Barbara McClintock's Ideas in the Century of Genetics* (Cold Spring Harbor, N.Y.: Cold Spring Harbor Press, 1992); Evelyn Fox Keller, "McClintock's Maize," *Science 81,* no. 8 (Oct. 1981): 54–59; and idem, *A Feeling for the Organism: The Life and Work of Barbara McClintock* (San Francisco: W. H. Freeman, 1980).

18. On Russell, see *AMWS,* 15th ed., 1982, 6:350; "Of Mice and AAUW Women," *AAUWJ* 47 (1954): 254; Elizabeth S. Russell, "A History of Mouse Genetics" (semi-autobiographical), *Annual Review of Genetics* 19 (1985): 1–28; and Edna Yost, *Women of Modern Science* (New York: Dodd, Mead, 1959), ch. 4. Russell's then husband William also worked at the Jackson Laboratory for a time (see *AMWS,* 17th ed., 1989, 6:374). On Green, see ibid., 13th ed., 1976, 2:1620. See also Jean Holstein, *The First Fifty Years at the Jackson Laboratory,* ed. William L. Dupuy [1929–1979] (Bar Harbor: n.p., 1979]); and Claire Fulcher, "In Memoriam [Margaret Dickie]," *AAUWJ* 63, no. 1 (Oct. 1969): 25.

19. See S. E. A. McCallan, "Lela Viola Barton, 1901–1967," *Contributions from the Boyce Thompson Institute* 24 (1967): 1–2; and John A. Small, "Lela Viola Barton, 1901–1967," *Bulletin of the Torrey Botanical Club* 95 (1968): 103–10, with a lengthy bibliography. See also William Crocker, *Growth of Plants: Twenty Years of Research at the Boyce Thompson Institute* (New York: Reinhold, 1948). Barton's lifelong friend Clyde Chandler (female) joined the staff in 1948. On Pfeiffer, see obituary, *NYT,* Sept. 12, 1989, B-10.

20. On the Lewises, see *AMS,* 10th ed., 1961, 3:2411, 2413. See also Margaret P. O'Neill Davis, "Wistar Institute of Anatomy and Biology," in *Research Institutions and Learned Societies,* ed. Joseph C. Kiger (Westport, Conn.: Greenwood, 1982), 481–85. The Warren H. Lewis Papers are in the library of the APS.

21. On McCulloch, see Erin O'Malia, "Emeritus Professor McCulloch Dies at 101,"

Daily Trojan (University of Southern California), June 10, 1987, 1. (I thank Dorothy Soule, director of the Irene McCulloch Foundation and herself a notable Hancock biologist, for this and several other clippings about McCulloch.) On Hartman, see *AMS,* 7th ed., 1944, 757; and correspondence between John Garth and Waldo Schmitt, Apr.–May 1966, nominating her for several awards. See also Richard C. Brusca, "The Allan Hancock Foundation of the University of Southern California," *A[ssociation of] S[ystematics] C[ollections] Newsletter* 8, no. 1 (Feb. 1980): 1–7, plus list of errata by John S. Garth. (I thank John Garth of the Hancock Foundation for sending me copies.)

22. On Sears, see *AMWS,* 14th ed., 1979, 5:4545; and Mary Sears and Daniel Merriman, eds., *Oceanography: The Past* (New York: Springer-Verlag, 1980), 22, 43, 44n, 505. On Bunce, see *AMWS,* 16th ed., 1986, 1:808; and oral history at Archives, Marine Biology Laboratory, Woods Hole, Mass. On Simpson, see Edward F. Taylor, "Joanne Simpson: Pathfinder for a Generation," *Weatherwise* 37 (Aug. 1984): 182–83, 206–7; Joanne Simpson, "Meteorologist," *Successful Women in the Sciences: An Analysis of Determinants,* ed. Ruth B. Kundsin, *Annals of the New York Academy of Sciences* 208 (1973): 41–45; and "The *Bulletin* Interviews: Dr. Joanne Simpson," *W[orld] M[eteorological] O[rganization] Bulletin,* 35, no. 1 (Jan. 1986): 3–14. There is a small collection of her papers with more biographical material at SLRC. Her full name was Joanne Gerould Starr Malkus Simpson.

23. On Wood, see *AMS,* 11th ed., 1967, 6:5958; "Elizabeth A. Wood" (biographical statement), Bell Laboratories, Whippany, N.J. (I thank Deirdre La Porte for a copy); "Citations for 1970," *American Journal of Physics* 38 (1970): 683; Elizabeth A. Wood, "Crystallography at Bell Laboratories," in McLachlan and Glusker, *Crystallography in North America,* 61–66; idem, "Address at ACA Meeting, 1975," ibid., 170–73 (reminiscences); and idem, "Memorial to Ida Helen Ogilvie," GSA *Bulletin* 75 (Feb. 1964): 35–39. Also at Bell Labs were mathematician Florence MacWilliams, a pioneer in digital coding in telecommunications from 1957 to 1982 (obituary, *NYT,* May 31, 1990, D-23), and Vera Compton (not in *AMWS* but cited in interview with Margaret Geller in Lightman and Brawer, *Origins,* 360).

24. On Goldman, see *NAWMP;* and [Juliet A. Mitchell, ed.], *A Community of Scholars: The Institute for Advanced Study, Faculty and Members, 1930–1980* (Princeton: Institute for Advanced Study, 1980), 26.

25. On Wallace, see *AMWS,* 13th ed., 1978, S&B, 1244; obituary, *NYT,* Jan. 13, 1993, A-19. Schwartz is not listed in the *AMWS* or *WWAW,* but see the introduction to Michael D. Bardo, ed., *Money History and International Finance: Essays in Honor of Anna J. Schwartz* (Chicago: Univ. of Chicago Press, 1989); and Anna J. Schwartz, *Money in Historical Perspective* (Chicago: Univ. of Chicago Press, 1987). (I thank Geoffrey Carliner of the National Bureau of Economic Research for his assistance.)

26. In astronomy and natural history women volunteers have been the backbone of certain organizations. See Marguerite Ainley, "D'assistantes anonymes a chercheures scientifiques: Une rétrospective sur la place des femmes en sciences," *Cahiers de recherche sociologique* 4 (Apr. 1986), 55–71. See also Mary Desmond Rock, *Museum of Science, Boston: The Founding and Formative Years, The Washburn Era, 1939–1980* (Boston: Museum of Science, 1989), ch. 14 and appendix.

27. Yet Mead and the AMNH in a way suited each other: it made relatively few

demands upon her, and she brought it a certain amount of publicity. Jane Howard captures this tradeoff well in *Margaret Mead: A Life* (New York: Simon & Schuster, 1984), 100–101, 336–37.

28. On Hyman, see *NAWMP;* Rachel D. Fink, "Hyman, Libbie Henrietta," in *Dictionary of Scientific Biography* 17 (1990): 442–43; "Scientists in the News," *Science* 131 (1960): 1599; Horace W. Stunkard, "In Memoriam, Libbie Henrietta Hyman, 1888–1969," in *Biology of the Turbellaria,* ed. Nathan W. Riser and M. Patricia Morse (New York: McGraw-Hill, 1974), ix–xii; William K. Emerson, "Bibliography of Libbie H. Hyman," ibid., xv–xxv; and Roman Kenk, "History of the Study of Turbellaria in North America," ibid., 17–22. The *Journal of Biological Psychology* 12, no. 1 (1970), was entirely devoted to her memory. There are a few letters to and from Hyman in the Elisabeth Deichmann Papers at the Museum of Comparative Zoology Archives at Harvard University. Deichmann, in a letter to Aida Martinez of Venezuela, May 13, 1967, praised Hyman's *Echinodermata* as "one of the most valuable books which has been published in this century." Her review in *Science* 123 (1956): 592 was only slightly less effusive.

29. On Bliss, see *AMWS,* 16th ed., 1986, 1:561; obituary, *NYT,* Jan. 2, 1988, 28; *Radcliffe Quarterly* 74, no. 2 (1988); *Harvard Magazine* 90 (May–June 1988): 129; and Linda H. Mantel, "Dorothy E. Bliss (1916–1987)," *Journal of Crustacean Biology* 9 (1988), 706–9. On Weitzner, see *NYT,* Apr. 9, 1988, 12; and Geoffrey Hellman, *Bankers, Bones, and Beetles: The First Century of the American Museum of Natural History* (Garden City, N.Y.: Natural History Press, 1969), 228, 242. On Messina, see Brooks F. Ellis, "Memorial to Angelina Rose Messina, 1910–1968," *Proceedings Volume of the Geological Society of America, Inc., for 1968* (Boulder, Colo.: GSA, 1971), 212–14. On Salmon, see Miriam E. Phelps, "Memorial to Eleanor Seely Salmon, 1910–1984," GSA *Memorials* 17 (1987).

30. On Wiley, see obituary, *NYT,* Nov. 19, 1986, D-27, and Hellman, *Bankers, Bones, and Beetles,* 220 and 253; on Heilbrun, see *NYT,* Oct. 28, 1987, B-11; and on Shaw, *AMWS,* 14th ed., 1979, 6:4604. See also Douglas J. Preston, *Dinosaurs in the Attic: An Excursion into the American Museum of Natural History* (New York: St. Martin's, 1986).

31. On Naumberg, see *NCAB* 41 (1956): 500–501; and Hellman, *Bankers, Bones, and Beetles,* 212–13. The first woman on the AMNH's board of trustees was Millicent Carey McIntosh, dean of Barnard College in 1952 (*AAUWJ* 45 [1952]: 167). In 1965 Laura Cabot Hodgkinson, a Smith College alumna and a past chairman of Smith's board of trustees, as well as a prominent fund-raiser, was the first woman elected president of the Boston Museum of Science—just in time to head its $8 million expansion campaign ("The Smith Medal," *SAQ* 57 [1965–66]: 22; Rock, *Museum of Science, Boston,* 157–58).

32. Constance Holden, "Ruth Patrick: Hard Work Brings Its Own (and Tyler) Reward," *Science* 188 (1975): 997–99; Bill Mandel, "Ralph Nader of Water Pollution Hasn't Given Up Yet," *Philadelphia Inquirer,* Feb. 23, 1975, reprinted in *Biography News* 2 (Mar.–Apr. 1975): 407; Marion Steinmann, "Rivers of America: The Source is Ruth Patrick," *R[ockefeller] F[oundation] Illustrated,* June 1983, 14–16. See also Jan Schmitz, "Women in Science [at the Academy of Natural Sciences]," *Academy News* 7 (fall 1984): 5–7. The Carnegie Museum of Natural History in Pittsburgh also had two women in the junior ranks of its curatorial staff: ornithologist Mary Heimerdinger Clench, Ph.D., as research assistant and then assistant and associate curator (*AMWS,* 16th ed., 1986,

2:265), and herpetologist Olive Brown Goin, with a master's degree, who served as assistant to the curator in the museum's Laboratory of Mammalogy (Olive Brown Goin, "The Marked Frog," *WAM* 50 [1966]: 21, 42).

33. See Margaret Titcomb, "The Bishop Museum and Library," *Special Libraries* 44 (1953): 398–400. On Neal, see *AMS,* 10th ed., 1961, 3:2935; Edwin H. Bryan, Jr., "Marie Catherine Neal," *Bulletin of the Torrey Botanical Club* 93 (1966): 199–200; and Constance E. Hartt and Barbara Peterson, "Marie Catherine Neal," in *Notable Women of Hawaii,* ed. by Barbara Bennett Peterson (Honolulu: Univ. of Hawaii Press, 1984), 282–85. On Titcomb, see Cynthia Timberlake, "Margaret Titcomb," ibid., 381–84.

34. On Eastwood, see *NAWMP;* and on McClintock, see *AMWS,* 18th ed., 1992, 5:26, and Carol Holleuffer, "Elizabeth McClintock, California Academy of Sciences Curator, Ornamental Plant Specialist," 1985, Regional Oral History Office, Bancroft Library, University of California, Berkeley.

35. William C. Steere, "Research and Education at the New York Botanical Garden," in *Current Topics in Plant Science,* ed. James E. Gunckel (New York: Academic, 1969), 253–61. On Allen, see *AMS,* 11th ed., 1965, 1:59; on Hervey, *AMWS,* 14th ed., 1979, 3:2151; on Anchel, ibid., 17th ed., 1989, 1:116; on Barksdale, Lindsay S. Olive, "Alma Whiffen Barksdale, 1916–1981," *Mycologia* 74 (1982): 359–62; and on Hall, Linda Yang, "Got a Question? Just Ask Miss Hall," *NYT,* Mar. 31, 1988, C-12, and obituary, ibid., Apr. 21, 1989, A-19. See also John F. Reed, "The Library of the New York Botanical Garden," *Garden Journal* 19 (May–June 1969): 77–88; and, more generally, on women science librarians, abstractors, editors, and the like, Margaret W. Rossiter, "Women and the History of Scientific Communication," *Journal of Library History* 21 (1986), 39–59..

36. On Horner and Lange, see Biographical File, Missouri Botanical Garden Archives, St. Louis, Missouri. (I thank Jane Miller for much help in St. Louis.)

37. See "The Smith Medal," *SAQ* 56 (1964–65): 24; and Linda Yang, "The Name of the Rose is Elizabeth Scholtz," *NYT,* June 23, 1988, C-12.

38. On Crane, see *AMS,* 11th ed., 1965, 1:1031; Yost, *Women of Modern Science,* ch. 8; and Jocelyn Crane, *Fiddler Crabs of the World (Ocypodidae: Genus Uca)* (Princeton: Princeton Univ. Press, 1975). On Beebe, see *AMS,* 10th ed., 1960, 1:249; and Robert Henry Welker, *Natural Man: The Life of William Beebe* (Bloomington: Indiana Univ. Press, 1975), esp. 145–46, 150–51, 160, 164, 211. (Beebe's first wife, Mary Blair Beebe, better known as Blair Niles, was a founder of the SWG.) See also William Bridges, *Gathering of Animals: An Unconventional History of the New York Zoological Society* (New York: Harper & Row, 1974), 427–30 and ch. 26. On Griffin, see *AMWS,* 17th ed., 1989, 3:327.

39. On Benchley, see *Who Was Who in America* 7 (1977–81): 44; *CBY, 1940,* 68–70; Victor Boesen, "The Zoo Director Is a Lady," *Coronet,* Dec. 1952, 30–33; and Margaret Poynter, *The Zoo Lady: Belle Benchley and the San Diego Zoo* (Minneapolis: Dillon, 1980). Benchley also wrote several autobiographical works about her animals: *My Friends: The Apes* (Boston: Little, Brown, 1942); *My Life in a Man-Made Jungle* (Boston: Little, Brown, 1943); and *My Animal Babies* (Boston: Little, Brown, 1943). See also Bernard Livingston, *Zoo: Animals, People, Places* (New York: Arbor House, 1974), 99; and Joan Morton Kelly, "Bringing Up Gorillas," *WAM* 37 (1952–53): 211–14. Another woman scientist at a city zoo was Evelyn Tilden, curator of laboratories at the Chicago Zoologi-

cal Park from 1954 until 1963 (*AMS*, 11th ed., 1967, 6:5413). Some of her earlier work as an assistant to Hideyo Noguchi of the Rockefeller Institute is described in Isabel R. Plesset, *Noguchi and His Patrons* (Rutherford, N.J.: Fairleigh Dickinson Univ. Press, 1980), 173–75, 217, 225–28.

40. On Sears, see above, n. 22; on Shane, see Elizabeth Spedding Calciano, "The Lick Observatory: Oral History Interview with Mary Lea Heger Shane," 1969, 227–33, Archives, University of California, Santa Cruz, copy at Bancroft Library, Berkeley.

41. Harriet L. Rheingold, "The First Twenty-Five Years of the SRCD," *Monographs of the Society for Research in Child Development* 50 (1985): 138–39.

42. "In Memoriam: Margaret Kuenne Harlow," *Child Development* 42 (1971): 1313–14; "Memorial Resolution of the Faculty of the University of Wisconsin on the Death of Professor Margaret Kuenne Harlow," University of Wisconsin Faculty Document 79, [1 Nov. 1971], in box 26, f. 88, *NAWP*, SLRC. Similarly, at the Association of American Geographers, a woman became its first paid employee, an "office manager," in 1950. Her duties increased, but because the association's officers felt unable to pay her a full-time salary, she resigned. By 1963, however, the rapidly growing association had hired an "executive officer" (a male Ph.D.) and four assistants, including two full-time women, one with the title "office manager" (Preston E. James and Geoffrey J. Martin, *The Association of American Geographers: The First 75 Years, 1904–79* [n.p.: Association of American Geographers, 1978], 122–23).

43. On Fairchild, see *WWAW* 7 (1972–73): 262, and *AMWS*, 12th ed., S&B, 1973, 1:659; on Hyslop, *WWAW* 7 (1972–73): 435; on Creagh, *Foremost Women in Communication* (New York: Foremost Americans Publishing, 1970), 142, and Edwin B. Eckel, *The Geological Society of America: Life History of a Learned Society,* GSA Memoir 155 (Boulder, Colo.: GSA, 1982): x, 38, 54, 57–58, 110; and on Clabaugh, Patricia Sutton Clabaugh, "Women and Science . . . Yesterday, Today and Tomorrow," *SAQ* 58 (1967): 4–5. For the many women on the staff of *Science,* see Dael Wolfle, *Renewing a Scientific Society: The American Association for the Advancement of Science from World War II to 1970* (Washington, D.C.: AAAS, 1989), 73–80, 86; on those at the ACS, see Herman Skolnik and Kenneth M. Reese, eds., *A Century of Chemistry: The Role of Chemists and the American Chemical Society* (Washington, D.C.: ACS, 1976), 108.

44. On Taeuber, see *AMWS*, 12th ed., S&B, 1973, 2:2425; and on Hyslop, *WWAW* 7 (1972–73): 435. On Atherton, see ibid., 27; Pauline A. Atherton, "Toward National Information Networks, Part 5. An Action Plan for Indexing," *Physics Today* 19 (Jan. 1966): 58, 60; and Eugene Garfield, "The ASIS Outstanding Information Science Teacher Award: Pauline Atherton Cochrane Wins the Second Award," *Current Contents,* no. 49 (Dec. 7, 1981), reprinted in his *Essays of an Information Scientist,* vol. 5 (Philadelphia: ISI, 1981–82), 331–34. On Lerner, see *AMWS*, 17th ed., 1989, 4:713.

45. Rita G. Lerner, "The Professional Society in a Changing World," *Library Quarterly* 54 (1984): 36–47; and Skolnik and Reese, *A Century of Chemistry,* ch. 4.

46. Charles A. Browne and Mary E. Weeks, *A History of the American Chemical Society* (Washington, D.C.: ACS, 1952), 336–67; E. J. Crane, ed., *CA Today: The Production of Chemical Abstracts* (Washington, D.C.: ACS, 1958); Skolnik and Reese, *A Century of Chemistry,* 126–43; Dale B. Baker, Jean W. Horiszny, and Wladyslaw V. Metanomski, "History of Abstracting at Chemical Abstracts Service," *Journal of Chemical Information*

and Computer Science 20 (1980): 193–201. Although Crane professed to believe that women offered special talents for abstracting (E. J. Crane, "Women Chemists" [letter to the editor], *C&EN* 25 [1947]: 394), few of the longtime volunteers were women (idem, "Chemical Abstracts," in Browne and Weeks, *A History of the American Chemical Society,* 362).

47. On Magill, see *AMWS,* 12th ed., 1972, 4:4023; "CA's Mary Magill Honored," *Journal of Chemical Documentation* 4 (Apr. 1964): ii; and Gail Tabor, "Outstanding Career Woman of 1963: Success in a 'Man's Field,'" *Columbus (Ohio) Citizen-Journal,* Feb. 3, 1964, 1, 9. E. J. Crane explained the "special difficulties" of organic indexing in *CA Today,* 67f.

48. On Parkins, see *AMWS,* 12th ed., 1972, 5:4779. See also William C. Steere, *Biological Abstracts/BIOSIS: The First Fifty Years of a Major Science Information Service* (New York: Plenum, 1976), xi, 132–34. Far less happy and less successful at *Biological Abstracts* was Nellie Payne, who worked there intermittently in the 1920s and 1930s. On Payne, see *AMS,* 12th ed., 1972, 5:4819, and some of her correspondence in the Entomology and Economic Zoology Papers of the Division of Agriculture, University of Minnesota Archives, Minneapolis. Payne preferred research work to library and literature tasks, which were nearly invisible and lacked promotions, but she was unable to obtain an academic post, perhaps because one confidential letter of recommendation of 1927 reported that she was not very careful about her personal appearance.

49. Edmund R. Arnold, "The American Geographical Society Library, Map and Photograph Collection: A History, 1951–1978" (Ph.D. diss., University of Pittsburgh, 1985). For the earlier years, see Margaret W. Rossiter, *Women Scientists in America: Struggles and Strategies to 1940* (Baltimore: Johns Hopkins Univ. Press, 1982), 259–60, and occasional items in the Isaiah Bowman Papers, Special Collections, Eisenhower Library, The Johns Hopkins University, Baltimore, Md. See also Sandra E. Belanger, "History of the Library of the Marine Biological Laboratory, 1888–1973," *Journal of Library History* 10 (1975): 260; and Mary Sears, "Memorial: Priscilla Braislin Montgomery," *Biological Bulletin* 113 (1957): 9–10.

50. Keturah E. Baldwin, *The AHEA Saga* (Washington, D.C.: AHEA, 1949); Mary Hawkins, ed., *The American Home Economics Association, 1950–1958: A Supplement to the AHEA Saga by Keturah Baldwin* (Washington, D.C.: AHEA, 1959); Frances Zuill, "Fifty Years of Achievement: The American Home Economics Association," *JHE* 51 (1959): 519–25; Helen Pundt, *AHEA: A History of Excellence* (Washington, D.C.: AHEA, 1980). But the most informative items about events at headquarters were the occasional pieces in the *JHE* about the comings and goings of major staff personnel, e.g., "Field Secretary and Editor for AHEA," *JHE* 40 (1948): 367, and "AHEA Appoints New Field Secretary," ibid. 47 (1955): 346.

51. On Riley, see *WWAW* 14 (1985–86): 673. She was also a professor at Rutgers at the time.

52. See Lucy Rathbone, "Introducing AHEA's New Executive Secretary," *JHE* 39 (1947): 66; "Mildred Horton, Executive Secretary, 1947–1960," ibid. 52 (1960): 811–12; "Jane L. Rees to be AHEA Executive Director," ibid. 57 (1965): 758; "Doris Hanson Named Executive Director," ibid. 59 (1967): 340; and *Ninth International Congress on Home Economics, Held at the University of Maryland, College Park Maryland, USA, July 28–*

August 2, 1958 (Washington, D.C.: AHEA, 1958). See also Hawkins, *American Home Economics Association,* 5–9.

53. On Creagh, see items in n. 43 above. Edwin B. Eckel, *The Geological Society of America,* ch. 7, offers a full account of the staff's duties and responsibilities at headquarters. This is rare in histories of scientific societies. Other women executives of scientific societies included Ernesta Drinker Ballard, director of the Pennsylvania Horticultural Society (90 percent men) ("Horticulturalist," in *Successful Women in the Sciences: An Analysis of Determinants,* ed. Ruth B. Kundsin, *Annals of the New York Academy of Sciences* 208 [1973]: 32–36) and Catherine Borras, longtime "secretary" of the AAAS (but not in any biographical directories).

54. Simon Baatz, *Knowledge, Culture, and Science in the Metropolis: The New York Academy of Sciences, 1817–1970, Annals of the New York Academy of Sciences* 584 (1990): ch. 6.

55. For a list of social scientists who moved from wartime social science projects to positions as foundation executives, see Jean M. Converse, *Survey Research in the United States: Roots and Emergence, 1890–1960* (Berkeley and Los Angeles: Univ. of California Press, 1987), 235–36. See also Carol Brown, "Sexism and the Russell Sage Foundation," *Feminist Studies* 1 (1972): 25–44.

56. On Anderson, see *WWAW* 8 (1974–75): 20; obituary, *NYT,* Dec. 20, 1985, D-26; and Isabel Grossner, "Carnegie Corporation Project—Florence Anderson," 1966 and 1967, Oral History of Florence Anderson, Columbia University Oral History Research Office, New York. Anderson, an art history major at Mount Holyoke (class of 1931), served in the Marine Corps during World War II and worked for the Carnegie Corporation from 1934 until the mid-1970s.

57. On Paschal, see *AMWS,* 13th ed., S&B, 1978, 923; *WWAW* 6 (1970–71): 955; "In the Field of Educational Administration," *WAM* 45 (1960–61): 213; and Paul Woodring, *Investment in Innovation: An Historical Appraisal of the Fund for the Advancement of Education* (Boston: Little, Brown, 1970), 54.

58. On Chamberlain, see *AMWS,* 13th ed., S&B, 1978, 200, and *WWAW* 12 (1981–82): 129. See also Susan Glauberman, "A Conversation with Mariam Chamberlain and Red Crossland," *Change* 13 (Nov.–Dec. 1981): 32–37; and Richard Magat, *The Ford Foundation at Work: Philanthropic Choices, Methods, and Styles* (New York: Plenum, 1979).

59. On Parker, see Lula Thomas Holmes, "Pablo Is Eating Better Now," *IW* 31 (Nov. 1952): 331–32; Elvin C. Stakman et al., *Campaign against Hunger* (Cambridge: Harvard Univ. Press, 1967), 42, 48–49; Robert E. Chandler, Jr., *An Adventure in Applied Science: A History of the International Rice Research Institute* (Los Banos, Philippines: International Rice Research Institute, 1982), 68–72; and *WWAW* 7 (1972–73): 685. A collection of her papers is at the RAC.

60. See Teresa Jean Odendahl, Elizabeth Trocolli Boris, and Arlene Kaplan Daniels, *Working in Foundations: Career Patterns of Women and Men* (New York: Foundation Center, 1985), which is based on 1982 data.

61. On Apgar, see *AMWS,* 12th ed., 1971, 1:148–49, and *NAWMP.*

62. Kenneth E. Caster, "Memorial to Katherine Van Winkle Palmer, 1895–1982," *Journal of Paleontology* 57 (1983): 1141–44, reprinted in GSA *Memorials* 17 (1987); obituary, *Bulletin of Sigma Delta Epsilon* 46, no. 3 (fall 1982): 15; Kenneth E. Caster, "Presentation

of the Paleontological Society Medal to Katherine Van Winkle Palmer," *Journal of Paleontology* 47 (1973): 599–601, and Palmer's response, ibid., 601–2. See also Katherine Van Winkle Palmer, "Role Models" (brief autobiography), in *Women in Geology: Proceedings of the First Northeastern Women's Geoscientists Conference, St. Lawrence University, April 26–27, 1976*, ed. Susan D. Halsey et al., Department of Geology and Geography Monograph 5 (Canton, N.Y.: St. Lawrence University, 1976), 25–28; taped interview with Katherine Van Winkle Palmer [1975], KLCU; and Katherine Van Winkle Palmer, *Paleontological Research Institution: Fifty Years, 1932–1982* (n.p., 1982).

63. On Dayhoff, see *AMWS*, 15th ed., 1982, 2:548; obituary, *NYT*, Feb. 9, 1983, B-12; and "In Memoriam," *Computers in Biology and Medicine* 14, no. 1 (1984): 1–2 (I thank Robert S. Ledley for a copy). See also *Research Accomplishments, 1960–1970* (Washington, D.C.: [National Biomedical Research Foundation], 1973).

64. On Harwood, see *AMWS*, 12th ed., 1972, 3:2548. Thomas E. Drake, *A Scientific Outpost: The First Half Century of the Nantucket Maria Mitchell Association* (Nantucket, Mass.: Nantucket Maria Mitchell Association, 1968), covers the period before 1952. The fame of the museum and the association greatly increased after the publication of Helen Wright, *Sweeper in the Sky: The Life of Maria Mitchell, First Woman Astronomer in America* (Nantucket: Nantucket Maria Mitchell Association, 1959). On Hoffleit, see *AMWS*, 17th ed., 1989, 3:765; and Dorrit Hoffleit, "Some Glimpses from My Career," *Mercury* 21 (Jan.–Feb. 1992): 16–18.

65. On Clark, see *AMWS*, 12th ed., 1971, 1:1042; *CBY, 1953*, 120–22; and Anne La-Bastille, "Scientist in a Wetsuit" (interview with Clark), *Oceans* 14 (Sept.–Oct. 1981): 44–47, 49, 50. The two autobiographies are Eugenie Clark, *The Lady with a Spear* (New York: Harper & Brothers, 1953) and *The Lady and the Sharks* (New York: Harper & Row, 1969). On the Mote Marine Laboratory, see *Research Centers Directory* 12 (1988): 182.

66. On Ray, see *CBY, 1973*, 345–48; George A. W. Boehm, "Extraordinary First Lady of the AEC," *Reader's Digest*, July 1974, 81–85; Christine Russell, "Profile: Dixy Lee Ray," *Biosciences* 24 (1974): 489–91; Robert Gillette, "A Conversation with Dixy Lee Ray," *Science* 189 (1975): 124–26; "Women in Science," *Science Teacher* 40 (Dec. 1973): 15–16; and Louis R. Guzzo, *Is It True What They Say about Dixy? A Biography of Dixy Lee Ray* (Mercer Island, Wash.: Writing Works, 1980). The extensive collection of Dixy Lee Ray Papers at the archives of the Hoover Institution on War, Revolution and Peace in Stanford, California, is closed to researchers. There apparently is no history of the Pacific Science Center, but it is described in the *Official Museum Directory, 1988* (Washington, D.C.: American Association of Museums, 1987), 860, and in Paul Ashdown, "Seattle 1962: Seattle World's Fair (Century 21 Exposition)," in *Historical Dictionary of the World's Fairs and Expositions, 1851–1988*, ed. John E. Findling and Kimberly D. Pelle (New York: Greenwood, 1990), 319–21, which also discusses source materials.

67. "Mamie Phipps Clark" (autobiography), in O'Connell and Russo, *Models of Achievement*, 1:266–77.

68. On Carson, see *NAWMP*; on Fieser, "Mary Fieser, Garvan Medal," *C&EN* 48 (Dec. 14, 1970): 64, and Stacey Pramer, "Mary Fieser: A Transitional Figure in the History of Women," *JCE* 62 (1985): 186–91; and on Snell, *AMS*, 11th ed., 1967, 5:5029, and ch. 12, n. 14, below.

69. For a meteorologist, see "Marion G. Hogan," vita, box 724, WB Papers, NARA.

On Westcott, see *AMS,* 9th ed., 1955, 2:1211, and obituary, *NYT,* Mar. 23, 1983, B-5. See also Eugene Kinkead, "Profiles: Physician in the Flower Beds," *New Yorker,* July 26, 1952, 26-39; and her autobiography, *Plant Doctoring Is Fun* (Princeton: Van Nostrand, 1957). Her papers are at the KLCU (I thank Sara Tjossem for telling me of them).

70. "Alumnae in the News: Louise (Capen) Baker, '27, Seed Analyst," *Agnes Scott Alumnae Quarterly* 18, no. 1 (Nov. 1939): 8-9, and "Seed Analysis—Highly Specialized Career," ibid. 26, no. 3 (spring 1948): 29. On Mauro, see obituary, *Albany (N.Y.) Times-Union,* July 18, 1976, and other clippings, esp. Kathleen Condon, "Dr. Jacqueline Mauro—Cancer Detective," *Albany Knickerbocker-News,* July 29, 1971, 1C, in Biographical Files, New York State Department of Health Library.

71. Alfred and Elma Milotte won six Academy Awards while filming for Walt Disney studios from 1948 until 1959 (joint obituary, *NYT,* Apr. 27, 1989, B-16); and Lois and Cris Crisler wrote and photographed in the Arctic in 1952 and 1953 for Disney productions (Richard Dyer MacCann, "Crislers Take Disney Films: Animals Are Their Actors," *CSM,* [1952-53], in Lois Brown Crisler Papers, University of Washington Special Collections, Seattle). See also Lydia and Ray Jewell's autobiographical "Nomads of the North," *WAM* 44 (1959-60): 146-47, 163, 167; and "The Call of the Wild," *Newsweek,* Sept. 16, 1974, 92, about Fred and Elaine Meaders, who spent fifteen years in Alaska.

Chapter 12   Corporate Employment: Research and Customer Service

1. Frances M. Fuller and Mary B. Batchelder, "Opportunities for Women at the Administrative Level," *Harvard Business Review* 31 (Jan. 1953): 111-28, plus "In This Issue," 14. See also T. North Whitehead, "Management Training for Women," *Journal of College Placement* 15, no. 2 (Dec. 1954): 15-16, 18-19; Christine Hobart, "Administrative Opportunities for Women," ibid. 20, no. 3 (Feb. 1960): 24-27, 92, 94, 96; Jane Knowles, "Harvard-Radcliffe Program in Business Administration: Training Women for Business," *Radcliffe Quarterly* 73, no. 4 (Dec. 1987): 27-29.

2. See, e.g., U.S. Civil Service Commission Library, *Scientists and Engineers in the Federal Government,* Personnel Bibliography Series, 30 (Washington, D.C.: GPO, 1970), which included industry. See also Business and Professional Women's Foundation, *Women Executives: A Selected Annotated Bibliography* (Washington, D.C., 1970).

3. See, e.g., Donald C. Pelz and Frank M. Andrews, *Scientists in Organizations: Productive Climates for Research and Development* (New York: John Wiley & Sons, 1966).

4. On Conwell, see *AMWS,* 16th ed., 1986, 2:350, and *WWAW* 14 (1985-86): 159. On Ancker-Johnson, see *AMWS,* 15th ed., 1982, 1:108; and Betsy Ancker-Johnson, "Physicist" (autobiography), *Successful Women in the Sciences: An Analysis of Determinants,* ed. Ruth B. Kundsin, *Annals of the New York Academy of Sciences* 208 (1973): 24-28.

5. NSF, *American Science Manpower, 1956-58* (Washington, D.C.: GPO, 1961), 81, 83, and *1968* (1969), 32, 252. Industrial psychology might have become a kind of "women's work" but did not. See Robert E. Webber and Richard C. Arvey, "The Woman Industrial Psychologist: An Emerging Reality," *AP* 33 (1978): 963-65; Virginia E. Schein, "The Woman Industrial Psychologist: Illusion or Reality?" ibid. 26 (1971): 708-12; and Donald S. Napoli, *Architects of Adjustment: The History of the Psychological Profession in the United States* (Port Washington, N.Y.: Kennikat, 1981): 138-40.

6. *Women in Higher-Level Positions,* WB Bulletin 236 (Washington, D.C.: GPO, 1950), 50. See also Katharine Hamill's "Women as Bosses," *Fortune* 53 (June 1956), 104–08, 213–14, 216, 219–20; and "What Companies Think about Woman," *Management Methods* 13, no. 3 (Dec. 1957): 14–16.

7. Margaret Cussler, *The Woman Executive* (New York: Harcourt, Brace, 1958), 125. Cussler had a hard time locating her interviewees, fifty-five women earning over $4,000 and supervising more than three people. On Cussler, see *AMWS,* 13th ed., S&B, 1978, 266; *WWAW* 10 (1977–78): 198, and obituary in *Radcliffe Quarterly* 73, no. 4 (Dec. 1987): 47. The project was supported by the NFBPWC's new foundation ("Foundation's First Grant to Dr. Cussler," *IW* 35 [Aug. 1956]: 10, 29; "First Foundation Grant Produces," ibid. 37 [Apr. 1958]: 7 [and advertisement on 6]; Marguerite Rawalt, *History of the National Federation of Business and Professional Women's Clubs, Inc., 1944–1960* [Washington, D.C.: NFBPWC, (1969)], 196).

8. The ACS's only major salary study in this period—after 1943 and before the 1970s—was that of 1955: Andrew Fraser, "The 1955 Professional and Economic Survey of the Membership of the American Chemical Society," *C&EN* 34 (1956): 1731–81, which broke down almost all of its categories (specialties, salaries, and degree levels) by gender, allowing one to locate male and female librarians and literature chemists.

9. Claudine Carlton, "Women and Science" (letter to the editor), *Science* 145 (1964): 1123.

10. See Frederick P. Li, "Suicide among Chemists," *Archives of Environmental Health* 19 (1969): 518–20, on 115 deaths; and "Suicides High for Female Chemists," *Industrial Research* 11 (Dec. 1969): 31. See also ch. 15, n. 1.

11. See Rosina Fusco, "The College Trained Secretary," *Journal of College Placement* 15, no. 4 (May 1955): 13–18.

12. Else L. Schulze, "Wanted: More Library Chemists," *JCE* 23 (1946): 178. See also *The Outlook for Women in Occupations Related to Science,* WB Bulletin 223-8 (Washington, D.C., 1948); and Margaret W. Rossiter, "Chemical Librarianship: A Kind of 'Women's Work' in America" (forthcoming, *Ambix*).

13. See also Lura Shorb and Lewis W. Beck, "Opportunities for Chemists in Literature Service Work," *JCE* 21 (1944): 315–18; ACS, Division of Chemical Education, *Training of Literature Chemists: A Collection of Papers Comprising the Symposium on Training of Literature Chemists, Presented before the Division of Chemical Education and the Division of Chemical Literature at the 127th National Meeting of the American Chemical Society, Cincinnati, Ohio, March 1955,* Advances in Chemistry Series, no. 17 (Washington, D.C., 1956); as well as the several articles presented at the Symposium on the Literature Chemist in the Chemical Industry, published in the *Journal of Chemical Documentation* 2 (1962): 158–93, and at a second symposium on the Education of Literature Chemists, published in ibid., 195–209ff. Women chemists who became patent attorneys included Ruth Merling at Eastman Kodak as early as 1931 (*AMS,* 9th ed., P&B, 1955, 1:1310); E. Janet Berry at Esso and elsewhere starting in 1948 (*AMWS,* 16th ed., 1986, 1:477); and Pauline Newman at FMC Corporation from 1954 (ibid., 5:665).

14. On Snell, see *AMS,* 11th ed., 1967, 5:5029. A literature chemist and chemical writer for her husband's consulting firm, Snell wrote several articles between 1945 and 1955 in which she advised women how to adapt to the sex-typed nature of chemical

employment, e.g., "Women as Professional Chemists," *JCE* 25 (1948): 450–53; "Organic Analysis as a Tool for Women Chemists," ibid. 27 (1950): 138–41; "The Woman Chemist," *C&EN* 28 (1950): 3110–12, reprinted in ACS, *Careers in Chemistry and Chemical Engineering* (Easton, Pa.: ACS, 1951), 44–46; and "Literature Searchers, as Needed by the Chemical Consultant," in ACS, Division of Chemical Education, *Training of Literature Chemists*, 13–15.

15. Fraser, "The 1955 Professional and Economic Survey," 1746, 1750, 1765.

16. Alma C. Mitchill, *Special Libraries Association: Its First Fifty Years, 1909–1959* (New York: Special Libraries Association, 1959), esp. 99–105.

17. On Cole, see *Who Was Who in America* 5 (1969–73): 142.

18. On Strieby (Mrs. R. Norris Shreeve), see *WWAW* 11 (1979–80): 754. See also Irene M. Strieby, "The Chemical Librarian in Industry," *JCE* 30 (1953): 522–25; and idem, "The Problem of Literature Chemists in Industry," in ACS, Division of Chemical Education, *Training of Literature Chemists*, 3–12.

19. See Eleanor B. Gibson, "Mrs. Catherine Deneen Mack: Her Life and Contributions," typescript, 1966, plus obituaries, in Special Collections, Arthur A. Houghton, Jr., Library, Corning Community College, Corning, N.Y.

20. "Who Says It's a Man's World?" *Texaco Topics* 3, no. 2 (1970): 10, and photo 12, copy in JLP.

21. On Gramse, see "Erna Gramse Awarded New York Honor Scroll," *The Chemist* 54, no. 5 (Sept. 1977): 14; and Erna L. Gramse, "Women in Chemistry," ibid., 14–15.

22. One exception was Eleanor M. Ullman, "Women in the Chemical Industry," *Industrial and Engineering Chemistry* 50 (Feb. 1958): 111A–112A, 114A.

23. On Mellon, see *AMS*, 11th ed., 1966, 4:3565; M. G. Mellon, "Introduction," ACS, Division of Chemical Education, *Training of Literature Chemists*, 1 and, for another mention of the need for chemists of both sexes, 6.

24. See, e.g., *Journal of Chemical Documentation* 2 (1962): 158–93.

25. See, e.g., help-wanted advertisements in *Special Libraries* 50 (1959): 90, 137, 364.

26. On Cortelyou, see *AMS*, 11th ed., 1965, 1:1000; see also "Let's Get Acquainted," *Chemical Bulletin* (ACS Chicago section) 40, no. 9 (Nov. 1953): 17, 19.

27. Ethaline Cortelyou, "Counseling the Woman Chemistry Major," *JCE* 32 (1955): 197.

28. Fraser, "The 1955 Professional and Economic Survey," 1746; the median monthly salary for "editing, writing, advertising, and publicity" was given as $586 (1765).

29. Ethaline Cortelyou, "The Training of Chemists and Chemical Engineers for Technical Journalism," *JCE* 33 (1956): 64–67. Cortelyou was also cited in a later article, Evelyn L. Hankins, "Technical Writing: A New Field with Immediate Prospects for Women," *National Business Woman* 39, no. 3 (Mar. 1960, 7. See also Israel Light, "Technical Writing and Professional Status," *Journal of Chemical Documentation* 1, no. 3 (1961): 4–10; and R. E. Speers, "Technical Editing and Writing in the Chemical Industry," ibid. 2 (1962): 162–64.

30. Ethaline Cortelyou, "Utilizing Chemical Womanpower to Combat the Alleged Shortage of Chemists," *Chemical Bulletin* (ACS Chicago section) [44?] (June 1958): 18–19, copy in carton 4, f. EC 83, ACE-CEW Papers, SLRC. For more on Cortelyou, see ch. 15.

31. On Stafford, see *AMWS*, 12th ed., 1973, 6:6048; "We See by the Papers," *SAQ* 47 (1955–56): 216; "National Association of Science Writers, Jane Stafford, Transcript of an Interview conducted by Bruce V. Lewenstein, 6 February 1987" (I thank Lewenstein for letting me see a copy of this). Although Stafford generally minimized any problems that she had faced as a pioneering woman science journalist, she was angry about the unequal pay, which she felt was patently unfair (47–48).

32. On Bishop, see *CBY, 1957,* 56–58; Patricia Evers Glendon, "Scientist in the Land of Hocus," *Barnard Alumnae Magazine* 47 (Apr. 1958): 8–9; "People," *Financial World* 148 (Feb. 1, 1979): 25–26; as well as Hazel Bishop, "My Observations Concerning Career Opportunities for Women in Chemistry," *The Chemist* 49 (1972): 425–26.

33. On Sullivan, see *AMWS*, 13th ed., 1976, 6:4372.

34. On Simon, see ibid., 16th ed., 1986, 6:759; *WWAW* 9 (1975–76): 818; and Roul Tunley, "The Lady Knows Her Rockets," *Saturday Evening Post,* Nov. 14, 1955, 39, 115–19, which found it bizarre that Simon could be both a rocket scientist and a woman. See also Marcel LaFollette, "Eyes on the Stars: Images of Women Scientist in Popular Magazines," *Science, Technology, and Human Values* 13 (1988): 262–75.

35. On Elion, see *AMWS*, 16th ed., 1986, 2:867; "Garvan Medal, Gertrude B. Elion," *C&EN* 46 (Jan. 15, 1968): 65; and James Bordley III and A. McGehee Harvey, *Two Centuries of American Medicine* (Philadelphia: W. B. Saunders, 1976), 466–68.

36. On von Rümker, see *AMWS*, 16th ed., 1986, 7:354; and "FC Profile: Dr. Rosemarie von Rümker," *Farm Chemicals* 127 (June 1964): 50–51.

37. On Skala, see *AMWS*, 14th ed., 1979, 6:4707; and "People," *Chemical Week* 121 (Nov. 23, 1977): 48.

38. On Telkes, see *AMWS*, 15th ed., 1982, 7:54–55; *CBY, 1950,* 563–64; and Betsy Burke, "Woman Engineer Who Tames the Sun," *IW* 34 (July 1955): 10–12.

39. On Tesoro, see *AMWS*, 16th ed., 1986, 7:70; "Giuliana C. Tesoro, 20th Olney Medalist," *American Dyestuff Reporter* 52 (Oct. 14, 1963): 98–102; and Shirlee Sherkow, "Interview with Giuliana Tesoro," oral history for Project on Women Scientists and Engineers, MIT Archives, 1976–77.

40. On Pierce, see *AMWS*, 15th ed., 1982, 5:978; and obituary, *White Plains (N.Y.) Reporter Dispatch,* June 2, 1988 (I thank Mary Ellen Bowden for a copy).

41. On Rebstock, see *AMWS*, 14th ed., 1979, 6:4129; and Ritchie Calder, *Profile of Science* (London: George Allen & Unwin, 1953), 247.

42. On Root, see *AMWS*, 15th ed., 1982, 6:263; "We See That . . . ," *AAUWJ* 57 (1963): 44–45.

43. On Hobby, see *AMWS*, 15th ed., 1982, 3:729; obituary, *NYT,* July 9, 1993, D-19; Elizabeth Moot O'Hern, *Profiles of Pioneer Women Scientists* (Washington, D.C.: Acropolis, 1985), ch. 6; Simon Baatz, *Knowledge, Culture, and Science in the Metropolis: The New York Academy of Sciences, 1817–1970, Annals of the New York Academy of Sciences* 584 (1990): 227–28 (quotation on 227). See also Gladys L. Hobby, *Penicillin: Meeting the Challenge* (New Haven: Yale Univ. Press, 1985); and *VAM,* Apr. 1961, 36.

44. "Women in Research: At Du Pont They Are Making Impressive Contributions in Many Fields of Science," *Better Living* 14 (Jan.–Feb. 1960): 12–15, clipping in carton 4, f. EC 83, ACE-CEW Papers, SLRC.

45. On Hood, see *AMWS*, 12th ed., 1972, 3:2807.

46. On Kwolek, see ibid., 16th ed., 1986, 4:538; and Ethlie A. Vare and Greg Ptacek, *Mothers of Invention* (New York: William Morrow, 1988), 192–93. (I thank Mary Ellen Bowden for information about Kwolek.)

47. On Blodgett, see "Katharine Burr Blodgett," *Physics Today* 33 (Mar. 1989): 107; Kathleen A. Davis, "Katharine Blodgett and Thin Films," *JCE* 61 (1984): 437–39; *CBY, 1940,* 90–91, and *1952,* 55–57; "Problem-Solver Katharine Blodgett Wins Garvan Medal," *C&EN* 29 (1951): 1408 and cover of April 9, 1951, issue; numerous press releases, clippings, some speeches, and occasional correspondence in Biographical Files, General Electric Corporation, Schenectady, N.Y. GE also has old personnel rating books, kept by managers, that ranked the scientists in the laboratory. In one chart for March 1953 that recorded the ranks for 230 GE laboratory employees, Blodgett was ranked 27th of 28 in general physics, and Edith Boldebuck (see n. 49 below) was ranked 31st of 76 chemists. There are also occasional Blodgett letters in the Irving Langmuir Papers at the Manuscript Division, LC. For a recent assessment of her work, see Gareth Roberts, ed., *Langmuir-Blodgett Films* (New York: Plenum, 1990), 11–13, 321–24.

48. See Tape of Margaret Welsh Goldsmith, July 18, 1977, GE. Goldsmith, a GE employee from 1926 into the 1940s, could have been confusing Greta Garbo with Greer Garson, who played Marie Curie in the award-winning wartime movie *Madame Curie.*

49. On Boldebuck, see *AMWS,* 14th ed., 1973, 1:465; obituary, *Schenectady Gazette,* June 5, 1981, clipping in Biographical Files, GE. See also Herman A. Liebhafsky, *Silicones Under the Monogram: A Story of Industrial Research* (New York: John Wiley & Sons, 1978).

50. J. W. Dumas to R. A. Gleason, Jan. 17, 1964, GE. (I thank George Wise for much help at GE.) See also "Coeds' Opportunities Widen, but Many Jobs Remain Closed to Them," *WSJ,* Feb. 1, 1962, 1, 2.

51. Paul C. Wensberg, *Land's Polaroid: A Company and the Man Who Invented It* (Boston: Houghton Mifflin, 1987), 9, 16, 73,127, 128, 149, 161–62, 180, 219, 223; and "Back from the Home to Business," *Business Week,* Oct. 7, 1961, 99. On Morse, see obituary, *Boston Herald-Traveler,* July 30, 1969, and other biographical materials in box 21, f. 30, *NAWP,* SLRC; none of "the Princesses" are listed in the *AMWS,* however. On Hopkins, see *AMWS,* 17th ed., 1989, 3:821–22; and Esther A. H. Hopkins, "Alternative Development of a Scientific Career," in *Women in Scientific and Engineering Professions,* ed. Violet B. Haas and Carolyn C. Perucci (Ann Arbor: Univ. of Michigan Press, 1984), 137–46.

52. Some women mathematicians held unusual positions. A. Estelle Glancy, a mathematician with a doctorate in astronomy, long designed, despite her deafness, specialized lenses for the American Optical Company ("Recent Awards to Distinguished Alumnae," *WAM* 36 [1951–52]: 285; "The World's Only Woman Lens Designer," ibid. 37 [1952–53]: 275–76, 309), and mathematics major Nancy Schoonover, a 1957 graduate of Smith, was reported a few years later as specializing in the design of small airplane tires for B. F. Goodrich (*SAQ* 52 [1960–61]: 90).

53. On Hopper, see *AMWS,* 16th ed., 1986, 3:813; James W. Cortada, comp., *Historical Dictionary of Data Processing: Biographies* (Westport, Conn.: Greenwood, 1987), 132–34; "Grace Murray Hopper," Biographical Files, Naval Historical Center, Department of the Navy, Washington, D.C., which contains a bibliography of her writings; John H.

Cushman, Jr., "Admiral Hopper's Farewell," *NYT,* Aug. 14, 1986, B-6; obituaries, *NYT,* Jan. 3, 1992, A-17, and *Washington Post,* Jan. 4, 1992; Grace Murray Hopper, "The Education of a Computer" (1952), reprinted in *Annals of the History of Computing* 8 (1988): 271–81. There is a collection of her papers at the Archives Center, National Museum of American History, Smithsonian Institution, Washington, D.C. Less well known was Maria von Wedemeyer-Weller, who at the time of her death at age fifty-three was the highest-ranking woman mathematician at the Honeywell Corporation (*NYT,* Nov. 17, 1977, D-3).

54. On Sammet, see *AMWS,* 17th ed., 1989, 6:421; Cortada, *Historical Dictionary of Data Processing: Biographies,* 229–30; Rina J. Yarmish and Louise S. Grinstein, "Brief Notes on Six Women in Computer Development, Part II," *Journal of Computers in Mathematics and Science Teaching* 2, no. 3 (spring 1983): 26–27; Jean Sammet, "The Early History of COBOL," in *The History of Programming Languages,* ed. Richard Wexelblat (New York: Academic, 1981), 199–277; and Jean Sammet, *Programming Languages: History and Fundamentals* (Englewood Cliffs, N.J.: Prentice-Hall, 1969).

55. John T. Soma, "The Computer Industry," *An Economic-Legal Analysis of Its Technology and Growth* (Lexington, Mass.: D. C. Heath, 1976), 21; Franklin M. Fisher, James W. McKie, and Richard B. Mancke, *IBM and the U.S. Data Processing Industry: An Economic History* (New York: Praeger, 1983), 24–25; and F. G. "Buck" Rogers, *The IBM Way* (New York: Harper & Row, 1986), 150–52 and ch. 8.

56. On Ruth Leach Pollock, see *CBY, 1948,* 373–74; *WWAW* 6 (1970–71): 993; William Rodgers, THINK: *A Biography of the Watsons and IBM* (New York: Stein & Day, 1969), 125, 268; and Nancy Foy, *The IBM World* (London: Eyre, Methuen, 1974), 116. See also Ruth M. Leach, "Where Women Get a Chance," *IW* 23 (Feb. 1944): 52, in which she said that she employed four hundred women in systems service, and idem, "Training on the Job—After College," in "War and Post-War Employment and Its Demands for Educational Adjustment" (New London: Institute of Women's Professional Relations, 1944, mimeographed), 19–24.

57. McCabe evidently kept herself out of biographical dictionaries. See Rita McCabe, "Women in Data-Processing," *National Business Woman* 38 (Sept. 1959): 6–7, 28, and, based on this, Gordon P. Lovell, "Women and the Growing World of Data Processing," *Journal of College Placement* 20, no. 4 (Apr. 1960): 34–36, 80. In 1964 IBM employed about twelve hundred women, as well as twenty-five hundred male programmers and four thousand male systems engineers. See Vartanig G. Vartan, "Computers Are Getting Ideas from Women: IBM the Leader in Employing Girls as Programmers," *NYT,* Mar. 12, 1964, 47, 53; and Rita McCabe, "The Commitment Required of a Woman Entering a Scientific Profession," in *Women and the Scientific Professions: The M.I.T. Symposium on American Women in Science and Engineering,* ed. Jacquelyn A. Mattfeld and Carol G. Van Aken [Cambridge: MIT Press, 1965], 24–28.

The directors of placement offices at the women's colleges often reported the new opportunities open to mathematics majors. See, e.g., Alice Davis, "Employing the Woman Scientist," *SAQ* 48 (1956–57): 12–13; Joan Fiss Bishop, "Mathematicians Needed!" *WAM* 45 (1960–61): 305; and "Who Makes Computers Compute?" ibid. 48 (1963–64): 22–23, 31. By 1964 IBM had hired several Wellesley and other graduates as programmers directly out of college, and some of the "systems service" women had moved into this area (including star Ann Robinson, featured in some IBM literature, when a computer she had programmed to write in Braille was demonstrated at a show in Paris) (McCabe, "Women in Data-Processing," 6–7; Lovell, "Women and the Growing World of Data Processing," 34, 36).

58. In the first "directory" of programmers, "Who's Who in the Computing Machinery Field," published in *The Computing Machinery Field* (later *Computing and Automation*) 2, nos. 1 (Jan. 1953) and 3 (Apr. 1953) and 3, no. 1 (Dec. 1953), 6.9 percent, or 36 of 523, were women, to judge by first names. A rather extensive joint survey by the NSF and the Mathematical Association of America of jobs in professional mathematics did not refer to "programmers," although 23.2 percent of the jobholders responding worked for computing laboratories (NSF with BLS and Mathematical Association of America, *Employment in Professional Mathematical Work in Industry and Government: Report on a 1960 Survey* [Washington, D.C.: GPO, 1962], 15–17, 38). See also Gerald H. F. Gardner, "The Status of Women in the Field of Computing," *Computers and Automation* 19, no. 1 (Jan. 1970): 57–58.

59. BLS, *Automation and Employment Opportunities for Office Workers*, BLS Bulletin 1241 (Washington, D.C., 1958), 11.

60. Ibid.

61. T. C. Cowan, "The Recruiting and Training of Programmers," *Datamation* 4 (May–June 1958): 16–18.

62. M. Ostrofsky, "Woman Mathematicians in Industry," *AAUWJ* 57 (Mar. 1964): 114, 117–18.

63. Jackson W. Granholm, "How to Hire a Programmer," *Datamation* 8 (Aug. 1962): 32. Proper attire was not clear in these years before the "dress for success" movement. John T. Molloy, *Women: Dress for Success* (London: W. Foulsham, 1980), was based partly on research on women engineers (46–48, 116–18, 144–46).

64. Ostrofsky, "Woman Mathematicians in Industry," 114, 117–18.

65. *Employment Opportunities for Women Mathematicians and Statisticians*, WB Bulletin 262 (Washington, D.C., 1956), 34.

66. Mary K. Hawes to John B. Parrish, Apr. 13, 1961, John B. Parrish Papers, SLRC. It was not clear whether Hawes had a higher degree, which was often cited as a requirement for promotions in the computer field. RCA reportedly employed twenty-nine women engineers (Edward M. Tuft [RCA vice president], "Women in Electronics," *National Business Woman* 35 [Nov. 1956]: 10–11, 25–26).

67. On this program at Douglass College, see ch. 3, n. 32.

68. "Mixing Math and Motherhood," *Business Week,* Mar. 2, 1963, 86–87.

69. On Husted, see *CBY, 1949,* 286–87; Marjorie Child Husted, "Would You Like More Recognition?" *JHE* 40 (1948): 459–60; and Marjorie Child Husted Papers, SLRC. On Kelly, see obituary, *NYT,* May 31, 1958, 15. Adelaide Hawley Cumming was the radio and television personality for "Betty Crocker" from 1950 to 1964. There are many relevant items in her papers at SLRC. Also at SLRC is some material on Mary Barber, head of the home economics department at the Kellogg Company from 1924 until 1949 (box 22, *NAWP*).

70. On General Foods, see, on Marie Sellers, *WWAW* 3 (1964–65): 909; on Grace M. Gustafson, obituary, *NYT,* Aug. 31, 1951, 15; and on Helen Thackeray, *Omicron Nu* 27, no.

3 (spring 1951): 12–13. On Corning, see, on Lucy Maltby, *WWAW* 3 (1964–65): 641–42, and Lucy Maltby, "How Test Kitchens Function," *Omicron Nu* 26, no. 2 (spring 1947): 47–51. On Sealtest, see, on Mary Horton, Frances Maule, *Executive Careers for Women*, rev. and enl. ed. (New York: Harper, 1961), 46–47, and obituary, *NYT*, July 7, 1974, 37. On Frigidaire, see Mary E. Huck, *WWAW* 4 (1966–67): 561. And on First National Stores, see Margaret L. Ross, "1974 Recipient of the Marjorie Hulsizer Copher Award," *Journal of the American Dietetic Association* 65 (1974): 670–74.

71. For example, "Ambassador of American Living," *JHE* 49 (1957): 652–53, was about Barbara Sampson of General Foods Kitchens, the only woman representative of an American business at an international trade fair in Poland in the summer of 1957. She was particularly responsible for demonstrating frozen foods and small appliances.

72. On Herbert, see *CBY, 1954*, 333–34, and *Omicron Nu* 29, no. 4 (spring 1955): 26; see also, on Anna Rush, *WWAW* 6 (1970–71): 1069.

73. On Guy, see *WWAW* 10 (1977–78): 353.

74. On Fisher, see obituary, *NYT*, Mar. 17, 1958, 29; Herbert R. Maves, "Town Hall," *Good Housekeeping* 136 (Apr. 1953): 16–17, upon her retirement; and Katharine Fisher, "Wanted: More Home Economists for Business," *JHE* 39 (1947): 324–26. See also George S. Wham, "The Good Housekeeping Institute and Home Economics," *Omicron Nu* 36, no. 1 (fall 1967): 12–14; and, for a floor plan of the institute's laboratories, *Good Housekeeping* 181 (Aug. 1975): 28. The revised Good Housekeeping Seal is described in ibid., July 1975, 6.

75. On Rogers, see "Ballad of Willie Mae," *Business Week*, Feb. 22, 1969, 45–46, and "Editor's Notebook," *Good Housekeeping* 181 (Aug. 1975): 6, upon her retirement. Rogers's successor, Zoe Coulson, was already on the magazine's staff as food editor (ibid. 181 [July 1975]: 20). Also on the staff was Ruth Bien, with a bachelor of science degree in chemistry from Teachers College, who long headed the magazine's chemical laboratory. After her retirement in the late 1960s, she did similar evaluation work for Avon Products (*AMWS*, 12th ed., 1971, 1:482; Elizabeth Weston, "She's a Good Cook, Too!" *Good Housekeeping* 143 [Aug. 1956]: 4).

76. On the HEIBs, see Carolyn Goldstein, "Mediating Consumption: Home Economics and American Consumers, 1900–1940," Ph.D. diss., Univ. of Delaware, 1994, and Frances Maule, *Careers for the Home Economist: Fields Which Offer Openings to the Girl with Modern Training in the Homemaking Arts* (New York: Funk & Wagnalls, 1943), ch. 8. A directory of members started to appear in 1965. The 1970 volume contains a list of past national chairmen (*Directory, 1970, Home Economists in Business, Section of the American Home Economics Association, Official List of Members* [Kensington, Md., n.d.], 3).

77. Earl J. McGrath and Jack T. Johnson, *The Changing Mission of Home Economics: A Report on Home Economics in the Land-Grant Colleges and State Universities* (New York: Teachers College Press, 1968), 72–75.

78. See, e.g., "Jobs in Home Economics," *Journal of College Placement* 15, no. 3 (Mar. 1955): 53–56; Josephine Hemphill, "Home Economics Unlimited," *JHE* 47 (1955): 653–60; and "Shortages and Projected Needs in Important Areas of Home Economics," ibid. 51 (1959): 415–17.

79. On Horton, see Maule, *Executive Careers for Women*, 46–47; and obituary, *NYT*, July 7, 1974, 37.

80. Marjorie Child Husted, "A Critical Evaluation of Modern Home Service, Given before the Mid-West Regional Gas Sales Conference, Chicago, April 23, 1952," typescript, box 2, Marjorie Child Husted Papers.

81. "HEIBs Speak Out," *JHE* 65 (Apr. 1973): 32.

82. Gerda W. Bowman and N. Beatrice Worthy, "Are Women Executives People?" *Harvard Business Review* 43, no. 4 (July–Aug. 1965): 14–16, 19–20, 22, 24, 26, 28, 164, 166, 168–70, 172, 174–76, 178 (quotation from 164). Two years later another article tried to put a more positive spin on essentially the same data: Joseph Famularo, "Woman Power," *Journal of College Placement* 27 (Apr.–May 1967): 32–41.

83. Caroline Bird, *Born Female: The High Cost of Keeping Women Down* (New York: David McKay, 1968), ch. 5, captures the mentality and strategies of corporate women of the time.

Chapter 13   Government "Showcase"?

1. See also NSF, *Employment in Professional Mathematical Work in Industry and Government: Report on a 1960 Survey* (Washington, D.C.: GPO, 1962), though none of the data are broken down by gender or race. Yet at a meeting at the AAUW Educational Center in March 1962, demographer Irene Taeuber reported that the U.S. Census Bureau had been trying to determine how to pick successful computer programmers: "The fascinating thing is that two of their most successful programmers for the computers are colored married women with young children" ("College Change and Choice for the College Woman: Views and Re-views," *AAUWJ* 55 [1962]: 280).

2. See Arnold Thackray et al., *Chemistry in America, 1876–1976: Historical Indicators* (Dordrecht: Reidel, 1985), 125–34.

3. See Elizabeth Moot O'Hern, *Profiles of Pioneer Women Scientists* (Washington, D.C.: Acropolis, 1985), which devotes six chapters to NIH women; Margaret W. Rossiter, *Women Scientists in America: Struggles and Strategies to 1940* (Baltimore: Johns Hopkins Univ. Press, 1982), 229–30; and Barney G. Glaser, *Organizational Scientists: Their Professional Careers* (Indianapolis: Bobbs-Merrill, 1964), based on questionnaires completed by 332 NIH personnel (including 56, or 17 percent, women) in 1952. Thelma Dunn, "Intramural Research Pioneers, Personalities, and Programs: The Early Years," *Journal of the National Cancer Institute* 59, no. 2 (suppl., Aug. 1977): 605–16, describes the accomplishments of many other women at the NCI. On Dunn, see *AMWS*, 14th ed., 1979, 2:1273, and a file at the History of Medicine Division of the National Library of Medicine (NLM), Bethesda. On Sarah Stewart, who was also important there, see *AMWS*, 13th ed., 1976, 6:4317; Bernice E. Eddy, "Sarah Elizabeth Stewart, 1906–1976," *Journal of the National Cancer Institute* 59, no. 4 (1977): 1039–40; O'Hern, *Profiles of Pioneer Women Scientists*, ch. 14; biographical information and oral history interview with Wyndham Miles, Feb. 10, 1964, both in History of Medicine Division, NLM. For more on NIH scientists, see DeWitt Stetten, Jr., and W. T. Carrigan, *NIH: An Account of Research in Its Laboratories and Clinics* (New York: Academic, 1984); "Margaret J. Rioch" (autobiography), in *Models of Achievement: Reflections of Eminent Women in Psychology*, ed. Agnes N. O'Connell and Nancy Felipe Russo, vol. 1 (New York: Columbia Univ. Press, 1983), 173–88; Margaret J. Rioch, "Training the Mature Woman for a

Professional Role," *AAUWJ* 55 (1962): 236–39; and *Report of the Committee on Education to the President's Commission on the Status of Women* (Washington, D.C.: GPO, 1963), 6. On Neufeld, see "Academy of Sciences Elects Sixty Members," *NYT,* May 1, 1977.

4. Rossiter, *Women Scientists in America,* 223–29.

5. Among the women civilian branch chiefs in the Army were mathematician and geodesist Irene Fischer (*AMWS,* 18th ed., 1992, 2:1131), physicist Rita Sagalyn (ibid., 15th ed. 1982, 6:378), and geologist Katharine Mather, one of the many Bryn Mawr College graduates in that field, who was in 1973 president of the Clay Minerals Society (ibid., 16th ed., 1986, 5:256; "Highlights of Katharine Mather's Career," *Station Break* [Waterways Experiment Station newsletter], Oct–Nov. 1982, 7 [I thank Michele Aldrich for the latter]). Meteorologist Frances Whedon, an MIT graduate in the 1920s, was chief of a section in the Army's Signal Research Office from 1942 until 1959 (on Whedon, see *AMS,* 11th ed., 1967, 6:5800). Women civilian scientists at the Navy included chemist Kathryn Shipp (see ibid., 5:4885, and obituaries, *Washington Post,* Oct. 21, 1977, C-12, and *Washington Star,* Oct. 23, 1977, E-12) and physical chemist and meteorologist Florence van Straten, the only woman scientist to get a "supergrade" appointment in 1950 (*AMS,* P&B, 11th ed., 1967, 6:5558; Edna Yost, *Women of Modern Science* [New York: Dodd, Mead, 1959], 124–39; CSC, "Appointments under Section 3.2 of Civil Service Rule III," *Sixty-seventh Annual Report of the U.S. Civil Service Commission, Fiscal Year Ended June 30, 1950* [Washington, D.C.: GPO, 1951], 59).

6. Mark R. Finlay, "The Industrial Utilization of Farm Products and By-Products: The USDA Regional Research Laboratories," *Agricultural History* 64, no. 2 (spring 1990): 41–52. Several women scientists, especially chemists, at the USDA laboratories were outstanding, such as Ruth Benerito (see *AMWS,* 16th ed., 1986, 1:425, and *C&EN* 48 [Jan. 19, 1970]: 43); Helen Hanson ("Christie Research Award," *Poultry Science* 32 [1953]: 371–72); Allene Jeanes ("Allene R. Jeanes, Garvan Medal," *C&EN* 34 [1956]: 1984; Paul A. Sandford, "Allene R. Jeanes," *Carbohydrate Research* 66 [1978]: 3–5; "Civil Service Inventors, Part I," *Civil Service Journal* 3, no. 1 [July–Sept. 1962]: 13); and Masters (*AMWS,* 12th ed., 1972, 4:3991, Rosetta McKinney, "Women in Cereal Chemistry," *Cereal Science Today* 19 [1974]: 534–38).

7. Donald S. Napoli, *Architects of Adjustment: The History of the Psychological Profession in the United States* (Port Washington, N.Y.: Kennikat, 1981), ch. 7; "The Veterans Administration," *Clinical Psychologist* 32, no. 2 (1979): 4–5 and 8; Alan Cranston, "Psychology in the Veterans Administration: A Storied History, A Vital Future," *AP* 41 (1986): 990–95. Among the notable women psychologists at the VA were Mildred Mitchell, Janet Spence, and Anne Roe. On Mitchell, see "Mildred B. Mitchell" (autobiography), in O'Connell and Russo, *Models of Achievement,* 1:120–39, and Mildred B. Mitchell, "Careers in Applied Psychology," *Proceedings of the Conference on Science, February 21–23, 1947, Rockford, Illinois* (Rockford, Ill.: Rockford College, 1947), 80–85 (I thank Joan B. Surrey for sending me a copy). On Spence, see "Janet Taylor Spence" (autobiography), in *Models of Achievement Reflections of Eminent Women in Psychology,* vol. 2 (Hillsdale, N.J.: Lawrence Erlbaum, 1988): 191–203. And on Roe, see ch. 7, n. 49.

8. On Karle, see *AMWS,* 15th ed., 1982, 4:202; Isabella L. Karle, "Crystallographer," in *Successful Women in the Sciences: An Analysis of Determinants,* ed. Ruth B. Kundsin, *Annals of the New York Academy of Sciences* 208 (1973): 11–14; I. L. Karle and J. Karle,

"Recollections and Reflections," in *Crystallography in North America*, ed. Dan McLachlan, Jr., and Jenny P. Lusker (New York: American Crystallographic Association, 1983), ch. 18; Maureen M. Julian, "Isabella Karle and a New Mathematical Breakthrough in Crystallography," *JCE* 63 (1986): 66–67; Nina Matheny Roscher, "Chemistry's Creative Women," ibid. 64 (1987): 748–52 (quotation from 751); and "Naval Research Laboratory [Awardee]," *Journal of the Washington Academy of Sciences* 66 (1976): 161.

When a WB official inquired about jobs for women at the NRL in 1947, C. N. Mason, Jr., responded, "The successful use of women in scientific positions [during World War II] has served to dispell *[sic]* most of the former scepticisms, but, in general, there is still a tendency to be more critical of the qualifications of the female applicant. Nearly all representatives [persons consulted at the lab] were of the opinion that women could be used most successfully in a scientific organization only when their total number did not exceed 10% of the working group. A variety of reasons for this limitation were advanced." The three reasons given were that (1) because women tended not to stay long enough to become career employees at the lab, "there is a decided preference for males at the P-1 [junior entry] level"; (2) women are best suited for precision work, which also has a lot of monotony, and the lab had very few such positions; and (3) "female employment in excess of 10% tends to introduce social problems and creates undesireable distractions" (Mason to Mary Brilla, Jan. 8, 1947, box 598, WB Papers, NARA). Only a woman as highly qualified as Isabella Karle was likely to override this skepticism. By contrast, none of this caution was present in a contemporary recruitment talk in which it was claimed that the NRL was so eager to recruit good young chemists that it was offering special educational opportunities to allow them to complete doctorates on the job (P. Borgstrom, "Professional Development of Young Scientists in Naval Research," *JCE* 26 [1949]: 78–79).

9. On the Goldhabers, see *AMWS*, 15th ed., 1982, 3:189; "Women in Science," *Science Teacher* 40 (Dec. 1973): 15; and a chapter in Alma Payne, *Partners in Science* (Cleveland: World, [1968]). She should not be confused with her sister-in-law, Sulamith Goldhaber, a research physicist at the Lawrence Radiation Laboratory in Berkeley (*AMS*, 11th ed., 1965, 2:1851; "Sulamith Goldhaber," *Physics Today* 19 [Feb. 1966]: 101; Luis Alvarez, "Sulamith Goldhaber [1923–1965]," in *Advances in Particle Physics*, ed. R. L. Cool and R. E. Marshak, vol. 2 [New York: Interscience, 1968], vii–ix). There is a reference to a 1969 report on women at the Lawrence Radiation Laboratory by Miriam L. Machlis in the extensive "Report of the Subcommittee on the Status of Academic Women on the Berkeley Campus," in University of California, Academic Senate, Berkeley Division, "Report of the Committee on Senate Policy," May 1970, 77, copy in JLP. Among other physicist-wives of note at Brookhaven was Renate Weiner Chasman, who came in 1963 (John Blewett, "Renate Wiener Chasman," *Physics Today* 31 [Feb. 1978]: 64; obituary, *NYT,* Oct. 19, 1977, B-2).

10. On Liane Russell, see *AMWS*, 15th ed., 1982, 6:352; and Richard G. Hewlett and Francis Duncan, *Atomic Shield: A History of the U.S. Atomic Energy Commission, 1947–1952* (University Park: Pennsylvania State Univ. Press, 1969), 506–9. Another radiation biologist of note at Oak Ridge was Jane Setlow (*AMWS*, 15th ed., 1982, 6:610). On Oak

Ridge in general see William G. Pollard, *Atomic Energy and Southern Science* (Oak Ridge, Tenn.: Oak Ridge Associated Universities, 1966).

11. On Yalow, see *AMWS*, 15th ed., 1982, 7:776; Genevieve Millet Landau, "Rosalyn Sussman Yalow: An Interview," *Parent's Magazine* 53, no. 1 (Jan. 1978): 38–39, 70, 72; and Lisa Z. Cohen, "Madame Curie and Dr. Rosalyn Yalow: Two Women Who Overcame the Odds," *Lab World* 29 (July 1978): 16–13, 21–22, 24.

12. On Meggers, see *AMS*, 13th ed., 1976, 4:2950; Don D. Fowler, Gus W. Van Beek, and Mario Sanoja, "Obituaries: Clifford Evans, 1920–1981," *American Antiquity* 47 (1982): 545–56; Payne, *Partners in Science;* and Ellis Yochelson, *The National Museum of Natural History: Seventy-five Years in the Natural History Building* (Washington, D.C.: Smithsonian Institution Press, 1985), 140.

13. Yochelson, *National Museum of Natural History,* 140; Richard C. Froeschner, Elsie M. L. Froeschner, and Oscar L. Cartwright, "Doris Holmes Blake, 1892–1978," *Proceedings of the Entomological Society of Washington* 83 (1981): 644–64; John Sherwood, "Doris Blake: The Courtly Coleopterist," *Washington Star,* Jan. 9, 1977, E-1, E-3; Joan B. Chapin, "Women in Systematics," *Bulletin of the Entomological Society of America* 33 (spring 1987): 13–15, on other Smithsonian and federal entomologists (I thank Edward Smith for a copy). The Doris Holmes Blake Papers are at the Smithsonian Institution Archives, Washington, D.C. Congress did not adopt antinepotism restrictions on federal employment until 1967, and then apparently to prevent a future president from appointing his relatives to important positions (Public Law 206, 90th Cong., 1st sess. [Dec. 16, 1967]).

14. On Wang, see *AMS*, 10th ed., 1961, 4:4279.

15. On Cochrane, see obituary, *Radcliffe Quarterly* 61, no. 4 (Dec. 1975): 37; on her husband, Chappelle Cochrane, see *AMWS*, 12th ed., 1971, 1:1082. Only about 2 percent of all Negroes working for the federal government in 1956 were at GS-8 or higher, and most of these were in Washington, D.C. (The President's Committee on Government Employment Policy, *A Five-City Survey of Negro-American Employees of the Federal Government* [Washington, D.C.: GPO, 1957]).

16. Thomas G. Alexander, "Alma Levant Hayden," *Journal of the Association of Analytical Chemists* 50 (1967): 1381; *AMS*, 11th ed. P&B, 1966, 3:2179. Hayden's husband Alonza was also a federal biochemist (ibid., suppl. 4, 1968, 261).

17. E. Katcher, memo on interview with Mrs. Boardman of FDA, Dec. 9, 1946, box 110, WB Papers.

18. On Underhill, see *CBY, 1954,* 617–19; Pat Paton, "Ruth Underhill Remembered," *Colorado Heritage,* no. 1 (1985): 14–21; and, in the Western History Department of the Denver Public Library, a taped interview with Underhill in 1962 (#CH-101), as well as an extensive clipping file.

19. Chester A. Thomas, "Jean McWhirt Pinkley, 1910–1969," *American Antiquity* 34 (1969): 471–73.

20. On Gerry, see *NAW;* Lida W. McBeath, "A Woman of Forest Science, Eloise Gerry," *Journal of Forest History* 22 (1978): 128–35; Harriett M. Grace, "Blazing a Trail in Woodland Research," *International Altrusan,* Nov. 1944, 10–11, 20; and Charles A. Nelson, *History of the U.S. Forest Products Laboratory, 1910–1963* (Madison, Wis.: Forest

Products Laboratory, 1971). On Richards, see John E. McDonald, "We Present Dr. C. Audrey Richards," *Journal of Forestry* 49 (1951): 918–19; on Bomhard, W. A. Dayton, "Miriam Lucile Bomhard," *Journal of the Washington Academy of Sciences* 43 (1953): 136; and on Duncan, *NCAB* 54 (1973): 480–81 and Ellis B. Cowling, "Catherine Gross Duncan, 1908–1968," *Phytopathology* 59 (1969): 1777. There are clippings on McBeath in box 19, f. 16, *NAWP*, SLRC.

After a WB investigator interviewed Fred Miller of the U.S. Forest Service on July 8, 1947, she reported the Forest Service's reasons for having no women in the field and rarely in the office: "The Forest Service now [unlike earlier in wartime] has no women in field work at all. The work environment makes it almost impossible to assign women to field work. The work is isolated and involves much travel; also all foresters are subject to fighting forest fires, which is very hard work. The foresters travel alone or in small groups, often have to camp overnight in all sorts of places. There are no facilities for women. Office jobs are in the higher grades and are filled by promotion. Because field experience is a prerequisite for advancement, there is little opportunity for women even in this kind of work (i.e. desk jobs)" ("Report of interview with Fred Miller, of the U.S. Forest Service Personnel Office, July 8, 1947," box 110, WB Papers). A later recruitment booklet, *Women in the Forest Service,* USDA Miscellaneous Publication 1058 (Washington, D.C.: GPO, 1967), took the opposite tack, however, and described in upbeat tones the many opportunities for women in the service.

21. Rossiter, *Women Scientists in America,* 229; Ruth O'Brien, "BHNHE [Bureau of Human Nutrition and Home Economics] Celebrates a Quarter Century of Service," *JHE* 40 (1948): 293–96. On Stanley, see *NAWMP;* and Helen T. Finneran, "Louise Stanley: A Study of the Career of a Home Economist, Scientist, and Administrator, 1923–1953" (M.A. thesis, American University, 1965). On Stiebeling, see *AMS,* 11th ed., 1967, 6:5182–83; *CBY, 1950,* 548–50; and obituary, *Washington Post,* May 20, 1989. See also T. Swann Harding, *Two Blades of Grass* (Norman: Univ. of Oklahoma Press, 1947), ch. 12. There is room for a full history of the bureau, which touched many aspects of American life.

22. Hazel Stiebeling, "The Institute of Home Economics—Present and Future Programs," in National Association of State Universities and Land-Grant Colleges, Division of Home Economics, *Selected Papers Presented to Seventy-first–Seventy-fifth Annual Meetings, 1957–1961* ([Washington, D.C.], 1962), 19–24; "Research Division Closing Deplored by AHEA," *JHE* 57 (1965): 166; "Research in Clothing and Housing in USDA to be Phased Out," ibid., 172; "Excerpts from AHEA Statement Supporting Continuation of Clothing and Housing Research in the U.S. Department of Agriculture," ibid., 457; U.S. Congress, Senate, Committee on Appropriations, Subcommittee on Agriculture and Related Agencies, *Hearings on Department of Agriculture, Elimination of Agricultural Research Stations and Lines of Research,* 89th Cong., 1st sess. (Washington, D.C.: GPO, 1965), 383–402, 579–603.

23. Julia I. Dalrymple, "Contributions of the United States Office of Education to Home Economics Teacher Education," in *Home Economics Teacher Education: Sixty Significant Years,* ed. Elizabeth M. Ray, AHEA Teacher Education Section Yearbook, 1 (Bloomington, Ill.: McKnight, 1981), 187–225. Lena Bailey and Beulah Sellers Davis, eds., *Home Economics Teacher Education: Seventy Significant Leaders,* AHEA Teacher

Education Section Yearbook, 2 (Bloomington, Ill.: McKnight, 1982), has short biographies of several of the USOE home economists. Virginia Thomas (ibid., 266–72) was particularly interesting, for she started out when home economics was segregated and rose to a high state position in West Virginia black home economics groups (among other things, she started the New Homemakers of America, a black group, which later merged with the Future Homemakers of America). Then when integration started, Thomas found a place among white agency officials at the USOE, earned a doctorate, and then served for many years on the faculty at Iowa State University. See also Helen Pundt, *AHEA: The History of Excellence* (Washington, D.C.: AHEA, 1980), 336 ("Resolutions"); and Hugh Davis Graham, *The Uncertain Triumph: Federal Education Policy in the Kennedy and Johnson Years* (Chapel Hill: Univ. of North Carolina Press, 1984), 97–102.

24. Judith Sealander, *As Minority Becomes Majority: Federal Reaction to the Phenomenon of Women in the Work Force, 1920–1963* (Westport, Conn.: Greenwood, 1983), ch. 7. On Zapoleon, see *Who's Who in the South and Southwest,* 16th ed. (Chicago, 1978), 815; on Peterson, *Who's Who in America,* 43d ed. (1984–85): 2:2574. See also Carole Levin, "Women's Bureau," in *Government Agencies,* ed. Donald R. Whitnah (Westport, Conn.: Greenwood, 1983), 623–29.

25. On Rees, see *AMWS,* 12th ed., 1972, 5:5155; "Who's Who in Naval Research," *ONR Research Reviews,* Feb. 1952, 13–14; *CBY, 1957,* 453–55; "Award for Distinguished Service to Mathematics," *American Mathematical Monthly* 69 (1962): 185–87; Mina Rees, "The Computing Program of the Office of Naval Research, 1946–1953," *Annals of the History of Computing* 4 (1982): 102–20; and Fred D. Rigby, "Pioneering in the Federal Support of Statistics Research," in *On the History of Statistics and Probability,* ed. D. B. Owen (New York: Marcel Dekker, 1976), 403–18. There are two oral history interviews with her: Uta Merzbach, "Interview with Mina Rees, March 19, 1969," transcript in Archives, NMAH, Smithsonian Institution, Washington, D.C.; and an interview by Rosamund Dana and Peter J. Hilton in *Mathematical People: Profiles and Interviews,* ed. Donald J. Albers and G. L. Alexanderson (Boston: Birkauser, 1985), 255–67. There is some correspondence in box 34 of the Alan T. Waterman Papers, Manuscript Division, LC, and there is a letter from Rees to Miss Frieda Miller, Apr. 17, 1947, on *The Outlook for Women in Mathematics* (eventually Bulletin 223-4) in box 116a, WB Papers.

Statistician Dorothy Gilford later headed this same division (*AMWS,* 12th ed., 1972, 2:2134; W. Allen Wallis, "'The President's Column," *American Statistician* 19 [Apr. 1965]: 2; Gertrude M. Cox to Mrs. Leon Gilford, Nov. 9, 1965, Gertrude Cox Papers, North Carolina State University Archives, Raleigh). Evelyn Pruitt headed its geography program from 1959 to 1973 (*AMWS,* 13th ed., 1976, 5:3554). See also Harvey M. Sapolsky, *Science and the Navy: The History of the Office of Naval Research* (Princeton: Princeton Univ. Press, 1990).

26. NIH, *Scientific and Administrative Personnel in the Division of Research Grants, NIH* (Bethesda, Md., 1963).

27. Marian W. Kies, "It Takes More Than Luck," in *Women Scientists: The Road to Liberation,* ed. Derek Richter (London: Macmillan, 1982), 24–43 (quotation from 38). On Kies, see *AMWS,* 16th ed., 1986, 4:314.

28. J. Merton England, *A Patron for Pure Science: The National Science Foundation's*

Formative Years, 1945–57 (Washington, D.C.: NSF, 1982), 215–16; Green then had a long career at the Roscoe Jackson Memorial Laboratory in Bar Harbor, Maine (*AMWS*, 13th ed., 1976, 2:1620). On Hogg, see ibid., 16th ed., 1986, 3:766; Yost, *Women of Modern Science*, 31–47; and David DeVorkin, "Helen Sawyer Hogg, Transcript of an Interview, 17 August 1979," 11–12, NBL.

29. On Anderson, *WWAW* 4 (1966–67): 35. On Embrey, see ch. 15, n. 46; on Sides, ch. 10, n. 30. The names and titles of NSF personnel can be found in the NSF's *Annual Reports* (Washington, D.C., 1951–). Similarly, at the program level, although titles were revealing of institutional hierarchies, they could be misleading, and some seemingly subordinate "assistant program officers" made considerable contributions, such as Estelle ("Keppie") Engel in biochemistry (thanked in Martin D. Kamen, *Radiant Science, Dark Politics: A Memoir of the Nuclear Age* [Berkeley and Los Angeles: Univ. of California Press, 1985], 275); Josephine Doherty, long in ecology (Forest Stearns, "Distinguished Service Citation," *Bulletin of the Ecological Society of America* 67 [1986]: 42–43); and Mary Bostain Greene Seymour in anthropology ("Fellowships and Awards: American Anthropological Association," *Chronicle of Higher Education,* Dec. 2, 1987, A8).

30. On Olson, see "Lois Olson, 1899–1977," *Professional Geographer* 29 (1977): 248.

31. On Watt, see *AMWS*, 13th ed., 1976, 6:4733; "1980 Awardees," *Journal of Nutrition* 110 (1980): 1723 (she had earlier won the USDA's Distinguished Service Award). See also Margaret C. Schindler, "The Preparation of the *Bibliography of Agriculture,*" in *Bibliographic Organization*, ed. Jesse H. Shera and Margaret E. Egan (Chicago: Univ. of Chicago Press, 1951), 226–35; and Carl F. W. Muesebeck, "Obituary: Luella Walkley Muesebeck, 1905–81," *Proceedings of the Entomological Society of Washington* 84 (1982): 864–66.

32. On Way, see *AMWS*, 16th ed., 1986, 7:465, and Leona Marshall Libby, *The Uranium People* (New York: Crane Russak, Charles Scribner's Sons, 1979), 116; for additional material on Way, see T. Lauritsen Papers, California Institute of Technology Archives, Pasadena, and the Papers of the Physics Department, Duke University Archives, Durham, N.C. On Sitterly, see obituary, *Washington Post*, Mar. 2, 1990, C-6; *AMWS*, 16th ed., 1986, 6:788; *CBY, 1962,* 391–93; and Rexmond C. Cochrane, *Measures for Progress: A History of the National Bureau of Standards* (Washington, D.C.: Department of Commerce, 1966), 471 n. 118. See also David DeVorkin, "Charlotte M. Sitterly, Transcript of an Interview, June 15, 1978," NBL; and "The Eistophos Science Club of Washington, D.C., The 75th Anniversary Papers, 1893–1968," typescript in Papers of the Eistophos Club, Rare Book Room, LC.

33. On the Deignans, see *AMS*, 11th ed., P&B, 1965, 2:1174; see also William H. Fitspatrick and Monroe E. Freeman, "The Science Information Exchange: The Evolution of a Unique Information Storage and Retrieval System," *Libri* 15 (1965): 127–37.

34. Very informative is Jane H. Wallace's "Women in the Survey," *Geotimes* 24 (Mar. 1979): 34 (I thank Michele Aldrich for a copy). On Hooker, see *AMWS*, 13th ed., 1976, 3:1987–88; "Marjorie Hooker," *Journal of the Washington Academy of Sciences* 66 (1976): 162–63; and Anna Jespersen, "Memorial to Marjorie Hooker, 1908–1976," GSA *Memorials* 8 (1978). On Jespersen, see *AMS*, 11th ed., P&B, 1966, 3:2576–77; and Phyllis Dedekam, "Memorial to Anna Jespersen, 1895–1989," GSA *Memorials* 23 (1993), 33. See

also Jewell J. Glass to Dr. Frieda S. Miller, May 6, 1947, and Eleanora Bliss Knopf to Miss Anne Larrabee, Mar. 18, 1947, both in box 117, WB Papers.

35. Mary E. Dowse, "Memorial to Alice Mary Dowse Weeks, 1909–1988," GSA *Memorials* 20 (1990), 28; *AMWS*, 13th ed., 1976, 6:4753.

36. On Krieger, see *AMWS*, 14th ed., 1979, 4:2767, and Wallace, "Women in the Survey." Krieger and her camp surely merit further study. The frequently proclaimed rigors of fieldwork may have been a bit overstated at times, for one woman paleontologist was later praised for discovering certain things at her slow pace that others rushed by: "Although afflicted by a rheumatic heart throughout her professional life Helen [Duncan] undertook fieldwork in the Rockies and Great Basin, and by finding diagnostic fossils (when she sat down for her much needed rest) in a formation hitherto considered unfossiliferous, she demonstrated conclusively the folly of hurrying a paleontologist" (Rousseau H. Flower and Jean M. Berdan, "Memorial to Helen Duncan, 1910–1971," GSA *Memorials* 5 [1977]). Duncan's papers are at the Smithsonian Institution Archives.

37. On Ruth Hooker, see *WWAW* 11 (1979–80): 382.

38. On Lincoln, see *AMWS*, 16th ed., 1986, 4:776; J. Virginia Lincoln, "Radio Warning . . . A Woman in a Man's World," *WAM* 48 (1963–64): 90–91, 109; and idem,, "Activities at World Data Center–A for Solar-Terrestrial Physics," *Telecommunications Journal* 42 (1975): 730–33.

39. One of the few sources on the modern period is a short article by Louis Lasagna, "1933–1968: The FDA, the Drug Industry, the Medical Profession, and the Public," in *Safeguarding the Public: Historical Aspects of Medicinal Drug Control*, ed. John B. Blake (Baltimore: Johns Hopkins Press, 1970), 171–79. On Bowman, see *AMWS*, 12th ed., 1971, 1:622. The Frances W. Bowman Papers are in the East Carolina Manuscript Collection at East Carolina University, Greenville, N.C. At her job she dealt with Margaret Pittman of NIH and several women scientists in industry.

40. On Eddy, see *AMWS*, 14th ed., 1979, 2:1307; O'Hern, *Profiles of Pioneer Women Scientists*, ch. 13; oral history of Bernice Eddy by Wyndham Miles, Nov. 3, 1964, and folder of clippings, including 1953 award nomination, History of Medicine Division, NLM; Sen. Abraham Ribicoff, "Vaccine Safety," *Congressional Record* (Oct. 15, 1971), 117, pt. 28:36, 369–77; "Vaccine Regulations," ibid. (Dec. 8, 1971), 117, pt. 35:45, 386–98. Nicholas Wade covered the hearings in *Science* between February and April 1972. See also Bert Spector, "The Great Salk Vaccine Mess," *Antioch Review* 38 (1980): 291–303. For more on whistleblowers, see Deena Weinstein, *Bureaucratic Opposition: Challenging Abuses at the Workplace* (New York: Pergamon Press, 1979). Women may possibly be more likely than men to become "whistleblowers," but just how to investigate this is not clear.

41. On Kelsey, see *AMWS*, 16th ed., 1986, 4:268; *CBY, 1965*, 218–20; John Lear, "The Unfinished Story of Thalidomide," *Saturday Review*, Sept. 1, 1962, 35–40; Will Jonathan, "The Feminine Conscience of FDA: Dr. Frances Oldham Kelsey," ibid., 41–43; and "President's Awards, 1962," *Civil Service Journal* 3, no . 2 (Oct.–Dec. 1962): 4. See also Margaret Truman, *Women of Courage* (New York: William Morrow, 1976), 219–39.

42. CSC, Office of Career Development, Incentive Awards Office, *Awards and Honors for Scientists and Engineers* (Washington, D.C., 1963); CSC Library, *Productivity,*

Motivation, and Incentive Awards, Personnel Bibliography Series, no. 16 (Washington, D.C., 1965). See also Robert W. Van de Velde, *The Rockefeller Public Service Awards* (Princeton: Princeton Univ. Press, 1967); and the program for the final awards presentation ceremony in December 1981, which has a retrospective list of all the winners (I thank Earle E. Coleman, Princeton University archivist, for his help). On Campbell, see *AMWS,* 16th ed., 1986, 2:26; and Eleanor Swent, "Ian and Catherine Campbell, Geologists: Teaching, Government Service, Editing," 1988, xi, Bancroft Library, University of California at Berkeley.

43. On Price, see *AMWS,* 13th ed., 1976, 5:3541; and [Ruth Roettinger], *Idealism at Work: Eighty Years of AAUW Fellowships, A Report by the American Association of University Women* (Washington, D.C.: AAUW, 1967), 72.

44. On Gunderson, see *WWAW* 3 (1964–66): 408; and CSC Library, *Fifty United States Civil Service Commissioners: Biographical Sketches, Biographical Sources, Writings* (Washington, D.C., 1971); press release announcing the Federal Woman's Award, Oct. 11, 1960, and flyer of eligibility requirements, f. "Federal-W," box 1009, Alphabetical File, Dwight D. Eisenhower's Papers as President, Dwight D. Eisenhower Library, Abilene, Kansas; Frances Lide, "Awardees' Event Is Expected to 'Find' Women Leaders," *Washington Star,* Oct. 12, 1960, clipping in carton 4, f. EC8b, ACE-CEW Papers, SLRC; and Barbara Bates Gunderson, "The Federal Woman's Award," *Civil Service Journal* 1, no. 3 (Jan.–Mar. 1961): 2–4, 12. See also Jeannette Cheek, "Interview with Ida Craven Merriam, Nov. 20, 22 and 23, 1982," 138–39, SLRC. Some regions of the CSC and other agencies created their own women's awards; for example, Catherine Campbell despite her doctorate and her married status, won the "Miss Federal Employee" award for the Los Angeles region of the CSC in 1960 (Swent, "Ian and Catherine Campbell," xi, 69).

45. "Confirmation of Oveta Culp Hobby," in U.S. Congress, Senate, Committee on Finance, *Hearing on Nominations of George M. Marshall, Secretary of the Treasury–Designate and Oveta C. Hobby, Federal Security Administrator–Designate,* 83d Cong., 1st sess., 1953, 23–28. There are many clippings about her at the Eisenhower Library, Abilene.

46. A 1947 study of 730 top federal employees (P-4 and above) found that women were "less than 8 per cent" of the total but 39.3 percent of those at the Department of Labor (because of its WB and the BLS) and 28.4 percent of those at the Federal Security Agency, as HEW was known then (because of its Children's Bureau and the Social Security Administration). At that time the CSC had no systematic data on the sex distribution of employees by agency and grade level. As those data became more widely available in the 1950s, the sex inequities became more visible. (Frances T. Cahn, *Federal Employees in War and Peace: Selection, Placement, Removal* [Washington, D.C.: Brookings Institution, 1949], 176–77, 184–85).

W. Lloyd Warner, Paul P. Van Riper, Norman H. Martin, and Orvis F. Collins, *The American Federal Executive* (New Haven: Yale Univ. Press, 1963), ch. 11, and idem, "Women Executives in the Federal Government," *Public Personnel Review* 23 (1962): 227–34, were based on 145 women at GS-14 or higher (just 1.12 percent of the 12,929 persons studied). See also John J. Corson and R. Shale Paul, *Men Near the Top: Filling Key Posts in the Federal Service* (Baltimore: The Johns Hopkins Press, 1966); Mary M. Lepper, "A Study of Career Structures of Federal Executives: A Focus on Women," in *Women in Politics,* ed. Jane S. Jaquette (New York: John Wiley & Sons, 1974), 116–17; and Dorothy E. Walt, "The Motivation for Women to Work in High-Level Professional Positions" (Ph.D. diss., American University, 1962), on 50 federally employed women.

47. On Switzer, see *NAWMP;* and Martha L. Walker, *Beyond Bureaucracy: Mary Elizabeth Switzer and Rehabilitation* (Lanham, Md.: Univ. Press of America, 1985).

48. On Eliot, see *NCAB* 60 (1981): 228–29; and William H. Schmidt, "Martha May Eliot, M.D., 1891–1978," *American Public Health Association News* 68 (1978): 696–700. See also Dorothy E. Bradbury, *Four Decades of Action for Children: A Short History of the Children's Bureau* (Washington, D.C.: GPO, [1956]). There is an extensive collection of Eliot's papers at SLRC.

49. On Ottinger, see U.S. Congress, Senate, Committee on Labor and Public Welfare, *Hearing on Nominations,* 85th Cong., 1st sess., 1957, 1–4; and Polly Weaver Crone, "Children's Bureau Chief Interviewed," *SAQ* 58, no. 1 (Feb. 1967): 11–14.

50. Frieda Miller was already chief of the WB. On Wickens, see obituary, *NYT,* Feb. 8, 1991, B-5; *CBY, 1962,* 462–64; "The Reminiscences of Mrs. A. J. Wickens," 1957, Columbia University Oral History Research Office; and George Martin, *Madam Secretary: Frances Perkins* (Boston: Houghton Mifflin, 1976), 453–55. There is a collection of her papers at SLRC. On Beyer, see *WWAW* 5 (1968–69): 107; obituary, *NYT,* Sept. 28, 1990, A-18; and the large collection of her papers at SLRC. See also Joseph P. Goldberg and William T. Moye, *The First Hundred Years of the Bureau of Labor Statistics* (Washington, D.C.: GPO, 1985).

51. On Merriam, see *AMWS,* 13th ed., S&B, 1978, 820; and Jeannette Cheek, "Interview with Ida Craven Merriam, November 20, 22, and 23, 1982," SLRC; see also Martha Derthick, *Policy-Making for Social Security* (Washington, D.C.: Brookings Institution, 1979), 19.

52. On Wood, see *AMWS,* 13th ed., S&B, 1978, 1305.

53. On Aitchison, see ibid., 13; there is a small collection of her papers at SLRC.

54. On Stinson, see James H. Winchester, "Lady Engineer," *(Senior) Scholastic* 66 (Feb. 23, 1955): 4; Jerri Cuthbertson, "Katharine Stinson: The Right Stuff," *North Carolina State Alumni Magazine* 56 (Mar. 1984): 10–11; and Kenneth Leish, "The Reminiscences of Katherine *[sic]* Stinson," July 1960, Columbia University Oral History Office.

55. See Mary Finch Hoyt, *American Women of the Space Age* (New York: Atheneum, 1966), which was aimed at a juvenile audience. On Roman, see *AMWS,* 16th ed., 1986, 6:266–67; "Dr. Nancy Roman: Transcript of an Interview by David DeVorkin, August 19, 1980," NBL; Joseph N. Tatarewicz, *Space Technology and Planetary Astronomy* (Bloomington: Indiana Univ. Press, 1990), 42–43, 46, 48. See also Nancy Roman, "The Role of Women in the Space Program" (address at Marymount College, Tarrytown-on-the Hudson, N.Y., Jan. 31, 1963), press release, copy in Library, Business and Professional Women's Foundation, Washington, D.C., 13–14.

56. On Townsend, see Constance Holden, "NASA Satellite Project: The Boss Is a Woman," *Science* 179 (1973): 48–49; and Maria Purl, "One Giant Step for Womankind:

Equal Opportunity Gets Off the Gound," *Working Woman* 5 (May 1980): 54–58, 60–61, and 109. Her papers are at the archives at the Virginia Polytechnic Institute and State University, Blacksburg.

57. On Gill, see *AMWS*, 14th ed., 1979, 3:1727; Jocelyn R. Gill, "The Exploration of Space," *WAM* 46 (1962): 276–79; and "A Cornerstone," in *Wellesley After-Images* (Los Angeles: Wellesley Club of Los Angeles, 1974), 102–4.

58. On Anderson, see *AMS*, 11th ed., 1965, 1:96. Others worked for the space program in agencies other than NASA. For example, in 1959 psychologist Mildred Mitchell left her job at the VA hospital in Dayton, Ohio, to join a project at Wright-Patterson Air Force Base testing future astronauts. On Mitchell, see "Mildred B. Mitchell" (auto-biography); Mitchell, "Careers in Applied Psychology"; and Hoyt, *American Women in the Space Age*, 74–79.

59. U.S. Congress, House, Committee on Science and Astronautics, Special Sub-committee on the Selection of Astronauts, *Hearings on Qualifications for Astronauts, July 17 and 18, 1962*, 87th Cong., 2d sess., 1962; Ken Hechler, *Toward the Endless Frontier: A History of the House Committee on Science and Technology, 1959–1979* (San Diego: American Astronautical Society, 1982); Erik Bergaust, "Is Space a Place for the Ladies?" *This Week Magazine*, Sept. 2, 1962, 12, 14 (in microfilm of *Detroit News*); "Astronauttes?" *Newsweek*, July 30, 1962, 17–18; Louis Lasagna, "Why Not 'Astronauttes' Also?" *New York Times Magazine*, Oct. 21, 1962, 52–64. On Cobb, see *CBY, 1961*, 106–7; *WWAW* 8 (1974–75): 176; and Jerrie Cobb with Jane Rieker, *The Jerrie Cobb Story* (Englewood Cliffs, N.J.: Prentice-Hall, 1963). See also W. Joan McCullough, "13 Who Were Left Behind," *Ms.* 2 (Sept. 1973): 41–45 (I thank Deborah Giannoni for a copy.); and Roger Wheeler and Philip Snowdon, "American Women in Space," *Journal of the British Interplanetary Society* 40 (1987): 81–88. (I thank Sharon Valiant for sharing her material on this topic.)

One of the women tested in 1961 was electronics engineer Janet Guthrie, who gained greater fame in 1978 as the first woman race car driver to compete at the Indianapolis 500. On Guthrie, see *CBY, 1978*, 182–85; Red Smith, "Fast Woman," *NYT*, Apr. 16, 1978, Sports section, 3; and Robert Amon, "Engineering Talent in Short Supply," *Industrial Bulletin* (New York State Department of Labor) 41 (June 1962): 2–5. On Tereshkova, see Valentina Tereshkova-Nikolayeva, "Women in Space," *Impact of Science on Society* 20 (Jan.–Mar. 1970): 5–12.

60. See, e.g., Roman, "The Role of Women in the Space Program," 13–14. See also "Resolutions Adopted at 1963 National Convention," *National Business Woman* 42 (Aug. 1963): 23; and "Resolutions Adopted at the 1964 National Convention," ibid. 43 (Sept. 1964): 29–30. See also Northern California chapter president Nita Ladewig's "NITA and NASA: A Case History of Male Chauvinism in the Space Program," *NOW Acts* 2, no. 1 (winter–spring 1969): 16–19.

61. For a full account of the commission, see Cynthia Harrison, *On Account of Sex: The Politics of Women's Issues, 1945–1968* (Berkeley and Los Angeles: Univ. of California Press, 1988); on the Committee on Federal Employment, see ibid., 142–46. See also Sealander, *As Minority Becomes Majority*, ch. 7; Judith Paterson, *Be Somebody: A Biography of Marguerite Rawalt* (Austin, Tex.: Eakin, 1986), chs. 5, 6; and Patricia G. Zelman, *Women, Work, and National Policy: The Kennedy-Johnson Years* (Ann Arbor: UMI Research Press, 1982), 28–33.

Officials of Sigma Delta Epsilon and the AAUW were among the many women's groups that offered their support to the new commission (Ernestine B. Thurman to Mrs. Eleanor Roosevelt, Jan. 29, 1962 and reply; and Thelma M. Herlin to President John F. Kennedy, Nov. 5, 1962, box 1, PCSW Papers, John F. Kennedy Presidential Library, Boston). So did Alan T. Waterman of the NSF (Alan T. Waterman to Eleanor Roosevelt, Sept. 7, 1962, enclosing his 1959 speech [see ch. 15, n. 46], Alan T. Waterman Papers).

62. On Hickey, see *WWAW* 11 (1979–80): 366, and *CBY, 1944,* 293; her papers and scrapbooks are at the Archives, University of Missouri at St. Louis. On Harrison, see *WWAW* 5 (1968–69): 517. Harrison wrote numerous articles on the commission and working women: "The Quiet Revolution," *Civil Service Journal* 3, no. 2 (Oct.–Dec. 1962): 5–7, 22–24; "Facts, Not Fancy, about Women in the Federal Service," ibid. 4, no. 2, (Oct.–Dec. 1963): 21–25; "The Working Woman: Barriers in Employment," *Public Administration Review* 24 (1964): 78–85; "Talent Search for Womanpower," *AAUWJ* 58 (1965): 99–101; and interview in "CSC Official Says Women Occupy High Positions in Government," *Federal Times,* Dec. 14, 1966, 13. Harold Leich mentioned Harrison's recent retirement in "Opportunities for Women Engineers in the Federal Service," in *Women in Engineering: Bridging the Gap Between Science and Technology,* ed. George Bugliarello et al. (Chicago: University of Illinois at Chicago Circle, 1972), 64–69.

63. Lucy S. Howorth to Miss Margaret Hickey, Mar. 5, 1963, and reply, Mar. 15, 1963, both in addenda, carton 3, f. 101, Somerville-Howorth Papers, SLRC.

64. John Macy, Jr., "Employment Policies and Practices of the Federal Government (Report No 2)," Apr. 9, 1962, reports the appointing officers' explanations of why they sought only male candidates (box 5, PCSW Papers, including the chart "Requests for Certificates Specifying Sex"). See also John W. Macy, Jr., *Public Service: The Human Side of Government* (New York: Harper & Row, 1971), ch. 7. Each committee published its own report, e.g., *Report of the Committee on Federal Employment to the President's Commission on the Status of Women* (Washington, D.C.: GPO, 1963), with several useful appendixes prepared by Evelyn Harrison's Bureau of Programs and Standards at the CSC.

65. "PSAC Panel on Scientific and Technical Manpower, Minutes of Meeting, December 19, 1962," 4–5, box 190, Office of Science and Technology Papers (RG 359), NARA. Evidently there had been some interest in women in science at the White House the previous month, since on November 21, 1962, Alan T. Waterman had written Jerome B. Wiesner, special assistant to the president for science and technology, "Following our appointment at the White House yesterday, it occurred to me that the President might be interested in a copy of our report, *Women in Scientific Careers.* I am enclosing a couple of copies in the event you wish to bring it to his attention." (box 265, ibid.).

66. Lady Bird Johnson, *A White House Diary* (New York: Holt, Rinehart & Winston, 1970), 51–62 passim; "Women are Appointed to Key Posts at AEC, Export-Import Bank," *WSJ,* Mar. 30, 1964, 3. On May, see *WWAW* 5 (1968–69): 785. See also "Virginia Mae Brown Dead at 67; First Woman to Head the I.C.C.," *NYT,* Feb. 27, 1991, B-12. The list "Major Female Appointments of President Johnson," app. A to a printout of Johnson appointees in the Office Files of John Macy, gives names of only ten women appointed to full-time jobs at GS-17 level or above (I thank Mary Knill of the Lyndon

Baines Johnson Library, Austin, Tex., for sending me copies). See also Alan Whitney, "Ladies Gaining Top Jobs, Prestige in Government," *Federal Times,* Oct. 13, 1965, 6, 7; Philip Abelson, "A Special Opportunity," *Science* 145 (1964): 115; and Macy, *Public Service,* ch. 7.

Anthropologist Margaret Mead was the only woman to serve in 1964–66 on the White House's "central group of domestic experts," which advised Eric Goldman on civil rights, poverty, unemployment, Vietnam, and women's issues. In particular she advised Goldman on how to publicize the top women's appointments and Mrs. Johnson's trip and speech at Radcliffe College in June 1964 (box E166, MMP).

67. *WSJ,* June 29, 1965, 1, col. 5.

68. Federal Woman's Award Study Group on Careers for Women, *Progress Report to the President, March 3, 1967* (Washington, D.C., 1967). See also Cheek, "Interview with Ida Craven Merriam," 138; and Charlotte Moore Sitterly, "Women in Science—The Changing Picture," in "The Eistophos Science Club of Washington, D.C., The 75th Anniversary Papers, 1893–1968," 68–76; press release, Backup File, Oct. 13, 1967, Lyndon Baines Johnson Presidential Papers, Lyndon Baines Johnson Library.

69. J. Philip Bohart, "The Federal Women's Program," *Civil Service Journal* 10, no. 1 (July–Sept. 1969): 16–18; Helene S. Markoff, "The Federal Women's Program," *Public Administration Review* 32 (1972): 144–51; CSC, Statistics Section, *Study of Employment of Women in the Federal Government [1966 and] 1967* (Washington, D.C.: GPO, 1968).

70. The NSF's National Register counted only 1,084 federally employed women scientists in 1954, compared with the 4,303 given in the comprehensive CSC survey cited in table 13.1. Thus it is hard to know how to interpret the Register's probably equally incomplete 1968 data, which showed large increases for women in statistics, where they jumped from just 8 in 1956–58 to 101 in 1968, and in the social sciences, where there was a hundredfold increase, from 2 to 197.

71. Hugh Heclo, *A Government of Strangers: Executive Politics in Washington* (Washington, D.C.: Brookings Institution, 1977), esp. 61–64; Arthur J. Gartaganis, "Trends in Federal Employment, 1958–1972," *Monthly Labor Review* 97 (Oct. 1974): 17.

72. BLS (for NSF), *Employment of Scientific and Technical Personnel in State Government Agencies: Report on a 1959 Survey* (Washington, D.C.: GPO, n.d.); *Employment of Scientific and Technical Personnel in State Government Agencies, 1962,* BLS Bulletin 1412 (Washington, D.C., 1964); *Employment of Scientific, Professional, and Technical Personnel in State Governments, January 1964,* BLS Bulletin 1557 (Washington, D.C., 1967); NSF, *Research and Development in State Government Agencies, Fiscal Years 1967 and 1968,* Surveys of Science Resources Series (Washington, D.C.: GPO, 1970); Edith Wall Andrew and Maurice Moylan, "Scientific and Professional Employment by State Governments," *Monthly Labor Review* 92 (Apr. 1969): 40–45.

73. Anna M. Sexton, *A Chronicle of the Division of Laboratories and Research, the New York State Department of Health: The First Fifty Years, 1914–1964* (Lunenberg, Vt.: Stinehour, 1967); "Polio's Little Brother," *Time,* Oct. 19, 1959, 61; J. R. Paul, *History of Poliomyelitis* (New Haven: Yale Univ. Press, 1971), 397–99. On Hazen, see *NAWMP* and O'Hern, *Profiles of Pioneer Women Scientists,* ch. 8; on Brown, see Yost, *Women of Modern Science,* 64–79; and on both Hazen and Brown, Richard S. Baldwin, *The Fungus Fighters: Two Women Scientists and Their Discovery* (Ithaca: Cornell Univ. Press, 1981).

74. On Moore, see *AMS,* 10th ed., 1961, 3:2840; on Kirkbride, Sexton, *Chronicle of the Division,* 168–69; on Gilbert, ibid., 167; on Andrus, *AMWS,* 12th ed., S&B, 1973, 1:57, and Jane D. Hildreth and Carolyn L. Konold, eds., *American Psychological Association Directory, 1951* (Washington, D.C.: APA, [1951]), 11; on Cornell, vita and clippings in Ethel Cornell Papers, KLCU (see also *The New York State Education Department, 1900–1965* [Albany: University of the State of New York, State Education Department, Division of Research, 1967]); and on Goldring, *NAWMP.* See also Charles E. Davis, "A Survey of Veterans' Preference Legislation in the States," *State Government* 53 (1980): 188–91.

75. On Kendrick, see O'Hern, *Profiles of Pioneer Women Scientists,* ch. 19; *AMS,* 11th ed., 1966, 3:2748; and clippings in Pearl Kendrick Biographical File, Bentley Historical Library, Ann Arbor, Mich. On Gillette, see "Obituaries: Helen Hawthorne Gillette," *A[merican] S[ociety of] M[icrobiology] News* 41, no. 1 (Jan. 1975): 50–51; and on Hirschberg, *AMWS,* 16th ed., 1986, 3:727, and Nell Hirschberg, "'Lab' Inspector," *SAQ* 42 (1950–51): 144–45. Greenfield is mentioned with Kendrick and others in Elizabeth D. Robinton, "A Tribute to Women Leaders in the Laboratory Section of the American Public Health Association," *American Journal of Public Health* 64 (1974): 1006–7.

76. On Griffin, see Katharine M. Banham to Dorothy Park Griffin, Feb. 28, 1956, Katharine M. Banham to Mary G. Clarke (letter of recommendation for Griffin), Jan. 22, 1969, and press release, Sept. 3, 1971, on fiftieth anniversary of North Carolina Department of Social Services, all in box 13, Katharine Banham Papers, Duke University Archives. On Turner, see Robert V. Guthrie, *Even the Rat Was White: A Historical View of Psychology* (New York: Harper & Row, 1976), 146–48 with photo.

77. For example, Charlotte Pratt in New York State ("Fellows-Elect of American Society for Horticultural Science, Class of 1981," *HortScience* 16 [1981]: 710); Edith C. Higgins in North Dakota (clipping, "New Building Enhances State Seed Department Service," *Fargo Forum,* Mar. 8, 1957, in file on North Dakota State Seed Laboratory, Institute for Regional Studies, North Dakota State University, Fargo); and Pearl Swanson in Iowa (*AMS,* 11th ed., 1967, 6:5277).

78. For example, Erna Gunther (see *AMS,* 11th ed., S&B, 1968, 1:619; there is a substantial collection of her papers at the University of Washington Archives in Seattle and a small one in the Alaska and Polar Regions Department, Ramussen Library, University of Alaska at Fairbanks); and Dorothy Jensen (see *AMS,* 11th ed., S&B, 1968, 1:782; Francis P. Conant, "Dorothy Cross Jensen, 1906–1972," *American Anthropologist* 76 [1974]: 80–82; and New Jersey Department of Civil Service, Career Development Office, *Women in New Jersey State Government* [Trenton: Department of Civil Service, 1974], 11).

79. For example, Louise Jordan in Oklahoma (see *AMS,* 11th ed., 1966, 3:2644; and Elizabeth A. Ham, *A History of the Oklahoma Geological Survey, 1908–1983,* Oklahoma Geological Survey Special Publication 83-2 [Norman, 1983], 36–37); Virginia Kline in Illinois (see Margaret O. Oros, "Memorial to Virginia Harriett Kline, 1910–1959," *Proceedings Volume of the Geological Society of America, Inc., for 1960* [Boulder, Colo.: GSA, (?1960)], 114–17); and Josie McGlamery in Alabama (see Charles W. Copeland, "Memorial to Josie Winifred McGlamery, 1887–1977," GSA *Memorials* 9 [1979]).

80. Louise Gohdes, "Lady Engineers: Professional Role of Distaffers in Highway Work Increasing," *California Highways and Public Works* 42 (Jan.–Feb. 1963): 46–49.

81. Among those who worked for state conservation agencies were Charles and Elizabeth Schwartz in Missouri (*Something about the Author* 8 [1976]: 184–86); and Ruth Hine in Wisconsin (*Journal of Soil and Water Conservation* 23 [July 1968]: 151). On the Hamerstroms, see *AMWS*, 15th ed., 1982, 3:440; Frances Hamerstrom, *Strictly for the Chickens* (Ames: Iowa State Univ. Press, 1980); and idem, *Is She Coming Too? Memoirs of a Lady Hunter* (Ames: Iowa State Univ. Press, 1989).

82. On Clark, see *AMWS*, 12th ed., 1971, 1:1042; and "Dr. Frances Clark Retires as Laboratory Director," *Pacific Fisherman* 54 (May 1956): 55. She is mentioned in Elizabeth Noble Shor, *Scripps Institution of Oceanography: Probing the Oceans, 1936 to 1976* (San Diego: Tofua, 1978), 44, 67; and Arthur F. McEvoy, *The Fisherman's Problem: Ecology and Law in the California Fisheries, 1850–1980* (Cambridge: Cambridge Univ. Press, 1986), 158–63, 198–201. (I thank Harry Scheiber for calling Clark to my attention.)

83. On Howard, see *AMWS*, 16th ed., 1986, 3:844; Kenneth E. Campbell, Jr., ed., *Papers in Paleontology Honoring Hildegarde Howard*, published as Natural History Museum of Los Angeles County, *Contributions in Science*, no. 330 (Sept. 15, 1980), esp. vii–xxv; and [Gretchen Sibley], *Sixty-five and Still Growing: Anniversary Annals of the Museum*, published as *Terra* 16 (winter 1978) (I thank Dr. Sibley for her help). Howard may deserve a full biography.

84. On Aberle, see *AMWS*, 16th ed., 1986, 1:8. Her papers are at the Harry S. Truman Presidential Library in Independence, Missouri.

85. On Kidd, see "Presenting Two Who Need No Introduction," *IW* 33 (1954): 409; see also David M. Schneider, "Opportunities for Statistical Work in State and Local Governments," *Journal of the American Statistical Association* 40 (1945): 62–70.

86. On Goldfeder, see Anna Goldfeder, "An Overview of Fifty Years in Cancer Research: Autobiographical Essay," *Cancer Research* 36 (1976): 1–9; *AMWS*, 15th ed., 1982, 3:188; and obituary, *NYT*, Feb. 18, 1993, B-11.

87. On Ratner, see "Garvan Medal, Sarah Ratner," *C&EN* 39 (1961): 103; Sarah Ratner, "A Long View of Nitrogen Metabolism" (autobiography), *Annual Reviews of Biochemistry* 46 (1977): 1–24 (quotation from 18). She was the first woman to be invited to present her autobiography in the *Annual Reviews*. In 1983 friends published a festschrift in her honor: Maynard Pullman, ed., *An Era in New York Biochemistry: A Festschrift for Sarah Ratner, Transactions of the New York Academy of Sciences*, 2d ser., 41 (Aug. 1983); of the thirty-nine contributors eleven were women.

88. On Baumgartner, see *CBY, 1950*, 22–24; "Sedgwick Memorial Medal for 1964," *American Journal of Public Health* 54 (1964): 2090–91; "Two Women Appointed to Cabinet of New York's New Mayor," *IW* 33 (Feb. 1954): 52; "She Guards the City's Health," ibid. 33 (Mar. 1954): 86–88, 119; O'Hern, *Profiles of Pioneer Women Scientists*, ch. 9; and Milton J. E. Senn, "Interview with Leona Baumgartner, April 15, 1974," in Milton Senn Child Study Oral Histories, History of Medicine Division, NLM. She is mentioned in Tom Rivers, *Reflections on a Life in Medicine and Science*, ed. Saul Benison (Cambridge: MIT Press, 1967), 393–96. She may merit a full biography, especially since there is a large collection of her papers at the Harvard Medical School's Countway Library. See also Peter Allan, "Career Patterns of Top Executives in New York City Government," *Public Personnel Review* 33 (1972): 114–17.

89. On Wormington, see obituary, *NYT*, June 2, 1994, D-23; and Cynthia Irwin-

Williams, "Women in the Field: The Role of Women in Archaeology before 1960," in *Women of Science: Righting the Record*, ed. G. Kass-Simon and Patricia Farnes (Bloomington: Indiana Univ. Press, 1990), 29–33. On MacLean, see Jeane Kenworthy, "Big City Traffic Planner," *National Business Woman* 40 (Feb. 1961): 9. On Flinn, see Helen Flinn Papers, Archives of the History of American Psychology, University of Akron, Akron, Ohio; and Lawrence S. Rogers, "Psychologists in Public Service and How They Grew," *AP* 12 (1957): 232–33, which, unfortunately, is marred by many arithmetical errors. On Untermann, see Alice White, "Obituary: Billie R. Untermann, 1906–1973," *Society of Vertebrate Paleontology News Bulletin*, no. 99 (1973): 65–66.

Any survey based, as the National Register was, on membership in national professional organizations would have seriously undercounted scientists at the local level, since many of them would have belonged only to state or regional groups. Perhaps disproportionately many of them would have been women. See ch. 5, n. 5.

Chapter 14   Invisibility and Underrecognition

1. The usual starting point for the history of scientific societies is Ralph S. Bates, *Scientific Societies in the United States*, 3d ed. (Cambridge: MIT Press, 1965), chs. 6 and 7. See also the section on women's scientific societies in the bibliographical essay below.

2. See, e.g., NSF, *Dues and Membership in Scientific Societies: Report of a Survey Conducted by the Office of Science Information Service* (Washington, D.C., 1960).

3. AAAS, *A Brief History of the Association, Its Present Organization and Operation, Summarized Proceedings, 1940–1948, and A Directory of Members as of December 31, 1947* (Washington, D.C., 1948), 500–515.

4. Andrew Fraser, "The 1955 Professional and Economic Survey of the Membership of the American Chemical Society," *C&EN* 34 (1956): 1746. Among the 29,237 chemists responding to the survey were 2,275 women. No sex was given, however, for 8,779 chemical engineers and 7,305 others who also belonged to the ACS. The overall proportion was probably closer to 5 percent ("Summary of the Report," ibid., 1731).

5. Margaret W. Mayall to Mrs. L. B. Zapoleon, Dec. 10, 1946, citing a May 1945 membership list, box 116, WB Papers (RG 86), NARA.

6. Mildred B. Mitchell, "Status of Women in the American Psychological Association," *AP* 6 (1951): 195.

7. G. E. Erikson, "A Statistical and Graphical Array of Data and Their Analyses Bearing on the History of the American Association of Anatomists," appendix to *The American Association of Anatomists, 1888–1987: Essays on the History of Anatomy in America and A Report on the Membership—Past and Present*, ed. John E. Pauly (Baltimore: Williams & Wilkins, 1987), 250.

8. Kay E. Leffler, "The Role of Women in Geography in the United States" (master's thesis, University of Oklahoma, 1965), 3, based on data in directories of the Association of American Geographers. See also Preston E. James and Geoffrey J. Martin, *The Association of American Geographers: The First Seventy-Five Years, 1904–1979* (Washington, D.C.: Association of American Geographers, 1978), 118–19.

9. Kenneth E. Clark, *American Psychologists: A Survey of a Growing Profession* (Washington, D.C.: APA, 1957), esp. 21, 71–72, as well as 160–61, 172, 182, 191–92. See also idem,

"The APA Study of Psychologists," *AP* 9 (1954): 17–20; Samuel Karson, Lenin A. Baler, and Herbert A. Carroll, "A Question of Values," ibid. 13 (1958): 243; and Kenneth E. Clark, "A Reply," ibid., 243–44.

10. Anselm L. Strauss and Lee Rainwater, *The Professional Scientist: A Study of American Chemists* (Chicago: Aldine, 1962), vi, 16–26, 136–37 (about the traits a chemist would like his son to have), 232–33, 270. Interviews with fifty-two other chemists were similarly excluded because the subjects were retired, foreign-born, or foreign-educated. See also "Board Committee Reports . . . Public, Professional and Member Relations," *C&EN* 38 (Oct. 31, 1960): 65–66; correspondence about it in the box 40, ACS Papers, Manuscript Division, LC; and several negative reviews (*Book Review Digest, 1963* [New York: H. W. Wilson, 1964], 973).

11. Mitchell, "Status of Women," 195; Roger G. Stewart, "Status of Fellows and Associates," *AP* 12 (1957): 273.

12. "Annual Report [of Marine Biological Laboratory for 1959]," in *Biological Bulletin* 119 (1960): 35–54. An even higher proportion of the associates were women (105 of 173, or 60.7 percent).

13. The other three were Ethel Browne Harvey of Princeton, New Jersey, 1951–54, Hope Hibbard of Oberlin College, 1952–59, and Mary Sears of Woods Hole Oceanographic Institution, 1955–62. See "[Annual] Report [of the Marine Biological Laboratory]," *Biological Bulletin* 79 (1940) through 139 (1970). The report also lists the names of those who taught in the laboratory's famous summer courses. Among these over the years were Hannah Croasdale of Dartmouth College Medical School, Madelene Pierce of Vassar, Mary D. Rogick of the College of New Rochelle, Marion Pettibone of the University of New Hampshire and after 1963 the Smithsonian Institution, Muriel Sandeen of Duke, and several others. See also "Opening Doors: Women, Science, and the MBL," *The Collecting Net* 6, no. 2 (fall 1988): 12–14, and the brief obituary of Hope Hibbard in the same issue, 28 (I thank Anne Simon Moffat for a copy); the brief account by Anne G. Maher in *Research Institutions and Learned Societies,* ed. Joseph C. Kiger (Westport, Conn.: Greenwood, 1982), 332–36; and Jane Maienschein's pictorial *One Hundred Years Exploring Life, 1888–1988: The Marine Biological Laboratory at Woods Hole* (Boston: Jones & Bartlett, 1989).

14. Edwin B. Eckel, *The Geological Society of America: Life History of a Learned Society,* GSA Memoir 155 (Boulder, Colo.: GSA, 1982), 34–39. A list of the names of the new fellows and members was published in its annual *Proceedings* until 1969.

15. Harold H. Williams, *History of the American Institute of Nutrition: The First Fifty Years,* chs. 5 and 6 and app. 2, published as part of *Fiftieth Anniversary of the Journal of Nutrition, September 27, 1978* (Bethesda, Md.: AIN, 1978).

16. On Clarke, see *NAWMP;* on Kellems, *CBY, 1948,* 340–42; on Rockwell, Alice C. Goff, *Women CAN Be Engineers* (Youngstown, Ohio: privately printed, 1946), 94–112; and on Rand, *NAWMP.*

17. On Rousseau, see *WWAW* 1 (1958–59): 627 (under her maiden name, Hutchinson); "Miss Chemical Engineer of 1955," *C&EN* 33 (1955): 3504; and Mary Dakis, Membership Department, AIChE Research, telephone conversation with author, July 5, 1994. On Burford, see L. N. Liggett, "Topside Coal Miner," *IW* 26 (1947): 125–26.

18. Elected to the SEP in 1970 was Dorothea Jameson Hurvich, a research scientist at

the University of Pennsylvania and the wife of another longtime member of the society, raising marriage to a member, as Laurel Furumoto has suggested, almost to a qualification of membership for a woman experimentalist (see Laurel Furumoto, "Shared Knowledge: The Experimentalists, 1904–1929," in *The Rise of Experimentation in American Psychology,* ed. Jill G. Morawski [New Haven: Yale Univ. Press, 1988], 108; list of SEP members, 1959–60, in LCP; *AMWS,* S&B, 13th ed., 1978, 601; and "Leo M. Hurvich and Dorothea Jameson" (joint autobiography), in *A History of Psychology in Autobiography,* ed. Gardner Lindzey, vol. 8 [Stanford: Stanford Univ. Press, 1989], 156–206).

19. "A Woman in the Chicago Club!" *Chicago Daily Times,* Nov. 12, 1942, 1, and "Signs of Times—Women at Chicago Club," *Chicago Sun,* Nov. 13, 1942, clippings in box 28, Icie Macy Hoobler Papers, Bentley Historical Library, Ann Arbor, Michigan (I thank Karen M. Mason for her help). See also Icie Macy Hoobler, *Boundless Horizons: Portrait of a Pioneer Woman Scientist* (Smithtown, N.Y.: Exposition Press, 1982), 161–62, where she recounts an (undated) incident at the New York City University Club in which she not only had to enter through a back hallway but also had to eat alone.

20. Jack Craig, "Inquiring Photographer," *The Percolator: Bulletin of the Chemists' Club,* Sept.–Oct. 1958, 11. Accompanied lady guests had long been allowed in certain parts of the building at specified hours. The Chemists' Club hired a female chemical librarian in 1967. See Paul B. Slawter, Jr., "The Chemists' Club at 75," CHEMTECH 3 (Dec. 1973): 718. See also *The Chemists' Club Yearbook for 1910–11* (New York: Chemists' Club, 1910); and D. H. Killeffer, *Decades of The Chemists' Club* (New York: Chemists' Club, 1957), which describes the premises in detail (I thank Mary Ellen Bowden and Arnold Thackray for assistance).

21. Elinor Langer, "Science Goes to Lunch," *Science* 146 (1964): 1145–49. In 1945, when the National Association of Science Writers, of which Jane Stafford was then president, held a dinner for a visiting Englishman at the Cosmos Club, she managed by virtue of her office to get in and preside ("National Association of Science Writers, Jane Stafford, Transcript of an Interview conducted by Bruce V. Lewenstein, 6 February 1987," 58–59 [I thank Lewenstein for letting me see a copy]).

22. Alden H. Emery, "Women Chemists as Section Chairmen," *C&EN* 30 (1952): 2526; Herman Skolnik and Kenneth M. Reese, *A Century of Chemistry: The Role of Chemists and the American Chemical Society* (Washington, D.C.: ACS, 1976), 406–41. There is also some relevant material in the Frances C. Brown Papers at the Duke University Archives, Durham, N.C. Brown headed the North Carolina section in 1956. In some fields certain local sections or societies were of particular importance. Among petroleum geologists, for example, Houston was central, and several women were active for a while in its local geological society (Alva C. Ellisor, *Rockhounds of Houston: An Informal History of the Houston Geological Society* [Houston: Houston Geological Society, 1947]).

23. James L. Pater, ed., *No Small Part: A History of Regional Organizations in American Psychology* (Washington, D.C.: American Psychological Assn., 1993), pp. 36, 48–49, 84.

24. "AAAS Officers, Committees, and Representatives for 1960," *Science* 131 (1960): 507. On Shields, see ch. 9 n. 27.

25. Skolnik and Reese, *A Century of Chemistry*, 401–6, has lists of chairs of divisions. The three women who headed the Division of Chemical Literature (later the Division of Chemical Information) were Hanna Friedenstein of the Cabot Corporation of Boston in 1959 (*AMWS*, 16th ed., 1986, 2:1195), Harriet Geer of Parke, Davis and Company in 1965 (ibid., 12th ed., 1972, 2:2081), and Helen Ginsburg of Schering Company and later of Abbott Laboratories in 1967 (ibid., 2:2149). See also W. V. Metanomski, *Fifty Years of Chemical Information in the American Chemical Society, 1943–93* ([Washington, D.C.]: ACS, Division of Chemical Information, 1993). Virginia Bartow chaired the Division of the History of Chemistry in 1952–54. Her papers at the University of Illinois Archives in Urbana have some relevant materials.

26. This is only a partial list. On Goodenough, see *NAWMP;* on McCarthy, Eileen Maxwell Canty to Barbara Sicherman, Mar. 8, 1977, box 26, f. 88, *NAWP;* on Duffy, *WWAW* 5 (1968–69): 335, and "Deaths," in University of North Carolina at Greensboro *Alumni News,* spring 1971, 31–32, copy in box 26, f. 88, *NAWP;* and on Roe, Robert L. Wrenn, "The Evolution of Anne Roe," *Journal of Counseling and Development* 63 (1985): 274. Roe had apparently been nominated for president of the APA and been narrowly defeated, but when asked to run a second time she had had to decline because of recent heart trouble (see also ch. 7, n. 49).

27. *C&EN* 30 (1952): 1514. There is a brief account of the earliest Gordon Conferences in AAAS, *A Brief History of the Association,* 67; Dael Wolfle, *Renewing a Scientific Society: The American Association for the Advancement of Science from World War II to 1970* (Washington, D.C.: AAAS, 1989), 115–17; and W. Gordon Parks, "Gordon Research Conferences: A Quarter Century on the Frontiers of Science," *Science* 124 (1956): 1279–81.

28. [Catharine Borras], "Women Officers of AAAS, 1885–1976" (I thank her for a copy). Cohn had also been chairman of the Women's Service Committee of the ACS, 1958–63 (*NCAB* 51 [1969]: 10; see also ch. 15). She deserves a biography.

29. Wolfle, *Renewing a Scientific Society,* ch. 7.

30. Ibid., 53–55; Carleton Mabee, "Margaret Mead's Approach to Controversial Public Issues: Racial Boycotts in the AAAS," *The Historian* 48 (1986): 191–208.

31. On both Taeubers, see *AMWS,* 12th ed., S&B, 1973, 2:2425; and see Irene Taeuber's obituary, *NYT,* Feb. 26, 1974, 40.

32. *American Journal of Physics* 8 (1940)–38 (1970); Robert H. Carleton, *The NSTA Story* (Washington, D.C.: National Science Teacher Association, 1976), app. 2; "SOPHE Presidents," list, 1950–89, provided by Society for Public Health Education, Nov. 21, 1989. See, e.g., Mamie L. Anderzhon, "In Memoriam: Mary Viola Phillips," *Journal of Geography* 78 (1979): 6; and Mary Viola Phillips, "Edith Putnam Parker: Her Work and Contributions to Geographic Education" (Ed.D. diss., Columbia University, 1964). The feminized world of geographic education, so evident in the pages of the *Journal of Geography,* merits further study. It was much more personal than the highly quantified view presented in Leffler, "The Role of Women in Geography in the United States."

33. *American Biology Teacher* 1 (1938)–32 (1970). At the AHEA, many (but by no means all) of whose members were schoolteachers, all of the presidents between 1940 and 1970 were women, despite a rising male membership in these decades. For the presidents of the AHEA, see Keturah E. Baldwin, *The AHEA Saga* (Washington, D.C.:

AHEA, 1949); Mary Hawkins, ed., *The American Home Economics Association, 1950–1958: A Supplement to the AHEA Saga by Keturah Baldwin, 1949* (Washington, D.C.: AHEA, 1959); and Helen Pundt, *AHEA: A History of Excellence* (Washington, D.C.: AHEA, 1980).

34. See C. M. Christensen, *E. C. Stakman: Statesman of Science* (St. Paul, Minn.: American Phytopathological Society, 1984), 85.

35. Dorothy Stimson, informal remarks, fiftieth-anniversary celebration of the History of Science Society, Norwalk, Conn., Oct. 25, 1974; John C. Greene, "Fiftieth Anniversary Celebration of the Society," *Isis* 66 (1975): 443.

36. On Anderson, see *NAWMP;* and Ronald L. Kathren and Natalie E. Tarr, "The Origins of the Health Physics Society," *Health Physics* 27 (1974): 419–28.

37. Judith Schachter Modell, *Ruth Benedict: Patterns of a Life* (Philadelphia: Univ. of Pennsylvania Press, 1983), 300–303; Margaret Mead, *Ruth Benedict* (New York: Columbia Univ. Press, 1974), 68, 70; Margaret M. Caffrey, *Ruth Benedict: Stranger in This Land* (Austin: Univ. of Texas Press, 1989), 332–33, esp. n. 9; Ruth Benedict to Father John Cooper, Jan. 9 and 22, 1947, Ruth Benedict Papers, Special Collections, Vassar College Library, Poughkeepsie, N.Y.

38. M. Patricia Faber of Barat College was Sigma Xi's first woman president, in 1984–85. On chapters at women's colleges, see Margaret Farrand Thorp, *Neilson of Smith* (New York: Oxford Univ. Press, 1956), 203; Gladys K. McCosh, "Sigma Xi, Wellesley College Chapter, 1938–1948," *Wellesley Magazine* 33 (Oct. 1948): 15–16; Mary Mrose and Elisabeth Deichmann, "The Radcliffe Chapter of the Society of the Sigma Xi, from 1943 to 1953" [1954], plus other lists and correspondence in box 17, Papers of the Office of the Graduate Dean; petition papers for establishing chapters at Vassar (1959) and Mount Holyoke (1965) in their respective college archives.

39. See, e.g., Dael Wolfle to Margaret Mead, Dec. 6, 1955, and reply, Dec. 13, 1955, box 14; Wolfle to Mead, Aug. 22, 1960, and reply, Sept. 7, 1960, and Mead to Chauncey Leake, Sept. 8, 1960, box 20; and Mead to Paul Fajon, Jan. 11, 1961, box E164 [Wenner Gren Foundation File], all in MMP; and Wolfle to Hans Nussbaum, Mar. 16, 1983, copy in possession of author.

40. On McBride, see obituary, *NYT,* June 4, 1976, D-16. The only previous woman president of the ACE, established in 1918, had been Virginia Gildersleeve of Barnard College in 1926 (pamphlet on the history of the ACE, c. 1970, Doe Library, University of California, Berkeley).

41. On White, see *NAWMP;* Elaine Schrecker, *No Ivory Tower: McCarthyism and the Universities* (New York: Oxford Univ. Press, 1986); and "The McCarthy Era," *Academe: The Bulletin of the AAUP* 75, no. 3 (May–June 1989): 27–30, but White is curiously invisible.

42. See obituary of Cora Hennel in "Necrology," *Proceedings of the Indiana Academy of Sciences* 57 (1947): 3–4; Winona Welch Papers, DePauw University Archives, Greencastle, Ind.; and Frances Brown Papers, Duke University Archives. On officers, see *Bulletin of the AAUP,* 1940–70. Florence Lewis, professor of mathematics at Goucher College, served as the AAUP's national treasurer from well before 1940 until 1955, and three other women—mathematician Jewell Bushey Hughes of Hunter College, psychologist Ethel Sabin-Smith of Mills College, and mathematician Marie J. Weiss of

Newcomb/Tulane—served as second vice-president between 1944 and 1949. A clipping and memorial (1952) on Marie J. Weiss are in the box 23, f. 56, *NAWP*. Committee W (on women) did not exist in these years.

43. NRC, *The Invisible University: Postdoctoral Education in the United States* (Washington, D.C.: National Academy Press, 1969) indicates that women and foreigners held their postdoctoral fellowships longer than men, partly because so few faculty positions or other forms of employment were offered to them (103–4, 117–18). On chemists, see also Barbara Reskin, "Sex Differences in Status Attainment in Science: The Case of the Postdoctoral Fellowship" *American Sociological Review* 41 (1976): 597–612.

44. Among the biomedical scientists listed in 1950 and 1951 by their first initial only were such later recognizable women as Harriet Dustan in cardiology, Elizabeth Weisburger in oncology, Estelle Ramey in endocrinology, and Helen Van Vanukis in biochemistry (*Research Grants Awarded by the Public Health Service, 1950*, PHS 63 ([Washington, D.C.?]: FSA, PHS, 1951), 54–59, and *1951*, PHS 164 (1952), 66–71. See also *The Invisible University*, 29–33).

45. See NSF, *Annual Reports*, fiscal years 1952–63 (Washington, D.C., [1952?]–64); and NSF, *Grants and Awards*, fiscal years 1964–71 (Washington, D.C., 1965–72). See also *The Invisible University*, 33–34. Terence L. Porter to author, Oct. 17, 1989, explains that from 1956 to 1966 the October awards were inadvertently omitted from the published annual report.

46. When the Nixon administration cut the NSF's total fellowship budget, the agency suspended the senior postdoctoral program for the fiscal year beginning in 1969 rather than cut the regular postdoctoral program (see *The Invisible University*, 248–49). The Elsa Allen Papers are in KLCU.

47. John Simon Guggenheim Memorial Foundation, *Reports of the President and Treasurer*, 1941 to 1970 (New York, [1942]–1971). Only two of the fifty-one awards to Canadians went to women. The report of the president for 1968 and especially its appendixes contain cumulative data on all fellowships awarded since 1925 by field and four-year time periods. My own taxonomy and calculations, presented in table 14.5, may differ, since many research topics were at the intersection of two or more fields.

There were several women on the Guggenheim Advisory Board in these years (unlike the pattern at NSF): historian and historian of science Marjorie Hope Nicolson, 1932–66 (with gaps); anatomist Florence Sabin, 1936–47; art historian and archaeologist Gisela Richter, 1947–51; sociologist and demographer Dorothy Swaine Thomas, 1951–67; poet Louise Bogan, 1955–70; historian Caroline Robbins, 1959–65; film critic Pauline Kael, 1967–72; and French professor Germaine Bree, 1969–74.

Another set of prestigious private postdoctoral awards in these years were those of the Sloan Foundation, started in 1955, and designed, as one adviser summed it up, to "place emphasis primarily on men—not projects or institutions as such—but that the climate of the educational institution for research should be given consideration. In seeking men, particular heed should be paid to younger men who offer marked promise." Accordingly not until 1963–64, after 487 such men had been awarded fellowships, was the first one given to a woman, Charlotte Froese (later Fischer) of the University of British Columbia in applied mathematics. The second went to physicist Laura Roth in 1966 (D. Stanley Tarbell and Ann Tracy Tarbell, *Roger Adams: Scientist and Statesman*

[Washington, D.C.: ACS, 1981], 187; Alfred P. Sloan Foundation, *Annual Report, 1953–70* [New York, (1954–71)]).

48. On Kober, see *NAWMP*.

49. On Carson, see ibid.

50. [Juliet A. Mitchell, ed.], *A Community of Scholars: The Institute for Advanced Study, Faculty and Members, 1930–1980* (Princeton: Institute for Advanced Study, 1980). On Goldman, see *NAWMP*. Even after her retirement several archaeologists came to work with Hetty Goldman on the results of recent excavations in Greece and Italy.

51. List, dated Jan. 23, 1989, of fellows at the Center for Advanced Study in the Behavioral Sciences, 1955–70 (I thank Margaret Amara, Center librarian, for sending this). These data differ slightly from those given in the Center's *Annual Report for 1988,* 10–12, which are given by field and five-year intervals. See also John Walsh, "Behavioral Sciences: The View at the Center for Advanced Study," *Science* 169 (1970): 654–58; and Arnold Thackray, "CASBS: Notes toward a History," *Annual Report for 1984,* 59–71. One of its founders was Bernard Berelson, who, as mentioned in chapter 4, was quite dismissive of women scholars' potential.

Siegel's study grew out of an earlier project funded by a grant from the Elizabeth McCormick Memorial Fund to the AAUW. See Alberta Engvall Siegel et al., "Dependence and Independence in the Children of Working Mothers," *Child Development* 30 (1959): 533 n. 1. See also Alberta Engvall Siegel, "The Working Mother," *AAUWJ* 55 (1962): 233–35. For more on women at the Center, see ch. 16, n. 17.

52. E. B. Wilson to Harlow Shapley, Nov. 23, 1942, box Wic–Z, Harvard College Observatory, Director's Correspondence, 1940–50, HUA (I thank Nathan Reingold for calling this file to my attention and the director of the Harvard College Observatory for permission to read it).

53. Recently elected fellows were occasionally listed in the *Proceedings of the American Academy of Arts and Sciences* (renamed *Daedalus* in 1958). Starting in 1958–59, the Academy published a separate *Records of the American Academy of Arts and Sciences,* which listed all current members and usually (but not in 1965) the recently elected ones (I thank Alexandra Oleson for help). On Kittrell, see *AMWS*, 14th ed., 1979, 4:2669. For a comparison of the twentieth-century growth patterns of the American Academy and with those of the National Academy, see Walter Rosenblith, "Bicentennial Address: Focus on Institutions of Knowledge," *Bulletin of the American Academy of Arts and Sciences* 35, no. 3 (May 1982): 9–27.

54. Daniel S. Greenberg, "The National Academy of Sciences: Profile of an Institution (II)," *Science* 156 (1967): 362, lists nine men since 1950 who had won the Nobel prize without first being elected to the National Academy; seven were eventually elected.

55. *Report of the National Academy of Science* (later *Annual Report of the National Academy of Science, National Academy of Engineering, and National Research Council*), fiscal years 1939–40 to 1969–70 (Washington, D.C.: GPO, 1941–73); Rexmond C. Cochrane, *The National Academy of Science: The First Hundred Years, 1863–1963* (Washington, D.C.: NAS, 1978), 524–25 and 563–64. Because any remarks about nominees in the National Academy's own archives are closed to researchers, the correspondence files of academy members elsewhere, which sometimes contain old ballots and related correspondence, can be useful. See, for example, the A. H. Sturtevant Papers for the zoology

and anatomy sections and the Lee DuBridge Papers for the physics section in the 1960s (both in California Institute of Technology Archives, Pasadena). On Mead's election in 1975, see box E84, MMP; and on Gilbreth's to the NAE, Lillian Gilbreth Papers, Special Collections, Purdue University Library, West Lafayette, Ind.

56. Public Law 86–209, 86th Cong., 1st sess. (Aug. 25, 1959), *An Act to Establish a National Medal of Science;* "National Medal of Science Winners," *Science* 171 (1971): 464. On Hyman, see Donald F. Hornig to Dr. James Oliver, Feb. 9, 1967, box 4, Donald F. Hornig Personal Papers, Lyndon Baines Johnson Library, Austin, Tex.

57. American Philosophical Society *Yearbook,* 1940–70 (Philadelphia, 1941–71). The APS's first woman member was Ekaterina Dashkova, elected in 1789.

58. Howard M. Jones to Mead, Oct. 13, 1948 and her reply Oct. 29, 1948, both in box 1, Organization series, MMP. In 1958 Mead delivered an address to the wives' luncheon at the annual meeting of the AAAS entitled "On Bringing Up Children in the Space Age" (Mary Waterman to Mead, Sept. 9, 1958, Jan. 4, 1959, and clippings, box 19, MMP).

In these years some scientific societies had a separate "ladies' program," as the elaborate ones of the GSA and the American Physical Society meetings were called, where the "ladies," presumably all wives of male scientists attending the sessions, were taken on special tours, treated to fashion shows, or otherwise diverted and entertained for several days while the men were busy at the meetings. See the reports on the annual meetings in the *Proceedings of the GSA* in the 1950s and 1960s; and the scrapbook of the Duke University physics department (in Duke University Archives) containing pictures and clippings of the ladies' day participants for March 1953 and June 1964, when national meetings of the American Physical Society were held in Raleigh-Durham, N.C. Phyllis K. Beranek listed among her accomplishments in *WWAW* 4 [1966–67]: 96 that she was chairman of the ladies committee of the International Congress of Acoustics, Cambridge, 1958, and of the 1966 meeting of the Acoustical Society of America, Boston.

In some associations an organized "women's auxiliary," led by the wives of prominent men in the field, ran the events themselves, holding teas or coffee hours where they could meet each other, electing each other to office, and even, in the tradition of philanthropic lady bountifuls, performing related good works. The Women's Auxiliary of the American Society of Mechanical Engineers (ASME), for example, was proud to report in 1962 that it had over the years loaned a total of $125,000 to about 350 students and awarded almost $40,000 in scholarships to fifty, usually foreign male, students ("Woman's Auxiliary to the ASME," *Mechanical Engineering* 85 [Jan. 1963]: 111–12). Lillian Gilbreth was proud to be a member of both the ASME and its women's auxiliary (Lillian Gilbreth, "The Auxiliary of the American Society of Mechanical Engineers," typescript, Jan. 4, 1965, box A, Lillian Gilbreth Papers).

At times, however, accompanying wives were put to physical labor to help the men's meetings run efficiently, as Betty Armstrong (later Wood) recalled nearly thirty years later had happened at the 1946 crystallography meetings. Held in the preseason at Lake George, New York, where the resort staff was not yet in place, the seemingly idle wives saved the day by reporting for voluntary kitchen duty. Armstrong, then unmarried, had felt no conflict and stayed with the men discussing the latest research, but she still thought the episode worth recalling many years later. (Elizabeth A. Wood, "Address at

ACA Meeting, 1975," in *Crystallography in North America,* ed. Dan McLachlan, Jr., and Jenny P. Glusker [New York: American Crystallographic Association, 1983], 172).

Mrs. Barry Commoner, herself an elementary science editor, suggested in 1962 that the AAAS do more for children at its annual meeting (Commoner to Dr. Thomas Park, Jan. 5, 1962, copy in box 23, MMP).

59. Colin MacLeod to Dr. Rebecca Lancefield, Dec. 22, 1955, box 6, f. 4, Rebecca Lancefield Papers, RAC, North Tarrytown, N.Y.

60. Ralph W. Burhoe to "Sir," Apr. 21, 1959, and Burhoe to Dr. Hilda Geiringer, Apr. 27, 1959, both in Hilda von Mises Papers, HUA.

61. See, e.g., Meg Greenfield, "Science Goes to Washington," *The Reporter* 29 (Sept. 23, 1963): 20–26; Bernice T. Eiduson, "Scientists as Advisors and Consultants in Washington," *Bulletin of Atomic Scientists* 22 (Oct. 1966): 26–31; Thomas E. Cronin and Norman C. Thomas, "Federal Advisory Processes: Advice and Discontent," *Science* 171 (1971): 771–79; and Martin L. Perl, "The Scientific Advisory Systems: Some Observations," ibid. 173 (1971): 1211–15.

62. Avery Leiserson, "Scientists and the Policy Process," *American Political Science Review* 59 (1965): 408–16, cited in Nicholas C. Mullins, "Power, Social Structure, and Advice in American Science: The United States National Advisory System, 1950–1972," *Science, Technology, and Human Values* 7, no. 37 (fall 1981): 4–19 (quotation from 4).

63. [Detlev W. Bronk], *The Science Committee: A Report by the Committee on the Utilization of Young Scientists and Engineers in Advisory Service to Government,* 2 vols. (Washington, D.C.: NAS-NRC, 1972), is essentially a personal statement by possibly the consummate committeeman of his generation. One of the appendixes is a short history of scientific advice with a lengthy bibliography. For some discussion of Bronk's "reliance on friends, many of them of long standing," see John Walsh, "The Rockefeller University: Science in a Different Key," *Science* 150 (1965): 1692–95.

64. *The Food and Nutrition Board, 1940–1965: Twenty-five Years in Retrospect* (Washington, D.C.: NAS-NRC, [1965]), addenda, i–ii. On Goldsmith, see *AMWS,* 12th ed., 1972, 2:2205; *NAWMP,* which curiously omits her service on the Food and Nutrition Board; and Biographical File, Tulane University Archives, New Orleans, La.

65. *International Educational Exchange: The Opening Decades, 1946–1966, A Report of the Board of Foreign Scholarships* (Washington, D.C.: [GPO], 1966), 38–39. Helen White chaired the board in 1950. See also Walter Johnson and Frances J. Colligan, *The Fulbright Program: A History* (Chicago: Univ. of Chicago Press, 1965), 347–48.

66. For a list of the home economists on the Committee of Nine, see ch. 8, n. 24. See also Vivian Wiser and Douglas E. Bowers, *Marketing Research and Its Coordination in the U.S. Department of Agriculture: A Historical Approach,* USDA Economic Research Service, Agricultural Economic Report No. 475 (Washington, D.C., 1981), 55; Ruth O'Brien, "The Flanagan-Hope Act in Relation to Home Economics Research," *JHE* 39 (1947): 150–52; Morgan in *NAWMP;* Agnes Fay Morgan Papers, Bancroft Library, University of California, Berkeley; and Catherine Personius Papers, KLCU.

67. The NSF *Annual Reports* list the members of the National Science Board. There is occasional correspondence with the female board members in the Alan T. Waterman Papers, Manuscript Division, LC, and, more interesting, one letter to his successor

discussing the need for "women, Catholics, negros, etc." on the board (Waterman to Leland J. Haworth, Jan. 2, 1964, box 29). The Sophie Aberle Brophy Papers are in the Harry S. Truman Presidential Library, Independence, Mo. On Gerty Cori, see n. 69.

68. On Jane Russell (Wilhelmi), see Jane Russell Faculty File, Special Collections, Emory University Library, Atlanta, Ga. The oral histories of Mina Rees (Archives of the NMAH, Smithsonian Institution, Washington, D.C.) and Mary Bunting (SLRC) have very little to say on this aspect of their lives. For members of NSF advisory panels, see its *Annual Reports*. See also Lyle Groenveld, Norman Koller, and Nicholas Mullins, "The Advisers of the United States National Science Foundation," *Social Studies of Science* 5 (1975): 343–54 for 1950–72.

69. On Cori, see Joseph Fruton, "Cori, Gerty Theresa Radnitz," in *Dictionary of Scientific Biography,* vol. 3 (New York: Charles Scribner's Sons, 1971), 415–16, and *NAWMP* (both with bibliographies); Carl F. Cori, "Gerty Theresa Cori, 1896–1957," in *American Chemists and Chemical Engineers,* ed. Wyndham Miles (Washington, D.C.: ACS, 1976), 94–95; Joseph Lardner, "Gerty, Theresa Cori, 1896–1957," in *NAS Biographical Memoirs* 61 (1992): 111–35; and Edna Yost, *Women of Modern Science* (New York: Dodd, Mead, 1959), 1–16. There is a small amount of material about her in box 6 of the Philip A. Shaffer Papers and box 4 of the Joseph Erlanger Papers, both in Washington University Medical Center Archives, St. Louis, Mo. Erlanger had won the Nobel prize in 1944, and he nominated the Coris for 1946 (Erlanger to The Nobel Committee for Physiology and Medicine, Jan. 3, 1946) and again for 1947 (Erlanger to the Nobel Committee of the Royal Caroline Institute, Nov. 20, 1946).

70. On Mayer, see *NAWMP;* "At Home with Maria Mayer," *Science Digest* 55 (Feb. 1964): 30–36; Mary Harrington Hall, "Maria Mayer: The Marie Curie of the Atom," *McCall's* 91, no. 10 (July 1964): 38, 40, 124; Dorothy Nelkin, *Selling Science: How the Press Covers Science and Technology* (New York: W. H. Freeman, 1987), 18–20; and many clippings in the Maria Goeppart Mayer Papers, Mandeville Department of Special Collections, University of California, San Diego.

71. Rita Levi-Montalcini, "Reflections on a Scientific Adventure," in *Women Scientists: The Road to Liberation,* ed. Derek Richter (London: Macmillan, 1982), ch. 7.

72. On Wu, see *AMWS,* 17th ed., 1989, 7:782; Edna Yost, *Women of Modern Science* (New York: Dodd, Mead, 1959), 80–93; Gloria Lubkin, "Chien-Shiung Wu: The First Lady of Physics Research," *Smithsonian,* Jan. 1971, 52–57; and Fay Ajzenberg-Selove, *A Matter of Choices: Memoirs of a Female Physicist* (New Brunswick, N.J.: Rutgers Univ. Press, 1994), 115. There is a large collection of material about Wu in the Physics Department Papers, Special Collections, Columbia University Library, New York City.

73. James B. Sumner mentioned Graham in "The Story of Urease," *JCE* 14 (1937): 257 but not in his Nobel prize acceptance speech, "The Chemical Nature of Enzymes," *Les Prix Nobel en 1946* (Stockholm: P. A. Norstedt & Söner, 1948), 185–92. Graham spent most of her later career at Florida State College for Women (*AMWS,* 11th ed., 1965, 2:1909; I thank Nell Monday for bringing Graham to my attention).

74. *CBY, 1959,* 251–52.

75. See Walter Eckhart, "The 1975 Nobel Prize for Physiology or Medicine," *Science* 190 (1975): 650; and on Vogt, *AMWS,* 13th ed., 1976, 6:4650.

76. On Schwartz, see the introduction to Michael D. Bardo, ed., *Money History and*

International Finance: Essays in Honor of Anna J. Schwartz (Chicago: Univ. of Chicago Press, 1989); and Anna J. Schwartz, *Money in Historical Perspective* (Chicago: Univ. of Chicago Press, 1987).

77. "3 Nobels in Science," *NYT,* Oct. 17, 1985, 17.

78. Judith Walzer, "Interview with Ruth Hubbard," July 1981, typescript, 98–102, MCRC; and Patricia Farnes, "Women in Medical Science," in *Women of Science: Righting the Record,* ed. G. Kass-Simon and Patricia Farnes (Bloomington: Indiana Univ. Press, 1990), 289.

Chapter 15   Women's Clubs and Prizes: Partial Palliatives

1. "Suicides High for Female Chemists," *Industrial Research* 11 (Dec. 1969): 31; Frederick P. Li, "Suicide among Chemists," *Archives of Environmental Health* 19 (1969): 518–20. See also J. S. Mausner and R. C. Steppacher, "Suicide in Professionals: A Study of Male and Female Psychologists," *American Journal of Epidemiology* 98 (1973): 436–45. The reasons proffered for the preponderance of suicides were chemical for the chemists (they had ready access to cyanide, the poison of choice) and psychological for the psychologists (they were maladjusted, professional women at a time when not many white, middle-class American women had careers). An unnamed woman professor of political science at Penn State also committed suicide (mentioned in Robert C. Bannister, *Jessie Bernard: The Making of a Feminist* [New Brunswick, N.J.: Rutgers Univ. Press, 1991], 144). For others, see the index. A seemingly high proportion (perhaps 50 percent) of the women interviewed in the mid-1970s by Shirlee Shakow's oral history project on women in science and engineering at MIT reported that they had been receiving some sort of psychotherapy around 1970.

2. See John Robson, ed., *Baird's Manual of American College Fraternities,* 18th ed. (Menasha, Wis.: George Banta, 1968), 663; this book is the essential source on the whole fraternity movement. See also Helen L. Horowitz, *Campus Life* (New York: Alfred A. Knopf, 1987), 202.

3. See, e.g., Grace M. Henderson, "For What Purpose, Home Economics in Higher Education[?]" *Omicron Nu* 36, no. 3 (fall 1968): 7–10, 37–39; idem, "Your Degree Yields Opportunity," ibid. 37, no. 3 (fall 1970): 18–21, 25–27; and Helen LeBaron, "What Price Honoraries?" ibid. 32, no. 1 (fall 1959): 16–18. There is a set of *Omicron Nu* in Mann Library, Cornell University. See esp. "A Review of 50 Years from the Records," ibid. 33, no. 2 (spring 1962): 13–18, 29–32, and the constitution, "Omicron Nu Society, Incorporated, National Constitution and By Laws," ibid. 32, no. 1 (fall 1959): insert between pp. 18 and 19.

4. Flemmie P. Kittrell, "Dear Omicron Nu Members," ibid. 27, no. 3 (spring 1951): 16–17 (from Baroda, India); "Dr. Flemmie P. Kittrell," ibid. 28, no. 2 (fall 1952): 6; Flemmie P. Kittrell, "New Horizons for Omicron Nu," ibid. 29, no. 1 (fall 1953): 18–22; idem, "Challenge of Home Economics—An International Common Denominator in Research and Service," ibid. 34, no. 1 (fall 1963): 9–11, 35; Bernice N. Bridgeport, "New Chapter . . . Alpha Phi," ibid., 15; Margaret E. Tillett, "Alpha Phi Chapter, Dr. Kittrell Honored," ibid. 38, no. 2 (spring 1972): 14.

5. Robson, *Baird's Manual,* 564–65, 567, with lists of chapters. On the formation of a

three-way "coordinating council" with Phi Upsilon Omicron and Kappa Omicron Phi, see "Omicron Nu in Action," *Omicron Nu* 37, nos. 3 (fall 1970): 4 and 4 (spring 1971): 4–5.

6. Robson, *Baird's Manual*, 730–31. When in 1968–69 Phi Lambda Upsilon voted to admit women members, many graduate students voted to join, but the alumnae preferred their separate Iota Sigma Pi chapters. This meant that before long there were few but alumnae left in Iota, and some chapters, such as Vanadium in New York City, became metropolitan chapters (Sister Mary Rose Stockton, *A History of Iota Sigma Pi* [Indianapolis: n.p., 1980], 65).

7. Robson, *Baird's Manual*, 658. See also Aaron J. Ihde, *Chemistry, as Viewed from Bascom Hill* (Madison: Department of Chemistry, University of Wisconsin, 1990), 373–81.

8. Stockton, *History of Iota Sigma Pi*, updates earlier histories by Agnes Fay Morgan, most recently that of 1963. ISP started a newsletter in 1941, *The Iotan*, but no library in the United States has a complete set. Occasional issues, however, exist along with other ISP materials in some of the collections of papers of former members, such as the Virginia Bartow Papers, University of Illinois Archives, Urbana; Mary Willard Papers, Penn State University Archives, Penn State Room, Pattee Library, Pennsylvania State University, University Park; Agnes Fay Morgan Papers, Bancroft Library, University of California at Berkeley; Gertrude Perlmann Papers, RAC, N.Y.; Icie Macy Hoobler Papers, Bentley Historical Library, University of Michigan, Ann Arbor; Mercury Chapter Records, University of Minnesota Archives, Minneapolis; H. Marjorie Crawford Biographical File, Special Collections, Vassar College Library, Poughkeepsie; and the Denver Public Library's clipping file on Essie White Cohn.

In September 1971 four of Alpha Chi Sigma's chapters admitted eighteen women, including Mary Willard, long-retired professor of chemistry at Penn State University (Robson, *Baird's Manual*, 475–77; Daniel A. Pelak, "Mary L. Willard, First Professional Woman Initiate," *Hexagon*, fall 1971, 2–3; "Four Chapters Initiate Eighteen Collegiate Women," ibid., 3–4). There was also a men-only fraternity for geologists (Sigma Gamma Epsilon, established at the University of Kansas in 1915), which had fifty-nine active chapters across the nation by 1968, but Chi Upsilon, the only equivalent group for women in geology, had only one chapter—at the University of Oklahoma (Robson, *Baird's Manual*, 671–72).

9. Agnes Fay Morgan to Clara A. Storvick, Feb. 14, 1966, box 3, Agnes Fay Morgan Papers.

10. Agnes Fay Morgan, "Women in Chemistry or Allied Fields and Professional Opportunities: The Real Purpose of Iota Sigma Pi," *Iotan* 27 (Apr. 1968): 13–16 (quotation from 16), copy in Agnes Fay Morgan Papers.

11. Naomi F. Goldsmith, "Women in Science: Symposium and Job Mart," *Science* 168 (1970): 1124–27.

12. On the SWG, see Megan Murray, "A Club That Celebrates Women of Adventure," *Ms.* 12 (Dec. 1983): 20 (I thank Jacqueline Goggin for a copy). In general the best source on the SWG, and often the only source, is its own publications, including the *Bulletin of the Society of Women Geographers*, twenty-fifth anniversary issue, Dec. 1950, which contains historical pieces; and the fiftieth-anniversary issue, spring 1975, which contains a directory of all SWG members from 1925 to 1975 by occupational group. The *Bulletin* of November 1974 discussed the society's past and present activities and con-

tained a list of active members and their current activities (I thank Helen Loerke of SWG for copies). See also Gretchen Smith, "And Life Still Unfolds . . . ," *IW* 33 (1954): 446–48, on Irene Wright, SWG president, 1954–57. SWG teas and fellowships were mentioned on occasion in the *Geographical Review,* put out by the American Geographical Society (see its annual index). There are a few items about the SWG, including Margaret Mead's 1942 speech "The Role of Women Geographers in Winning the War and the Peace," in boxes E126 and E127, MMP. Mead's comment about the Museum was in Jane Howard, *Margaret Mead: A Life* (New York: Simon & Schuster, 1984), 132. On founders Harriet Chalmers Adams and Blair (Beebe) Niles, see Marion Tinling, ed., *Women into the Unknown: A Sourcebook on Women Explorers and Travelers* (New York: Greenwood, 1989), 3–8 and 197–202.

13. See ch. 3, nn. 3–4; "What Do You Know about the Beginnings of SWE?" *SWE Newsletter* 21 (May–June 1977): 18; Marta Navia Kindya with Sudha Dave, *Four Decades of the Society of Women Engineers* (n.p.: 1990), 11–21; "Women's Council, Western Society of Engineers," *Journal of the Western Society of Engineering* 53, no. 2 (June 1948): 77–79; "Girls Studying Engineering See Future for Women in These Fields," *CSM,* Apr. 16, 1949, 14; Beatrice C. Horneman, "Engineers Wanted," *IW* 29 (July 1950): 213–14; and "Women Engineers See Field Widening," *NYT,* Mar. 11, 1951, p. 47, col. 1. The exact membership of SWE in any given year is hard to determine. In 1953 it was reported as ca. 350 (*World Directory of Women's Organizations* [London: World Directory, (1953)]).

14. See, e.g., "Dr. Gilbreth Speaks on Some Problems That Challenge Engineers Today," *Midwest Engineer* 5 (Oct. 1952): 5–6, 16; and numerous *SWE* publications, such as program, SWE Celebration Banquet, Nov. 8, 1961, copy in Lillian Gilbreth Papers, Special Collections, Purdue University Library, West Lafayette, Ind.; "Dr. Gilbreth Honored," *SWE Newsletter* 11, no. 4 (Nov. 1964): 1, 5; "'Dr. Gilbreth Speaks on Engineering: PNW Section and ASME Sponsor Career Program," ibid. 12, no. 6 (Jan.–Feb. 1966): 1–2. There are also numerous letters from SWE officers and members (e.g., Beatrice Hicks, Irene Peden, and Olive Salembier) thanking her for her advice and "inspiration" in Lillian Gilbreth Papers.

15. In Elsie Eaves, "SWE—25 Years in Perspective," *SWE Newsletter* 22, nos. 3–4 (spring 1976): 15; used with permission of the Society of Women Engineers, New York, N.Y.

16. The best sources of information about SWE are ibid., 14–15, 19; Kindya with Dave, *Four Decades of the Society of Women Engineers;* and the newsletters themselves. For information on the fund drive in 1958–61, see Pearl F. Clark, "You Asked for It," *SWE Newsletter* 5, no. 4 (Nov. 1958): 3–4. On Clark herself, see Howard P. Emerson and Douglas C. E. Naehring, *Origins of Industrial Engineering: The Early Years of a Profession* (Atlanta: Institute of Industrial Engineering, 1988), 78.

17. SWE, *Report on Women Undergraduate Students in Engineering: Biennial Survey, 1959–69* (New York, 1971); Patricia L Brown, ed., *Women in Engineering* (New York: SWE, 1955). Brown was later profiled in Aline de Grandchamp, "Engineer Likes Fixing, Heads SWE, From South," *CSM,* Sept. 5, 1961, 5.

18. *Employment Opportunities for Women in Professional Engineering,* WB Bulletin 254 (Washington, D.C., 1954); "Employment Characteristics of Women Engineers," *Monthly Labor Review* 79 (May 1956): 551–54. An earlier draft, "Survey of Women Engi-

neers, 1955," which is based on the same 874 replies, including 625 from employed women, is in Alice Leopold Papers, SLRC.

19. *Proceedings of the First International Conference of Women Engineers and Scientists: Focus for the Future, Developing Engineering and Scientific Talent* (New York: SWE, 1964); "First International Conference of Women Engineers and Scientists," *The Woman Engineer: Journal of the Women's Engineering Society* (U.K.) 9, no. 12 (spring 1964): 24; "International Conference of Women Engineers and Scientists," *ASHRAE Journal* 6 (Aug. 1964): 44, 91. There are several letters about the conference in the Lillian Gilbreth Papers; and there is a letter from Donald Hornig, presidential science adviser, to Beatrice Hicks, president of SWE, June 3, 1964, as well as memo from Hornig to Fred Holborn, June 12, 1964, enclosing the presidential welcoming remarks in Donald F. Hornig Papers, box 1, Lyndon Baines Johnson Library, Austin, Tex.

20. "The Talk of the Town—Cyberculture and Girls," *New Yorker,* July 4, 1964, 22–23. The meeting fared better in Elinor Nelson, "Women Engineers Meet," *CSM,* June 24, 1964, 4. On Cavanagh, the SWE president, who presided over the conference, see *WWAW* 5 (1968–69): 211; Elinor Nelson, "Women Engineers Stress Youth: Aileen Cavanagh Heads Drive for New Recruits," *CSM,* June 10, 1964, 12; and interview in Sidney Feldman, "Increasing Role for Women in Electronic Engineering," *Electronic Industries* 23 (Feb. 1964): 46–50. There is a biographical sketch of Ruth Shafer, who ran the 1964 conference, in *Proceedings of the Second International Conference of Women Engineers and Scientists, Cambridge, U.K., 1–9 July 1967* (London: Women's Engineering Society, 1968), 1. A somewhat critical account of this second international conference is Molly Neal, "The World Around—Technical Women," *International Science and Technology,* no. 69 (Sept. 1967): 13, 16, 18, 20.

21. Eaves, "SWE—25 Years in Perspective," 19. On Hicks, see "Career Women," *Coronet,* Apr. 1953, 118–19; *CBY, 1957,* 255–27; Robert C. Toth, "Couple May Beautify 'The Ugly American,'" *New York Herald Tribune,* June 18, 1959, clipping in carton 4, f. 86, ACE-CEW Papers, SLRC; and "Outstanding Engineer, Manager, and Consultant Is a Woman," *Product Engineering,* Apr. 8, 1968, 137–38.

22. Essie White Cohn, "[Annual Report of the] Women's Service [Committee for 1958]," *C&EN* 37 (May 25, 1959): 74.

23. "Garvan Medal: Gertrude B. Elion," ibid. 46 (Jan. 15, 1968): 65. For a list of Garvan medalists, see Herman Skolnik and Kenneth M. Reese, eds., *A Century of Chemistry: The Role of Chemists and the American Chemical Society* (Washington, D.C.: ACS, 1976), 444. On the early days of the Garvan medal, see Margaret W. Rossiter, *Women Scientists in America: Struggles and Strategies to 1940* (Baltimore: Johns Hopkins Univ. Press, 1982), 308.

24. See Perlmann's obituary in *NYT,* Sept. 10, 1974, 44.

25. Skolnik and Reese, *A Century of Chemistry,* 441–45, lists many ACS award winners to 1976. At the national level, the only ACS prize to be awarded to women was the James Bryant Conant Prize in High School Chemistry Teaching, established in 1967. At the divisional level, four women are listed as winning prizes between 1940 and 1970: two won the distinguished service award in environmental chemistry (Louise Lamphere in 1965 and H. Gladys Swope in 1968), and two won the Dexter Chemical Corporation Award in the history of chemistry (Eva Armstrong in 1958 and Mary E. Weeks in 1967),

but this list omits Clara D. Craver, who won the Carbide and Carbon Award of the ACS division on organic coatings and plastics chemistry in 1956 ("ASTM's New Technical Committee Chairman," *ASTM Standardization News* 7, no. 3 [Mar. 1979]: 34, 62). Women did a little better at the local level, with Gerty and Carl Cori winning the Midwest Award of the St. Louis section of the ACS in 1946 ("Midwest Award to the Coris," *C&EN* 24 [1946]: 893–96) and DuPont's Stephanie Kwolek winning an award from the Wilmington, Delaware, section (on Kwolek, see *AMWS*, 16th ed., 1986, 4:538; and Ethlie A. Vare and Greg Ptacek, *Mothers of Invention* [New York: William Morrow, 1988], 192–93; I thank Mary Ellen Bowden for information about Kwolek).

26. Annual reports of the Women's Service Committee were published with those of the ACS governing boards in the *C&EN*.

27. Marjorie J. Vold, "[Annual Report of the] Women's Service [Committee for 1950]," *C&EN* 29 (1951): 2184; idem, "[Annual Report of the] Women's Service [Committee for 1949]," ibid. 28 (1950): 1844; Hoylande D. Young, "[Annual Report of the] Women's Service [Committee for 1948]," ibid. 27 (1949): 1306. See also "Women at Work," ibid. 26 (1948): 2872–73.

28. See ch. 14, n. 58, for mention of "ladies programs" in other scientific societies.

29. H. Gladys Swope, "[Annual Report of the] Women's Service [Committee for 1963]," *C&EN* 42 (June 1, 1964): 85.

30. Florence H. Forziati, "[Annual Report of the] Women's Service [Committee for 1966]," ibid. 45 (June 5, 1967): 86.

31. In 1967 the Women's Service Committee, perhaps sensing the need for some more private means of communication than annual reports in the *C&EN,* started a separate bulletin for women members of the ACS; unfortunately, no library in America seems to have kept a set. The ACS Library in Washington, D.C., has a set dating from 1972 on. The women's response to the 1969 report on suicide would have been interesting.

The angry letter was Roberta J. Schilit and Eleanor D. Edge, both of Wilmington, Delaware, "Status of Women Chemists," *C&EN* 47 (June 16, 1969): 8. On Edge, see *AMS,* 11th ed., 1965, 2:1359.

32. See also ch. 1, nn. 22, 25, and ch. 2; Lillian G. Portenier, ed., *The International Council of Psychologists, Inc.: The First Quarter Century, 1942–1967* (n.p., [1967]) and Mary Roth Walsh, "Academic Professional Women Organizing for Change: The Struggle in Psychology," *Journal of Social Issues* 41, no. 4 (Winter 1985): 19–23. Among the several relevant collections at the Archives for the History of American Psychology at the University of Akron are the Margaret Ives Papers (longtime archivist of NCWP), the Doris Twitchell-Allen Papers, and the Leah Gold Fein Papers.

33. There are several Mitchell letters about membership in the Katharine Banham Papers, Duke University Archives, Durham, N.C. (Mitchell to Banham, May 15, 1951, Oct. 31, 1954, and Mar. 6, 1955), and her frequent reports as membership committee chairman are in the Margaret Ives Papers. But Mitchell did not mention this activity in her autobiography, "Mildred B. Mitchell," in *Models of Achievement: Reflections of Eminent Women in Psychology,* ed. Agnes N. O'Connell and Nancy Felipe Russo, vol. 1 (New York: Columbia Univ. Press, 1983), 121–39. When Banham was asked to succeed Mitchell as chairman of the membership committee in 1959 she accepted only reluctantly

(Banham to Emma Layman, Oct. 1, 1959, Katharine Banham Papers) and served only one year (Banham to C. Ruth Russell, June 3, 1960, ibid.; in both of these letters she recommended that a man be added to the committee in order to help recruit male members).

34. Virginia B. Sanderson, "Progress Report: Committee on the Status of Women Psychologists," Feb. 22, 1955, box 489, Doris Twitchell-Allen Papers.

35. Walter M. Nielsen to Banham, May 21, 1952, her reply, May 23, 1952, and his response July 9, 1952; Nielsen to Banham, Dec. 2, 1954, and her reply Dec. 4, 1954; Banham to Gertrude Reiman (ICWP secretary), June 20, 1955; and Banham to Harriet O'Shea, July 6, 1955, all in Katharine Banham Papers.

36. Portenier, *The International Council of Psychologists, Inc.*, 17–18, 34; Leah Gold Fein, "President's Column: Membership Trends in ICP," typescript of newsletter article, [1974], in Leah Gold Fein Papers. See also Dan W. Dodson, "Is It Desireable for Women's Organizations to Maintain Their Autonomy?" *Journal of the National Association of Women Deans and Counselors* 32 (fall 1968): 40–45. These were the very same years (1960–69) for which epidemiologists found the suicide rate for women psychologists to be three times that expected for women overall and 3.8 times that for male psychologists, who killed themselves only three-quarters as often as predicted (see above, n. 1).

Riess had earlier (1952) lost his position at Hunter College when he refused to tell a congressional committee whether he had ever been a member of the Communist Party (Elaine W. Schrecker, *No Ivory Tower: McCarthyism and the Universities* [New York: Oxford Univ. Press, 1986], 169, 284).

37. Robson, *Baird's Manual*, 745.

38. Mary Louise Robbins, ed., *A History of Sigma Delta Epsilon, 1921–1971: Graduate Women in Science* (n.p., 1971), 1. The administrative records of Sigma Delta Epsilon are located in KLCU. There are additional materials in the Winona Welch Papers, DePauw University Archives, Greencastle, Indiana; Mary Willard Faculty File; Lillian Gilbreth Papers; MMP; Rebecca Lancefield Papers, RAC; and Ruth Benedict Papers, Special Collections, Vassar College Library. See also the Mary E. Maver Papers, History of Medicine Division, National Library of Medicine, Bethesda, Md., and obituary of Mary E. Reid, f. 17, box 19, *NAWP*, SLRC.

39. Robbins, *A History of Sigma Delta Epsilon*, 11. See, e.g., Virginia Bartow, "Women in Chemistry: An Historical Survey," *Sigma Delta Epsilon News*, Feb. 1937; Julia Haber, "Women in the Biological Sciences" (State College, Pa., [1939], mimeographed); Helen Brewster Owens, "Early Scientific Work of Women and Women in Mathematics" (State College, Pa., [1940], mimeographed) and Mary Willard, "Pioneer Women in Chemistry" (State College, Pa., [1940], mimeographed).

40. Robbins, *A History of Sigma Delta Epsilon*; Cynthia Westcott, *Plant Doctoring Is Fun* (Princeton: Van Nostrand, 1957), 234–35; Dael Wolfle, *Renewing a Scientific Society: The American Association for the Advancement of Science from World War II to 1970* (Washington, D.C.: AAAS, 1989), 67. The New York meetings also tended to be among the largest. I thank Michele Aldrich, AAAS archivist, for information on SDE.

41. Barbara A. Sokolosky, ed., *AAUW Archives, 1881–1976: A Guide to the Microfilm Edition* (New York: Microfilming Corporation of America, 1980). In 1959 committee members showed interest in the activities of the American Council of Women in Sci-

ence, but its staff associate had other, higher-priority activities to attend to (Meeting of AAUW Status of Women Committee, "Minutes," Oct. 31–Nov. 1, 1959, 9 [reel 117, sec. 738]).

42. Bess Furman, "Flemming Backs Women in Science," *NYT,* Dec. 30, 1958, p. 18, col. 3.

43. "Interim Report, National Conference on the Participation of Women in Science," c. 1958, box 724, WB Papers (RG 86), NARA.

44. See a program for the meeting and an invitation in box 19 (AAAS), MMP; some correspondence about the session in box 19 (AAAS), President's Committee on the Development of Scientists and Engineers Papers, Dwight D. Eisenhower Library, Abilene, Kans.; a press release in carton 4, f. EC 83, ACE-CEW Papers; an invitation to speak (M. V. King to Elisabeth Deichmann, Nov. 3, 1958, and her decline, Dec. 12, 1958, in Elisabeth Deichmann Papers, Museum of Comparative Zoology Library, Harvard University, Cambridge, Mass.); as well as Sigma Delta Epsilon Papers, 1958, KLCU. See also "Womanpower Needed," *Science News Letter* 75 (Jan. 10, 1959), 22; and the more critical evaluation, " . . . Untapped Reservoir?" *America* (a Catholic magazine) 100 (Jan. 31, 1959): 508, which claimed, "Motherhood remains the prime natural function of woman . . . we do not accept the status of Soviet woman as the ideal to supplant our own."

45. "Women in Science," *Science* 129 (1959): 1117. "Minutes, Washington Area Discussion Meeting," Jan. 31, 1959; Mary Louise Robbins and Ethaline Cortelyou to Stella Manor, WB, May 22, 1959 (two letters of this date), enclosing "Minutes, National Council on the Participation of Women in Science, May 2, 1959"; and Mary Louise Robbins, "To all Persons Interested in the National Council on the Participation of Women in Science," June 20, 1959, all in box 724, WB Papers. See also related correspondence in "AAAS Women's Caucus" files in current Opportunities in Science Office Files, AAAS Headquarters, Washington, D.C. (I thank Michele Aldrich for telling me of this and sending copies).

46. Alan T. Waterman, "Scientific Womanpower—A Neglected Resource," *Science Education* 44 (1960): 207. On Embrey, see Milton Lomask, *A Minor Miracle: An Informal History of the National Science Foundation* (Washington, D.C. NSF, [1976]), 77; memo (about her duties), Embrey to the Director, July 9, 1963, box 29, Alan T. Waterman Papers, Manuscript Division, LC; Lee Anna Embrey, "Annual Report, Research Assistant to Director, July 9, 1963," 5, and Waterman to Embrey, July 15, 1964, both in ibid., about her activities. See also Ethaline Cortelyou to Waterman, Oct. 5, 1959, Sigma Delta Epsilon Papers; and Ethaline Cortelyou, "Encouraging Women to Select and to Advance in Scientific Careers (X2)," *Science* 131 (1960): 548.

47. Waterman, "Scientific Womanpower," 209.

48. Ibid., 210. The statistics did not come out until October 1961, in a booklet prepared by Bella Schwartz, *Women in Scientific Careers* (Washington, D.C.: GPO, 1961). More than half of its data, however, were drawn from non-NSF sources, such as WB, BLS, Census Bureau, USOE, HEW, and NEA. Although the report's appearance was a sign that how women were faring was of some interest to some persons at NSF, the pessimistic tone of the report and of its press release (copy in Collection PA 66-452, Rutgers University, Ford Foundation Archives, New York) was echoed in two journal

articles ("Few Women Choose Science as a Career," *Science News Letter* 80 [Nov. 4, 1961]: 305, and "Science Not Likely to Attract Many Women," *C&EN* 39 [Nov. 6, 1961]: 40–41). Once again, therefore, publicity purportedly trying to recruit women into science proved strangely counterproductive.

49. Waterman, "Scientific Womanpower," 211. A week later anthropologist David Aberle of the University of Michigan wrote Waterman that more effort should be put into changing the antinepotism rules than into trying to change basic social attitudes about women's brains. Alberle's wife, E. Kathleen Gough, a trained anthropologist also, was barred from holding a job "in the only academic location near her home" simply because she was married to him. Waterman replied that the nepotism rules "represent a curious anomaly in the present situation," but it was not "appropriate for the Federal Government to take a position" on university regulations: that was up to the university's board of regents or the state legislature (Aberle to Waterman, Jan. 6, 1960, and reply, Jan. 20, 1960, Director's Correspondence, NSF Papers [RG 307], NARA). See also ch. 6, n. 38.

50. Waterman, "Scientific Womanpower," 212.

51. Ibid., 213.

52. Program in box 20 (AAAS), MMP. The program chair was Ernestine Thurman. She and the council were mentioned in Ruth Dean, "NIH Scientist Excels, Mosquitoes to Millinery, Ernestine Thurman Is Master of All," *Washington Star,* Jan. 10, 1960 (clipping in carton 4, f. 86, ACE-CEW Papers),; and the council was mentioned in Maxine Cheshire, "Distaffers Cutting Capers on Horizons of Science," *Washington Post,* Jan. 17, 1960 (clipping in carton 4, f. EC 83, ACE-CEW Papers). Ernestine Thurman's own address to the National League for Nursing Convention in April 1961, "Women in Science" (copy in Sigma Delta Epsilon Papers), published as "Needed: More Women in Science," in *Nursing Outlook* 9 (1961): 633–35, also mentioned the council and Waterman's address.

53. This may have been the same person as Meta Ellis Heller of Seattle (and so Boeing?), who in 1963 wrote a congressional committee urging passage of the Equal Pay Act, which she saw as a way of encouraging young women to go into science and engineering. As a member of the local chapter of SWE, she had tried to present a science fair award to a qualified female, but there were none that year. See U.S. Congress, House of Representatives, *Equal Pay Act: Hearings before the Special Subcommittee on Labor of the Committee on Education and Labor, House of Representatives, 88th Congress, 1st session on HR. 3861 and Related Bills* (Washington, D.C.: GPO, 1963), 310.

54. Meta Ellis, chairman, "Annual Report of the Committee for Encouraging Women in Science," Dec. 1959, and idem, "Annual Report, Committee on Encouraging Women in Science," Dec. 1960, both in Sigma Delta Epsilon Papers. Winona Welch, past president of SDE and longtime professor of botany at DePauw University, responded by collecting materials on women in science and increasing her outreach efforts (box 498, Winona Welch Papers).

55. Robbins, *A History of Sigma Delta Epsilon,* 21; Beverly Henderson and Joanne Ingwall, "Progress Report on 'Women in Science' Workshop," in Sigma Delta Epsilon Papers. See also Jennie T. Farley, "Women on the March Again: The Rebirth of Feminism in an Academic Community" (Ph.D. diss., Cornell University, 1970).

56. [Helen S. Hogg], "Notice, the Annie J. Cannon Prize," *Astronomical Journal* 71 (Aug. 1966): 544; Margaret A. Firth, ed., *Handbook of Scientific and Technical Awards in the United States and Canada, 1900–1952* (New York: Special Libraries Association, 1956), 25–26; Claire Walter, *The Book of Winners* (New York: Harcourt Brace Jovanovich, 1978), 358. For reactions of winners, see Emma Williams Vyssotsky to Harlow Shapley, Oct. 28, 1946, box U–V, Papers of the Director of the Harvard College Observatory, 1940–50, HUA; Helen Hogg to Margaret Harwood, Feb. 25, Mar. 16, Apr. 2, 1966 (with enclosure from G. C. McVittie, secretary of the American Astronomical Society), Apr. 17, 1967, all in Margaret Harwood Papers, SLRC. For Swope, see vita, in f. 17, box 1; Harwood to Swope, June 1, 1968, box 1; Emma [Vyssotsky] to Swope, May 30, 1968, box 2; and Swope to Edith Muller, n.d, box 2, all in Henrietta Swope Papers, SLRC. Swope also gave the American Astronomical Society an anonymous donation of $1,000 at some point (box 1).

57. "Margaret Burbridge to Head Greenwich Observatory," *Physics Today* 24 (Dec. 1971): 63; "The Stargazer," *Time,* Mar. 20, 1972, 38; A. Cowley et al., "Report to the Council of the AAS from the Working Group on the Status of Women in Astronomy— 1973," *Bulletin of the American Astronomical Society* 6 (1974): 412–23. The award is given biennially. Winners for 1974–80 are listed in "The Annie Jump Cannon Award," in *Idealism at Work: AAUW Educational Foundation Programs, 1967–1981, A Report by Doris C. Davies* (Washington, D.C.: AAUW Educational Foundation, 1981), 368.

58. On Meitner, see "Honor Dr. Meitner for Work on Atom," *NYT,* Feb. 10, 1946, 13. Ruth Leach of IBM and Esther L. Richards of Johns Hopkins were also honored in 1946. Otto Hahn, Meitner's former collaborator, who won the 1944 Nobel prize alone for their joint work, overestimated the value of her "Woman of the Year" award in the United States as a kind of female Nobel prize (Otto Hahn, *Otto Hahn: My Life, The Autobiography of a Scientist,* trans. Ernst Kaiser and Eithne Wilkins [New York: Herder & Herder, 1970], 182–83 and esp. 199).

59. On Cori and Brady, see Bess Furman, "Women Reporters Roam Wonderland," *NYT,* Apr. 4, 1948, 52; on Rebstock, "Press Women Plan to Honor Educator," ibid., Apr. 9, 1950, 67, and "Politics Satirized by Women of Press," ibid., Apr. 16, 1950, 27. See also Maurine Beasley, "The Woman's National Press Club: Case Study of Professional Aspirations," *Journalism History* 15, no. 4 (winter 1988): 119 (I thank Bruce Lewenstein for telling me of this).

60. "People . . . American Women of Achievement," *Life,* May 28, 1951, 34.

61. "Six Most Successful Women," *Woman's Home Companion,* Jan. 1955, 18–19. The six were Senator Margaret Chase Smith, bacteriologist Margaret C. Modin, engineer Lillian Gilbreth, aviatrix Jacqueline Cochran, tennis player Doris Hart, and medical philanthropist Mary Lasker.

62. On MacDougall, see Dorothy Calder, "Fame for 'Miss Mac' Balanced with Flair for Gentle Arts," *CSM,* Dec. 29, 1949, 12; on Russell, Marjorie Rutherford, "Emory Biochemist, Dr. Jane Russell, Is 1960 WOTY in the Professions," *Atlanta Constitution,* Jan. 18, 1961, and Margaret Turner, "WOTY in the Professions Tops in Scientific Field," unidentified clipping, both in Jane Russell Faculty File, Special Collections, Emory University Library, Atlanta; on Fisher, *WWAW* 1 (1958): 420; on Cohn, Lois Cress, "Woman of the Year," *Sunday Denver Post,* Oct. 1, 1961, 10, and file on Cohn in Denver

Public Library, Western History Department; on Taussky, Olga Taussky-Todd, "Olga Taussky-Todd" (autobiography), in *Mathematical People: Profiles and Interviews,* ed. Donald J. Albers and G. L. Alexanderson (Boston: Birkhauser, 1985), 335; on Mathias, Mary Ann Callan, "12 Women of the Year Honored by *Times* in Colorful Ceremonies," pt. 2, *Los Angeles Times,* Dec. 15, 1964, 2, Irving S. Bengelsdorf, "*Times* Woman of the Year, Dr. Mathias One of Nation's Top Botanical Taxonomists," pt. 5, ibid., Dec. 16, 1964, 1, 12, and "W[ashington] U[niversity] Alumna Cited Woman of the Year," *Washington University Alumni News,* Feb. 1965, 10, clippings, all in Mathias Biographical File, Hunt Institute for Botanical Documentation, Carnegie-Mellon University, Pittsburgh (I thank Anita Karg for her kind help); and on Magill, "The C-J Salutes," *Columbus (Ohio) Citizen-Journal,* Feb. 3, 1964, 1, and Gail Tabor, "Outstanding Career Woman of 1963: Success in a 'Man's Field,'" ibid., 9. Retired professor of chemistry Mary L. Willard was chosen Penn State's woman of the year in 1965 (Leaflet, "1965 Penn State Woman of the Year," 1965, and clipping, "Miss Willard Selected '65 Woman of the Year," *Centre Daily Times* [State College and Bellefonte, Pa.], May 20, 1965, both in Mary Willard Faculty File).

63. John H. Fenton, "Radcliffe Cites 8 in Special Honors," *NYT,* Dec. 4, 1954.

64. See, e.g., "Honorary Degrees," *SAQ* 57 (1965–66): 214; and "Honorary Degrees at the Eighty-Ninth Commencement," ibid. 58, no. 4 (Aug. 1967): 9–10.

65. On Nice, see Rossiter, *Women Scientists in America,* 277; on Gifford, *C&EN* 48 (Mar. 23, 1970): 83, upon her winning a James Bryant Conant Teaching Award from the ACS.

66. Rebecca Lancefield to President Herrick B. Young, Jan. 4, 1966, f. 6, and Oct. 30, 1968, f. 7, both in box 8, Rebecca Lancefield Papers. See also "Alumnae Achievement Awards," clipping from a Wellesley College publication, n.d., in f. 1, box 1.

67. On the Margaret Morse Nice Club, see Doris Huestis Speirs to Edward S. Morse, Jan. 31, 1975, published as app. 2 to Margaret Morse Nice, *Research Is a Passion with Me* (Toronto: Consolidated Amethyst Communications, 1979), 267–71; on the Lancefield Club, see press release, Mar. 3, 1981 (on Lancefield's death), f. 1, box 1, Rebecca Lancefield Papers; "The Eistophos Science Club of Washington, D.C.: The 75th Anniversary Papers, 1893–1968," typescript in Papers of the Eistophos Club, Rare Book Room, LC. For information on the Women's Research Club at the University of Michigan, see *The University of Michigan,: An Encyclopedic Survey,* vol. 1 (Ann Arbor: Univ. of Michigan Press, 1942), 407–10.

68. See Ruth W. Tryon, *Investment in Creative Scholarship: A History of the Fellowship Program of the American Association of University Women, 1890–1956* (Washington, D.C.: AAUW, 1957), on the 1,121 fellows to 1956, and [Ruth Roettinger], *Idealism at Work: Eighty Years of AAUW Fellowships, A Report by the American Association of University Women* (Washington, D.C.: AAUW, 1967), 14–19 and chs. 3 and 4, on 896 responses to a 1966 survey.

69. Helen C. White, "The AAUW Achievement Award, Presented to Florence Seibert, April 13, 1943," *AAUWJ* 36 (1943): 258–59. The mastermind behind the achievement award was Dorothy B. Atkinson Rood of Minneapolis, who also led the AAUW campaign to raise a million-dollar endowment ([Roettinger], *Idealism at Work,* 294–95; on Rood, see *WWAW* 1 [1958–59]: 1098).

70. "AAUW Achievement Award, 1950," *AAUWJ* 44 (1950): 38; Ruth Benedict to Hope Hibbard, May 28, 1946, Ruth Benedict Papers.

71. [Roettinger]. *Idealism at Work,* ch. 5; there is a list of winners, 1967–80, in *Idealism at Work . . . 1967–1981,* 363–66.

Chapter 16   The Path to Liberation: Consciousness Raised, Legislation Enacted

1. Dorothy W. Weeks, "Women in Physics Today," *Physics Today* 13 (Aug. 1960): 22–23. It came after Lise Meitner's historical account, "The Status of Women in the Professions," ibid., 16–21. Weeks took some of her data from her earlier "Woman Power Shortage in the Physical Sciences," in the *AAUWJ* 48 (1955): 146–49. On Weeks, see *AMWS,* 16th ed., 1986, 7:487; "Dorothy W. Weeks—1969 Achievement Awardee," *AAUWJ* 63 (Jan. 1970): 89; Katherine Sopka, "Dorothy Weeks: Transcript of an Interview on a Tape Recorder," July 19, 1978, NBL; obituary, *NYT,* June 8, 1990, D-16; and Dorothy Weeks Papers, Institute Archives, MIT.

2. Sylvia Fleis Fava, "The Status of Women in Professional Sociology," *American Sociological Review* 25 (1960): 271–76. On Fava, see *AMWS,* 13th ed., 1978, S&B, 363.

3. Betty Friedan, *The Feminine Mystique* (New York: W. W. Norton, 1963), ch. 6, n. 12. Friedan was even more enthusiastic about Eleanor Flexner's recently published book, *A Century of Struggle: The Woman's Rights Movement in the United States* (Cambridge: Harvard Univ. Press, 1959), which she thought every college girl in America should read (ch. 4, n. 1).

4. On Alpenfels, see *WWAW* 8 (1974–75): 15; "Deaths, Ethel J. Alpenfels," *Anthropology Newsletter* (American Anthropological Association), Feb. 1982, 3; and the New York University Archives, for a photograph and card record of her employment history (I thank Hope Lewis for her assistance). See also Ethel J. Alpenfels, "Women in the Professional World," in *American Women: The Changing Image,* ed. Beverly Benner Cassara (Boston: Beacon, 1962), 73–89; idem, "Academic Opportunity Seen Fading for Women," *New York University Alumni News* 8 (Oct. 1962): 1, 3; and idem, "The World of Ideas—Do Women Count?" *Educational Record* 44 (1963): 40–43.

5. Alpenfels, "The World of Ideas," 40.

6. Ibid., 43. Alpenfels's other writings show no earlier or later interest in women's issues. Alpenfels and Friedan were consultants for the PCSW on the portrayal of women by the mass media (PCSW, *Four Consultations: Private Employment Opportunities, New Patterns in Volunteer Work, Portrayal of Women by the Mass Media, Problems of Negro Women* [Washington, D.C.: GPO, 1963]).

7. On Friedan, see *CBY, 1970,* 146–48. See also Betty Friedan, "I Say: Women Are People Too!" *Good Housekeeping,* Sept. 1960, 59–61, 161–62; idem, "If One Generation Can Ever Tell Another," *SAQ* 52 (Feb. 1961): 68–70; idem, *The Feminine Mystique;* "Reactions to the Feminine Mystique," *SAQ* 55 (1963–64): 2, 32; "Letters," ibid. 56 (1964–65): 30; and Betty Friedan, *It Changed My Life: Writings on the Women's Movement* (New York: Random House, 1976).

8. Sylvia Fleis Fava reviewed *The Feminine Mystique* favorably in the *American Socio-*

logical Review 28 (1963): 1053–54. She thought it was too "heavily psychological," however, and once again urged sociologists to study women's social roles.

9. On Rossi, see *Contemporary Authors,* vol. 17, rev. (Detroit: Gale Research, 1986), 389–90; Alice S. Rossi, "Seasons of a Woman's Life," in *Authors of Their Own Lives: Intellectual Autobiographies by Twenty American Sociologists,* ed. Bennett M. Berger (Berkeley: Univ. of California Press, 1990), 301–22; vita c. 1971 and typescript of autobiographical talk, "Deviance and Conformity in the Life Goals of Women," delivered at Wellesley College, Mar. 12, 1970, both in box 7, JLP; Alice S. Rossi, "Women in the Seventies: Problems and Possibilities," keynote speech at Barnard College Conference on Women, New York, Apr. 17, 1970, published in *Discrimination against Women: Hearings before the Special Subcommittee on Education of the Committee on Education and Labor of the U.S. House of Representatives, 91st Congress, 2nd session, on Section 805 of H.R. 16098,* 2 vols. (Washington, D.C.: GPO, 1970–71), 2:1064–67. See also her "Landladies' Attitudes on Student Room Rental in Cambridge, Mass., with Special Reference to Attitudes toward Renting to Negro Students," 1956, in box 16 ("Housing"), Papers of the Office of the Graduate Dean, RCA. Rossi's husband Peter had also been on the PCSW (*Contemporary Authors,* vol. 19, rev. [Detroit: Gale Research, 1987], 410–11).

10. Alice S. Rossi, "Equality between the Sexes: An Immodest Proposal," *Daedalus* 93, no. 2 (spring 1964): 607–52. Though lengthy, this may have been a shortened version of the "monograph" that she wrote Jessie Bernard she had submitted (Rossi to Bernard, Jan. 20, 1964, in box 7, Jessie Bernard Papers, Historical Collections and Labor Archives, Pattee Library, Pennsylvania State University, University Park, Pa.).

11. Alice S. Rossi, "The Case against Full-time Motherhood," *Redbook,* Feb.–Mar. 1965, 51, 129–31, 145.

12. The decision to retain coeducation and build a women's dormitory grew out of the whole concern in the 1950s at MIT and elsewhere to "humanize" (male) engineering students by broadening the curriculum and expanding the range of available social activities, including both organized ones, such as athletics, arts, student government, and publications, as well as informal ones, such as relaxed conversation over dinner in the dining hall. At MIT this "education of the whole man" was then incorporated into future building plans for a student center, an art museum, and a chapel, as well as dormitories and dining halls. Yet the big leap was in extending this new vision of "gracious living" to the women, for even when the women's housing situation was discussed and found to be demonstrably worse than the men's, their needs were given lower priority than the men's and often were postponed for still further study. But around 1956 MIT President James Killian, who had been a trustee of Mount Holyoke College and was possibly influenced by plans for the construction of Cronkhite Hall across town at the Radcliffe Graduate School, by the personal interest of Mrs. Karl Compton, his mentor's widow, by his wife, a Wellesley graduate, or, most likely, by the hint of a large future donation by Katherine Dexter McCormick, an MIT alumna (1904), decided to build a woman's dormitory on a central site. Unfortunately there would be some delay because McCormick, widow of the wealthy heir to the McCormick reaper fortune, a former suffragist, and still in the 1950s a militant feminist, was heavily committed to supporting research on the birth control pill. When that tapered off with the development of the oral contraceptive, she finally made her large gift to

MIT for the new women's dormitory in 1960. She was certainly one woman who made a big difference.

There is considerable material about this decision in boxes 81 and 83, President's Papers (AC 4), Institute Archives, MIT, e.g., "Housing for Women Students," in "Report of the Committee on Student Housing, MIT, June 1956," 61–62; J. A. Stratton, "A Statement of Policy on Women Students," Jan. 24, 1957; Killian to Dean Pietro Belluschi, Mar. 9, 1954, "Confidential"; and in the Dorothy Weeks Papers, also at MIT (esp. McCormick to Weeks, Apr. 25, Oct. 10, 1960). Weeks was on the Committee for the Second Century Fund at MIT, and McCormick's gift counted toward its total.

The December 1963 issue of *Technology Review* (vol. 66) was devoted to the dedication of the new women's dormitory in October 1963 and had celebratory articles about it ("A New Home for MIT Coeds," 15–16) as well as about the history of coeducation at MIT (Beth Bogie, "How Coeducation Came About," 17–18, 40, a sanitized account that drew a nearly continuous line from Ellen Swallow Richards to the new dorm, omitting the Hamilton Report and the crisis of 1955–56). It also discussed McCormick's earlier charities at MIT: she had given furniture to the first MIT women's residence, the apartment house on Bay State Road in Boston, in the 1940s and for years had maintained a famous taxi fund that transported the women students across the river on stormy days. For more on McCormick, see *NAWMP; A Tribute to Katharine Dexter McCormick, 1875–1967*, pamphlet (1968); James R. Killian, *The Education of a College President: A Memoir* (Cambridge: MIT Press, 1985), 298–302; R. Christian Johnson, "Feminism, Philanthropy, and Science in the Development of the Oral Contraceptive Pill," *Pharmacy in History* 19 (1977): 63–78; Loretta McLaughlin, *The Pill, John Rock, and the Church: The Biography of a Revolution* (Boston: Little, Brown, 1982); and James Reed, *From Private Vice to Public Virtue: The Birth Control Movement and American Society since 1830* (New York: Basic Books, 1978), ch. 26. On Elizabeth Parks Killian, see "New Members of the Alumnae Association Board," *WAM* 35 (1950–51): 273; and Lenice Ingram Bacon, "Courageous Heart—Elizabeth Parks Killian, '29," ibid. 44 (1959–60): 14–15, 32 (about Killian after her two debilitating strokes); and Killian, *Education of a College President*, 383–85.

13. On Nachmias, see *AMWS*, 18th ed., 1992, 5:639; Vivianne T. Nachmias, "Panelist," in *Women and the Scientific Professions: The M.I.T. Symposium on American Women in Science and Engineering*, ed. Jacquelyn A. Mattfeld and Carol G. Van Aken (Cambridge: MIT Press, 1965), 29–34. There is considerable correspondence about the symposium— its preparations, attendees, publicity, and publication—as well as the building of McCormick Hall in the Dorothy Bowe Papers (Women at MIT Collection [AC-220]), Institute Archives. See esp. J. A. Mattfeld to Nachmias, Oct. 29, 1964, box 3, f. 6; Cynthia Parsons, "Women in Sciences Spar at Conclave," *CSM*, Nov. 3, 1964, 6; and "Should Science Be for Men Only?" *Technology Review* 67 (Dec. 1964): 28, 46.

14. Alice S. Rossi, "Barriers to the Career Choice of Engineering, Medicine, or Science among American Women," in Mattfeld and Van Aken, *Women and the Scientific Professions*, 51–127. See also John Lear, "Will Science Change Marriage?" *Saturday Review*, Dec. 5, 1964, 75–77, on Rossi's talk; [Mrs. Ankeny], "American Women in Science and Engineering: A Symposium" (in box 3, f. 16), and Mattfeld to Mr. Carroll Bowen of MIT Press, Mar. 2, 1965 (box 3, f. 17), on the book manuscript, both in Dorothy Bowe

Papers (Women at MIT Collection); Alice S. Rossi, "Women in Science: Why So Few?" *Science* 148 (1965): 1196–1202.

15. "Women—In Science or Out?" *Science* 149 (1965): 707–8. See, e.g., Miriam Lipschutz Yevick, "Some Thoughts on Women in Science," *Technology Review* 72 (July–Aug. 1970): 43–46; letters to the editor, ibid., 73 (Feb. 1971): 76–77; and Rita Levi-Montalcini, "Women Scientists and the Women's Liberation Movement," *Proceedings of the Third International Conference of Women Engineers and Scientists,* vol. 1 (Turin: Associazione Italiana Donne Ingegneri e Architetti, 1971): 1, 11–14.

16. On Kundsin, see *AMWS,* 16th ed., 1986, 4:521; and Ruth Kundsin, "Why Nobody Wants Women in Science," *Science Digest* 58 (Oct. 1965): 60–65, which cites Rossi. See also the flippant tone in Molly Neal, "The World Around—Technical Women," *International Science and Technology,* no. 69 (Sept. 1967): 18, which ridicules some of John B. Parrish's "Future Role of Women in U.S. Science and Engineering," in *Proceedings of the Second International Conference of Women Engineers and Scientists, Cambridge, England, 1–9 July 1967,* vol. 3 (London: Women's Engineering Society, 1967). Parrish's essay was also published as John B. Parrish and Jean S. Block, "The Future for Women in Science and Engineering," *Bulletin of the Atomic Scientists* 24 (May 1968), 46–49.

17. Ann Fischer and Peggy Golde, "The Position of Women in Anthropology," *American Anthropologist* 70 (1968): 337–44. To a certain extent the article reflected its place of origin. It was conceived as essentially a data-collection exercise with some assertions about discriminatory behaviors. But despite the authors' prime location at the "center" of the behavioral sciences in the mid-1960s, the whole topic was far from consideration in "central" circles at the time. Their article did not refer to such essential works on the topic as Fava's 1960 report, Rossi's 1964 *Daedalus* essay, or Rossi's MIT speech in *Science* in 1965. Nevertheless, this collaborative project inspired Golde, a research associate and lecturer in the Stanford anthropology department and an assistant professor of psychiatry at Stanford's medical school, to collect and edit *Women in the Field: Anthropological Experiences* (Chicago: Aldine, 1970). See also Katherine Spencer Halpern, "Obituaries: Ann Fischer, 1919–1971," *American Anthropologist* 75 (1973): 292–94; and Munro S. Edmonson, "Ann Kindrick Fischer, 1919–1971," in *Women Anthropologists: Selected Biographies,* by Ute Gacs et al. (Urbana: Univ. of Illinois Press, 1989), 88–94.

18. Geraldine R. Mintz [pseud.], "Some Observations on the Function of Women Sociologists at Sociology Conventions," *American Sociologist* 2 (1967): 158–59. The author might have been Alice Rossi, Sylvia Fava, or maybe Martha S. White, who within a few years published "Psychological and Social Barriers to Women in Science," *Science* 170 (1970): 413–16.

19. On the formation of NOW, see Jo Freeman, *The Politics of Women's Liberation: A Case Study of an Emerging Social Movement and Its Relation to the Policy Process* (New York: Longman, 1975), ch. 3. On O'Hanrahan, see Virginia Carabillo, "Inka O'Hanrahan" (obituary), *NOW Acts* 3, no. 1 (winter 1970): 11; on Hacker, Barton C. Hacker, "Sally Hacker (1936–1988)," *Technology and Culture* 31 (1990): 927; on Epstein, Cynthia Fuchs Epstein, "Personal Reflections with a Sociological Eye," in Berger, *Authors of Their Own Lives,* 356. Northern California chapter president Nita Ladewig published

her fruitless correspondence with NASA officials in "NITA and NASA: A Case History of Male Chauvinism in the Space Program," *NOW Acts* 2, no. 1 (winter–spring 1969): 16–19. On Millett, see *CBY, 1971,* 271–74; and Kate Millett, *Token Learning,* pamphlet published by the New York City branch of NOW in 1968.

20. NSF, *American Science Manpower, 1966* (Washington, D.C.: GPO, 1968), 171, 200–201, and *1968* (Washington, D.C.: GPO, 1969), 202, 252–53. There was also a substantial rise (from 1,200 to 1,300, or 8.3 percent) in the number of women biologists working for federal agencies, particularly the Departments of Agriculture and Commerce, between 1969 and 1970, as perhaps many of these same subfaculty women left academia ("Federal Scientific, Technical, and Health Personnel in 1970," *NSF Science Resources Studies Highlights,* Feb. 22, 1972, 3, 4).

21. On Weisstein, see Naomi Weisstein, "Adventures of a Woman in Science," *Federation Proceedings* 35, no. 11 (1976): 2226–31; Sara Evans, *Personal Politics: The Roots of Women's Liberation in the Civil Rights Movement and the New Left* (New York: Alfred A. Knopf, 1979), 115–223 passim; "Women Scholars: Stymied by System," *Washington Post,* c. 1969, described in *WAM* 54 (winter 1970): 80; and Wellesley College Class of 1961, "Twenty Year Reunion Book" (1981) (I thank Jean Berry at the Wellesley College Archives for her assistance). Weisstein married radical historian Jesse Lemisch in 1965, and both moved on to SUNY-Buffalo in 1973. In 1980 she developed chronic fatigue syndrome, and she has been bedridden ever since (Jesse Lemisch, "Do They Want My Wife to Die?" *NYT,* Apr. 15, 1992, A-27). *Feminism & Psychology* 13, no. 2 (1993), reappraises her "Psychology Constructs the Female" twenty-five years later; her "Response and Afterword: Power, Resistance, and Science: A Call for a Revitalized Feminist Psychology," appears on pp. 239–45.

Weisstein's writings of the period include: "Kinder, Küche, Kirche, as Scientific Law: Psychology Constructs the Female," in *Sisterhood Is Powerful: An Anthology of Writings from the Women's Liberation Movement,* ed. Robin Morgan (New York: Random House, 1970), 205–20, reprinted in *Discrimination against Women,* 1:286–92, revised and expanded into "Psychology Constructs the Female; or The Fantasy Life of the Male Psychologist (with Some Attention to the Fantasies of His Friends, the Male Biologist and the Male Anthropologist)," *Social Education* 35 (Apr. 1971): 362–73; "Woman as Nigger," *Psychology Today* 3 (Oct. 1969): 20, 22, 58, reprinted in *Voices from Women's Liberation,* ed. Leslie B. Tanner (New York: New American Library, 1970), 296–303, and in *Discrimination against Women,* 2:885–88; "How Can a Little Girl Like You Teach a Great Big Class of Men? The Chairman Said: I Was Doing Just Fine Until You Asked Me That Question, I Replied: Or Women/Scientists, Social Expectation, and Self Esteem" (Millbrae, Calif.: Les Femmes, 1974), reprinted with a slightly different title in *Working It Out: Twenty-three Women Writers, Artists, Scientists, and Scholars Talk about Their Lives and Work,* ed. Sara Ruddick and Pamela Daniels (New York: Pantheon, 1977), 241–50.

22. See, in addition to n. 9 above, Alice S. Rossi, "Status of Women in Graduate Departments of Sociology, 1968–1969," *American Sociologist* 5 (1970): 1–12; her preface in *Academic Women on the Move,* ed. Alice S. Rossi and Ann Calderwood (New York: Russell Sage Foundation, 1973), xi; Kay Klotzburger, "Political Action by Academic

Women," ibid., 379–84; and Pamela Roby, "Women and the ASA: Degendering Organizational Structures and Processes, 1964–1974," *American Sociologist* 23 (spring 1992): 18–48.

23. Arlie Hochschild and Alice Rossi, "Postscript" to Rossi, "Status of Women in Graduate Departments of Sociology," 11–12; "Women and Professional Advancement [excerpt from Rossi's report]," *Science* 166 (1969): 356; "Rossi Rouses ASA Woman Power," *NOW Acts* 3, no. 1 (winter 1970): 14. The women's caucus of the ASA later became the independent Sociologists for Women in Society. See also Jo Freeman, "The Revolution Is Happening in Our Minds," *College and University Business* 48 (Feb. 1970): 67–68. This was part of a special issue on women on campus, notable for its early date, wide coverage, and even sympathetic viewpoint.

24. The petition was published as "Equality for Women in Science," *Science for the People* 2, no. 2 (Aug. 1970), 10–11. See also James K. Glassman, "AAAS Boston Meeting: Dissenters Find a Forum," *Science* 167 (1970): 36–38; W. G. Berl, "The 1969 Meeting of the AAAS: A Brief Appraisal," ibid. 1157–58; and Dael Wolfle, "AAAS Council Meeting, 1969," *Science* 167 (1970): 1151–53, which omitted any mention of the petition. Wolfle later reported that a group called WITCHES ("Women's International Terrorist Conspiracy from Hell") had put a hex on some of the sessions (Dael Wolfle, *Renewing a Scientific Society: The American Association for the Advancement of Science from World War II to 1970* [Washington, D.C.: AAAS, 1989], 259, 260). See also Stuart Blume, *Toward a Political Sociology of Science* (New York: Free Press, 1974), 166–67.

25. Robert Reinhold, "Women Criticize Psychology Unit," *NYT*, Sept. 6, 1970, p. 28, col. 1; Ann de Wolfe, "Rebel Women Shake-up APA," *NOW Acts* 3, no. 1 (winter 1970): 13; Klotzburger, "Political Action by Academic Women," 387; "First 'Town Meeting' Focuses on APA Discrimination of Women," *APA Monitor* 1, no. 1 (Oct. 1970): 1, 7; "AWP Seeks $1 Million Reparations from APA," ibid., 3; "Status of Women in Psychology to Be Studied by Task Force," ibid. 1, no. 2 (Nov. 1970), 3; "Astin Appointed to Chair Task Force on Women," ibid. 2, no. 2 (Feb. 1971), 8; "Task Force on Women Presses for Changes in Psychology," ibid. 2, nos. 8–9 (Aug.–Sept. 1971): 4; "Women's Task Force Sets Goals, Disbands," ibid. 3, nos. 9–10 (Sept.–Oct. 1972): 7; and Mary Roth Walsh, "Academic Professional Women Organizing for Change: The Struggle in Psychology," *Journal of Social Issues* 41, no. 4 (winter 1985): 17–27.

26. See also Harland G. Bloland and Sue M. Bloland, *American Learned Societies in Transition: The Impact of Dissent and Recession* (New York: McGraw-Hill, 1974).

27. Several of these reports are in JLP. There are partial bibliographies of these campus "status of women" reports in Rossi and Calderwood, *Academic Women on the Move*, 230–38; Lewis C. Solmon, *Male and Female Graduate Students: The Question of Equal Opportunity* (New York: Praeger, 1976), 124–27; and Linda A. Harmon, comp., *Status of Women in Higher Education, 1963–1972: A Selective Bibliography*, Series in Bibliography, 2 (Ames: Iowa State University Library, 1972). Some of these committees' archives are becoming available, such as that for Harvard's Committee on the Status of Women in the Faculty of Arts and Sciences, 1970–71, at SLRC.

Meanwhile, in 1967–72 the faculties and boards of trustees of several major men's colleges announced their decision to become coeducational institutions as soon as possible. Starting with Cal Tech in 1967 and Wesleyan University in 1968, the movement spread to such Ivy League bastions as Yale, Princeton, and Dartmouth; the Jesuit colleges of Georgetown, Holy Cross, and Boston College; the military and naval service academies; and later such holdouts as Amherst and Williams. This voluntary move toward undergraduate coeducation quickly became intertwined with issues on the status of faculty women, since men-only colleges had had few or no women faculty.

28. Bryce Nelson, "A Surplus of Scientists? The Job Market Is Tightening," *Science* 166 (1969): 582–84; "Suddenly Ph.D's are 'A Glut on the Market,'" *NYT,* Jan. 4, 1970, sec. 4, p. 9, col. 6; Sandra Blakeslee, "Young Physicists Find Fewer Jobs: State of Emergency for New Ph.D's Reported," ibid., Apr. 26, 1970, 50; H. William Koch, "On Physics and Employment of Physicists in 1970," *Physics Today* 24 (June 1971): 23–27; "Federal Scientific, Technical, and Health Personnel in 1970," *NSF Science Resources Studies Highlights,* Feb. 22, 1972, 1–4; Alice S. Rossi, "Discrimination and Demography Restrict Opportunities for Academic Women," *College and University Business* 48 (Feb. 1970): 74–78; idem, "Women in the Seventies," in *Discrimination against Women,* vol. 2, esp. 1069–72; Alan Wolfe, "Hard Times on Campus," *Nation* 210 (May 25, 1970): 623–27, predicting that women and radicals would bear the brunt of the cutbacks; Office of Scientific Personnel, NRC, "Employment Status of Recent Recipients of the Doctorate," *Science* 168 (1970): 930–39, which minimized the problem, but did not break its extensive data down by gender; Earl F. Cheit, *The New Depression in Higher Education: A Study of Financial Conditions at Forty-one Colleges and Universities* (New York: McGraw-Hill, 1971). See also Dael Wolfle and Charles V. Kidd, "The Future Market for Ph.D.s," *Science* 173 (1971): 784–93; Wolfle, *Renewing a Scientific Society,* ch. 12; Frank J. Newman, "The Era of Expertise: The Growth, the Spread, and Ultimately the Decline of the National Commitment to the Concept of the Highly Trained Expert, 1945 to 1970" (Ph.D. diss., Stanford University, 1981), esp. 108–9.

29. See ch. 13, n. 68. A press release about the signing (Office of the White House Press Secretary, Oct. 13, 1967) and the transcript of the accompanying news conference (#1010-A), as well as a copy of the executive order, are in box 79, President's Appointment File, box 79, Lyndon Baines Johnson Library, Austin, Tex.

30. "An Uppity Woman," *Time,* July 10, 1972, 92. See also Anne Ingram, "An Oral History of the Woman's Equity Movement: University of Maryland, College Park, 1968–1978," *Maryland Historian* 9 (fall 1978): 1–25.

31. In October 1970, WEAL also filed formal charges of sex discrimination against all the medical schools in the United States (Judith Randal, "Med School Bias Charge Justified," *Washington, D.C. Star,* Oct. 29, 1970 and WEAL press release, both in box 7, JLP). The lawsuit had some impact, for a year later, in October 1971, Congress passed the Public Health Service Act, the first to outlaw sex discrimination in medical and other health professional schools (Bernice Sandler, "A Little Help from Our Government: WEAL and Contract Compliance," in Rossi and Calderwood, *Academic Women on the Move,* 439).

32. Robert C. Alberts, *Pitt: The Story of the University of Pittsburgh, 1787–1987* (Pittsburgh: Univ. of Pittsburgh Press, 1986), 394–98.

33. See, e.g., Malcolm G. Scully, "Women in Higher Education: Challenging the Status Quo," *Chronicle of Higher Education* 4, no. 18 (Feb. 9, 1970): 2–5; Nancy Gruchow, "Discrimination: Women Charge Universities, Colleges with Bias," *Science*

168 (1970): 559–61; Robert J. Bazell, "Sex Discrimination: Campuses Face Contract Loss over HEW Demands," ibid. 170 (1970): 834–35; Francine Achbar and Pam Bishop, "Harvard Probed for Discrimination against Women," *Boston Herald Traveler,* Apr. 5, 1970, 56; Daniel Zwerdling, "Sex Discrimination on Campus: The Womanpower Problem," *New Republic* 164 (Mar. 20, 1971): 11–13; "Michigan U. Bars Raise for Woman," *NYT,* June 6, 1971, 43; Nancy Hicks, "Women on Faculties Pressing for Advances in Pay and Status," *NYT,* Nov. 21, 1971, 41; Deborah Shapley, "University Women's Rights: Whose Feet Are Dragging?" *Science* 175 (1972): 151–54; Constance Holden, "Women in Michigan: Academic Sexism under Siege," ibid. 178 (1972): 841–44; idem, "Women in Michigan: Parlaying Rights into Power," ibid., 962–65; Gertrude Ezorsky, "The Fight over University Women," *New York Review of Books,* May 16, 1974, 32–40; and Joan Abramson, *The Invisible Woman* (San Francisco: Jossey-Bass, 1975). See also Sandler, "A Little Help from Our Government," 439–62.

34. In the fall of 1971 the ACE reportedly formed a committee, to be chaired by Derek Bok, newly appointed president of Harvard University, to keep HEW off campus. KNOW, Inc., a branch of NOW in Pittsburgh, organized a petition campaign against the committee and its goal (folder on ACE, box 1, f. 1, Jessie Bernard Papers).

35. See Lawrence A. Simpson, "A Study of Employing Agents' Attitudes toward Academic Women in Higher Education" (Ed.D. diss., Pennsylvania State University, 1968); Jo Tice Bloom, "Equal Treatment for Academic Women," *National Business Woman* 50, no. 5 (May 1969): 8–10, about Simpson and his wife; and Lawrence A. Simpson, "A Myth Is Better Than a Miss: Men Get the Edge in Academic Employment," *College and University Business* 48 (Feb. 1970): 70–71; L. S. Fidell, "Empirical Verification of Sex Discrimination in Hiring Practices in Psychology," *AP* 25 (1970): 1094–98; and Arie Y. Lewin and Linda Duchan, "Women in Academia: A Study of the Hiring Decision in Departments of Physical Science," *Science* 173 (1971): 892–95.

36. The big battles were in the Senate, since the House had long since approved most of the changes ("Senate Passes Equal Jobs Bill after Ending Filibuster," *Congressional Quarterly* 30 [1972]: 454–57; "Equal Jobs: Approval of Court Enforcement Approach," ibid., 601–2. See also U.S. Congress, Senate, Committee on Labor and Public Welfare, Subcommittee on Labor, *Legislative History of the Equal Employment Opportunity Act of 1972 (H.R. 1746, P.L. 92-261), Amending Title VII of the Civil Rights Act of 1964* [Washington, D.C.: GPO, 1972]).

37. Freeman, *The Politics of Women's Liberation,* 209–21. Birch Bayh (D.-Indiana) and Marlow Cook (R.-Kentucky) led the move in the Senate, while Martha Griffiths (D.-Michigan), the House's leading feminist, led it there.

38. *Discrimination against Women,* vols. 1 and 2, title page. Soon thereafter Catharine R. Stimpson edited a shortened version, *Discrimination against Women: Congressional Hearings on Equal Rights in Education and Employment* (New York: Bowker, 1973). See Freeman, *The Politics of Women's Liberation,* 223. On Astin, see *AMWS,* 13th ed., S&B, 1978, 43; and Helen S. Astin, *The Woman Doctorate in America: Origins, Career, and Family* (New York: Russell Sage Foundation, 1969). There is abundant material about these hearings (and about discrimination against women in education in general) in the Edith Green Papers, recently opened but only partially arranged, at the Oregon Histori-

cal Society, Portland, Oregon. Green merits a full biography (see her obituary, *NYT*, Apr. 23, 1987, D-31; and *CBY, 1956*, 225–27).

39. Elizabeth Chipman, *Women on the Ice: A History of Women in the Far South* (Melbourne: Melbourne Univ. Press, 1986), ch. 9; unidentified memo, May 20, [1965?], plus clipping in file entitled "Women in Antarctica," Central Subject Files, Office of Antarctic Programs, NSF Papers (RG 307), NARA (I thank Dr. Frank Burch for bringing this file to my attention). On Allen, see *AMWS*, 12th ed., 1971, 1:77; on Walls, ibid., 17th ed., 1989, 7:408; and on McWhinnie, ibid., 14th ed., 1979, 5:3355. See also M. A. McWhinnie, "Antarctic Research," in *Marine Technology Society Journal* 8, no. 5 (June 1974): 3–7; and Jennie Darlington, *My Antarctic Honeymoon: A Year at the Bottom of the World* (London: Frederick Muller, 1956).

40. "Six Women to Work at Antarctic Bases," *NYT*, Sept. 14, 1969, p. 94, col. 8; Walter Sullivan, "Antarctic, a No-Woman's Land, to Get 6 Females," ibid., Oct. 1, 1969, p. 24, col. 1. See also Michael Satchel, "Women Who Conquer the South Pole," *Washington Post, Parade Magazine*, June 5, 1983, 16–17; and Barbara Land, *The New Explorers: Women in Antarctica* (New York: Dodd, Mead, 1981), esp. chs. 1 and 2. The sixth woman was a New Zealander.

41. Quotation in Bob Corbett, "Female Aquanauts Cite Contributions," *San Diego Evening Tribune*, Aug. 5, 1970, B-5. See also Richard B. Lyons, "5 Women Named Aquanaut Team," *NYT*, Mar. 3, 1970, C-27; "5 Women Scientists Start 2-Week Stay under Sea," ibid., July 7, 1970, C-20; "Female Underwater Team May Lead Women to Space," *San Diego Union*, July 28, 1970, A-1; Don Bane, "14 Days on the Ocean Floor," *Los Angeles Herald-Examiner*, Aug. 5, 1970, A-6, plus numerous other clippings in box entitled "Submersibles: Tektite II," in Records of the Public Information Office, 1970, SIO Archives (I thank Deborah C. Day for bringing this file to my attention). See also Sylvia A. Earle, "Tektite II: Part Two, All-Girl Team Tests the Habitat," *National Geographic Magazine*, Aug. 1971, 291–96. On Mead, see *CBY, 1972*, 319–22.

42. Christopher Lydon, "Role of Women Sparks Debate By Congresswoman and Doctor," *NYT*, July 26, 1970, 35; "Embattled Doctor Quits," ibid., Aug. 1, 1970, 5; James M. Naughton, "Washington Notes . . . ," ibid., Aug. 18, 1970, 23; "Does ♀ = ♂? Doctor Stirs a Medical Row," *Medical World News* 11, no. 33 (Aug. 14, 1970): 18–20; "Endocrinology: A Woman Physiologist Vies (as a Man?) with Dr. Berman," ibid., no. 38 (Sept. 18, 1970): 6; Estelle Ramey, "Well, Fellows, What Did Happen at the Bay of Pigs? And Who Was in Control?" *McCall's* 98, no. 4 (Jan. 1971): 26, 81–83; and idem, "Sex Hormones and Executive Ability," in *Successful Women in the Sciences: An Analysis of Determinants*, ed. Ruth B. Kundsin, *Annals of the New York Academy of Sciences* 208 (1973): 237–45. On Ramey, see *AMWS*, 15th ed., 1982, 6:35; and Michael Kolbenschlag, "Dr Estelle Ramey: Reclaiming the Feminine Legacy," *Human Behavior* 5 (July 1976): 25–27. Patsy Mink has donated a box of materials on the episode to the Sophia Smith Collection, Smith College, Northampton, Mass.

43. Copy of Alice Rossi's memo to members of Committee W, Status of Women, AAUP, Aug. 1970, enclosed with Vera Kistiakowsky to M. G. Mayer, Apr. 16, 1971, Maria Goeppart Mayer Papers, Mandeville Department of Special Collections, University of California, San Diego; Alice S. Rossi, "Report of Committee W, 1970–71,"

AAUP Bulletin 57 (1971): 215–20; "Faculty Appointment and Family Relationship," *Liberal Education* 57 (1971): 305; press release, n.d., box 7, JLP. See also Heather Sigworth, "The Legal Status of Antinepotism Regulations," *AAUP Bulletin* 58 (1972): 31–34; idem, "Issues in Nepotism Rules," in *Women in Higher Education,* ed. W. Todd Furniss and Patricia Albjerg Graham (Washington, D.C.: ACE, 1974), 110–20; "Anti-Nepotism Policies Discriminate against Women in Higher Education," *AAUWJ* 64 (Nov. 1970): 40–41. There is also some correspondence in the box 1, Davida Teller Papers, University of Washington Archives, Seattle, on attempts around 1970 to change the rules there. The AAUW had sought in 1966 to do a joint study with the AAUP on the status of academic women but got minimal response ("Minutes of Committee on Standards in Higher Education, Oct. 27, 1966," reel 101, sec. 1172; "Report of Committee on Standards in Higher Education to Board of Directors," p. 3, reel 101, secs. 1199–1200).

44. Advertisement by the Australian Commonwealth Scientific and Industrial Research Organization, *Science* 169 (1970): 1115; "Equal Work—Unequal Pay," ibid. 170 (1970): 1358; "Intrepid Analysis of Female Scientists," ibid. 171 (1971): 521–22; "Male Bias and Women's Fate," ibid. 172 (1971): 514–15; and Susan Ervin-Tripp, "Women with Ph.D.'s," ibid. 174 (1971): 1281. Ervin-Tripp had earned her Ph.D. in 1955.

45. On these later women's caucuses, see Klotzburger, "Political Action by Academic Women"; Ruth Oltman, "Women in the Professional Caucuses," *American Behavioral Scientist* 15 (1971): 281–302; Anne Briscoe, "Phenomenon of the Seventies: The Women's Caucuses," *Signs* 4 (1978): 152–58; and Bloland and Bloland, *American Learned Societies in Transition,* 36–38, 59–62, 99. See also Gloria B. Lubkin, "Women in Physics," *Physics Today* 24 (Apr. 1971): 23–27; and Lenore Blum, "A Brief History of the Association for Women in Mathematics: The Presidents' Perspectives," *Notices of the American Mathematical Society* 38 (1991): 738–54. Another founder of AWIS was Anne Briscoe, who received a doctorate in biochemistry from Yale in 1949. She recorded her twenty-year odyssey of service work for the clinical departments of the Cornell and Columbia medical schools in "Diary of a Mad Feminist Chemist," *International Journal of Women's Studies* 4, no. 4 (1981): 420–30; and idem, "Scientific Sexism: The World of Chemistry," in *Women in Scientific and Engineering Professions,* ed. Violet B. Haas and Carolyn C. Perucci (Ann Arbor: Univ. of Michigan Press, 1984), 155.

46. Press release, Apr. 16, 1971, and list of AWIS charter members, both in box 7, JLP; Phyllis Harber, "Registry of Women Scientists," *Science* 172 (1971): 1192. On Apter, see *AMWS,* 14th ed., 1979, 1:123; Julia T. Apter, "Increasing the Professional Visibility of Women in Academe: A Case Study," in Furniss and Graham, *Women in Higher Education,* 104–9; Apter to Dr. Perlmann, Dec. 15, 1971, and Perlmann to Apter, Dec. 20, 1971, enclosing a vita, box 6, f. 7, Gertrude Perlmann Papers, RAC, North Tarrytown, N.Y.

47. "Women's Lib Hits Club after 72 Years," *The Chemist* 48 (June 1971): 142; Paul B. Slawter, Jr., "The Chemists' Club at 75," CHEMTECH 3 (Dec. 1973): 720–21. On Hazel Bishop, see ch. 12, n. 32.

48. "Directory of Scientists Will Now List Women," *NYT,* Nov. 23, 1971, 7. The article was mis-headlined: the *AMS* had always included women; it was the title that was new.

49. John Walsh, "AAAS Meetings: Pro Forma Protest and Constitutional Reform,"

Science 175 (1972): 42–44; William Bevan, "AAAS Council Meeting, 1971," ibid., 798–99; Mary E. Clutter and Virginia Walbot, "AAAS Meeting" (letter to the editor), ibid., 944–45.

50. Maria Goeppart Mayer to Vera Kistiakowsky, [May?] 1971, Maria Goeppart Mayer Papers. See also Arlie Hochschild, "Making It: Marginality and Obstacles to Minority Consciousness," in Kundsin, *Successful Women in the Sciences,* 179–84.

51. "Garvan Medal, Birgit Vennesland," *C&EN* 42 (Feb. 17, 1964): 98; and Birgit Vennesland, "Reflections and Small Confessions," *Annual Review of Plant Physiology* 32 (1981): 1–20, esp. 15–16. Vennesland's mother had been a suffragist, and she considered her two years on an International Federation of University Women fellowship (at Harvard because of World War II) as a key period in her career; yet having "made it" with their help, she considered herself uninterested in women's issues in science.

52. Louise Daniel, taped interview, 1976, KLCU.

53. Gwendolyn S. Safier, "Jessie Bernard: Sociologist" (Ph.D. diss., University of Kansas, 1972), 269–71. See also Jessie Bernard, "Sexism and Discrimination," *American Sociologist* 5 (1970): 374–75; and idem, "My Four Revolutions: An Autobiographical History of the ASA," *American Journal of Sociology* 78 (1973): 773–91. Thereafter Bernard publicly endorsed women's liberation, wrote tenure letters for the burgeoning number of younger women sociologists, and even testified at hearings and in lawsuits for other embattled ones at a time when not many other senior sociologists were particularly interested in or sympathetic to their research (numerous letters in Jessie Bernard Papers; and Robert C. Bannister, *Jessie Bernard: The Making of a Feminist* [New Brunswick, N.J.: Rutgers Univ. Press, 1991]).

54. Judith Walzer, "Interviews with Ruth Hubbard," July and August 1981, typescript, 88, 96–97, 114–16, 123–25, MCRC. See also Madeline Drexler, "Ruth Hubbard An Outsider Inside," *Boston Globe,* May 16, 1990, 73, 78–79 (I thank Clark A Elliott for a copy).

55. Eleanor Jack Gibson, taped interview, 1976, KLCU.

56. On Public Law 92-318, see *U.S. Statutes at Large* 86 (1972): 373–75. See also *Higher Education Amendments of 1971: Hearings before the Subcommittee on Education of the Committee on Education and Labor, House of Representatives, 92nd Congress, 1st Session* (Washington, D.C.: GPO, 1971), pt. 2:1113–32, where Congresswoman Patsy Mink inserted into the record Ruth Oltman's *Campus 1970: Where Do Women Stand?* a report on sex discrimination in academia prepared for the AAUW.

BIBLIOGRAPHICAL ESSAY

Pursuing the history of women in science from 1940 to 1972 has been a challenging odyssey. The period is still rarely examined by historians, who are all too content to end their tales in 1940, and the topic includes fields and institutions not often studied by historians of science, such as nutrition, psychology, governmental bureaus, women's colleges, scientific societies, technical libraries, and nonprofit institutions. The subject therefore requires one to have not only a knowledge of the history of mainstream political events (wars and presidential administrations) and the major scientific institutions, events, and prominent male personalities but also an awareness of women's history broadly defined. But first one must locate the women scientists, who, though often present, are rarely recorded even in footnotes of the official accounts or institutional histories. Instead they can be found in a separate, more scattered literature comprising obituaries, biographical dictionaries, histories and publications of women's organizations, and a wide variety of other pieces. All this requires a lot of reference work, which stretches not only one's mind but also many libraries' resources.

General Works

Among the useful general works are George Q. Flynn, *The Mess in Washington: Manpower Mobilization in World War II* (Westport, Conn.: Greenwood, 1979); Susan M. Hartmann, "Prescriptions for Penelope: Literature on Women's Obligations to Returning World War II Veterans," *Women's Studies* 5, no. 3 (1978): 223–40; Jessie Bernard's *Academic Women* (State College: Pennsylvania State Univ. Press, 1964); and Frank J. Newman, "The Era of Expertise: The Growth, the Spread, and Ultimately the Decline of the National Commitment to the Concept of the Highly Trained Expert, 1945 to 1970" (Ph.D. diss., Stanford University, 1981). Three accounts of specific fields that are of particular relevance to women scientists are Donald S. Napoli, *Architects of Adjustment: The History of the Psychological Profession in the United States* (Port Washington, N.Y.: Kennikat, 1981); Dan McLachlan, Jr., and Jenny P. Glusker, eds., *Crystallography in North America* (New York: American Crystallographic Association, 1983); and NRC, Food and Nutrition Board, *The Food and Nutrition Board, 1940–1965: Twenty-five Years in Retrospect* (Washington, D.C.: NAS-NRC, [1966?]), which has some useful essays.

Most of the recent work on Big Science is on physics and has little on women. In fact a recent review of J. L. Heilbron and Robert W. Seidel's *Lawrence and His Laboratory: A History of the Lawrence Berkeley Laboratory,* vol. 1 (1989) put it well: "Women appear only in the margins of the narrative, where, significantly, they figure only as objects to be married or irradiated by men, never as voices" (Andrew Pickering, "The Rad Lab and the World," *British Journal for the History of Science* 25 [1992]: 249–50).

Autobiographies

There are what might appear to be a surprising number of autobiographies by women scientists but none by women engineers. Some are whole volumes, and others are shorter pieces, many of them presented as part of a panel, career day, or retrospective session during the 1970s. Among the former the outstanding ones are Icie Macy Hoobler's *Boundless Horizons: Portrait of a Pioneer Woman Scientist* (Smithtown, N.Y.: Exposition, 1982); *Cecilia Payne-Gaposchkin: An Autobiography* (Cambridge: Cambridge Univ. Press, 1984), edited by her daughter Katherine Haramundanis and with an introduction by historian Peggy Kidwell; Cynthia Westcott's *Plant Doctoring Is Fun* (Princeton: Van Nostrand, 1957); Hortense Powdermaker's *Stranger and Friends: The Way of an Anthropologist* (New York: W. W. Norton, 1966); and Frances Hamerstrom's, *Strictly for the Chickens* (Ames: Iowa State Univ. Press, 1980), along with Eugenie Clark's two aimed at children, *Lady with a Spear* (New York: Harper & Brothers, 1953) and *The Lady and the Sharks* (New York: Harper & Row, 1969). Two prominent women deans have also left autobiographies of interest: Virginia Gildersleeve's *Many a Good Crusade: Memoirs of Virginia Crocheron Gildersleeve* (New York: Macmillan, 1954) and Bernice Brown Cronkhite's *The Times of My Life* ([Cambridge, Mass.]: privately printed, 1982). To these can be added the nearly eighty oral histories listed below.

OTHER RELEVANT AUTOBIOGRAPHIES

Ajzenberg-Selove, Fay. *A Matter of Choices: Memoirs of a Female Physicist.* New Brunswick, N.J.: Rutgers Univ. Press, 1994.

Bernard, Jessie. "My Four Revolutions: An Autobiographical History of the ASA." *American Journal of Sociology* 78 (1973): 773–91.

———. "A Woman's Twentieth Century." In *Authors of Their Own Lives: Intellectual Autobiographies by Twenty American Sociologists,* edited by Bennett M. Berger. Berkeley and Los Angeles: Univ. of California Press, 1990.

Bunting, Mary I. "From *Serratia* to Women's Lib and a Bit Beyond." *American Society for Microbiology News* 37, no. 3 (1971): 46–52.

Cobb, Jerrie, with Jane Rieker. *The Jerrie Cobb Story.* Englewood Cliffs, N.J.: Prentice-Hall, 1963.

Hopper, Grace Murray. "The Education of a Computer." 1952. Reprinted in *Annals of the History of Computing* 8 (1988): 271–81.

Libby, Leona Marshall. *The Uranium People.* New York: Charles Scribner's Sons, 1979) (semi-autobiographical).

Rossi, Alice S. "Seasons of a Woman's Life." In Berger, *Authors of Their Own Lives,* 301–22. See Bernard, "A Woman's Twentieth Century," above.

Sandler, Bernice. "A Little Help from Our Government: WEAL and Contract Compliance." In *Academic Women on the Move,* edited by Alice S. Rossi and Ann Calderwood, 439–62. New York: Russell Sage Foundation, 1973.

Wax, Rosalie H. *Doing Fieldwork: Warnings and Advice.* Chicago: Univ. of Chicago Press, 1971.

Weisstein, Naomi. "Adventures of a Woman in Science." *Federation Proceedings* 35, no. 11 (1976): 2226–31.

————. "How Can a Little Girl Like You Teach a Great Big Class of Men? The Chairman Said, and Other Adventures of a Woman in Science." In *Working It Out: Twenty-three Women Writers, Artists, Scientists, and Scholars Talk about Their Lives and Work,* edited by Sara Ruddick and Pamela Daniels, 241–50. New York: Pantheon, 1977.

COLLECTIVE AUTOBIOGRAPHIES

Some collective autobiographies contain essays by women in several fields, such as Violet B. Haas and Carolyn C. Perucci, eds., *Women in Scientific and Engineering Professions* (Ann Arbor: Univ. of Michigan Press, 1984); *Successful Women in the Sciences: An Analysis of Determinants,* ed. Ruth B. Kundsin, *Annals of the New York Academy of Sciences* 208 (1973); and the international Derek Richter, ed., *Women Scientists: The Road to Liberation* (London: Macmillan, 1982). Others focus on women in a particular field, such as Peggy Golde, ed., *Women in the Field: Anthropological Experiences* (Chicago: Aldine, 1970); and Agnes N. O'Connell and Nancy Felipe Russo, eds., *Models of Achievement: Reflections of Eminent Women in Psychology,* 2 vols. (New York: Columbia Univ. Press, 1983; Hillsdale, N.J.: Lawrence Erlbaum, 1988). Still others, devoted to a particular field, include essays on or by women, such as Donald J. Albers and G. L. Alexanderson, eds., *Mathematical People: Profiles and Interviews* (Boston: Birkauser, 1985), and Donald J. Albers et al., *More Mathematical People: Contemporary Conversations* (Boston: Harcourt Brace Jovanovich, 1990); T. S. Krawiec, ed., *The Psychologists: Auto-biographies of Distinguished Living Psychologists,* 3 vols. (Brandon, Vt.: Clinical Psychology, 1978); Gardner Lindzey, ed., *A History of Psychology in Autobiography,* vols. 7 (San Francisco: W. H. Freeman, 1980) and 8 (Stanford: Stanford Univ. Press, 1989); and Alan Lightman and Roberta Brawer, *Origins: The Lives and Worlds of Modern Cosmologists* (Cambridge: Harvard Univ. Press, 1990).

Biographies

Biographies of women in science vary in quality. The older ones tilt toward juvenilia, a genre that has influenced several otherwise serious early works about women in science, or even humor, such as the two hilarious books on Lillian Gilbreth by her children Frank B. Gilbreth, Jr., and Ernestine Gilbreth Carey, *Cheaper by the Dozen* and *Belles on Their Toes* (New York: Thomas Y. Crowell, 1948 and 1950, respectively), both of which were made into movies. Later biographies are more likely to be serious works, even dissertations or university press publications, such as the three on the anthropologist Ruth Benedict and two on sociologist Jessie Bernard. Most biographers of women scientists have been their relatives, former students, journalists, or, as in the case of Rachel Carson, agent. Richard S. Baldwin's *Fungus Fighters: Two Women Scientists and Their Discovery* (Ithaca: Cornell Univ. Press, 1981) is unusual, for books about governmental scientists are rare.

Bannister, Robert C. *Jessie Bernard: The Making of a Feminist.* New Brunswick, N.J.: Rutgers Univ. Press, 1991.
Brooks, Paul. *The House of Life: Rachel Carson at Work.* Boston: Houghton Mifflin, 1972.

Caffrey, Margaret M. *Ruth Benedict: Stranger in This Land*. Austin: Univ. of Texas Press, 1989.

Cohen, Lisa Z. "Madame Curie and Dr. Rosalyn Yalow: Two Women Who Overcame the Odds." *Lab World* 29 (July 1978): 16–13, 21–22, 24.

Eagles, Juanita A., Orrea F. Pye, and Clara M. Taylor. *Mary Swartz Rose, 1874–1941: Pioneer in Nutrition*. New York: Teachers College Press, 1979.

Guzzo, Louis R. *Is It True What They Say about Dixy? A Biography of Dixy Lee Ray*. Mercer Island, Wash.: Writing Works, 1980.

Hill, Edward. *My Daughter Beatrice: A Personal Memoir of Dr. Beatrice Tinsley, Astronomer*. New York: American Physical Society, 1986.

Howard, Jane. *Margaret Mead: A Life*. New York: Simon & Schuster, 1984.

Jennings, Bojan Hamlin. "The Professional Life of Emma Perry Carr." *JCE* 63, no. 11 (1986): 923–27.

Keller, Evelyn Fox. *A Feeling for the Organism: The Life and Work of Barbara McClintock*. New York: W. H. Freeman, 1983.

Mead, Margaret. *Ruth Benedict*. New York: Columbia Univ. Press, 1974.

Modell, Judith Schachter. *Ruth Benedict: Patterns of a Life*. Philadelphia: Univ. of Pennsylvania Press, 1983.

Poynter, Margaret. *The Zoo Lady: Belle Benchley and the San Diego Zoo*. Minneapolis: Dillon, 1980.

Safier, Gwendolyn S. "Jessie Bernard, Sociologist." Ph.D. diss., University of Kansas, 1972.

COLLECTIVE BIOGRAPHIES

Some early collective biographies, such as Edna Yost's *American Women of Science* (Philadelphia: J. B. Lippincott, 1943; rev. ed., 1955) and *Women of Modern Science* (New York: Dodd, Mead, 1959), were a muted form of vocational guidance, portraying the lives of women in science to inspire the young as well as inform the public. Similarly, Mary Finch Hoyt's *American Women of the Space Age* (New York: Atheneum, 1966) can be interpreted as showing girls of the time all the interesting things they could do even if they could not be astronauts. Now these works constitute doubly interesting historical documents. As some of the biobibliographical projects of the 1970s have been completed, the genre of collective biographies of women scientists (and some engineers) has greatly expanded. The more recent works range from factual compendia to insightful essays on a whole field, such as the several essays on archaeology, geology, crystallography, mathematics, and other fields in G. Kass-Simon and Patricia Farnes, eds., *Women of Science: Righting the Record* (Bloomington: Indiana Univ. Press, 1990).

Bailey, Lena, and Beulah Sellers Davis, eds. *Seventy Significant Leaders: Home Economics Teacher Education*. Teacher Education Section Yearbook No. 2. Bloomington, Ill.: McKnight, [1982].

Gacs, Ute, et al. *Women Anthropologists: A Biographical Dictionary*. Westport, Conn.: Greenwood, 1988. Reprinted as *Women Anthropologists: Selected Biographies*. Urbana: Univ. of Illinois Press, 1989.

Grinstein, Louise S., and Paul J. Campbell, eds. *Women of Mathematics: A Biobibliographic Sourcebook*. New York: Greenwood, 1987.

Kenschaft, Patricia Clark. "Black Men and Women in Mathematical Research." *Journal of Black Studies* 18 (1987): 182–84.

Land, Barbara. *The New Explorers: Women in Antarctica*. New York: Dodd, Mead, 1981.

McKelway, Ben. "Women in Oceanography." *Oceanus* 25, no. 4 (1982–83): 75–79.

Miles, Wyndham, ed. *American Chemists and Chemical Engineers*. Washington, D.C.: ACS, 1976.

O'Connell, Agnes N., and Nancy Felipe Russo, eds. *Women in Psychology: A Bio-bibliographic Sourcebook*. New York: Greenwood, 1990.

O'Hern, Elizabeth Moot. *Profiles of Pioneer Women Scientists*. Washington, D.C.: Acropolis, 1985.

Payne [Ralston], Alma. *Partners in Science*. Cleveland: World, 1968.

Ray, Elizabeth M., ed. *Home Economics Teacher Education: Sixty Significant Years*. AHEA, Teacher Education Section Yearbook No. 1. Bloomington, Ill.: McKnight, 1981.

Robinton, Elizabeth D. "A Tribute to Women Leaders in the Laboratory Section of the American Public Health Association." *American Journal of Public Health* 64 (1974): 1006–7.

Satchel, Michael. "Women Who Conquer the South Pole." *Washington Post, Parade Magazine,* June 5, 1983, 16–17.

Shapley, Virginia B. "Science in Petticoats: In Wartime and in Peace." *AAUWJ* 39 (spring 1946): 148–50.

Stevens, Gwendolyn, and Sheldon Stevens. *The Women of Psychology.* Cambridge, Mass.: Schenckmann, 1982.

Tinling, Marion, ed. *Women into the Unknown: A Sourcebook on Women Explorers and Travelers*. New York: Greenwood, 1989.

Tolman, Ruth. "Some Work of Women Psychologists in the War." *Journal of Consulting Psychology* 7 (1943): 127–31.

Vare, Ethlie A., and Greg Ptacek. *Mothers of Invention*. New York: William Morrow, 1988.

Wallace, Jane H. "Women in the Survey." *Geotimes* 24, no. 3 (1979): 34.

Wheeler, Roger, and Philip Snowdon. "American Women in Space." *Journal of the British Interplanetary Society* 40 (1987): 81–88.

Williams, Barbara. *Breakthrough: Women in Archaeology*. New York: Walker, 1981.

Yarmish, Rina J., and Louise S. Grinstein. "Brief Notes on Six Women in Computer Development, Part I." *Journal of Computers in Mathematics and Science Teaching* 2, no. 2 (winter 1982): 38–39.

———. "Brief Notes on Six Women in Computer Development, Part II." *Journal of Computers in Mathematics and Science Teaching* 2, no. 3 (spring 1983): 26–27.

Vocational Guidance

Books and articles purporting to give young and other women advice about job opportunities have been an invaluable source of material for this study. Each is, of course, a product of its time and reflects some political stance regarding why women were being recruited or discouraged at that particular moment. Even an article bearing the bland title "Opportunities for Women in Science" can reveal a broad range of attitudes and messages. Helpful bibliographic guides to this literature include Florence S. Hellman, comp., "Women's Part in World War II: A List of References" (Washington, D.C.: LC, Division of Bibliography, 1942, mimeographed); Louise Moore, comp., *Occupations for Girls and Women: Selected References, July 1943–June 1948*, U.S. Department of Labor, WB Bulletin 229 (Washington, D.C., 1949); and LC, Science and Technology Division, "Scientific Personnel: A Bibliography" (Washington, D.C., 1950, mimeographed). For the later period the *Applied Science Index* has had entries under "women as."

Some of the most balanced and most judicious materials of the time were publications, often containing a bibliography, by the federal Women's Bureau (WB), especially its series of bulletins, which focused especially on scientific and technical careers in the period 1948–61. These can be supplemented by the actual materials used to create them, preserved in the Papers of the Women's Bureau (RG 86) at NARA in Washington, D.C. A useful listing of these bulletins, as well as a list of the box numbers of related materials, is appended to Mabel Deutrich and Virginia Purdy, eds., *Clio Was a Woman: Studies in the History of American Women* (Washington, D.C.: Howard Univ. Press, 1980), 329–42.

The many publications by the Bureau of Labor Statistics (BLS) on the occupational outlook for certain fields usually included some mention of women's prospects. Articles in its *Monthly Labor Review* are usually authoritative. For factors affecting federal scientists, one should start with U.S. Civil Service Commission Library, *Scientists and Engineers in the Federal Government*, Personnel Bibliography Series, no. 30 (Washington, D.C., 1970). Many articles on women's careers in chemistry were published in the *Journal of Chemical Education (JCE)* and are listed in its index volumes. Useful for providing some of the flavor of the times are such works as Evelyn Steele's *Careers for Girls in Science and Engineering* (New York: E. P. Dutton, 1943) and her *Wartime Opportunities for Women* (New York: E. P. Dutton, 1943). In the 1950s and 1960s there were occasional articles in the young woman's magazine *Mademoiselle*, as well as in books such as Frances Maule's *Executive Careers for Women* (New York: Harper, 1961) and several by Marguerite Zapoleon, including *The College Girl Looks Ahead to Her Career Opportunities* (New York: Harper, 1956) and *Occupational Planning for Women* (New York: Harper, 1961). In addition, numerous publications by members of the SWE urged women to become engineers.

American Chemical Society, Division of Chemical Education, *Training of Literature Chemists: A Collection of Papers Comprising the Symposium on Training of Literature Chemists, Presented before the Division of Chemical Education and the Division of Chemical Literature at the 127th National Meeting of the American Chemical Society, Cincinnati, Ohio, March 1955*. Advances in Chemistry Series, no. 17. Washington, D.C., 1956.

Goff, Alice C. *Women CAN Be Engineers*. Youngstown, Ohio: privately printed, 1946.

McCabe, Rita. "Women in Data-Processing." *National Business Woman* 38 (Sept. 1959): 6–7, 28.

Snell, Cornelia. "Organic Analysis as a Tool for Women Chemists." *JCE* 27 (1950): 138–41.

———. "The Woman Chemist." *C&EN* 28 (1950): 3110–12. Reprinted in *Careers in Chemistry and Chemical Engineering*. Easton, Pa.: ACS, 1951, 44–46.

————. "Women Professional Chemists." *JCE* 25 (1948): 450–53.

Strieby, Irene M. "The Chemical Librarian in Industry." *JCE* 30 (1953): 522–25.

U.S. Department of Labor, Women's Bureau. *Careers for Women as Technicians.* WB Bulletin 282. Washington, D.C., 1961.

————. *Careers for Women in the Biological Sciences.* WB Bulletin 278. Washington, D.C., 1961.

————. *Careers for Women in the Physical Sciences.* WB Bulletin 270. Washington, D.C., 1959.

————. *Employment Opportunities for Women in Professional Engineering.* WB Bulletin 254. Washington, D.C., 1954.

————. *Employment Opportunities for Women Mathematicians and Statisticians.* WB Bulletin 262. Washington, D.C., 1956.

————. *The Outlook for Women in Science.* WB Bulletin 223. Washington, D.C., 1948–49 (8 parts on specific fields).

Woodford, Lois. "Opportunities for Women in Chemistry." *JCE* 19 (1942): 536–38.

————. "Trends in the Industrial Employment of Women Chemists." *JCE* 22 (1945): 236–38.

Biographical Reference Works

For short biographical data, *American Men of Science (AMS,* since 1972 *American Men and Women of Science [AMWS])* is useful for those listed in it, despite occasional errors, but its selection criteria for the period after 1950 are not clear. Too many important persons are left out. A 1983 index for the first fourteen editions is useful in locating the last, and thus fullest, entry for an individual, although in many cases later entries, in the interests of economy and brevity, have omitted the details of a lifetime climb up a ladder. The occasional geographic indexes are helpful for identifying those couples who are both listed and live in the same town. (Likewise, the detailed geographic and employer indexes to more specialized membership directories, such as those of the APA, can provide useful starting places for studies of local employment patterns of those persons who bothered to join.) The *Who's Who of American Women (WWAW),* started in 1958, contains additional personal information, such as names of children and spouses. The *Current Biography Yearbook* has two- to three-page sketches of persons currently in the news, such as award winners, presidential appointees, best-selling authors, and controversial figures. Since its volumes go back to 1940, it and its index are useful for now historical figures. Obituaries, always a generous genre, can be particularly informative, because they are often written by a friend or acquaintance and provide personal details. Yet locating them can be difficult, since each field seems to have its own way of publishing such material. For a reference to an obituary or other biographical notice, such as an award presentation, the *Biography Index* can be helpful, though its coverage omits several important sources of scientific obituaries, such as newsletters and bulletins by professional societies, including the *A[merican] S[ociety] of M[icrobiology] Newsletter* and the *Notices of the American Mathematical Society,* and even the series of memorials by the GSA. Substantial and well-researched short biographies of prominent women in a number of fields who died between 1951 and 1975 are contained in *NAWMP,* edited by

Barbara Sicherman and Carol Hurd Green (Cambridge: Harvard Univ. Press, 1980). For particular subgroups there are Marion Tinling, comp., *Women into the Unknown: A Sourcebook on Women Explorers and Travelers* (New York: Greenwood, 1989), and Barbara Bennett Peterson, ed., *Notable Women of Hawaii* (Honolulu: Univ. of Hawaii Press, 1984).

Data and Statistics

Indispensable for work in the area of data and statistics are certain bibliographic guides, such as Malcolm C. Hamilton's *Directory of Educational Statistics: A Guide to Sources* (Ann Arbor: Pierian, 1974); Carolyn Shaw Bell, "Women in Science: Definitions and Data for Economic Analysis," *Successful Women in the Sciences: An Analysis of Determinants,* published as *Annals of the New York Academy of Sciences* 208 (1973): 134–42, on the merits and weaknesses of several sets of manpower statistics; and NSF, Division of Science Resources Studies, *A Listing of Publications, 1953–77* (Washington, D.C., [1978]) and subsequent editions.

The home economists have their own sets of data: USOE, Vocational Division, "Home Economics in Degree-granting Institutions," Misc. 2557 (Washington, D.C., mimeographed), revised biennially from 1939–40 to 1963–64 and then continued by the AHEA. Similarly, though less extensively, the AAUW has published its own data on its fellows and other award winners, many of whom were scientists.

As for the actual data sets, they are all flawed in one way or another. Some have vast coverage but few details; others have restricted scope but more details. Many of the data collected were not broken down by gender, and even when they were to some extent, direct comparisons with male scientists were rarely possible. Even what is probably the best series, *American Science Manpower,* 1954–70, of the National Register of Scientific and Technical Personnel, based on responses to questionnaires circulated through large scientific societies, changed over time and often admitted in its "technical notes" that its results did not constitute a valid statistical sample. In many ways the nongovernmental data, such as those collected by the NEA and by John Parrish of the University of Illinois, were the most useful, because they illuminated the very structural and institutional patterns that the others surveys obscured in their aggregates.

The several quantitative studies of scientists, based upon various data sets, are also quite revealing of attitudes of the time. Frequently authors include women in their data but bury them in or omit them from the text. The classic studies of this type are R. H. Knapp and H. B. Goodrich, *Origins of American Scientists* (Chicago: Univ. of Chicago Press, 1952); James A. Davis, *Stipends and Spouses: The Finances of American Arts and Science Graduate Students* (Chicago: Univ. of Chicago Press, 1962); and Anselm L. Strauss and Lee Rainwater, *The Professional Scientist: A Study of American Chemists* (Chicago: Aldine, 1962). In addition, the numerous studies of "career patterns" and the like that have taken sex differences into account, chiefly those by the Office of Scientific Personnel at the NRC, too long have taken the place of real studies. They ask those uninteresting questions to which their limited data can provide the answers, such as the number of years between degrees, but leave out and thus exonerate the whole institutional setting and its practices, which shaped the women's careers or lack thereof. More interesting and informative are those quantitative studies with a more feminist slant that used their data to answer interesting and important questions, such as Helen S. Astin, *The Woman Doctorate in America: Origins, Career, and Family* (New York: Russell Sage Foundation, 1969); Louise Dolnick Solomon, "Female Doctoral Chemists: Sexual Discrepancies in Career Patterns" (Ph.D. diss., Cornell University, 1972); and Barbara Reskin, "Sex Differences in the Professional Life Chances of Chemists" (Ph.D. diss., University of Washington, 1973). See also Marcel C. LaFollette, "Eyes on the Stars: Images of Women Scientist in Popular Magazines," *Science, Technology, and Human Values* 13 (1988): 262–75.

American Home Economics Association. *Home Economics in Institutions Granting Bachelor's or Higher Degrees, 1963–64: Tables Selected from Material Prepared by the U.S. Office of Education.* Washington, D.C., 1967.

Bureau of Labor Statistics. *Employment, Education, and Earnings of American Men of Science.* BLS Bulletin 1927. Washington, D.C., 1951.

———. *Employment of Scientists and Engineers, 1950–1970.* BLS Bulletin 1781. Washington, D.C., 1973.

———. *Personnel Resources in the Social Sciences and Humanities: A Survey of the Characteristics and Economic Status of Professional Workers in 14 Fields of Specialization.* BLS Bulletin 1169. Washington, D.C., 1954.

David, Deborah S. *Career Patterns and Values: A Study of Men and Women in Science and Engineering.* Washington, D.C.: U.S. Department of Labor, Manpower Administration, 1971.

Gartaganis, Arthur J. "Trends in Federal Employment, 1958–1972." *Monthly Labor Review* 97 (Oct. 1974): 17–25.

Gorman, J. C., and Laura J. Harper. *Home Economics in Institutions Granting Bachelors or Higher Degrees, 1968–69.* Washington, D.C.: AHEA, 1970.

Harmon, Lindsey R. *Career Achievements of NSF Graduate Fellows: The Awardees of 1952–1972.* Washington, D.C.: NAS, 1977.

———. *Career Achievements of the National Defense Education Act (Title IV) Fellows of 1959–1973: A Report to the U.S. Office of Education.* Washington, D.C.: NAS, 1977.

———. *Profiles of Ph.D's in the Sciences: Summary Report on Follow-up of Doctorate Cohorts, 1935–1960.* NAS-NRC Publication No. 1293. Washington, D.C., 1965.

———, comp. *A Century of Doctorates: Data Analyses of Growth and Change.* Washington, D.C.: NAS, 1978.

Harmon, Lindsey R., and Herbert Soldz, comps. *Doctorate Production in United States Universities, 1920–1962.* NAS Publication No. 1142. Washington, D.C., 1963.

National Education Association. "Teacher Supply and Demand in Degree-granting Institutions, 1954–55." *NEA Research Bulletin* 33, no. 4 (1955): 127–63 (whole issue).

———. Research Division. *Teacher Supply and Demand in Universities, Colleges, and Junior Colleges, 1957–58 and 1958–59.* Higher Education Series, Research Report 1959-R10. Washington, D.C.: NEA, 1959.

National Research Council. Office of Scientific Personnel. *Careers of PhD's, Academic*

versus Nonacademic: A Second Report on Follow-up of Doctorate Cohorts, 1935–1960.
NAS-NRC Publication No. 1577. Washington, D.C., 1968.

National Science Foundation. *Graduate Student Enrollment and Support in American
Universities.* Washington, D.C.: GPO, 1957.

———. *Scientific Manpower in the Federal Government, 1954.* Washington, D.C., 1957.

———. "Scientists and Engineers from Abroad: Trends of the Past Decade, 1966–1975."
NSF Reviews of Data on Science Resources, no. 28 (Feb. 1977): 1–3.

———. *Women in Scientific Careers.* Washington, D.C.: GPO, 1961.

National Science Foundation, with Bureau of Labor Statistics and Mathematical Association of America. *Employment in Professional Mathematical Work in Industry and
Government.* Washington, D.C.: GPO, 1962.

Parrish, John L. "Professional Womanpower as a National Resource." *Quarterly Review
of Economics and Business* 1 (1961): 54–63.

———. "Top Level Training of Women in the United States, 1900–1960." *Journal of the
National Association of Women Deans and Counselors* 25 (1962): 67–73.

———. "Women in Top Level Teaching and Research." *AAUWJ* 55 (1962): 99–103, 106–
7.

[Roettinger, Ruth.] *Idealism at Work, Eighty Years of AAUW Fellowships: A Report by the
American Association of University Women.* Washington, D.C.: AAUW, 1967.

Tryon, Ruth. *Investment in Creative Scholarship: A History of the Fellowship Program of the
American Association of University Women, 1890–1956.* Washington, D.C.: AAUW,
1957.

U.S. Civil Service Commission. Statistics Section. *Study of Employment of Women in the
Federal Government [1966 and] 1967.* Washington, D.C.: GPO, 1968.

Serials

The chief scientific serials of value here were those that presented news and commented
on the people in their profession, such as *Science, American Psychologist (AP),* and
Chemical and Engineering News (C&EN). These also printed letters to the editor, so that
the historian can see what alternative views were held at the time. The *JCE* published
recruitment articles, and the *Journal of Engineering Education (JEE),* educational statistics.

Others of interest were the organs of women's organizations, such as the *Journal of
the American Association of University Women, Independent Woman (IW,* later *National
Business Woman),* of the National Federation of Business and Professional Women, the
Journal of Home Economics (JHE), and during World War II, *Women's Work and Education.* The newsletters of various women's scientific "fraternities," committees, and other
organizations are invaluable, but they are difficult to locate because librarians have often
discarded them as ephemera.

Alumnae magazines also constitute a useful source of articles by and about a college's
unusual and high-achieving graduates. I examined those for Smith and Wellesley colleges, 1940–70, simply because they were easily available at the Boston Public Library.
There is an index to the *Wellesley Alumnae Magazine* in the Wellesley College Archives.

Among national newspapers, the *Christian Science Monitor* had the most extensive and most sympathetic treatment of women scientists and engineers, including feature articles, for example, on the leaders of the SWE. The *Wall Street Journal*, indexed since 1955, occasionally mentioned women in articles on corporate employment or spring recruitment at the colleges. The *New York Times Index* did not list articles about women under "Women" until quite late, limiting the usefulness of that newspaper to confirming what one had already found elsewhere.

Higher Education

Women's experience in higher education in World War II and after has not received the attention it deserves, but see Mabel Newcomer, *A Century of Higher Education for American Women* (New York: Harper & Brothers, 1959); and Barbara Miller Solomon, *In the Company of Educated Women: A History of Women's Higher Education in America* (New Haven: Yale Univ. Press, 1985), ch. 12. Contemporary works that extoll recent institutional upgrading are E. Alden Dunham, *Colleges of the Forgotten Americans: A Profile of State Colleges and Regional Universities* (New York: McGraw-Hill, 1969), and Harold L. Hodgkinson, *Institutions in Transition* (New York: McGraw-Hill, 1971). More analytical is David Riesman and Christopher Jencks, *The Academic Revolution* (Chicago: Univ. of Chicago Press, 1968). Although most traditional histories of universities focus on the higher administration and leave women out almost entirely, even in the case of coeducational institutions, they can be helpful if they mention a particular episode, such as the coming of coeducation, the first woman faculty member, the struggle to get accredited, the dispute over AAUW acceptance, or the changeover to human ecology. Very useful in general and especially for determining when a particular college became coeducational or a teachers college became a state college or university are John F. Ohles and Shirley M. Ohles, comps., *Private Colleges and Universities*, 2 vols. (Westport, Conn.: Greenwood, 1982); and idem, *Public Colleges and Universities* (Westport, Conn.: Greenwood, 1986). Also helpful was Walter Crosby Eells and Ernest V. Hollis, comps., *Administration of Higher Education: An Annotated Bibliography*, USOE Bulletin 1960, no. 7.

Departmental histories are rarer, but they are more valuable because they are more likely to reveal something about the presence and place of women in a university's daily life or annual academic year. Among these are Aaron J. Ihde, *Chemistry, As Viewed from Bascom's Hill* (Madison: University of Wisconsin, Department of Chemistry, 1990); Robert Rakes Shrock's *Geology at M.I.T., 1865–1965: A History of the First Hundred Years of Geology at Massachusetts Institute of Technology* (Cambridge: MIT Press, 1982); Saunders MacLane, "Mathematics at the University of Chicago, A Brief History," in *A Century of Mathematics in America, Part II*, edited by Peter Duren, vol. 2 (Providence, R.I.: American Mathematical Society, 1989), 127–54; and especially Mary Cookingham's inspired "Social Economists and Reform: Berkeley, 1906–1961," *History of Political Economy* 19 (1987): 47–65.

There are a few institutional histories of schools and colleges of home economics in the period since 1940, such as Agnes Fay Morgan, "The History of Nutrition and Home Economics in the University of California, Berkeley, 1914–62" ([1962], mimeographed); Flora Rose et al., *A Growing College: Home Economics at Cornell University*

(Ithaca: New York State College of Human Ecology, Cornell University, 1969); and Ercel Sherman Eppright and Elizabeth Storm Ferguson, *A Century of Home Economics at Iowa State University* (Ames: Iowa State University Home Economics Alumni Association, 1971). Earl J. McGrath and Jack T. Johnson's *Changing Mission of Home Economics: A Report on Home Economics in the Land-Grant Colleges and State Universities* (New York: Teachers College Press, 1968) was prepared with a certain negative viewpoint in mind, and Theodore Vallance's "Home Economics and the Development of New Forms of Human Service Education," in *Land-Grant Universities and Their Continuing Challenge,* edited by G. Lester Anderson (East Lansing: Michigan State Univ. Press, 1976), 79–103, boasted revealingly of recent progress in upgrading the field. Maresi Nerad is the first historian of higher education to focus on the topic. Her "Gender Stratification in Higher Education: The Department of Home Economics at the University of California, Berkeley, 1916–1962," *Women's Studies International Forum* 10 (1987): 157–64, was based on her recent dissertation, "'Home Economics Has to Move': The Disappearance of the Department of Home Economics from the University of California, Berkeley" (Ph.D. diss., University of California, Berkeley, 1986), and forthcoming book.

The topic of graduate education for women in the 1950s and 1960s is relatively unexplored. One can begin with such period pieces as *Graduate Education for Women, the Radcliffe Ph.D.: A Report by a Faculty-Trustee Committee* (Cambridge: Harvard Univ. Press, 1956); Bernice Brown Cronkhite's autobiography, *The Times of My Life,* mentioned above; and, lest one forget the nuns, Reverend Mother M. Regina, I.H.M., "Can the Mother General Afford to Send Sisters on to Graduate Study?" *NCEAB* 58 (1961): 140–43. Bernard Berelson's *Graduate Education in the United States* (New York: McGraw-Hill, 1960) can only infuriate.

Some mention of the variety of other historical works that proved useful on particular topics may stimulate more work in the many underdeveloped areas of educational history: Edna May Turner, "Education of Women for Engineering in the United States, 1885–1952" (Ph.D. diss., New York University, 1954); Geraldine Joncich Clifford, *Ed School* (Chicago: Univ. of Chicago Press, 1988); Carlos E. Kruytbosch and Sheldon L. Messinger, "Unequal Peers: The Situation of Researchers at Berkeley," *American Behavioral Scientist* 11 (May–June 1968): 33–43; Gerard M. McGrath, "Collegiality at Columbia: The Origin, Development, and Utility of Two Faculty Clubs" (Ed.D. diss., Columbia University, Teachers College, 1983); and Jennie Farley, "Women on the March Again: The Rebirth of Feminism in an Academic Community" (Ph.D. diss., Cornell University, 1970).

OTHER WORKS ON HIGHER EDUCATION

Ford Foundation. *The Pay of Professors: A Report on the Ford Foundation Grants for College-Teacher Salaries.* New York, 1962.

Lloyd, Alice. "Women in the Postwar College." *AAUWJ* 39 (spring 1946): 131–34.

McGuigan, Dorothy Gies. *A Dangerous Experiment: One Hundred Years of Women at the University of Michigan.* Ann Arbor: University of Michigan Center for Continuing Education of Women, 1970.

Rossi, Alice. "Discrimination and Demography Restrict Opportunities for Academic Women." *College and University Business* 74 (Feb. 1970): 74–78.

Ruml, Beardsley, and Sidney G. Tickton. *Teaching Salaries Then and Now: A Fifty-Year Comparison with Other Occupations and Industries.* New York: FAE, 1955.

Shaffer, Harry G., and Juliet P. Shaffer. "Job Discrimination against Faculty Wives: Restrictive Employment Practices in Colleges and Universities." *Journal of Higher Education* 37 (1966): 10–15.

Sigworth, Heather. "The Legal Status of Antinepotism Regulations." *AAUP Bulletin* 58 (1972): 31–34.

Snelling, W. Rodman, and Robert F. Boruch. *Science in the Liberal Arts College: A Longitudinal Study of Forty-nine Selective Colleges.* New York: Columbia Univ. Press, 1972.

Wolfe, Alan. "Hard Times on Campus." *The Nation* 210 (May 25, 1970): 623–27.

Women's Colleges

Useful starting places for the history of women's colleges in these years are Mabel Newcomer's 1959 book *Century of Higher Education,* mentioned above; Elaine Kendall, *"Peculiar Institutions": An Informal History of the Seven Sister Colleges* (New York: Putnam, 1976); and Liva Baker, *I'm Radcliffe, Fly Me!: The Seven Sisters and the Failure of Women's Education* (New York: Macmillan, 1976), which is better than its title and has an excellent bibliography. The last two are, however, more interesting as views of the women's colleges in the mid-1970s than they are adequate as history. Mary J. Oates introduces a widespread but little-studied type of woman's college in "The Development of Catholic Colleges for Women, 1895–1960," *Higher Education for Catholic Women: An Historical Anthology* (New York: Garland, 1987), reprinted in the *U.S. Catholic Historian* 7 (1988): 413–28.

Often used in evaluating the women's colleges is information on the subsequent careers of their graduates, for example, how many were later listed in *Who's Who in America* or *WWAW* or who later earned doctorates. It seems unfair to judge or rank a college in terms of what later happens to its alumnae, for a variety of reasons, including a lack of graduate fellowships for women or job discrimination, but see M. Elizabeth Tidball and Vera Kistiakowsky, "Baccalaureate Origins of American Scientists and Scholars," *Science* 193 (1976): 646–52, and Mary J. Oates and Susan Williamson, "Women's Colleges and Women Achievers," *Signs* 3 (1978): 795–806.

Several institutional histories get to the post-1940 period, including Florence M. Read, *The Story of Spelman College* (Atlanta: privately printed, 1961); Grace E. Hawk, *Pembroke College in Brown University: The First Seventy-five Years, 1891–1966* (Providence: Brown Univ. Press, 1967), and Polly Welts Kauffman, ed., *The Search for Equity: Women at Brown University, 1891–1991* (Hanover, N.H.: Univ. Press of New England for Brown Univ. Press, 1991); Jean Glasscock et al., *Wellesley College, 1875–1975: A Century of Women* (Wellesley, Mass.: Wellesley College, 1975); Bridget S. Pieschel and Stephen Pieschel, *Loyal Daughters: One Hundred Years at Mississippi University for Women, 1884–1984* (Jackson: Univ. Press of Mississippi, 1984); and most interesting of all, Joyce Thompson, *Marking a Trail: A History of the Texas Woman's University* (Denton: Texas Woman's Univ. Press, 1982). Virginia Gildersleeve's autobiography, mentioned above, reminds one of a bygone academic woman's Anglophilic culture. Here too histories of particular departments are a very useful genre, given sufficient sources either in the alumnae magazines or in the college archives, for example, Beale W. Cockey, "Mathematics at Goucher, 1888–1979" (undergraduate honors project, 1980, typescript).

Institutional Histories

Although histories of particular scientific institutions constitute an invaluable resource for the details and atmosphere of the working lives and careers of innumerable women scientists, only those of the nonprofit institutions are particularly valuable. Often written by former employees or associates, many of them the mainstays of their organizations, which often had unusual origins and later transformations, they can provide unexpected insights and are fun to read as well; there ought to be more of them. Lawrence J. Friedman, *Menninger: The Family and the Clinic* (New York: Knopf, 1990), which is a critique of a family, a field, and a cluster of institutions, sets a new standard for the genre. Among the most helpful or enjoyable are Jean Holstein, *The First Fifty Years at the Jackson Laboratory* [1929–79], ed. William L. Dupuy (Bar Harbor, Maine: n.p., [1979]); Elizabeth Noble Shor, *Scripps Institution of Oceanography: Probing the Oceans, 1936 to 1976* (San Diego: Tofus, 1978); Katherine Van Winkle Palmer, *Paleontological Research Institution, Fifty Years, 1932–1982* (n.p., 1982); William Campbell Steere, *Biological Abstracts/ BIOSIS: The First Fifty Years of a Major Science Information Service* (New York: Plenum, 1976); Thomas E. Drake, *A Scientific Outpost: The First Half Century of the Nantucket Maria Mitchell Association* (Nantucket, Mass.: Nantucket Maria Mitchell Association, 1968); William Bridges, *Gathering of Animals: An Unconventional History of the New York Zoological Society* (New York: Harper & Row, 1974); Alexander Bunts and George Crile, *To Act as a Unit: The Story of the Cleveland Clinic* (Cleveland: Cleveland Clinic Foundation, 1971); and [Juliet A. Mitchell, ed.], *A Community of Scholars: The Institute for Advanced Study, Faculty and Members, 1930–1980* (Princeton: Institute for Advanced Study, 1980). Harold Orlans, *The Nonprofit Research Institution: Problems and Prospects* (New York: McGraw-Hill, 1972), is more analytical of the pressures on all of them.

Foundations

The impact the foundations have had on higher education and opportunities for women in American society is incalculable. Accounts of particular foundations, such as Paul Woodring, *Investment in Innovation: An Historical Appraisal of the Fund for the Advancement of Education* (Boston: Little, Brown, 1970), and Richard Magat, *The Ford Foundation at Work: Philanthropic Choices, Methods, and Styles* (New York: Plenum, 1979), provide some information about their internal workings in these years. Annual reports of particular foundations reveal limited information on those grants that were actually made and omit those that were rejected. The oral history of Florence Anderson, at the Oral History Office at Columbia, provides a useful perspective. The most provocative critiques were by angry outsiders, such as Carol Brown, "Sexism and the Russell Sage Foundation," *Feminist Studies* 1 (1972): 25–44, and Peter Seybold, "The Ford Foundation and Social Control," *Science for the People,* May–June 1982, 28–31, and "The Ford

Foundation and the Triumph of Behavioralism in American Political Science," in *Philanthropy and Cultural Imperialism: The Foundations at Home and Abroad,* ed. Robert Arnove (Boston: G. K. Hall, 1980), 269–302. Yet even they fail to convey what opportunities and directions were missed because of the foundations' biases. Harold M. Keele and Joseph C. Kiger, eds., *Foundations* (Westport, Conn.: Greenwood, 1984), provides helpful background information.

Governmental Agencies

For an introduction to some issues surrounding scientific employment in the federal government from 1945 to 1965, see Margaret W. Rossiter, "Setting Federal Salaries in the Space Age," *Osiris* 7 (1992): 23–42; and Sharon Gibbs Thibodeau, "Science in the Federal Government," in *Historical Writing on American Science: Perspectives and Prospects,* ed. Sally Gregory Kohlstedt and Margaret W. Rossiter (Baltimore: Johns Hopkins Univ. Press, 1986), 81–96. For a statistical overview of high-ranking women, see W. Lloyd Warner, Paul P. Van Riper, Norman H. Martin, and Orvis F. Collins, "Women Executives in the Federal Government," *Public Personnel Review* 23 (1962): 227–34. One study of women executives at an unnamed federal scientific agency is Dorothy E. Walt, "The Motivation for Women to Work in High-Level Professional Positions" (Ph.D. diss., American University, 1962).

The few histories of scientific agencies vary greatly in sophistication and value. Of special importance here is the NSF, which has been well served. J. Merton England's *A Patron for Pure Science: The National Science Foundation's Formative Years, 1945–57* (Washington, D.C.: NSF, 1982) is as clear and helpful as one is likely to find, but it ends just as this important agency is gearing up for the post-Sputnik age. For a lot of specific information, such as who served on which boards at the time, one still needs to consult the NSF's *Annual Report* (Washington, D.C., 1951–). Also useful are Lyle Groenveld, Norman Koller, and Nicholas Mullins, "The Advisers of the United States National Science Foundation," *Social Studies of Science* 5 (1975): 343–54, and, more generally, Nicholas C. Mullins, "Power, Social Structure, and Advice in American Science: The United States National Advisory System, 1950–1972," *Science, Technology, and Human Values* 7, no. 37 (1981): 4–19. Less is written about the history of the National Institutes of Health (NIH), which was even more important for women in science than the NSF, but see Barney G. Glaser, *Organizational Scientists: Their Professional Careers* (Indianapolis: Bobbs-Merrill, 1964), for a 1952 survey of NIH personnel (17 percent women); and Thelma Dunn, "Intramural Research Pioneers, Personalities, and Program: The Early Years," *Journal of the National Cancer Institute* 59, no. 2 (suppl., 1977): 605–16.

Judith Sealander's *As Minority Become Majority: Federal Reaction to the Phenomenon of Women in the Work Force, 1920–1963* (Westport, Conn.: Greenwood, 1983) treats the WB in terms of an age cohort of single women who ran the agency for decades. One could do the same for the Bureau of Home Economics in the Department of Agriculture or the Home Economics Branch of the Division of Vocational Education at the USOE. The closest one can get to this for women scientists is Jane H. Wallace's "Women in the Survey," *Geotimes* 24, no. 3 (1979): 34, on the USGS. Valuable because, like the histories of the nonprofit institutions discussed above, it is more personal and even more affectionate than the usual history of a governmental agency is Ellis Yochelson, *The National Museum of Natural History: Seventy-five Years in the Natural History Building* (Washington, D.C.: Smithsonian Institution Press, 1985). The very few accounts of military and naval laboratories that exist tend to be administrative histories that give much attention to the unit's many reorganizations and omit discussion of the activities of its scientists, especially the many civilians employees. One interesting exception, as notable for what it leaves out (a discussion of wife swapping) as for what it contains is Evelyn Glatt, "Professional Men and Women at Work: A Comparative Study in a Research and Development Organization" (Ph.D. diss., Case Institute of Technology, 1966), on the Naval Ordnance Laboratory at remote China Lake, California.

On federal policy on women's issues in these years, see Patricia G. Zelman, *Women, Work, and National Policy: The Kennedy–Johnson Years* (Ann Arbor: UMI Research Press, 1982); and Cynthia Harrison, *On Account of Sex: The Politics of Women's Issues, 1945–1968* (Berkeley: Univ. of California Press, 1988), which gives more importance to the PCSW than that temporary and transitional group may merit.

At the state level, Anna M. Sexton, *A Chronicle of the Division of Laboratories and Research, New York State Department of Health: The First Fifty Years* (Lunenberg, Vt.: Stinehour, 1967), remains a classic on a very important agency. It can be supplemented by Richard Baldwin's dual biography of the chemist Rachel Brown and the bacteriologist Margaret Hazen, mentioned above in the section on biographies. For data, see NSF, Surveys of Science Resources Series, *Research and Development in State Government Agencies, Fiscal Years 1967 and 1968* (Washington, D.C.: GPO, 1970).

Scientific Societies

Given the importance of scientific societies and their central role in facilitating the practice of science and the recognition of scientists, it is surprising how little sophisticated work has been done on the topic. Usually such histories have been mere chronicles or lists of officers and changes in bylaws, but a few recent efforts have provided more detail and even some drama; see, for example, Dael Wolfle, *Renewing a Scientific Society: The American Association for the Advancement of Science from World War II to 1970* (Washington, D.C.: AAAS, 1989); Simon Baatz, *Knowledge, Culture, and Science in the Metropolis: The New York Academy of Sciences, 1817–1970,* published as *Annals of the New York Academy of Sciences* 584 (1990); and Carleton Mabee, "Margaret Mead's Approach to Controversial Public Issues: Racial Boycotts in the AAAS," *Historian* 48 (1986): 191–208. Others of interest include Harriet L. Rheingold, "The First Twenty-five Years of the SRCD," *Monographs of the Society for Research in Child Development* 50 (1985): 126–40; and Edwin B. Eckel, *The Geological Society of America: Life History of a Learned Society,* GSA Memoir 155 (1982), which offers a full account of the staff's duties and responsibilities at headquarters. This is rare in histories of scientific societies. Essential reference works for any work in this area are Joseph C. Kiger, *American Learned Societies* (Washington, D.C.: Public Affairs Press, 1963) and idem, ed., *Research Institutions and Learned Societies* (Westport, Conn.: Greenwood, 1982). There was very little on women's participation in scientific societies before the many "status of women" reports of 1968–

72, to be discussed below, but Kay E. Leffler, "The Role of Women in Geography in the United States" (master's thesis, University of Oklahoma, 1965), is an exception.

Women's Scientific Societies

Two recent articles have attempted to analyze the activities of groups of women psychologists: Mary Roth Walsh, "Academic Professional Women Organizing for Change: The Struggle in Psychology," *Journal of Social Issues* 41, no. 4 (winter 1985): 17–28; and James H. Capshew and Alejandra C. Laszlo, "'We Would Not Take No for an Answer': Women Psychologists and Gender Politics during World War II," ibid. 42, no. 1 (spring 1986): 157–80, to which Alice I. Bryan responded in "A Participant's View of the National Council of Women Psychologists: Comment on Capshew and Laszlo," ibid., 181–84. Both articles are based on the rich materials at the Archives of the History of American Psychology at the University of Akron in Akron, Ohio.

Less analytical are the several chronicles or chronologies of women's scientific societies, such as that by Sister Mary Rose Stockton, *A History of Iota Sigma Pi* (Indianapolis: n.p., 1980), which updates earlier efforts by Agnes Fay Morgan; that by Mary Louise Robbins, ed., *A History of Sigma Delta Epsilon, 1921–1971: Graduate Women in Science* (n.p., 1971); another edited by Lillian G. Portenier, on the well-studied ICWP, entitled *The International Council of Psychologists, Inc.: The First Quarter Century, 1942–1967* (n.p., [1967]); and recently, Marta Navia Kindya with the collaboration of Sudha Dave, *Four Decades of the Society of Women Engineers* (n.p., 1990).

Neither a scientific society nor a women's group, the AHEA has been well served by three chronicles: Keturah Baldwin, *The AHEA Saga* (Washington, D.C.: AHEA, 1949); Mary Hawkins, ed., *The American Home Economics Association, 1950–1958: A Supplement to the AHEA Saga by Keturah E. Baldwin, 1949* (Washington, D.C.: AHEA, 1959); and Helen Pundt, *AHEA: A History of Excellence* (Washington, D.C.: AHEA, 1980).

Meetings

To a certain extent, the history of women in science between 1940 and 1972 can be glimpsed through the proceedings of a series of meetings, starting with one during World War II to urge college women to join the war effort and ending with "Successful Women in the Sciences," a New York Academy of Sciences special symposium in May 1972. In between were others at which the nation's postwar needs were discussed: "Women in the Defense Decade," in 1951 (*Women in the Defense Decade*, ACE Studies, ser. 1, no. 52, ed. Raymond F. Howes [Washington, D.C., 1952]); a meeting on the need for women in the physical sciences, held at Bryn Mawr College in 1954 (summarized in "Conference on the Role of Women's Colleges in the Physical Sciences, Held at Bryn Mawr College, Bryn Mawr, Pennsylvania, on June 17–18, 1954" [1954, mimeographed]); and meetings in 1958–60 hosted by the National (or American) Council of Women in Science. Meetings in the 1960s included the bizarre one in 1962 at the University of Pittsburgh at which some speakers openly ridiculed and even defied the prevailing "womanpower" message that women were needed in engineering (*Women in Profession-*

al Engineering: A Conference Held under the Auspices of the Executive Office of the President of the United States and Sponsored by the University of Pittsburgh and the Society of Women Engineers, April 23 and 24, 1962 at the University of Pittsburgh [New York: SWE, 1962]). This episode might be interpreted as one of the worst of the innumerable "career days" of the 1950s and 1960s exhorting women but discouraging them as well. It was followed by the historic 1964 meeting at MIT, where an illustrious cast of speakers represented a kind of watershed in old and new thinking about women's place in science (Jacquelyn A. Mattfeld and Carol G. Van Aken, eds., *Women and the Scientific Professions,: The M.I.T. Symposium on American Women in Science and Engineering* [Cambridge: MIT Press, 1965]). In the same year the SWE hosted the first international conference of women in engineering as a kind of signpost of women's emergence in that field (*Proceedings of the First International Conference of Women Engineers and Scientists: Focus for the Future, Developing Engineering and Scientific Talent* [New York: SWE, 1964]). Finally, one could end the era with the 1972 conference in New York City hosted by the New York Academy of Sciences, where a series of women speakers "found their voice" and spoke out rather honestly and rather angrily about what science had done to them over the years (*Successful Women in the Sciences: An Analysis of Determinants,* ed. Ruth B. Kundsin, *Annals of the New York Academy of Sciences* 208 [1973]).

Status of Women Reports

Status of women reports constitute a political bellwether that goes back to the 1890s. As documents of overt political criticism, they mark a certain discontent and level of ferment among the women either enrolled or employed at an institution, such as a university, or among members of a scientific society. To produce and publish such a report requires a certain "consciousness," anger, energy, and organization. Yet these reports are a varied lot. Some "let the numbers speak for themselves"; others add considerable editorial comment; a few, such as that on the University of California at Berkeley, contain considerable useful historical information. The archival records of the groups writing these reports should be of considerable interest when they are made ready for use. Some of the reports were mimeographed or pamphlets and are now hard to locate. Several are in JLP. A useful starting place is Linda A. Harmon, comp., *Status of Women in Higher Education, 1963–72: A Selective Bibliography,* Series in Bibliography, no. 2 (Ames: Iowa State University Library, 1972).

Bryan, Alice I., and Edwin G. Boring. "Women in American Psychology: Statistics from the OPP Questionnaire." *AP* 1 (1946): 71–79.

Fava, Sylvia Fleis. "The Status of Women in Professional Sociology." *American Sociological Review* 25 (1960): 271–76.

Fischer, Ann, and Peggy Golde. "The Position of Women in Anthropology." *American Anthropologist* 70 (1968): 337–44.

Mitchell, Mildred. "Status of Women in the American Psychological Association." *AP* 6 (1951): 193–201.

Rossi, Alice S. "Status of Women in Graduate Departments of Sociology, 1968–1969." *American Sociologist* 5 (1970): 1–12.

———. "Report of Committee W., 1970–71." *AAUP Bulletin* 57 (1971): 215–20.

Simpson, Lawrence A. "A Study of Employing Agents' Attitudes toward Academic Women in Higher Education." Ed.D. diss., Pennsylvania State University, 1968.

Weeks, Dorothy W. "Woman Power Shortage in the Physical Sciences." *AAUWJ* 48 (Mar. 1955): 146–49.

———. "Women in Physics Today." *Physics Today*, vol. 13, Aug. 1960, 22–23.

Suicides

There were two contemporary statistical studies of women chemists and psychologists who committed suicide: Frederick P. Li, "Suicide among Chemists," *Archives of Environmental Health* 19 (1969): 518–20; and J. S. Mausner and R. C. Steppacher, "Suicide in Professionals: A Study of Male and Female Psychologists," *American Journal of Epidemiology* 98 (1973): 436–45. A larger, later study was Judy Walrath et al., "Causes of Death among Female Chemists," *American Journal of Public Health* 75 (1985): 883–85, based upon 347 white women members of the ACS who had died between 1925 and 1979 (plus eleven nonwhite women who showed "no unusual mortality pattern"). For commentary on this later study, see "Women Chemists Mortality Study Finds High Suicide Rate," *C&EN* 62 (Apr. 23, 1984): 16–17; Molly Gleiser, "Suicide among Women Chemists," *Nature* 328 (July 2, 1987): 10; and Malcolm W. Browne, "Women in Chemistry: Higher Suicide Risk Seen," *NYT*, Aug. 4, 1987, 21. Aside from these, the topic has received little attention, and individual cases such as those of Klara Dan von Neumann Eckhart, Elsie Frenkel-Brunswik, and an unnamed professor of political science at Pennsylvania State tend to be covered up. Occasionally individuals or a woman's group, such as that of the ACS, deplores this lack of attention in an article or a letter to the editor of a scientific journal. The subject's importance, both for women scientists and for the sociopathology of scientific careers, merits more than statistical interest.

Other Women's Organizations

Beyond the publications listed above under serials, a few works about women's organizations have chronicled their activities to a certain extent. There have been two histories of the Business and Professional Women's Clubs: Geline M. Bowman, comp., *A History of the National Federation of Business and Professional Women's Clubs, Inc., 1919–1944* (New York: NFBPWC, 1944); and Marguerite Rawalt, *History of the National Federation of Business and Professional Women's Clubs, Inc., 1944–1960* (Washington, D.C.: NFBPWC, [1969]). The AAUW lacks a history of the period after 1930 but has published two accounts of its fellows and fellowships listed above in the section on data. See also Susan M. Hartmann, "Women's Organizations during World War II: The Interaction of Class, Race, and Feminism," in *Woman's Being, Woman's Place: Female Identity and Vocation in American History*, ed. Mary Kelley (Boston: G. K. Hall, 1979), 313–28; and Janice Leone, "Integrating the American Association of University Women, 1946–1949," *Historian* 51 (1989): 423–45.

Archival Collections

The personal papers of individuals and the archives or official records of organizations have proven invaluable for this study. They exist nearly everywhere. In fact the more one looks, the more one finds. The primary ones are listed below, the most useful being marked by an asterisk (*). Academic women are generally more fully documented than industrial, corporate, or governmental ones, who have occasionally been interviewed for oral histories. One collection of unexpectedly great value, because it has materials on the many lesser-known persons who did not quite merit inclusion in *Notable American Women: The Modern Period,* was the *Notable American Women* Papers, at the SLRC. Another particularly valuable collection there is the Papers of the Commission on the Education of Women, of the American Council on Education, c. 1952–61, which contains extensive subject files with clippings on women in science, among many other topics. The Jeanne Lowe Papers at Vassar College are useful for their subject files— status of women reports, interviews, and other material c. 1969–70 on women in the sciences, engineering, and medicine.

The numerous oral histories are a varied lot. Usually done rather late in life, some are quite informative, but others are narrowly factual. Some open up new topics not otherwise documented; others omit topics of interest entirely. Their usefulness depends largely on what else is available.

COLLECTIONS USED

Alaska
> Fairbanks. University of Alaska. Ramussen Library. Alaska and Polar Regions Department. Louise Boyd Collection. Erna Gunther Papers.

California
> Berkeley. University of California.
>> Bancroft Library. Oral History of Ian and Catherine Campbell. Patricia Durbin Reminiscences. Faculty Memorials and Necrology. Oral History of Emily Huntington (copy at UCLA). Oral History of Claude Hutchison. Oral History of Mildred Mathias. Oral History of Elizabeth McClintock. Centennial Oral History of Agnes Fay Morgan. *Agnes Fay Morgan Papers. Oral History of Knowles Ryerson. Oral History of Flora M. Scott. Oral History of Lois H. M. Stolz. Oral History of Harry Wellman.
>> Chancellor's Office. Chancellor's Files.
> Chico. California State University at Chico Archives. Meriam Library. Special Collections. Vesta Holt Papers. Yearbooks.
> La Jolla.
>> Scripps Institution of Oceanography Archives. Margaret Robinson Biographical File. Special Committee on Non-Senate Academic Personnel File. Tektite II File. Women in Oceanography Subject File.
>> University of California at San Diego. Mandeville Department of Special Collections. *Maria Goeppart Mayer Papers. Harold Urey Papers.

Pasadena. California Institute of Technology Archives. Biology Division Papers. Dan Campbell Papers. Lee DuBridge Papers. Jesse Greenstein Papers. Historical Files (on coeducation). T. Lauritsen Papers.

San Francisco. University of California at San Francisco Archives. Hooper Foundation Papers, Bernice U. Eddie File.

Santa Barbara. University of California at Santa Barbara. University Library. Special Collections. Katherine Esau Autobiography and Oral History. Kappa Omicron Phi Scrapbooks. Oral History of Hazel Severy.

Stanford. Stanford University Archives. Lewis Terman Papers.

Colorado

Denver. Denver Public Library. Western History Department. Clippings. Interview with Florence Sabin. Interview with Ruth Underhill.

Connecticut

New Haven. Yale University. Yale University Archives. Records of President Griswold. Records of the Provost. Graduate School Papers: Records of Dean John Perry Miller.

District of Columbia

American Council on Education. Archives.

Library of Congress.

Manuscript Division. American Chemical Society Papers. Ames-Ilg Papers. Irving Langmuir Papers. *Margaret Mead Papers. Oswald Veblen Papers. *Alan T. Waterman Papers.

Rare Book Room. Eistophos Science Club Papers.

National Anthropological Archives. Dorothea Leighton Papers.

National Archives and Records Administration. Bureau of the Budget (RG 51). National Science Foundation Papers (RG 307). National Security and Resources Board Papers (RG 304). Office of Science and Technology Papers (RG 359). President's Committee on Scientists and Engineers Papers (in RG 307). Women's Bureau Papers (RG 86).

National Federation of Business and Professional Women's Clubs. Library. Archives.

Smithsonian Institution.

Archives. Doris Mable Cochran Papers.

National Museum of American History. Archives. Oral Histories in Computing (D. Lehmer, E. Marden, Mina Rees, and Ida Rhodes).

Georgia

Atlanta

Emory University Library. Special Collections. Jane Russell Faculty File.

Georgia Institute of Technology Library. Clippings and correspondence files on coeds.

Spelman College Archives. *Spelman Messenger*.

Decatur. Agnes Scott College Archives. Alumnae magazine. Bradley Observatory File. Mary MacDougall File.

Illinois
  Urbana. University of Illinois Archives. *Virginia Bartow Papers. Faculty Files.
  Janice Smith Papers. Women Faculty Exhibit File.

Indiana
  Bloomington. Indiana University Archives. Housing File. Kate Heuvner Mueller
  Autobiography.
  Greencastle. DePauw University Archives. Oral History of Winona Welch. *Winona
  Welch Papers.
  West Lafayette. Purdue University Library. Special Collections. Lillian Gilbreth Papers.

Iowa
  Ames. Iowa State University Archives. Faculty Files.

Kansas
  Abilene. Dwight D. Eisenhower Library. Jacqueline Cochran Papers. Dwight D.
  Eisenhower Papers as President—Alphabetical File. Oral History of Eli Ginzberg.
  Oveta Culp Hobby Papers. Papers of the Office of the Special Assistant for Science
  and Technology. *President's Committee on the Development of Scientists and
  Engineers Papers. Papers of the President's Scientific Advisory Committee. White
  House Photos and Appointment Book.
  Lawrence. Spencer Research Library. University of Kansas Archives. Cora Downs
  Faculty Folder. Oral History of Cora Downs. Faculty Handbooks. Oral History of
  Fritz and Grace Heider. Records on antinepotism rule (Board of Regents and
  Psychology Department).
  Manhattan. Kansas State University Archives. Margaret Justin File.

Louisiana
  New Orleans. Tulane University Archives. Clara deMilt Papers. Faculty Files. Newcomb College Faculty List.

Maryland
  Bethesda. National Library of Medicine. History of Medicine Division. Biographical
  Files. Oral histories, especially Milton J. Senn Oral History Collection on Child
  Development.
  College Park.
    American Institute of Physics. Niels Bohr Library. Oral History of Mildred Allen.
    Oral History of Nannielou Dieter Conklin. Oral History of Esther Conwell.
    Oral History of Janet Guernsey. Oral History of Dorrit Hoffleit. Oral History of
    Helen Sawyer Hogg. Oral History of Cecilia Payne-Gaposchkin. Oral History
    of Nancy Roman. Oral History of Charlotte Moore Sitterly. Oral History of
    Dorothy Weeks.
    University of Maryland Archives. Mary Shorb Papers. Adele Stamp Papers.

Massachusetts
  Boston. John F. Kennedy Presidential Library. Awards File. John Kenneth Galbraith
  Papers. Oral History of John W. Macy. Papers of the President's Commission on
  the Status of Women. White House Central Subject Files.

Cambridge
 Harvard University
  Museum of Comparative Zoology Library. Archives. Elisabeth Deichmann Papers. Tilly Edinger Papers.
  Pusey Library. University Archives. *Bart J. Bok Papers. Edwin G. Boring Papers. Papers of the Director of the Harvard College Observatory. Faculty Club Photos. Wilbur K. Jordan Papers. Kirtley Mather Papers. S. S. Stevens Papers. Hilda von Mises Papers. Ralph Wetmore Papers.
 Massachusetts Institute of Technology. Institute Archives. Dorothy Bowe Papers (Women at MIT Collection). Margaret Compton obituaries. Oral History of Mildred Dresselhaus. Oral History of Ellen Henderson. Oral History of Christina Jansen. Oral History of Christine Jones. Oral History of Vera Kistiakowsky. Oral History of Nancy Kopell. President's Papers (AC 4). Oral History of Lisa Steiner. Tech Show playbills. Oral History of Giuliana Tesoro. Dorothy Weeks Papers. Oral History of Sheila Widnall.
 Polaroid Corporation Library. Clipping folders.
 Radcliffe College
  Henry A. Murray Center for Research on the Study of Lives. Oral History of Cora DuBois. Oral History of Ruth Hubbard. Oral History of Ruth Dixon Turner. Oral History of Emily Vermeule.
  Radcliffe College Archives. Oral History of Mary Bunting. Mary Bunting Presidential Papers. Oral History of Bernice Brown Cronkhite. Bernice Cronkhite Papers. *Papers of the Office of the Graduate Dean. Deceased Alumnae Records. Elisabeth Deichmann File. *Wilbur Jordan Presidential Papers. Radcliffe Graduate School materials.
  Schlesinger Library. Beatrice Aitchison Collection. *American Council on Education, Commission on the Education of Women Papers. American Council on Education, Papers of the Conference on Women in the Defense Decade. Sarah Gibson Blanding Papers. Helen M. Cam Papers. Elizabeth Cooper Papers. Adelaide Cumming Papers. DACOWITS Papers (Defense Advisory Committee on Women in the Services). Margaret Harwood Papers. Marjorie Child Husted Papers. Oral History of Flemmie Kittrell. Alice Leopold Papers. Oral History of Ida Craven Merriam. *Notable American Women Papers. *John B. Parrish Papers. Joanne Simpson Papers. Somerville-Howorth Papers. Henrietta Swope Papers.
Northampton. Smith College.
 Archives. Faculty Files. *Myra Sampson Papers.
 Sophia Smith Collection. Ernestine Gilbreth Carey Papers. Margaret Wooster Curti Papers. Dorothy Wrinch Papers.
South Hadley. Mount Holyoke College. Archives. Chemistry Department Files.
Wellesley. Wellesley College Archives. Oral History of Harriet Creighton. *President's Office, Statistical Notebook.
Woods Hole. Marine Biological Laboratory. Archives. Annual Reports. Photographs.

Michigan

Ann Arbor. University of Michigan. Bentley Historical Library. *Icie Macy Hoobler Papers. Women's Research Club Papers.

East Lansing. Michigan State University Archives. *Papers of College of Human Ecology. Marie Dye Papers.

Minnesota

Minneapolis. University of Minnesota Archives. Biographical Files. Division of Agriculture—Entomology and Economic Zoology Papers. *Ruth Eckert Papers. Faculty Files. Florence Goodenough Papers. Ruth Grout Papers. Papers of Iota Sigma Pi Chapter. Mildred Templin Papers.

Missouri

Independence. Harry S. Truman Presidential Library. William and Sophie Aberle Brophy Papers. *Records of the President's Scientific Research Board.

St. Louis

Missouri Botanical Garden. Biographical Files.

University of Missouri at St. Louis Archives. Margaret Hickey Papers.

Washington University Medical Center Archives. Gerty Cori Papers. Joseph Erlanger Papers. Helen Tredway Graham Papers. Tape by Ruth Silverberg. Margaret G. Smith Papers. Tape by Mildred Trotter. Mildred Trotter Papers.

New Jersey

Princeton. Princeton University Archives. Henry Norris Russell Papers.

New York

Albany. New York State Department of Health Library. Biographical Files.

Ithaca. Cornell University. Kroch Library. Rare and Manuscript Collections. Elsa Allen Papers. Oral History of Sarah Gibson Blanding. Ethel Cornell Papers. Faculty Files. Graduate Wives Newsletters. Hazel Hauck Papers. Oral History of Mary Henry. Helen Lyons Papers. Margaret Morse Nice Papers. Oral History of Catherine Personius. *Catherine Personius Papers. Sigma Delta Epsilon Papers. Statler Club Papers. Oral History of Elizabeth Lee Vincent. Cynthia Westcott Papers. Women Faculty Taped Interviews.

New York City

Columbia University. Oral History Research Office. Oral History of Florence Anderson. Oral History of Maria Goeppart Mayer. Oral History of Adelaide Oppenheim. Oral History of Katherine Phillips. Reminiscences of Katharine Stinson.

Ford Foundation Archives. National Manpower Council Files. Rutgers University—File on Retraining Women College Graduates.

North Tarrytown. Rockefeller Archive Center. Rebecca Lancefield Papers. File on the National Research Council—Personnel. Gertrude Perlmann Papers. Rockefeller University Faculty Files.

Poughkeepsie. Vassar College Library. Special Collections. *Alumnae in Science File. Alumnae magazine. Ruth Benedict Papers. Sarah Gibson Blanding Papers. Faculty Files. Helen Lockwood Correspondence. *Jeanne R. Lowe Papers.

Rochester. University of Rochester Medical School. Miner Medical Library. Estelle Hawley Papers.

Schenectady. General Electric Corporation. Biographical Files. Tape of Margaret Welsh Goldsmith.

Syracuse. Syracuse University. George Arendts Research Library. M. Eunice Hilton Papers.

North Carolina

Durham. Duke University Archives. Katharine Banham Papers. Frances C. Brown Papers. Department Clipping Files. Faculty Biographical Files. Muriel I. Sandeen Papers. Hertha Sponer items.

Greenville. East Carolina University Library. East Carolina Manuscript Collection. Frances Bowman Papers.

Raleigh. North Carolina State University Archives. Gertrude Cox Papers. Files on Gertrude Cox, Statistics Department, and Women.

Ohio

Akron. University of Akron. Archives of the History of American Psychology. Polsky Building. Dorothy Adkins (Wood) Papers. Margaret Wooster Curti Papers. Leah Fein Papers. Helen Flinn File. Margaret Ives Papers. Oral History of Emily Stogdill. Emily Stogdill Papers. Doris Twitchell-Allen Papers.

Pennsylvania

Philadelphia.

American Philosophical Society Library. Leonard Carmichael Papers. Warren H. Lewis Papers. Florence Seibert Papers. Rose Mooney Slater Papers. Items from T. Dobzhansky Papers, L. C. Dunn Papers, and J. C. Slater Papers.

University of Pennsylvania Archives. Althea K. Hottel Alumni File. Women's Faculty Club Papers.

Pittsburgh. Carnegie-Mellon University. Hunt Institute for Botanical Documentation. Biographical Files. Mildred Mathias Papers. Women in Science File.

University Park. Pennsylvania State University. Pattee Library.

Penn State Room. University Archives. Papers of the Dean of the College of Liberal Arts. Ben Euwema Vertifical File. Faculty Files. Grace Henderson Papers. Ellen Richards Institute Papers. Lowell Schoenfeld Vertical File. Mary P. Shelton Vertical File. Carolyn Sherif Vertical File.

Historical Collections and Labor Archives. *Jessie Bernard Papers.

Texas

Austin. Lyndon Baines Johnson Library. Aides' Files. Oral History of Donald F. Hornig. Donald F. Hornig Papers. Lyndon Baines Johnson Presidential Papers. President's Appointment File. White House Central Files.

Virginia

Blacksburg. Virginia Polytechnic Institute and State University Archives. Marjorie Rhodes Townsend Papers.

Washington
  Seattle
    University of Washington Archives. Oral History of Grace Denny. Viola Garfield
      Papers. Erna Gunther Papers. Florence Harris Papers. Thelma Kennedy Papers.
      Irene Peden Papers. Ruth Svihla Papers. Davida Teller Papers.
    University of Washington Special Collections. Lois Brown Crisler Papers.

Wisconsin
  Madison
    University of Wisconsin Archives. Course catalogs. Faculty Files. Report on E. B.
      Fred Fellows.
    University of Wisconsin Oral History Office. Oral History of Ruth Henderson.
      Oral History of James and Elizabeth Miller. Oral History of Dean Glenn Pound.
      Oral History of May Reynolds.

INDEX

Library of Congress Cataloging-in-Publication Data

Rossiter, Margaret W.
Women scientists in America : before Affirmative Action, 1940–1972
Margaret W. Rossiter.
p.   cm.
Includes bibliographical references and index.
ISBN 0-8018-4893-8
1. Women scientists—United States—History.   2. Women in science—
United States—History.   I. Title.
Q130.R683   1995
305.43′5′0973—dc20   95-2320